THE GREAT WAR

THE CAUSE OF THE QUARREL

ARCHIVES

d'Histoire et Militaire

ADELAIDE

Kaiser Wilhelm II, 1891 ~ Max Johann Bernhard

ISBN-10: 099437142X
ISBN-13: 978-0994371423

Diplomacy ~ Foreign Relations ~ Imperialism ~ World War 1914-18 Causes ~ Campaigns; Eastern Front, Western Front & Middle East

THIS EDITION IS LIMITED TO FIVE HUNDRED AND TEN COPIES OF WHICH FIVE HUNDRED ARE FOR SALE AND TEN FOR PRESENTATION

No..

To Arabella and Harald

CONTENTS

PART I

PREFACE

UNDERLYING CAUSES OF THE GREAT WAR

CONTENTS

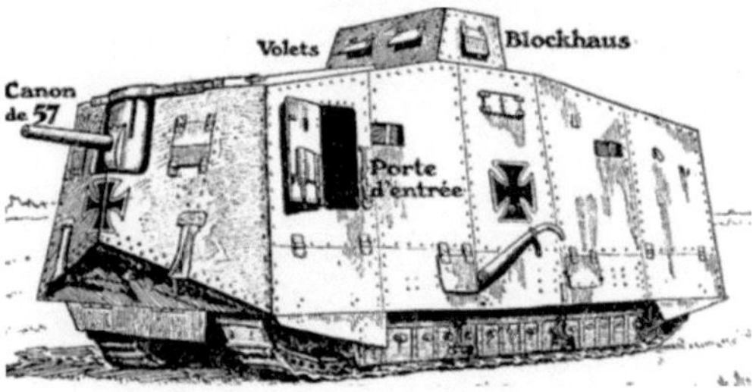

PART II

ENTR'ACTE

IMMEDIATE CAUSES OF THE GREAT WAR

British forces valiantly fending off a night attack in the trenches using star pistols and rifles.

CONTENTS

PART III

EXPANSION OF THE WAR

OTHER MOBILIZATIONS & DECLARATIONS OF WAR

LIST OF MAPS

Inspection of the Russian Black Sea Fleet ~ Ivan Ajvazovski

Kaiser Wilhelm II reviews the Prussian Guard ~ Carl Rochling

Conflicting interests of the European Empires in North Africa and the Near East

It is not easy to recapture the atmosphere of these years, but in general we are aware of a certain grandiosity invading the thought and action of all nations. Everywhere the demand was for the unparalleled and the colossal. The prevailing spirit infected governments and their military and naval advisers, who then set themselves to produce the largest armies, the most formidable battleships, the fastest destroyers, the most powerful guns, and the most destructive projectiles. The necessity of keeping ahead of the competition of possible enemies was urged by each government in turn, and in the circumstances generally with good reason; but the process developed a collective zeal in the competition, which carried it forward with a momentum that no government could control. Every collision between them was from this time onward attended by a press chorus of increasing volume. Such was the atmosphere in which the European groups pursued their rivalries J.A. Spender

PREFACE

More than one true thing may be said about the causes of the war, but the statement that comprises the most truth is that militarism and the armaments inseparable from it made the Great War inevitable. Armaments were intended to produce a sense of security in each nation ~ that was the justification put forward in defence of them. What they really did was to produce fear in everybody. Fear causes suspicion and hatred. It is hardly too much to say, that between nations, it stimulates all that is bad and depresses all that is good. Sir Edward Grey

When the Great War erupted, people asked as they still do, what was the cause of the quarrel? Many believed it impossible that the continent of Europe could be plunged into war through a local squabble in the Balkans. But while the greatest events, as Sophocles pointed out many centuries ago, may seem to spring from trivialities, the causes are always profound. The truth of this is abundantly clear in the origins of the Great War. They can be traced in various deep and different ways ~ and as far back as in the primitive days when Teutonic warriors sallied forth for a *place in the sun* which they regarded as theirs by *right of conquest* ~ and if we examine the various episodes that lead to the Great War, we soon become conscious of a tide of affairs that gathers force and carries all before it.

Acting on the historical assumption that war was a legitimate instrument of foreign policy, and on the whole the best way of clearing up the knots and tangles that had defied their best statecraft, most diplomats and statesmen of pre-war Europe were unaware that the collective result of their foreign policies and their armaments would all end so tragically. The wars that they had in mind were the limited and localized Prussian-Danish War of 1864 and the Franco-Prussian War of 1870-71. These masterfully planned short wars had both achieved their purpose without excessive sacrifice or the derangement of European society. Whilst the diplomacy previously practiced with its adoption of the ethics of war was well suited to limited localized wars, it became wholly unsuited to the hazards of multi-national alliances that only divided Europe into two hostile groups who became increasingly suspicious of one another, who steadily built up greater armies and navies, and who would all be drawn into the Great War which few wanted. It all came from their system of alliances and therefore remains a question whether this theme of history may be regarded as closed or whether it might be reopened by another oversight of misguided statecraft, and be repeated to an even worse conclusion.

If we search carefully, we can find everywhere underlying the epic struggles of recorded history, (and in those now occurring) an antagonism of some kind between two systems, visions or ideals, which in some particular contention were fundamentally opposed and could not be easily reconciled. State papers and the memoirs of diplomats are apt to confuse the real issues, by setting forth a diary of minor incidents and piquant details, not in their true proportions, but as they appeared at the moment of their occurrence to the awareness of harassed and suspicious officials. For example ; all the archival heritage of the American Civil War upholds the essential fact, that in this conflict a million lives were sacrificed by one of the most intelligent and practical nations of the mid-1800's for no other cause than an irreconcilable difference amongst them ~ for what St. Paul called *the substance of things hoped for.* On one side, there was the ideal of Union and a determination to make it prevail. On the Confederate side there was an ideal of Independence and an equal determination to defend it ~ whatever the cost. If war on such basis is possible within a single nation, nurtured in the same traditions and tongue, how futile is any assurance that material considerations can make war impossible between nations. Collisions of interest occur when one Power's interests venture crosswise to the course of another, so the essential question becomes ~ did any nation or Empire have possessions which appeared to obstruct the national ideals of others, and did these others come to believe that alone or in alliance they had the power to redress the perceived balance in triumph?

As the balance of power changed throughout Europe in military alliances and territorial disputes, new threats took form to set the scene for subsequent conflict. As Austria-Hungary, Germany and Italy established their Triple Alliance, France reached out to Russia to enter into a defensive Éntente with Britain and became diplomatic partners with their signing of the Anglo-Russian Éntente in 1907. Following the effort of the German group by a succession of shock foreign policy stunts to disintegrate the French-Russian-British Éntente ~ the subsequent focus of both groups on Austro-Russian rivalry in the East became the supreme test of strength for both Alliances. Russian ambition for dominance over the Balkans and Straits spurred southern Slavic dreams for a Greater Serbia to create their erratic Balkan League that during the Balkan Wars all but drove the Ottoman Turkish Empire from Europe, and foresaw the final forcing of the issue between the Alliances that thereafter, loomed ever nearer.

PREFACE

For years, problems connected with the Near East caused intense international rivalry for future domination over the Balkan Peninsula and Asia Minor, and were a continual danger to world peace. These localities, together with the waterways they controlled, formed the great corridor from east to west and from north to south, and constituted the natural thoroughfare from Central Europe to Asia and from Russia to the Mediterranean. Thus, since the advent of the German *Mittel-Europa* scheme, Berlin was just as determined to thrust open the Near-Eastern door via the Balkans to optimize their commercial vitality with the Ottoman Empire - as the Russians were to expand their influence through the same region towards warm water - and the Austrians were to secure their fracturous Empire.

Geo-political success would probably come to whichever power controlled the Balkan Peninsula. If the Tsar's regiments guarded the Bosporus and the Ægean, then Russia would have her window upon the world, and from her often proposed new capital at Constantinople ~ *Tsargrad*, Russia could control all Danube trade, and with her finger thus upon the artery of Vienna, would end the dream of the Germanic powers of stretching their dominion down to the Persian Gulf. Conversely, if Germany and Austria won the Bosporus and the Hellespont, then Russia, thrust back from the south and east, would lie like a giant bound among its enemies, a vast but an inferior power. Therefore, the Balkans remained the danger-zone of Europe, and the destinies nations were in the sights of strategic planners.

There could hardly be a worse motive for ordering the affairs of a civilized world, but under their Alliance system in a world organized for war, it became irresistible. While many historians seeking the origin of the war waste their ink when they look for it in one root when there are multiple webs, behind every contention were the massed Empires seeing tests of their power in which they could not afford to be vanquished, and within each limited decision matrix, those favouring the neutralist cause failed because there seemed no realization on the side of peace that the danger was most desperate. That they were on the edge of Armageddon… with all psychological momentum on the side of war.

This book examines this history and the condition of Europe with its component nationalities and provides significant new insights on the incidents that occurred before the eruption of the Great War, as it explains the deep-rooted causes of the conflict, the immediate events, and considers the glaring miscalculations that arose out of the myopic misunderstandings between the Chancelleries of Europe. Indeed, just as the Viennese protested that London only viewed A-H policy thru the distorted prism of its suspicion of Germany, according to German calculations : the British Empire would stand aside ~ but Britain took part. Italy would help her allies ~ but Italy refused. Serbia was a thing of naught ~ but Serbia shattered several A-H divisions. Belgium would not count ~ but brave Belgium by her actions counted for the loss of eight vital days, while by her sufferings she mobilized against the invaders the condemnation of the world. Similarly, British statesmen of the era were mostly unaware of the strength and intensity in the Russian mind of the often-balked ambition to reach Constantinople and to dominate the Straits. That was only revealed to them in full measure after the war had erupted.

The tragedy of the Great War has been ascribed to many different causes. It was of course, the natural outcome of foreign policies deliberately pursued, and behind the policy was the personality of those in whose hands were the destinies of Empires. Therefore, I have attempted to sketch the qualities of the leading actors in this tragedy, to bring forth some clear vision on this torment of world history which society still shudders to recall, to reveal how those who were responsible for the conduct of affairs tried to think in the years before the war, and how they applied their conclusions. All thru this whirlpool of history, the accent is on those individuals caught up in the prelude to war. Accordingly, and having plumbed the depths of archives for detail to bring this cast of characters to life, they can be sensed anew through their words to feel what they felt and knew at that fateful time, for a genuineness of feeling, thought and knowledge ~ uninfluenced by subsequent reaction or recantation. This timeless quality in the perception of events is especially esteemed in crystallizing the sequence of episodic scenes within, to help us fathom times past, and our shared humanity.

Rodney MacLean 2018

Le Petit Journal

ADMINISTRATION
61, RUE LAFAYETTE, 61
Les manuscrits ne sont pas rendus
On s'abonne sans frais dans tous les bureaux de poste

5 CENT. SUPPLÉMENT ILLUSTRÉ 5 CENT.
21me Année — Numéro 1.031
DIMANCHE 21 AOUT 1910

ABONNEMENTS

	SIX MOIS	UN AN
SEINE et SEINE-ET-OISE..	2 fr.	3 fr. 50
DÉPARTEMENTS..........	2 fr.	4 fr. »
ÉTRANGER	2 50	5 fr. »

LES AVIATEURS MILITAIRES A LA FRONTIÈRE DE L'EST

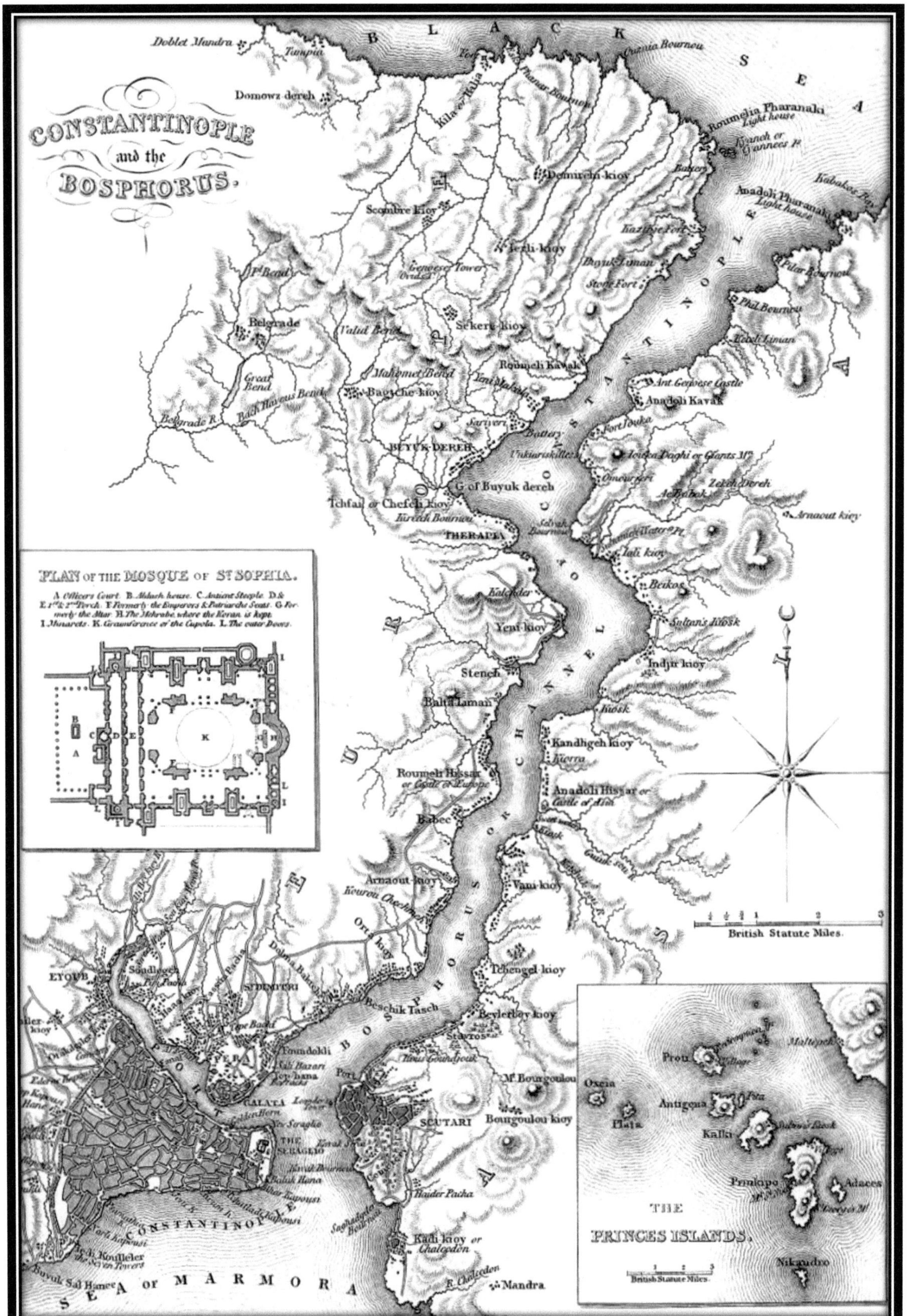

After the fall of Constantinople in 1453, Imperial Russia viewed itself as the last surviving part of Roman-Byzantium, and as the geo-political force to resurrect the lost leviathan ~ Third Rome. Such ambitions were supported by the Russian Orthodox Church and given legitimacy by the 1472 marriage of Ivan III to Sophia Palaiologina, relative of the last Byzantine Emperor. The centuries-long objective of the Tsars to recapture Constantinople ~ and to rename it as Tsargrad, City of Tsars, despite the many southern advances and expansions of Imperial Russia; was thwarted by opposition in the Crimean War, by the Austrian mobilization and British diplomatic pressure in 1878 (when Russian forces were 10km from Constantinople) and by the closing of the Turkish Straits in 1914.

PART I

UNDERLYING CAUSES OF THE GREAT WAR

During this early period, the threat of war or the use of war as an instrument of policy was throughout the principal weapon of all European Empires and the foundation of their instructions to their Ambassadors and attachés. From their use of war and the threat of war followed their alliances, armaments competitions, war scares, and their secret and disingenuous methods of diplomacy. As diplomacy that constantly kept war in view was under strong temptation to employ the ethics of war, the practice of deceit and the misleading of rivals were held to be justifiable and even necessary in the period leading up to the Great War. Thus, the leaders of the European Empires and other Kingdoms continued to play their compelling and dangerous game with the balance of power.

During the whole or the greater part of the period between the Crimean War and the Great War, there were four principal factors, the interplay of which determined the shape of politics in Europe. First was the fervent hostility of France and Imperial Germany. Second was the desire of Russia to gain full naval access to warm water ~ *to obtain the key of her house* ~ as Tsar Alexander II designated the control of the Straits of the Bosphorus and Dardanelles from the Black Sea through to the Ægean to ensure an outlet into European waters not blocked by winter ice or controlled by Baltic states. Third, the rising contention from 1900 between Britain and Germany over warships, colonies, and the perceived threat of the *Berlin-Baghdad Railway.* Fourth, as the old Ottoman Turkish Empire withered, an escalating alarm in Vienna over the rise of Serbian nationalism and an associated spread of Russian influence through the Balkan Peninsula, led to a view in the Austro-Hungarian Empire that their very existence was at stake. The key historical questions of how the war came to occur and who was responsible for it need to be answered if we are to learn any lessons from the mistakes of the past.

Russian foreign policy for a long period of time had been very aggressive. Ever since the time of Peter the Great, the Russians had looked to Constantinople, the once Holy City of their religion as the future capital of their Empire ~ *Tsargrad.* For three centuries Russia had been acquiring territory and expanding in a southerly direction toward the desired warm water ports of the Black Sea. From the strategic viewpoint of Tsar Alexander II, Russian tradition still demanded an advance to the Straits to achieve their dream of acquiring Constantinople and their control of the Straits. Russian Minister of War General Alexei Kuropatkin noted, *Our Tsar has grandiose plans in his head ~ to capture all Manchuria for Imperial Russia, and to annex all of Korea. He also desires to take the Bosphorus, and the Dardanelles.* Warm water ports that did not freeze over in the winter were essential for the development of Russian year-round trade and for their development of a strong navy. When Kaiser Wilhelm I of Prussia sent his own grandson Crown Prince Wilhelm to Imperial Russia with Bismarck on a conciliatory mission, *with direct instructions to offer Constantinople and the Dardanelles.* Tsar Alexander snapped sensationally; *If I wish to have Constantinople I shall take it whenever I like ~ without need of the approval of Prince Bismarck.*

Tsar Alexander II redeployed his forces to subdue the rebellious Tatars and Ukrainian Cossacks to secure the Ukraine, and then to acquire the Crimean Peninsula to ensure the Ottoman Empire had no buffer against Russian expansion. Russia and the Ottoman Empire then fell into direct conflict on this issue of *The Eastern Question* most briefly defined as the struggle of Europe for the heritage of the Ottoman Turkish Empire between the Danube and the Dardanelles, the Adriatic and the Black Sea, and the Balkan Peninsula in the far S.E. corner of Europe with its fertile plains of wild rich soil.

The Russians supposed that if they applied pressure that the Turks would give way and concede to their demands. However, the issue was complicated by Austria who had the vital interest in the Balkans to keep open the mouth of the Danube. Any threat by Russia to the Dardanelles would be viewed by Austria as a threat to their interests in the Balkans, while the Ottoman Empire saw that it would place Constantinople at the mercy of a Russian Fleet. Thus, British and French forces entered into the Crimean War of 1854 to confront Russia and prevent any such encroachment. The Crimean became the first war following the telegraphic connection of the capitals of Europe to benefit from recent advances in modern medicine, and to realize the full potential that steam power now offered.

Nicholas II
Tsar Emperor of Russia

The Grand Duchesses Olga & Tatiana
Daughters of Nicholas and Alexandra

Alexandra of Hesse-Darmstadt
Tsarina Empress Consort of Russia

Princess Sophia Hohenzollern of Prussia
Crown Princess of Greece
Younger Sister of Wilhelm II

Wilhelm II
Prince of Prussia & Kaiser of Germany

Princess Viktoria-Luise of Prussia
Daughter of Kaiser Wilhelm II

Prince Otto Eduard Leopold von Bismarck
First Chancellor of Imperial Germany

Archduke Franz Ferdinand d'Este
Heir Apparent to the A-H Dual Monarchy

Franz Josef von Hapsburg-Lothringen
Emperor of Austria & King of Hungary

King Haakon VII & Queen Maud of Norway
Maud of Wales

Alexandra of Denmark
Queen Consort of the United Kingdom

Edward VII
King of Great Britain & British Dominions

Princess Ena of Battenburg
Queen Victoria Eugenie of Spain

Mary Theresa Olivia Cornwallis-West
Princess of Pless

Princess Marie of Edinburgh
Queen Consort of Romania

Mîr-i livâ Ismail Enver Pasha
Minister of War ~ Ottoman Turkish Empire

Lord Horatio Herbert Kitchener
Field Marshal Earl Kitchener of Khartoum

Princess Katarzyna Radziwillowa
St Petersburg Courier

The German sentiment for the Near East and Africa was of some antiquity. The practical eye of Helmuth von Moltke in 1840 had discerned its potential when he conceived a German Principality of Palestine. Following the Franco-Prussian War, explorers like Eduard Sachau travelled in Asia Minor and Mesopotamia, and impressed their countrymen with the notion that these neglected regions, once so prosperous and fertile in ancient times could be reclaimed for civilization by German commerce. Paul Botticher, better known under his pen-name of Paul de Lagarde, wrote in the *National Zeitung; How often are we to repeat to our countrymen who emigrate to America, that they can find at their very doors magnificent countries only waiting to be exploited? We must create a Central Europe by conquering for German colonization large spaces to the east of our frontiers.* He foresaw a war would be necessary for this purpose, and urged his countrymen to prepare for it. Soon after Constantine Frantz suggested: *The establishment of a great Central European Confederation, extending from Holland to the Balkan States, and involving the retirement of Russia beyond the Dniester ~ Prussia must be able to wage an offensive war in the grand style.* Later writers carried their views further to include the entire Ottoman Empire. German Officials like Bernhard Dernburg, Professors Anton Sprenger, and Ernst Hasse pointed to the Near East as the most promising goal of German foreign interest and the new Pan-German League adopted their policies.

Between the nature of the German and that of the Turk, there is a profound relationship of character. The Turk, is the chief material and moral force of the East, there is therefore in his Empire, a sphere in which the German may have a great future, not only by way of material colonization, but also with a view to moral influence and political domination ~ even the prospect of using the Turkish Army as a spearhead against the English in Egypt, in case they should oppose German designs elsewhere. Paul Rohrbach

Immediately on his accession, the young Kaiser Wilhelm II having heard endless enthralling tales from Count Paul von Hatzfeldt, who for 14 years had been German Ambassador to Constantinople, was now so thoroughly convinced that the treasures of the East awaited German commerce in the vast territories of the Ottoman Empire, that he was ready to take full advantage of all opportunities that enabled him to further the economic progress of Germany, upon which his heart was truly set. Acting together in the summer of 1888, Hatzfeldt urged Kaiser Wilhelm to send discreet instructions to his Wilhelmstrasse Foreign Ministry in Berlin to embark on a more active foreign policy in the East and see their *Drang nach Osten ~ Drive to the East* transposed from the visionary phase into forward action.

Weltpolitik thus became the adopted foreign policy of Kaiser Wilhelm II and his diplomatic staff of the Wilhelmstrasse Foreign Ministry. The aim of *Weltpolitik* was to transform Germany into a global power through aggressive diplomacy and the acquisition of a series of strategic overseas colonies. German Chancellor, Prince Bernhard von Bülow made this abundantly clear in a speech of 1897.
Wir wollen niemand in den Schatten stellen, aber wir verlangen auch unseren Platz an der Sonne.
We wish to throw no one into the shade of the sun, but we demand our own Place in the Sun.

Thus the ground was prepared, and the favourable mental attitude for colonization induced. But a certain disillusionment followed. Colonies in fruitful and habitable portions of our world at that time were no longer to be had ~ at least without fighting for them. Yet the Ottoman Turkish Empire was a vast area, unoccupied or occupied only by wandering tribes or semi-civilized peoples, under the sway of an outdated, ineffective régime. Here the flourishing population of Germany might attempt promising settlements, where German finance could be invested in railway and irrigation projects, in mining and agriculture; and the increasing prosperity of these proposed colonies could in time provide markets for German produce. New land rendered fit for agriculture and possessing mineral wealth would also soon supply materials for manufacture, and food for the people at home.

To these economic incentives were added the political and other advantages to be drawn from a close connection with Turkey. Financial and economic control over Turkey, European and Asiatic, meant the control of S.E. Europe and the Eastern Mediterranean, the command of the Danube from source to mouth, the severance of Russia from the open sea, and an end to the Imperial Russian dream of celebrating an Orthodox Mass in Santa Sophia within Constantinople. More than that, German guidance of Turkey might also provide for the practical possession of Bagdad, and a rail route to the Persian Gulf, from where it would be easy to bring pressure to bear not only on Persia but also on Russian territory east of the Caspian, and even on British India.

Professor G.W. Prothero
King's College Cambridge

Kaiser Wilhelm paid his first visit to Constantinople in 1889, then to Jerusalem and Damascus in 1898 to unleash his powers of persuasion on Sultan Abd-Ul-Hamid II and declare; *The three hundred millions of Mohammedans who live scattered over the globe may be assured that I will be their friend at all times.* The eccentric Sultan, who was a devoted fan of Sherlock Holmes for insights into detection methods, asked his new friend Wilhelm to unwrap his gift of a German hunting rifle from behind the protection of a wrought-iron screen to reassure him that it did not contain an assassin's bomb. Berlin publicists grasped the importance of this visit to extol the Kaiser's activity and foresight, while Wilhelm attired in Eastern costume wrote to his cousin Tsar Nicholas II. *Dear Nicky, the Mahommedans are a tremendous card in our game, in case you or I are suddenly confronted by a war, with a certain meddlesome power.*

Queen Victoria, received reports of concern from the leaders of both Britain and Russia, who were anxious of the ambitions of Kaiser Wilhelm and the rise of German influence in the Middle East; in particular his adventurous partnership with Sultan Abd-Ul-Hamid II to forge a new trade route. The inspired German plan incorporated the existing rail routes from the Atlantic seaports of Bremen and Hamburg to connect with Berlin. Then to the Austrian capital of Vienna, from where the train would take the route of the *Orient Express,* down through the Carpathian Mountains to Budapest. Once there, the *Berlin-Konstantinopel* sleeping cars would be attached to continue on via Belgrade and Sofia, and over the Bosphorus to Constantinople. From Constantinople, a new line would be constructed S.E. to Baghdad, then down to a port terminus at Al Basra on the confluence of the Tigris and the Euphrates to enter the Persian Gulf thru the waters of the Shatt-al-Arab. This newly proposed *Berlin to Baghdad Railway* would open all Asia Minor, Mesopotamia, and Persia to German commerce.

Forward teams of German surveyors and accompanying geologists soon found evidence of oil between Mosul and Baghdad, and prolific deposits of antimony, lead, zinc, copper, chrome, and coal. In 1899, Sultan Abd-Ul-Hamid granted a rail concession to Kaiser Wilhelm ~ *My only friend in Europe,* in addition to a mining concession to the Baghdad Rail Co. consisting of oil and mineral rights to a parallel 20km on either side of the rail tracks across Asiatic Turkey all the way to the Persian Gulf. So, the new masterminds of Imperial Germany set forth on their *Drang nach Osten ~ Drive to the East.* As many Germans celebrated their reports of amazing prospects for mining and market potential, Kaiser Wilhelm unveiled a statue of his father at a Bremen banquet. *It is no empty world dominion to which I aspire, but to a worldwide dominion of the Hohenzollerns. God has destined Germany for a great future. We are the salt of the earth, but we must also prove ourselves worthy of this high calling.*

Other rail projects proposed by German engineers for construction were ; a branch line from the docks of Alexandretta to Aleppo to absorb the valuable trade of Damascus and Northern Syria, in addition to the Hejaz Railway to the south. As the Imperial Germans began to enjoy their economic and political power in Asia Minor following the completed sections of their *Berlin-Bagdad Railway,* their progress increasingly detracted from Russia's economic influence towards the Caucasian frontier and northern Persia where Russia had previously held dominant market share. With Constantinople's commercial interests realigning, the Ottoman Empire began closing their doors to Russian trade.

As German foreign policy could not be pursued without diminishing future Russian ambitions, Kaiser Wilhelm now required the complete co-operation of Emperor Franz Josef I of Austria-Hungary. Since the railway led through Belgrade and Sofia, A-H control of Serbia and Bulgaria became essential to Germany for the continued viability of the rail project that now formed the very core of Germany's eastward drive. Tsar Ferdinand of Bulgaria (as a cousin of the Kaiser) entered into a convenient alliance with Austria, but throughout Serbia, there was increasing animosity to Vienna.

To operate any railway that ran thru Serbia, both Germany and A-H required sway in Belgrade to secure the crossing of the River Danube. But, if Vienna was ever infringed upon, Berlin's dreams of their *Weltpolitik* route to Byzantium and Bagdad, and their entire Central European Customs Union, which was to encompass the Balkan States and Turkey, all faded into thin air. Therefore, the principal object of Berlin and Vienna's foreign policy became their effective domination of the Balkan Peninsula. By extension, while the establishment of German influence in S.E. Europe and in the Ottoman Empire became the fundamental focus of German policy, so the *Berlin-Baghdad Railway* began to play a rôle in British-German trade rivalry and their strategic manœuvrings. The railway therefore hastened a unity of suspicious powers against Imperial Germany, which led to a fear of encirclement within Germany, as Berlin's increasing involvement in the East via the Balkans, strengthened her alliance with Vienna.

During the time of the South African War (1899-1902), anyone who had visited Germany would have discerned several disturbing trends. Regular disparaging articles against Britain confirmed that German foreign policy had become basically anti-British, while repulsive reports of savage tales about Great Britain would have persuaded visitors that Germany was not a land of *Küche, Kinder und Kircher* as previously supposed, but rather a people preparing to put on spiked helmets. Kaiser Wilhelm's geopolitical ambition to realign Germany with the Ottoman and Hapsburg rulers of the Balkans, and to dominate in the Bosphorus, now saw his grandiose *Weltpolitik* policy of the *Berlin-Baghdad Railway* promise to provide Germany an opening of a land corridor to the Near East that could elude British sea power in the event of war, and threaten both the Suez Canal and ultimately British rule in India. The senior British military advisor to the Serbian Army, Major R.D.G. Laffan expressed his concerns.

If Berlin-Baghdad were achieved, a huge block of territory producing every kind of economic wealth, and unassailable by sea power would be united under German authority. This barrier would cut off Russia from her western friends, Great Britain and France. German and Turkish armies would be within easy striking distance of our Egyptian interests and the Persian Gulf, and our Indian Empire would be threatened. The port of Alexandretta and the control of the Dardanelles would also give Germany naval power in the Mediterranean. Then our vast, but slightly defended Empire would soon feel the shock of Germany's eastward thrust.

The British response came in 1901, to formulate and intrigue for a mutually protective agreement with Sheikh Mubarak Al-Sabah of Kuwait. British warships appeared off the Kuwaiti coast to dictate to the Ottoman Empire that they recognise the Persian Gulf port located just below the Shaat-al-Arab, controlled by the Anaza tribe to become a British Protectorate. The corrupt, ruthless Sheikh of Anaza, who had seized power in 1896 after murdering his two half-brothers as they slept in his palace, was offered generous gifts of British gold and fine rifles to encourage his signing of a lease in perpetuity. This acquisition of influence over Kuwait was then applied in the hope of blocking the completion of the *Berlin-Baghdad Railway* from an eventual access to the waters of the Persian Gulf. The consequence of this supposed strategic advantage was that Great Britain had once again contributed to the German perception of encirclement, and further polarized the now perilous diplomatic situation.

In 1902, Germany secured from Turkey their next concession for the *Berlin-Bagdad Railway* to extend the rail connection from the recently completed eastern terminus at Konia in Asia Minor. From March 1903, construction from Konia continued across the Anatolian tablelands, along the foothills of the Taurus Mountains and on to connect with Mosul-Baghdad-Basra in stretches of 200km at a time. By the autumn of 1904, the first 200km rail section from Constantinople to Bulgurli was declared open. In 1911, a large team of German railway engineers was engaged to press railway construction at the rate of a kilometre a day to achieve the German dream of the *Bremen-Berlin-Bosphorus-Bagdad-Bahn*. By 1914, when Baghdad was just another 650km distant across the harsh desert landscape, the railway had already done much to poison the international atmosphere. Germany had come to believe that Britain, Russia, and France were deeply suspicious of their plans in the Near East. In particular, Russia knew any advance of Austro-German interests into the Balkans and Turkey needed to be thwarted if her own intrigues for securing control of the Straits and Constantinople were not to be checked by an obstacle more fatal than a defacto Turkish dominion in the form of a Teutonic corridor from Berlin to Baghdad, and a steam-powered connection from the North Sea to a terminus on the Indian Ocean.

For Berlin to secure so great a prize, it was indispensable that the Austro-Hungarian Empire should dominate the Balkan Peninsula. The disadvantage of a close alliance with the Dual Monarchy induced some in Germany to advocate a rupture of the connection, but this view appears to be have been rare. The retired first Chancellor of Imperial Germany, Prince Otto von Bismarck once stated; *Germany's interests in the Turkish-Balkan area are not worth the bones of a single Pomeranian Grenadier.* Bismarck had always been determined not to become involved in an area where Austria and Russia were at odds, since Germany could not appease either nation without offending the other, and both were dangerous as enemies; so while previously considered not worth becoming involved, as German economic power flowed into the Ottoman Empire, the great scheme of the *Berlin-Baghdad Railway* could in no other way be carried out. As the route from Berlin to Byzantium and thence to Bagdad, ran through Vienna, Russian dreams of *Tsargrad* and the Straits were compromised, and some said;
The road to Constantinople is via Vienna, the road to Vienna, via Berlin.

* * *

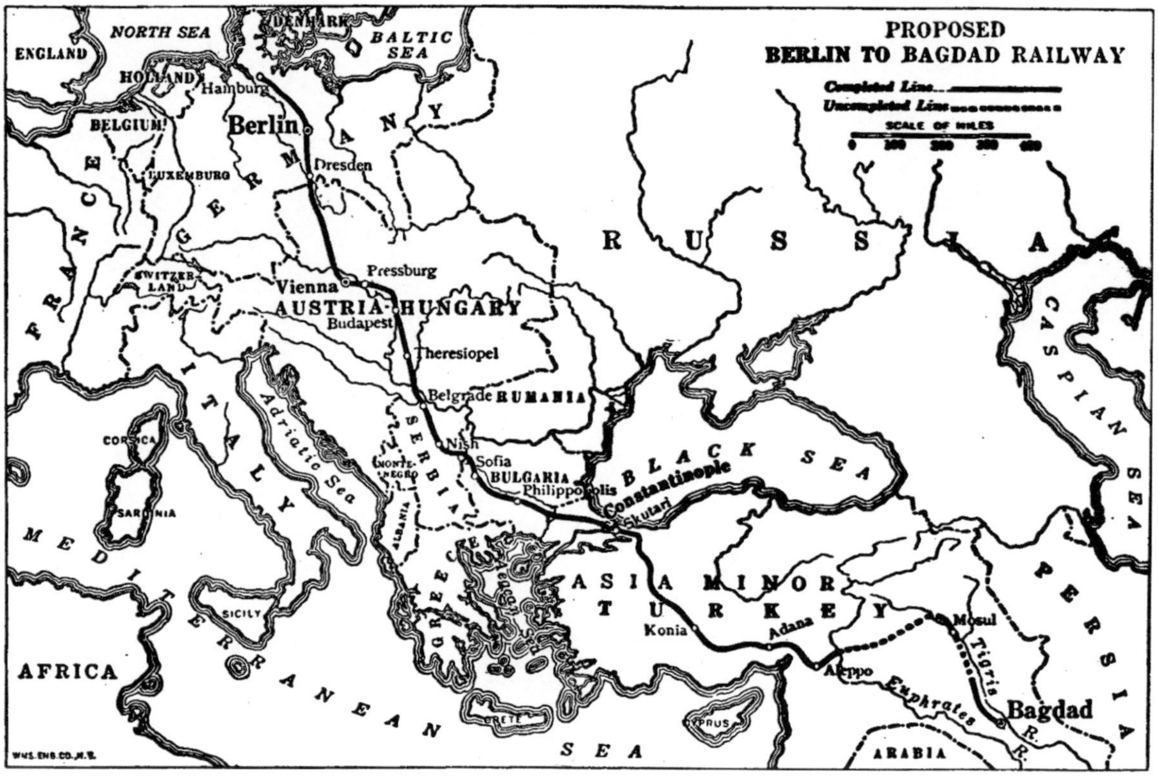

Smoking Lounge and Library

… a wash and a brush up

Interior of the Sleeping Car

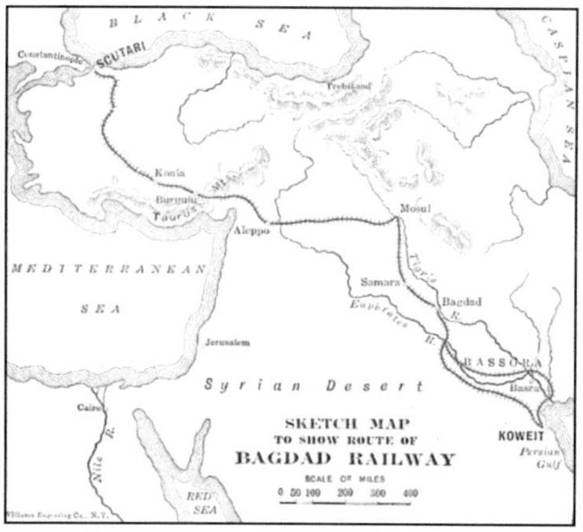

RISING CONTENTION BETWEEN THE EMPIRES OVER COLONIES

In 1870, when only 10% of Africa was under effective European control, surplus capital was often more profitably invested overseas where cheap materials, limited competition, and abundant raw materials made greater premiums possible. Added inducement for colonial adventures arose from the demand for raw materials unavailable in Europe; especially cotton, coffee, cocoa, palm oil, diamonds, sugar cane, tobacco, mahogany, rubber, ebony and ivory, for which European consumers had grown accustomed and upon which commerce had grown dependent. Thus began the *Scramble for Africa*. The travels of Livingstone and other explorers having revealed the *Dark Continent* as a new sphere for the expansion of the Empires moved many to peg out their claims. Great Britain secured the southern and eastern coasts of Africa for stopover ports on the route to Asia and her empire in India. In 1879, the King of Belgium claimed the Congo, while France and Britain established Dual Control in Egypt. France resumed her activities in West Africa and occupied Tunis in 1881. In 1882, Britain took full control of Egypt and Basutoland as Italy seized Eritrea on the Horn of Africa. In 1883, France claimed Upper Nigeria and Madagascar as Britain acquired Bechuanaland and Nyasaland.

As the *Scramble for Africa* intensified, Prince Otto von Bismarck convened the *Kongokonferenz* in Berlin during 1884 to discuss African issues. The diplomats condemned the slave trade, prohibited the sale of alcoholic beverages and firearms, and established rules of competition by which European Empires could be guided in seeking colonies. Through negotiation, Bismarck proclaimed German protectorates in arid South-West Africa, and sanctioned the occupation of Togoland and Wituland. German colonial expansion then became active in the South Pacific where many attractive islands were also acquired : German New Guinea, the Bismarck Archipelago, Bougainville, Nauru, Samoa, the Northern Solomons, and in April 1898, a further three island groups were purchased from Spain.

By a treaty which is not to be laid before the Spanish Cortes for ratification for months and must remain absolutely secret until then, the Spanish Government has handed to us the Caroline, Pellew, and Marianne Islands (without Guam) in return for the sum of 16 million marks and certain trade concessions. We shall thus become neighbours of the Americans who possess the Philippines. The value of the American market for our industry is immense. For the sake of our coming occupation of the Caroline Islands, we must be assured against disturbance by the Americans. So we must avoid any appearance of supporting the rebellious Tagals, and especially of any intention of extending our influence over the Philippines. Prince Bernhard von Bülow

Chancellor of Germany

Next, densely populated Qing Dynasty China came into view as a colonial venture for Germany, which despite her expanding colonial possessions and rapidly growing commerce still possessed no maritime coaling stations ~ and Germany needed coaling stations for her ships. To meet the demands of German commercial interests to open up China to international trade, Berlin resolved that under unimpaired Chinese sovereignty and after payment of the Chinese *likin* tax, a trading port with a maritime coaling station be founded, wherein German colonists would co-operate with the interests of Qing China to serve commerce, with a military presence limited to the security of their trading center.

Several places along the coast of China had been considered by Germany, but on more careful investigation most had either ; bad connection with the interior regions, were not promising from a commercial-political standpoint or were encumbered by privileges already granted to other nations. And so, it was agreed ~ because the reports of the Commander of the East Asiatic Cruiser Squadron, Admiral Alfred von Tirpitz, had drawn such a promising picture of the possibilities of the potential for development in Shandong ~ to found a settlement on the Bay of Kiao-chau. Further positive reports came in from many conversations among the talkative officers of the Russian China Division. It was learned that their Russian Admiral, had one winter anchored in the bay, but had found it too desolate and so atrociously lonesome ~ because there were no Japanese tea houses with geisha girls (which he deemed indispensable to winter quarters) thus the Russian Squadron would never return. Hence, the Russians had no intention of gaining a foothold there. This last piece of news arrived at the same time as their prerequisite answer from Tsar Nicholas. *I have no interest in the territory south of the Tientsin-Pekin line, which means that there is no reason why I should place obstacles in your path in Shandong. My interest is concentrated upon the territory on the Yalu, around Port Arthur. Now that the English have made difficulties for me at Mokpo; I would, in fact be pleased if Germany should locate herself in future on the other side of the Gulf of Chih-li as Russia's welcome neighbour.*

After the news arrived in Berlin of the murder of two German Catholic missionaries in 1897, by the *Big Sword Society* in Shandong, Admiral Ernst von Diederichs (now CO of the Cruiser Squadron) wired to Berlin; *May incidents be exploited in pursuit of further goals?* Upon receipt of the Diederichs cable, Kaiser Wilhelm intervened and replied; *Proceed immediately to Kiao-chau with entire squadron.* Diederichs; *Will proceed with greatest energy.* Compensation was demanded from the Qing Dynasty, and the acquisition of Jiao Zhou Bay in Kiao-chau province on the south coast of Shandong Peninsula was the successful outcome. In the small centre of Tsingtao, much was done to promote commerce and industry jointly with the Chinese. The flag of Chinese Empire was hoisted over the Custom House at Tsingtao, which soon became a prospering German commercial colony, where many Chinese people worked side by side with the German settlers to establish the Tsingtao Brewery and complete the Shandong rail link to the Trans-Siberian Railway to enable train travel from Tsingtao to Berlin. Germany now had her *Place in the Sun* aplenty. Though her acquisitions were small in area, these colonial possessions added much to her strength and attained the object for which they were acquired. They were a nucleus, which Imperial Germany now hoped to enlarge in time by diplomacy or arms. In a launch ceremony speech to the North German Regatta Association, Kaiser Wilhelm confirmed it.

In spite of the fact that we have no such fleet as we should have, we have conquered for ourselves a place in the sun. It will now be my task to see to it that this place in the sun shall remain our undisputed possession, in order that the sun's rays may fall fruitfully upon our activity and trade in foreign parts, that our industry and agriculture may develop within the state and our sailing sports upon the water, for our future lies upon the water. The more Germans go out upon the waters, whether it be in races or regattas, whether it be in journeys across the ocean, or in the service of the battle flag, so much the better it will be for us.

Chancellor Prince von Bülow, whose own mother-in-law said, Bernhard makes a secret out of everything, and who Friedrich von Holstein described as; Having read more Machiavelli than he could digest! will live on in German history as a man who resembled Bismarck in but one important particular; the gift of phrase-making. When he proclaimed that Germany demanded her ~ Place in the Sun ~ he flung into the fire fat which was to go sizzling down the age. It was worth its weight in precious gems to the dangerous blood-and-iron brigade. Bismarck's bluster in 1887 gave the War Party of that day its fillip. They snatched it up with alacrity, and being Germans, proceeded to exploit it with masterly efficiency and deadly thoroughness. A place in the sun forthwith inspired an entirely new German literature. It became the spiritual mother of this war. Lord Thomas Brassey

While some small German colonies proved disappointing ~ like barren German Southwest Africa where the only successful activity had been that of gathering small diamonds discovered in the sands of the beaches and in the desert running back from the sea ~ even if they had been more successful, they would hardly have satisfied surging German ambitions or been regarded as sufficient outlets or feeders of German trade. Commercial and industrial exigencies soon prompted other schemes and methods. Commercial treaties advantageous to Germany formed one of these methods. A notable example was the treaty forced upon Russia during the Russo-Japanese War, which gave valuable rights to German products whilst laying a heavy duty on Russian corn.

Commercial treaties were only a minor item in the Imperial German programme. The idea of an immense Central European Customs Union was re-launched of a vast protectorate extending from the Balkans to the Persian Gulf. Julius von Eckhardt wrote in *Berlin-Wien-Rom* proposing that *Mitteleuropaischer* be formed to embrace Austria-Hungary, Switzerland, Italy, and other small States, (which should be either forced to join or economically starved into adhesion). In *Germania Triumphans* published in 1895, the ultimate triumph of Germany was foretold as being accomplished in 1915 after a series of wars. With politics and trade regarded as inseparable, successful war was the means by which the double end could be achieved. Berlin militarists now aimed at the eventual absorption of Belgium and other countries or parts of countries, the conquest of the French and Belgian colonies, and the diminution of France and Russia. Such were the ideas of some German strategists of the 1890s.

Continual commercial and colonial expansion, going hand in hand, had another notable result : the German fleet. The two objects, for which the German fleet was originally formed, were the defence and extension of their oversea colonies, and the protection of oversea commerce. Without a fleet, neither colonies nor commerce could be regarded as secure, and both would be at the mercy of a *stronger naval power*. Such were the reasons alleged in the preamble to the First Navy Law. The reasons were good, the need was obvious, so the growing wealth of Germany supplied the means to meet it. The ulterior aims that were subsequently disclosed, were to serve German interests during peace through gunboat diplomacy, and in times of war through commerce raiding to protect German trade routes, and to disrupt hostile ones. A global network of coaled naval bases was a prerequisite for this.

The German Navy was the special creation of Kaiser Wilhelm, who began reforming it in the 1890s with Admiral Alfred von Tirpitz as its head. In 1898 the First Navy Law was passed to authorize the construction of 16 battleships in the following three years. Then during the South African War in 1900, British cruisers on patrol detained the *Bundesrath* and two other German steamers off the coast of Africa to search them on a firm suspicion of carrying Mauser rifles and Krupp artillery to the Boers. While all of Germany was outraged, Admiral Alfred von Tirpitz took advantage of their anger to pass a Second Navy Law to double the strength of the German fleet. An attached memorandum defined it. *To protect Germany's sea trade and colonies, in existing circumstances, there is only one means. Germany must have a battle-fleet so strong that, even for an adversary possessing the greatest seapower, a war against it would involve such dangers as to imperil his position in the world.* After this, the German Government refused to retard its accelerating rate of building warships ~ in the everclear expressive words of Kaiser Wilhelm;
The duty of watch-keeping officer in the ship of State has now devolved upon me!
The course remains as it was. Full steam ahead!

Although the creation of their fleet was welcomed throughout Germany with the Kaiser himself leading the campaign with phrases like ~ *Our future is on the water. Bitter is our need of a strong fleet* ~ the rapid expansion of the German Navy began to concern the British, who had long supposed that Russian designs on India were the main threat to their Empire, but from 1880 they came to believe that perhaps it was Imperial Germany, and not Russia perchance that now challenged the British Empire. The Earl of Selborne, as First Sea Lord of the British Admiralty was of the following opinion.

The naval policy of Germany is definite and persistent. The Kaiser seems determined that the power of Germany shall be used all over the world to push German commerce, possessions, and interests. Of necessity, it follows that German naval strength must be raised to compare more advantageously than at present with ours. The result of this policy will be to place Germany in a commanding position if ever we find ourselves at war. Naval officers who have seen much of the German Navy lately, are all agreed that it is as good as can be.

On 4th April 1897, Kaiser Wilhelm delivered a dramatic speech in Cologne in which he said; *Neptune with the trident is a symbol that we have new tasks to perform, and that trident must be in our hands.* This generality attained strategic focus in the Reichstag, when Admiral von der Goltz explained:

Let us consider the case of a war with England. In spite of what many people think, there is nothing improbable in it owing to the animosity which exists in our country towards England and to the sentiments of the British nation towards all Continental Powers and in particular against Germany. The opinion is generally held in this country that any resistance against England at sea would be impossible, and that all our naval preparations are but wasted efforts. It is time that this childish fear, which would put a stop to all our progress, should be pulled up by the roots and destroyed. The maritime superiority of Great Britain, overwhelming now, will certainly remain considerable in the future, but she is compelled to scatter her forces all over the world. In the event of war in home waters the greater part of her foreign squadrons would no doubt be recalled; but that would be a matter of time, and then all the stations overseas could not be abandoned. On the other hand, the German fleet, though much smaller, can remain concentrated in European waters. With the increases about to be made we will be in a position to measure our strength with the British naval forces in home waters.

On 10th February 1906, the British Royal Navy launched HMS *Dreadnought* as the culmination of practical experience gained in the 1905 Russo-Japanese War, and as a consequence of having asked: *What is the real object of successive developments in the German Navy, and can the reported facts be reconciled with official German communiqués that such naval expansion is merely an aspect of German foreign trade, and the duty of protecting her territorial interests over the seas?* The most comprehensive answer to this was revealed in a speech delivered by Winston Churchill to the Committee of Imperial Defence.

I should like to point out, that the repeated increases have occurred quite irrespective of what we had done ourselves. In fact, the ultimate scale of the German fleet is of the most formidable character. Of course, it is quite true that according to the German Navy Law, as a great many German speakers on the subject have always said, the German fleet does not exist in order to be a menace to the British fleet, and it does not contemplate anything of that character; it only exists for the protection of German trade and of German colonies. We are speaking here without the reserves which are necessary in public utterance, and I am bound to say, speaking on behalf of the Admiralty, that we find it very difficult to reconcile such statements with truth ~ very difficult indeed. The whole character of the German fleet shows that it is designed for aggressive and offensive action of the largest possible character in the North Sea or the North Atlantic ~ action, according to the to the memorandum accompanying their first Bill, against the strongest naval Power at some moment when that Power will not be able, owing to some duty which it may have to discharge to its colonies or to some other part of the Empire, to keep all its forces concentrated to meet the blow. The structure of the German battleships shows clearly that they are intended for attack and for fleet action. They are not a cruiser fleet designed to protect colonies and commerce all over the world. They have been preparing for years, and are continuing to prepare on an ever-larger scale a fleet, which, from its structure and character, can be proved by naval experts to have the central and supreme object of drawing out a line of battle for a great trial of strength in the North Sea or in the ocean. I will not go into technical details, but the gun position, the armament, the way the torpedo tubes are placed ~ all these things enable naval experts to say that this idea of sudden and aggressive action on the greatest scale against a great modern naval Power is undoubtedly the guiding principle of German naval policy.

When you go to the smaller types of vessels, the same principle can be traced. In their torpedo boats, speed has been the principle essentially that they have gone upon and that they have developed. Their torpedo boats are undoubtedly designed with a view to developing an attack upon the great ships of the navy that they may be opposed to. That again is a very significant fact. Now we come to the submarine. If there ever was a vessel in the world whose services to the defensive will be great, it is the submarine. But the German development of that vessel, from all the information we can obtain, shows that it is intended ~ to turn even this weapon of defence into one of offence that is to say, they are building not the smaller classes which will be useful for the defence of their somewhat limited coastline, but the large classes which would be capable of sudden operation at a great distance from their base across the sea. So, I think I am justified in saying that the German fleet exists for the purpose of fighting a great battle in the North Sea, both with battleships and with all ancillary vessels, against some other great naval Power which is not referred to by them.

Meanwhile, the escalating alarm of the Hapsburg Monarchy over the rise of Serbian nationalism as a threat to A-H control of the Balkans was being traced back to a series of events in old Venetia; inclusive of the city of Venice, Istria and Dalmatia that Napoleon had decreed Austrian possessions. On 2nd December 1848, the Archduke Franz Josef ascended the throne of the then Austrian Empire as Franz Josef I, but in this troubled year, his Hapsburg Monarchy was in grave danger of collapsing. Twice that year, revolution in Vienna put the Imperial Hapsburg family to flight. First to Innsbruck, then to the fortress city of Olomouc. Never, until its end in 1918, would the Hapsburg dynasty be harder pressed than in 1848. Franz Josef, once overwhelmed from the anxiety of such responsibility placed on his young 18-year old shoulders, knelt before his Uncle Ferdinand, who whispered to him… *God bless you, be good. God will protect you.* Austrian aristocrats, and army officers alike were confident that the dashing young and imposing Franz Josef I as their Emperor of Austria, King of Bohemia, Hungary, and Lombardy-Venetia, would be the savior of their turbulent Danubian Monarchy.

In the Italian provinces, then part of the Austrian Empire, increasing rebelliousness erupted into open revolution in Milan on 18th March 1848, when crowds clamored for the separation of Lombardy from the Austrian Empire. After the Austrian Governor rejected this, five days of subsequent street fighting forced Count Johann Josef Radetzky to withdraw his small garrison to Verona. During April, Franz Joseph left Vienna to join Count Radetzky's force in the field in Italy, to get a taste of army life. He showed a good aptitude, and it was prophetic that the first time he heard bullets whistle around his head in combat, it was in action against revolutionaries. Count Radetzky consolidated his force, and on 6th May facing heavy odds at Santa Lucia, his 19,000 men defeated 41,000 Piedmontese rebels. A series of Austrian victories by Radetzky soon compelled the Italians to conclude an Armistice.

When a Venetian revolt against the Austrians who controlled much of Italy erupted in 1849, Austrian forces retaliated by blockading Venice. When the siege was over, Count Radetzky wisely allowed the Venetian patriots to slip away, hoping to avoid any future commotion or serious ill will. The Austrians then set about administering Venetia in quite a different manner. By abandoning the traditional methods of the ancient *Venetocracy,* any image carved in stone on the walls or columns of the old city to indicate the presence of old Venice's power were chiseled away and destroyed forever. One of the great successes of Venetian rule, had been the peace and harmony of the relationships between the Italian, Venetian and pro-Venetian Slavic elements in Istria and Dalmatia. The Imperial Austrians, without realizing the damage they were doing ~ particularly to themselves ~ thoughtlessly rekindled old rancours. As the Italian community were drawn to the appeal of their national *Risorgimento* or Resurgence, so the Slav community responded to the appeal of pan-Slavism, looking towards Serbia and thinking of a union of southern Slavs, which in time would bring about the tragic synchronicity of Serbian nationalism to spark the Great War, and the end of the Hapsburg Monarchy.

Assassination Attempt. On 18th February 1853, as Franz Josef I was walking along the city walls of Vienna accompanied by his adjutant, Count Maximillian O'Donnell. As the two stopped to watch the soldiers drilling below, the Hungarian nationalist Janos Libényi rushed the Emperor from behind. Seeing the sneak attack coming, the shriek of a lady enabled Franz Josef to turn to avoid the full impact of Libényi's knife striking at his neck. While the heavy Austrian collar of Franz Josef's uniform protected his body from the assassin's blade, Count O'Donnell pounced on the threatening assassin. A young Viennese butcher, Josef Ettenreich was nearby, and he too rushed in to overwhelm Libényi. For his deed, he was elevated to the nobility by Franz Josef to become Baron Josef von Ettenreich. Libényi was led away and hanged. Thanksgiving services were held throughout the Empire, and the fine new church, *Der Votivkirche* was built in Vienna to commemorate the Emperor's deliverance. Libényi's assassination attempt had been instigated by Hungarian émigrés living in London, from whom Libényi had received money. The Austrians then overstepped the mark in both Budapest and the Hungarian countryside where people were arrested for wearing ribbons in the Hungarian national colours, and toy shopkeepers were questioned for displaying any dolls dressed in such colours.

In the summer of 1875, when a Serbian peasant rebellion erupted against the rule of the Ottoman Empire in their western province of Bosnia, the Austrian Foreign Minister Count Gyula Andrássy, followed the uprising with extreme attention as Vienna had identified the Near East as the area of particular interest for possible territorial and political expansion. With Ottoman forces soon unable to stop the uprising in Bosnia, Andrássy saw an opportunity to move Vienna's policy forward and after conferring with Berlin and St Petersburg, reached an understanding as to their attitude toward the Eastern Question. Their views were embodied in the dispatch known as the *Andrássy Note* sent on 30th December 1875 to their Austrian Ambassador to Great Britain, Count von Beust. The document outlined Vienna's vision for Bosnia-Herzegovina as a territory governed and administered by Austria. After obtaining assent from Britain and France, the document became the basis for future negotiations.

When the Principalities of Montenegro and Serbia similarly rose in rebellion and declared war on the Ottoman Empire, Vienna conferred with St Petersburg and came to the Reichstadt Agreement in which they mutually arranged that almost any result should be to their advantage. If Montenegro or Serbia were ever defeated, Russia was to intervene to save them from any disastrous consequences. If their nationalist uprising won, Austria was to get Bosnia-Herzegovina, and Russia was to recover Bessarabia, which she had lost in the Crimean War. If the Ottoman Empire collapsed, the two Empires were then to act together to create autonomous principalities ~ and to make Constantinople a free city.

All possible eventualities seemed to be worked out to the advantage of the two Empires, and for more than a year everything followed according to plan. In 1876 Russia intervened to save the Serbs when the Ottoman Turks had them at their mercy, and when the Bulgarians rebelled the following year, Tsar Alexander II of Russia made good his threat to act single-handed. The Austrians renewed the Reichstadt Agreement and gave him a free hand to do so.

However, the situation changed rapidly in January of 1878, when after an exhausting campaign, the Tsar approached the gates of Constantinople, and proposed to dictate the Treaty of San Stefano. The Tsar's brother, Grand Duke Nicholas wrote to him that it was necessary to take Constantinople. But fear of English and Austrian intervention restrained the Tsar. He ordered his brother to advance without taking either Constantinople or the Gallipoli Peninsula, which overlooked the Dardanelles. On 30th January, the Ottoman Turks accepted Russian conditions for peace negotiations, and signed an armistice with the Russians halted at a line that at its nearest point was 50km from Constantinople. With the Tsar's forces said to preparing the destruction of the Ottoman Empire, and for an emergence into the Mediterranean, Russia promptly demanded and obtained Turkish acquiescence in the occupation of San Stefano on the Sea of Marmara ~ only 10km from the outskirts of Constantinople. There the Russians and Turks signed the Treaty of San Stefano in March to create a Big Bulgaria, provide Russia with territorial gains in the Caucasus, territorial gains for Serbia and Montenegro, and recognize the full independence of Serbia, Montenegro, and Romania. While most were happy with the treaty, the Tsar anticipated that London and Vienna would find the treaty unacceptable.

The Austrians declared the San Stefano Treaty a breach of their pre-war agreement with Russia, and judged that the Russian proposal to establish a Big Bulgaria under Russian influence would more than neutralize the advantages they expected from the occupation of Bosnia-Herzegovina and diminish Austrian prestige and influence in the Balkans. British diplomats were soon in full eruption, when their Government also judged that the Big Bulgaria concept combined with the imminent Russian threat to Constantinople would be a blow to British interests that must be averted at all costs. Russia was serious, having already re-fortified Sebastopol and created a Black Sea Fleet. Austria and Britain joined in protests and when Prince Gorchakoff announced that Russia intended to occupy Constantinople, Austria mobilized her army and Britain ordered a fleet to the Sea of Marmara.

As Tsar Alexander could not confront the combined forces of the Austrian and British Empires, he succumbed to Austrian and British diplomatic pressure and agreed to some modifications with the concerned Powers agreeing to meet in Berlin to reach a mutually acceptable postwar settlement. The Congress of Berlin opened in June 1878 with Prince Gorchakoff as the head of Russia's delegation. His friend Count Shouvaloff, who mixed easily in fashionable circles, undertook the mission of maintaining whatever war gains were possible in the potentially hostile atmosphere of the conference. Bismarck, with whom Shouvaloff was on excellent terms, would keep the Congress sessions on track.

THE ENTRANCE TO THE DARDANELLES

THE CASTLE OF EUROPE

THE CASTLE OF ASIA

BRITISH FLEET STEAMING THRU THE ENTRANCE OF THE DARDANELLES

23RD JANUARY 1878 ~ DISRAELI ORDERED THE BRITISH FLEET TO THE DARDANELLES

8TH FEBRUARY 1878 ~ BRITISH FLEET ENTERED TURKISH WATERS AND ANCHORED OFF CONSTANTINOPLE. TSAR THREATENED TO OCCUPY CONSTANTINOPLE ~ DID NOT ACT

23RD MARCH 1878 ~ RUSSIA REJECTS LONDON'S PROPOSAL TO LAY THE PROPOSED TREATY OF SAN STEFANO BEFORE A SPECIAL SITTING OF A EUROPEAN CONGRESS

27TH MARCH 1878 ~ IN ANTICIPATION OF WAR WITH RUSSIA ~ DISRAELI MOBILIZED THE RESERVES AND DEPLOYED A CONTINGENT OF SPECIAL INDIAN REGIMENTS TO MALTA

As Ottoman Turkish rule in the Balkans faded to insignificance, their troops in Bosnia deserted in such numbers that 4000 deserters roamed the countryside in gangs terrorizing the people and seizing control of some areas. The Austrian consul in Sarajevo, Konrad von Wassitsch reported to Vienna: *Ottoman administrative bodies have no authority, and the population has lost its confidence in the government.* With this, Austrian Foreign Minister Gyula Andrássy accelerated his case for the Austrian occupation of Bosnia Herzegovina in a memorandum to the other European Empires arguing that an autonomous Bosnia within the Ottoman Empire lacked the means to overcome crime and internal ethnic divisions.

Count Andrássy first concurred with Berlin and London that the final adjustment of matters be submitted to a European congress. He then became the principal Austrian plenipotentiary and directed his efforts to diminish the gains of Russia ~ and to aggrandize the Hapsburg Monarchy. Before the Congress, talks between Andrássy and British Foreign Secretary, the Marquis of Salisbury had already concluded with Britain agreeing to all Austrian proposals relative to Bosnia-Herzegovina to come before the congress, while Austria would support all British demands ~ one of which was for Britain to administer the Ottoman-held island of Cyprus.

As a show of force and statement of intent simultaneous with the start of the treaty negotiations, the Austrian Army mobilized some 80,000 troops on its southeast border, ready to thrust into Bosnia. During this action, Vienna made a point of deploying Hapsburg subjects of a Southern Slav nature: ethnic Croats and Serbs, sensing that the Bosnian people would be less outraged by an Austrian occupation seeing their own kin amongst the invading troops. On 28th June, the Treaty of Berlin decreed that Austria receive a *mandate to occupy and administer the Bosnian Vilayet* ~ but not to annex it. Bosnia-Herzegovina remained part of the Ottoman Empire to be governed *de facto* as a part of Austria.

When it was over, Berlin had not been compelled to show her hand, and war had been averted. Whatever had been settled, had been settled by direct negotiations between the Empires directly concerned. While there were shouts of triumph in London and Vienna; there was disappointment for their Slav clients, both Bulgar and Serb who had got far less than they expected ~ there was also the shifting balance to Russia's disadvantage through the Austrian occupation of Bosnia-Herzegovina. There was above all the smothered disappointment at the miscarriage of her avowed aims ~ the possession of Constantinople as *Tsargrad,* and the mastery of the Straits. Prince Gorchakoff confessed; *The Treaty of Berlin was the darkest page in my life...* Alexander replied, *and in mine too.* The Treaty was a grave disappointment to Russian nationalists. The grateful Big Bulgaria they had hoped to see created was reduced in size. Although territorial gains, expanded rights for the Balkan peoples and a weaker Ottoman Empire remained real accomplishments, the nationalists were frustrated from compromises.

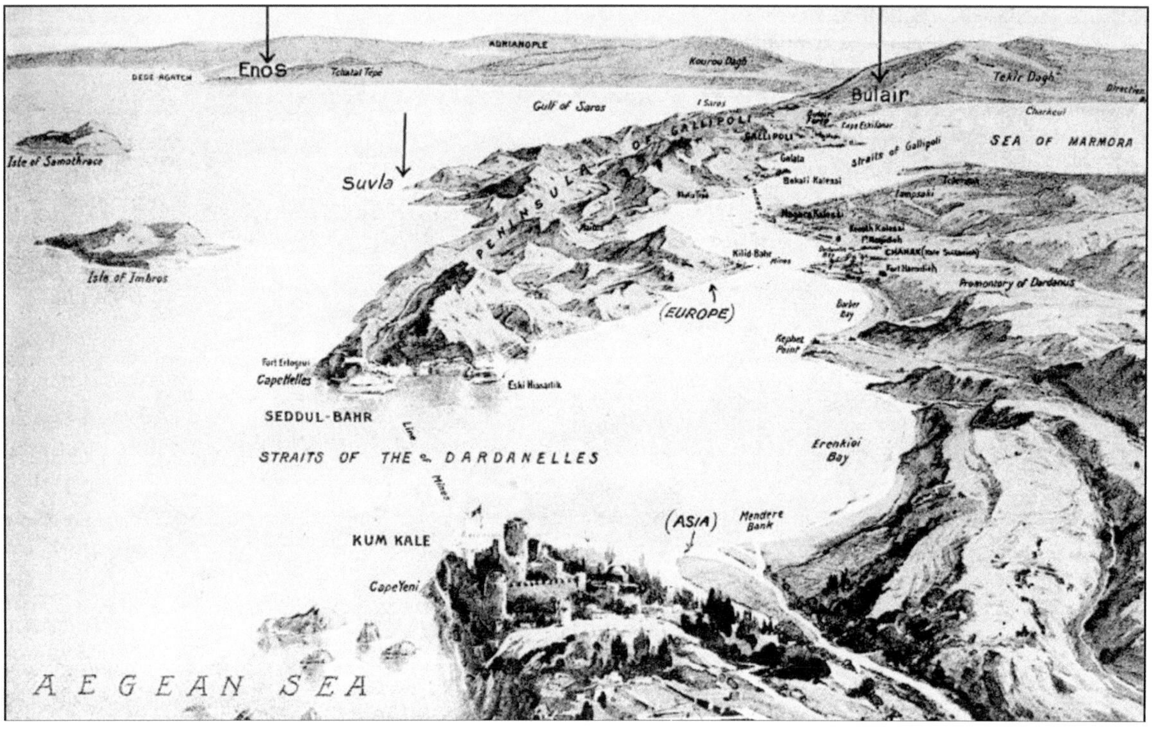

Balkan Border changes in the 1878 Treaty of Berlin saw territory gained by Serbia and Romania, some territory ceded by Romania to Russia, the new Principality of Bulgaria, and the autonomous province of Rumelia. The scene also saw Bosnia–Herzegovina occupied by Austria and Antivari given to Montenegro. Britain supported Austrian ambitions in Bosnia, but the smaller nations had objections. While satisfied to receive formal independence, Serbia opposed the Bulgarian expansion; Romania was disappointed with Russia breaking their agreements that guaranteed their country's territorial integrity ~ and Albania objected to a loss of territory to Serbia, Bulgaria, and Montenegro.

During the Berlin Congress, there were other ominous signs. France during the discussions had been suspiciously forthcoming to Russia, and it was reported from London that French Ambassador Comte de Saint-Vallier had broached the idea of a *rapprochement* between Britain and Imperial Russia. Already the future cast its shadow over the scene. During this time of peace, there continued to be elaborate and incessant Intelligence activity along the Franco-German border. Occasional deserters from the German Army were picked up by French agents in the fortress towns of Alsace and Lorraine and their papers were then used by agents of the *Deuxiéme Bureau* to get back into Germany. Meanwhile, Sarajevo was modernized by Austria who used the city as a test area for new inventions. In 1885, Sarajevo became the first city in Europe to have an electric tramway trialed before being installed in Vienna. Architects and engineers wanting to rebuild Sarajevo as a modern capital rushed to the city ~ then a fire that burned down part of the city centre, left more room for redevelopment.

After Queen Victoria closed her eyes forever in January of 1901, the way opened for the reign of King Edward VII who had spent much of his time as the Prince of Wales in Paris or on the Riviera. He spoke perfect French with ease, he had formed many warm attachments in France, and he ardently admired the people as a nation. King Edward was of the opinion that, as duels might be avoided in society by the exercise of tact, so wars might be avoided in Europe, if his art of kingcraft gave him the vision to see where to use his rank. Abandoning the old British policy of isolation, the fresh focus of Edwardian foreign policy turned to France. The King then personally planned to visit a number of European countries in the spring of 1903 during a Mediterranean cruise aboard the *Victoria and Albert.* King Edward saw it as a personal initiative, and on his own, quietly made it known to the French Government that on his return journey, it would give him pleasure to meet President Émile Loubet on French soil. British Ambassador Monson reported that; *The President received the invitation with delight. A visit from the King will in the present temper of France, do an amount of good, which is probably not realized in England.* On 28th April, the *Victoria and Albert* reached Marseilles and King Edward arrived in Paris the next day, where he paid a short informal visit to President Loubet at the Elysée. Edward VII was delighted by his reception and described his personal visit in the French phrase ~ *a success fou.*

It was a remarkable personal triumph for Britain's King Edward VII ~ *Edward the Peacemaker,* and exceeded all results that President Émile Loubet had predicted for it. With the ground prepared, the two foreign secretaries, the Marquess of Landsowne and Théophile Déclassé soon achieved a raft of colonial concessions in Central and West Africa, Siam, Madagascar, in the New Hebrides, and went on to formulate an agreement for the partition of Morocco between France and Spain. As Edward acquired exceptional popularity in France after his personal part in instigating the foreign policy of the Anglo-French Éntente Cordiallc, he was soon confirmed as the originator of the understanding by André Tardieu, who being a confidant of Théophile Déclassé and had reliable information said; *The English King was the initiator of the rapprochement. It was he, who both conceived and facilitated it while many still believed that the moment was premature. King Edward's visit in itself, proved to be an act of foreign policy.* After establishing his diplomatic position as a source of new strength in the affairs of state, Edward grasping the danger of Germany's rampant armaments program and naval expansion, took his role as the head of British armed forces seriously. He supported secret plans to provide the significant troop contribution of the British Expeditionary Force to support France, in the event of a war with Germany.

King Edward VII and his machinations became one of the chief themes of conversation between Kaiser Wilhelm and Tsar Nicholas in their extensive *Willy-Nicky* correspondence. The two nephews capped each other's stories about the iniquities of their uncle, *the arch mischief maker,* and planted the image of *Edward the Encircler* deeply within the German psyche, especially as further European situations surfaced to strengthen British influence in Europe. When the young King Alfonso XIII of Spain married Edward's niece, Princess Ena of Battenberg, German suspicions were again stirred. When Norway dissolved their union with Sweden in 1905, the Kaiser planned to install his younger son on the Norwegian throne. However, King Edward VII used his considerable diplomatic prowess to encourage his son-in-law, Prince Charles of Denmark to stand for the royal election. Prince Charles became King Haakon VII of Sweden, and Edward VII had once again contained German aspirations.

The rival Empires of Britain and Russia agreed to resolve their interests in Asia and elsewhere, in order to make an effective stand against the rising German influence within the Ottoman Empire. First the British began secret talks toward an Anglo-Russian Éntente, and then a wider Triple Éntente. In March of 1907, visiting Russian warships to Portsmouth foretold the Anglo-Russian alliance. Upon Edward VII's invitation, all of the visiting Russian officers and sailors were invited to visit London to be entertained as guests at the Admiralty and see an epic spectacle in their honour at the Hippodrome. After a grand banquet in the evening, there was a gala performance for them at the Alhambra Theatre. The German Ambassador said: *It has certainly never happened before, that an English Minister of Foreign Affairs has gone to a variety theatre to greet foreign guests.* The conclusion of the Anglo-Russian agreement (widely attributed to King Edward VII's policy) brought to a close the classic period of *The Great Game.* In November of the same year, when Kaiser Wilhelm II visited King Edward VII at Windsor Castle to improve his situation, he proclaimed at the Guildhall: *Blood is thicker than water, and that the main prop and base for the peace of the world is the maintenance of good relations between England and Germany.* However, the King's laughter at the Kaiser's alarm of the *Yellow Peril* was badly received by visitors. Edward declared: *I cannot recognize such a thing. The Japanese are an intelligent, brave, and knightly people.*

While 19th century Europe can be seen as an age of culture, before the terrible clash of civilizations it can also be seen as the original age of terror with many assassins targeting the leaders of the world. Continuous campaigns of bombings and assassinations terrorized the entire continent from the 1860s with the direct aim of unsettling the established social order. In these years, the revolutionary assassins of the Russian *People's Will* would succeed in killing Tsar Alexander II with a bomb, as fanatical Italian anarchists would stab President Carnot of France and Empress Elizabeth of Austria. Anarchists also assassinated King Umberto I of Italy, King Carlos and the Crown Prince Luis of Spain, Violent extremists represented a terminal threat to Europe's Empires and warrant particular attention to comprehend the context of the assassination of Franz Ferdinand ~ and Vienna's response to it.

The revolutionist is a dedicated man. He has no interests of his own, no affairs, no ties, and no possessions. Everything in him is swallowed up in a single passion ~ revolution. He knows only the science of destruction. Your every step now, should be taken in order to bring about the collapse of everything, both the state and its moral code. Only we shall remain, we who take power and ride over the fools. Sergei Nechayev

In 1849, the 28-year old Russian writer Fyodor Dostoyevsky was arrested for the crime of reciting such a revolutionary manifesto in public. Along with other members of his revolutionary entourage, he was sentenced 10 years in Siberian exile, to toil away doing hard labor. When he resumed writing, his realization of revolutionaries led him to become their outspoken foe. After noticing this, Dostoevsky was invited to visit Tsar Alexander II and he soon became a regular visitor to the Palace. Alexander II responded well to his critics. He freed the peasants, promoted self-governance, and established universities ~ and yet, he would be tracked, and targeted in seven assassination attempts.

FIRST: 4th April 1866. As Alexander was walking with his nephew through his Summer Palace garden in Saint Petersburg, a crowd of onlookers gathered to watch them… all peering over his fence. When their walk was done, as Alexander and the boy boarded their carriage, Dmitri Karakazov ran forward to fire a shot. Osip Komissarov, an apprentice hatter jostled Karakazov's elbow, just as his shot was fired to thwart the attempt. As the loyal crowd pounced on the failed assassin, the struggling Karakazov cried out, *Fools ~ I am doing this for you!* Tsar Alexander walked over to ask Karakazov, *Why are you shooting at me?* Dmitri Karakazov boldly answered ~ *Your Majesty, you hurt the peasants!* The hero of the hour; the apprentice hatter Osip Komissarov was embraced by Tsar Alexander II, ennobled and given a vacant country estate. Karakazov was executed, and in the Summer Garden in memory of saving Alexander, his family placed a chapel with an inscription; *Do not touch my Anointed.*

SECOND: The Paris International Expo of 1867. As Alexander II, his two sons, and Napoleon III rode by in an open carriage, an enraged Polish immigrant Anton Berezovsky surged from the admiring crowd to take aim and fire his double-barreled pistol. He succeeded only in killing a horse. A court swiftly sentenced Berezovsky, to transportation for life to the Pacific isles of New Caledonia.

THIRD: Morning of 20th April 1879. As Alexander walked through the grounds of his palace park, a man walked towards him. Alexander Soloviev then ran up to the Tsar till merely a few steps away. As soon as the Tsar saw a revolver in his hands, he evaded the assassin running in a zigzag pattern, dodging to the right ~ then to the left. Soloviev aimed and fired five times. Every shot missed, but one slug passed through the Tsar's sleeve. The failed assassin Soloviev was hanged before a large crowd.

FOURTH: December 1879. After the failure of three amateur marksmen, the revolutionaries of the *Narodnaya Volya People's Will* movement planned to ignite a social revolution by killing Alexander. The fanatical *People's Will* decided in November, to attack the Tsar and all his family with demolitions. They would mine the rail route on which the Russian royal family would make their return from their summer palace at Livadia on the south coast of Crimea as they journeyed north to St. Petersburg. The revolutionaries, led by the accomplished insider Sophia Perovskaya, knew that a freight train would precede the holiday entourage travelling in their private train. Fate would save the Tsar once more. Engine trouble with the royal freight train, meant that the first train became the one with Alexander. Not knowing this, the *People's Will* detonated their explosives beneath the wrong train ~ the freight train carrying the fruit, vegetables, and crates of the Tsar's favourite jam from his Crimean estate. After losing his groceries, Alexander asked his Guards ; *What do they have against me, these poor people? Why do they hunt me like a wild beast?* (Note: Just before this bomb derailment, Count Sergei Witte had a dispute with the Railway Minister and said, *the Tsar's trains are being driven at dangerous speeds!* Alexander II witnessing this, soon advanced Count Witte as his new Minister of Railways.

Tsar Alexander II

Vera Figner

FIFTH: The *People's Will* prepared another daring plan ~ to blow up Alexander in his own palace. Sophia Perovskaya, the daughter of a former Governor of Saint Petersburg, who was well placed to hear all manner of current goings on within the palace, soon learned thru her Imperial connections that the basement of the Winter Palace required renovation; including the wine cellar two floors below the formal dining room of the royal family where all visiting foreign dignitaries were entertained. Soon, the Winter Palace had a new carpenter, Stepan Khalturin, who had recently established the underground *Library of Revolution*. Once Khalturin began working everyday within the Winter Palace, he made a sketch of the palace room arrangement on the second floor to show Sophia Perovskaya. He marked the Imperial dining hall with a large *X* indicating the area he would target to be destroyed. The revolutionaries thought it a perfect plan. To assassinate the entire royal family, as they dined.

Every day, the carpenter Khalturin carried in among his tools a few packets of explosives that he secretly packed away in the middle of a wall, which he said he was plastering in the vast wine cellar. After three months of sly sneaking in explosives and clever covert storage, so great was the quantity of dynamite that the space of intervening floor between the dining room and the basement was of scant significance. Khalturin's frequent skin contact with the nitroglycerine was now causing his health to decline, and his appearance to degrade. He had almost constant severe headaches and could not sleep. With his appetite gone and his hair falling out, Stepan Khalturin knew he had to act soon.

When he heard that Tsar Alexander was expecting a visit from Prince Alexander of Hesse, on the evening of 17th February 1880, Khalturin set the timer on his demolition charge to detonate at 6.20 PM, but chance intervened. The train of Prince Alexander steaming from Berlin to St Petersburg was so slowed by a blizzard that it was overdue by 30 minutes. For the first time, dinner would be delayed. As Alexander's family left the drawing room for the dining room, the 228-lb bomb detonated with a shattering roar heard clear across St Petersburg. The hallway doors blew wide open and a shockwave of debris and dust swept over the Tsar and his guests. All the lights went out throughout the Palace. Beneath the catastrophic wreckage of the dining room, ten elite soldiers of the Imperial Finnish Guard were no more, and another 56 people were wounded in the explosion.

Stepan Khalturin, Vera Figner, and Sophia Perovskaya watched from the iron gates of the Winter Palace. After the shattering explosion had blasted two stories off the front of the Winter Palace and onto the square near to them, smoke and dust still obscured the scene for several minutes more. With the smell of the explosive detonation still strong in the air, the crowd began to see the damage. One of the Tsar's guards appeared to announce to the crowd. *A bomb attack on the Palace has taken place, and there are many casualties ~ but the Tsar and all his family are whole and well. God has saved the Tsar!* From the mostly loyal crowd, a heartfelt cheer erupted, *Ura! Ura! Thank God!* the crowd exclaimed. As Khalturin and his co-conspirators Sophia Perovskaya and Vera Figner slipped away undetected, he groaned ~ *All that work for nothing…* Vera Figner consoled him with expectations of the many new recruits that would join their growing cause. *An explosion in the Tsar's lair! Your deed will live forever. Don't grieve. We didn't succeed today ~ but we'll succeed tomorrow. Alexander II will be destroyed.*

After seeing their dramatic detonation of the Winter Palace, Vera Figner moved the *People's Will* store of explosives to her own apartment above a grocery. While this was done, Sophia Perovskaya organized an *Observation Brigade* to study the Tsar's movements, so they could ambush him on a street over which he habitually passed. Her observers learnt that the best place to target the Tsar was on the garden side of the Catherine Canal embankment ~ as he returned from his Sunday troop inspections.

After lunch on 13th March 1881, as Alexander II was known to do every Sunday for many years, he went to the Mikhailovsky Palace for a military review. He travelled in a closed carriage preceded by six mounted Cossacks, followed by two horse sleighs. The first sleigh carried Police Chief Colonel Adrian Dvorzhitsky and three policemen. In the rear sleigh rode Captain Eduard Kokh of the Palace Guard with two guards. When they reached the Mikhailovsky parade ground, the Tsar was pleased.

After a perfect military review, Alexander and his small entourage stopped by at the actual Mikhailovsky Palace to visit his cousin, the Grand Duchess Ekaterina Romanova who resided there. As Grand Duchess Ekaterina openly opposed his reforms ~ particularly Alexander's draft of a Russian constitution that he had been refining for six months (and had now signed all the necessary papers to begin its instillation to provide for an elected assembly with legislative powers) Tsar Alexander tried to settle their differences. After his brief visit to the Grand Duchess Ekaterina he divulged:
I am doing all that I have within my power to do, and that, no one on earth can demand more of me…

SIXTH: The return route was via the Catherine Canal, and then over the small Pevshesky Bridge. As the Tsar's convoy rode along the Catherine Canal embankment, Sophia Perovskaya thereupon deployed two pairs of bomb throwers 50m apart on the embankment. Across the canal, Sophia prepared to give the signal ~ when she saw the Tsar's entourage turn onto the embankment, she would take out her white handkerchief and blow her nose into it. The wheels of Alexander II's armoured coach screeched and crackled, as it turned on the ice-covered paving stones of the canal embankment. Seeing this, Sophia took out her white handkerchief and blew her revolutionary nose. Nikolai Rysakov waited until the rear sleigh came nearer… *After a moment's hesitation, I threw the bomb under the horse's hooves to blow up under the carriage. The strong explosion knocked me hard into the fence.* The bomb detonated with a sharp B-A-N-G under the rear axle of Alexander's armoured carriage thrusting it into the air. The rear Cossack and his horse were blown over onto the ice-covered street. As his once fine mount bawled in mortal agony, the Imperial Cossack lay lifeless…

While the explosion had only scratched Alexander's armoured carriage (a gift from Napoleon III), his coach horses panicked by the sudden blast and the dying squeals of the fallen horse bolted down the embankment at their best speed. The Tsar yelled out to his coachman to halt the coach, which he managed to do after covering 100m. As Police-Chief Dvorzhitsky's sleigh approached the Tsar's coach, he leapt out, ran over, and opened the left-side door. The Tsar appeared unhurt, stepped down onto the frozen pavement, and sincerely crossed himself. Dvorzhitsky, expecting that the Tsar should have been promptly taken back to the Winter Palace, asked the coachman ~ *Why have you stopped?* When Alexander explained that he wished to walk back to the bombsite to see if anyone was hurt, Dvorzhitsky protested as much as Alexander persisted. The explosion had attracted quite a crowd, who followed the Tsar as he walked back along the embankment to where his Cossack guards were restraining Rysakov. When Captain Kokh asked Rysakov ~ *Did you make that blast?* He replied: *It was I.*

SEVENTH: When Alexander reached the bombsite, he comforted a screaming child who was now seen to be badly injured and tried to comfort the Mother. Alexander pointed to the corpse of his fallen Cossack guard to enquire ~ *Have they caught the culprit?* Colonel Adrian Dvorzhitsky, seeing Captain Kokh had the criminal in custody said, *Exactly so, your Majesty. Shouldn't we set off back to the Palace?* Then Alexander walked across to the large group surrounding Rysakov, and asked ~ *Is this the one?* Captain Kokh replied, *He is the very one, your Majesty.* Alexander's brother, the Grand Duke Mikhail who next arrived on the scene, and not immediately able to see the Tsar enquired, *What about the Tsar?* Alexander answered happily ~ *I am whole, thank God.* On hearing that, the failed assassin Rysakov who saw the approach of his fellow assassin, Ignati Grinevitsky with another grenade, shouted to the Tsar; *Don't thank God yet!* Grinevitsky, who had been standing by the canal fence, thrust himself through the crowd to throw his grenade from a distance of two metres ~ directly at the Tsar's feet.

I was deafened by the new explosion, burned, wounded, and thrown to the ground. Suddenly, amid the smoke and snowy fog, I heard His Majesty's weak voice cry ~ Help! Gathering what strength I had, I rose up and rushed to the Tsar. His Majesty was half-lying, half-sitting, leaning on his right arm. Thinking he was merely wounded heavily, I tried to lift him but the Tsar's legs were shattered, and the blood poured out of them. Twenty people, with wounds of varying degree, lay on the sidewalk and on the street. Some managed to stand, others to crawl, still others tried to get out from beneath bodies that had fallen on them. Through the snow, debris, and blood, you could see fragments of clothing, epaulets, sabres, and bloody chunks of human flesh.

Colonel Adrian Dvorzhitsky ~ Imperial Police

Alexander mumbled… *I'm cold, so cold. Home to the Palace, to die there.* Colonel Dvorzhitsky and Captain Kokh lifted the Tsar onto the police sleigh. The sleigh, preceded by a mounted Cossack raced toward the Palace. When they arrived Tsar Alexander noticed that Colonel Dvorzhitsky's arm was bleeding. *Are you hurt?* Dvorzhitsky replied, *We do not matter, your Majesty. We are concerned about you.* When they arrived at the Saltykov entrance, they carried Alexander up the stairs of the marble staircase to his study, where 20 years before, he had signed the Emancipation Edict to free the serfs. Now Alexander II was bleeding to death, with his face mutilated and his belly torn open. With the Grand Duke Mikhail in tears, standing at the head of the bed in full parade uniform, as other members of the Romanov family came rushing to the scene, the dying Tsar was given Communion and Last Rites. That day, Grand Duke Mikhail was to take his sons ice-skating with Alexander's 13-year-old grandson, Nicky (as the future Tsar Nicholas II was referred to in the Romanov family). Now Nicky, in a sailor suit was ascending the marble steps and trying to avoid stepping in his grandfather's blood. At 3.30 PM, the standard (personal flag) of Tsar Alexander II was lowered.

To my left stood the new Tsar, Alexander III. A strange change came over him that instant. This was not the same Tsarevich Alexander who liked to amuse the small friends of his son, Nicky, by tearing a deck of cards in half or bending an iron rod into a knot. In the course of five minutes he was completely transformed. Something incomparably greater than the simple consciousness of the duties of the monarch illuminated his heavy figure. A fire burned in his eyes. It was the regal gaze, the heavy pitiless gaze, they had all waited for that they all believed would return peace and great power to the country. Grand Duke Mikhail

Tsar Alexander II's death caused a seismic setback for his proposed reforms. His draft plans for an elected parliament or Duma, completed the day before his assassination, were not as yet released to the Russian people. In a matter of only 48 hours, Alexander II had planned to release his plan for the Duma to the Russian people. Had he lived, Russia might have seen a path to constitutional monarchy. Instead, the first action Alexander III took after his coronation was to tear up all of these fine plans. After experiencing significant social progress under the reign of Alexander II ; whose assassination was witnessed first-hand by his son Alexander III and his grandson Nicholas, both of whom vowed not to have the same fate befall them. Both used the *Okhrana* Secret Service Guards to arrest protestors and uproot suspected rebel groups that only created further suppression of freedom for the people. The great folly of the *Narodnaya Volya* was that thru their assassination of Alexander II they had only ensured strong and perpetual autocratic oppression. After the assassination of his father, Alexander III would take no chances with reform or liberal ideas, and the peril of the *Narodnaya Volya* convinced the ruling Romanov family, that they must use a very firm hand to save their monarchy. Such were the constant considerations of the crowned heads of Europe, and what influenced many of their decisions. Even so, Alexander III inherited a white page in his country's political history and an exceedingly healthy condition of affairs within his own domains. But, externally, matters did not look so bright. Austria was avidly interested in the Balkan States, and Russo-Germanic relations were uncertain.

Assassination attempt on Queen Victoria at Windsor Station ~ 2nd March 1882

Sante Caserio

Luigi Lucheni

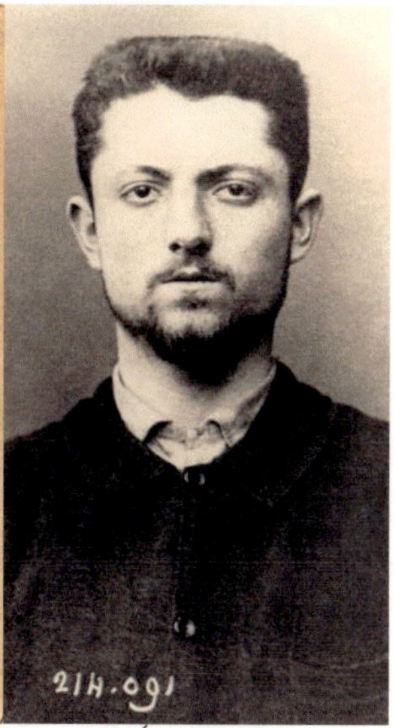

Émile Henry

The similar terror bombing and assassination tactics of French and Italian anarchists can be traced back to the Paris bomb scare of 1892-94 when *Ravachol* (the pseudonym of François Koenigstein) attempted to assassinate several judges, but instead injured other people and was betrayed to the police by a waiter from Restaurant Véry. Théodule Meunier took revenge and bombed the restaurant. Unable to feed his family, Auguste Vaillant then threw a bomb into France's Chamber of Deputies, wounding several to call attention to the heartfelt plight of the poor. On 12th February 1894, shortly after the trial and execution of Vaillant, a young anarchist Émile Henry vowed to avenge the death of Vaillant. The experienced Émile Henry, who one year before had used a bomb to kill five gendarmes, now walked up the Avenue de l'Opera, with a bomb hidden under his coat. Henry paused at several elegant cafés ~ and moved on. They were not full enough. He entered the Café Terminus (still) near Gare St Lazare, to order two beers and a cigar. With his cigar, Henry lit the fuse of his homemade bomb and threw it into the crowded café. Amid thick smoke, marble tables, chairs, and shattered café mirrors, the screams and shouts of the wounded shrilled through the smoke. Émile Henry ran away, before being caught and wrestled to the ground after a fierce struggle. In the Parisian Café Terminus, 22 people had been wounded, some critically, three of whom would die that day.

In early 1894, the Italian agitator Sante Caserio was arrested and promptly thrown into prison for distributing anarchist pamphlets outside the barracks in Milan. There he remained for five months. On his release, he went to Switzerland and worked with a baker, until he argued and was dismissed. Caserio then bought a dagger with a hilt of copper gilt in a scabbard of black and red striped velvet. The young anarchist then decided to kill French President Carnot to avenge the recently guillotined anarchists Vaillant and Henry. When Caserio learned of Carnot's coming visit to Lyons, he went to Lyons, bought a paper containing a programme of the Presidents' visit and joined the crowd to see Carnot's carriage travel from a banquet along the Rue de la Republique ~ a young man ran toward Carnot's carriage appearing to present a petition. Caserio then fatally stabbed President Sadi Carnot. Caserio was caught, convicted, and informed that the Lyon Fête guillotine would execute him.

Anatole Diebler, the headsman of Paris arrived at 3 o'clock, bringing with him the well-known guillotine. The van containing the guillotine was escorted by 15 gendarmes on horseback. The operation of erecting the machine occupied about an hour. There was a heavy shower during the night, which had the effect of driving away nearly all the curiosity seekers, who gathered only to witness the execution. At 3 o'clock there were hardly 100 persons waiting. Sometime before daylight the weather improved, and the curious began to gather again. Then the 15 mounted gendarmes returned to form a piquet d'execution, facing the guillotine. A few minutes before 5 o'clock the condemned man was led from his cell to the guillotine. His arms were bound behind him. Deibler's assistants whipped off his jacket. When the attendants seized him to lay him under the blade, he struggled fiercely to free himself. At 4.55, all was ready. In front of the guillotine, Caserio shouted to the crowd, Coraggio cugini, evviva l'anarchia! ~ Courage, cousins, long live anarchy! The click of the guillotine was heard, and at 5 o'clock precisely, the blade fell. Caserio's head dropped into the basket. The blood from the trunk spurted high as the body revolved into the basket. Deibler himself picked up the head from the sawdust and threw it viciously into the basket, with the body of the dead Anarchist. *The New York Times*

François Claudius Koënigstein, aka Ravachol ~ Service Régional d'Identité Judiciaire de Paris.
Photograph by Alphonse Bertillon ~ an influential French police officer and inventor of the mug shot.

Although the photographing of criminals began in the 1840s very soon after the invention of photography, it was not until 1888 that Alphonse Bertillon standardized the process. Bertillon also applied the anthropological technique of anthropometry to law enforcement creating an identification system based on physical measurements. Anthropometry was the first scientific system used by police to identify criminals. Before that time, criminals could only be identified by name or photograph. The method was eventually supplanted by fingerprinting. Bertillon also created other forensics techniques, including forensic document examination, the use of galvanoplastic compounds to preserve footprints, ballistics, and the dynamometer used to determine the degree of force used in breaking and entering. Today Ravachol remains an important cult hero among French anarchists.

On the night of 29th March 1886 Ravachol committed his first murders. He set out to rob Rivollier, who lived in the outskirts of the village of La Varizelle. Ravachol broke into the house armed with a hatchet, and surprising Rivollier in bed, killed him by splitting his skull. His female servant, alarmed for her life, fled up the road. Ravachol pursued and killed her. He then went back to the house for Rivollier's money. He then desecrated the grave of Baroness of Rochetaillé in 1891 to find her jewellery, then robbed and murdered an elderly recluse. With the police on his tail, he fled to Spain where he then justified his actions on the basis of anarchistic theories, learned bombmaking in Barcelona, and returned to Paris. On 11th March 1892, he bombed the home of Judge Benoit at 136 Boulevard Saint Germain. Harming no one in this first major act of French anarchist terrorism. Ravachol next made a bigger bomb and visited the home of Public Prosecutor Bulot on Rue de Clichy, this time wounding five and causing much damage. While eating his dinner at du Restaurant Véry, Ravachol foolishly gave himself away talking too freely with the waiter Jules Lérot, who turned him in a few days later to be arrested in mid-April of 1892.

Troisième année. — N° 73. Huit pages : CINQ centimes. Dimanche 8 Mai 1892.

LE PROGRÈS ILLUSTRÉ

On s'abonne sans frais dans tous les bureaux de poste.

Supplément littéraire du « PROGRÈS DE LYON »

Les manuscrits non insérés ne sont pas rendus

ABONNEMENTS

	SIX MOIS	UN AN
Lyon, Rhône et limitrophes.	2f »	3 50
Hors de ces départements.	2f50	4 50

ADMINISTRATION ET RÉDACTION
10, Place de la Charité, 10
ADRESSER LES CORRESPONDANCES ET ABONNEMENTS
à M. Léon DELAROCHE, administrateur

ANNONCES
Les annonces sont reçues exclusivement :
A LYON : à l'Agence V. Fournier, 14, rue Confort, et dans ses succursales de Grenoble, Saint-Etienne, Mâcon et Dijon
A PARIS : à l'Agence Havas, place de la Bourse, 8.

L'EXPLOSION DU RESTAURANT VÉRY

Théodule Meunier was a French anarchist who, along with Émile Henry and August Vaillant, was responsible for a series of Paris bombings during 1892. The three-targeted civilian and government buildings bombing boulevard cafés, and the homes of some magistrates. During the trial of Ravachol on 25th April 1892, the day before he was to be sentenced, Restaurant Véry in which Ravachol was arrested was targeted by Meunier who disguising his face and embellishing it with a thick false beard, next placed some dynamite cartridges in a wooden box, pierced with holes for a fuse or wick, and with his box in a bag, Meunier set out for M. Very's restaurant. Before entering it, however, he went into a urinal and there lighted his wick. Then, going to the restaurant, he called for a petit verve of rum ~ price three sous. Théodule Meunier drank it, left his bag near the door, and hurriedly departed. It was 9.25 in the evening when the huge explosion occurred. The owner was killed; so was one of his customers named Hamonod, and several other persons were injured, some of them very grievously. Fleeing to England, he lived in London before his arrest in 1894. Extradited to France, Meunier was tried and sentenced to life imprisonment in a penal colony in Cayenne showing no remorse ~ I only did what I had to do. If I could start over again, I would do the same thing.

The severe actions of Ravachol continued to inspire others. On the evening of 12th November, 1893, Léon-Jules Léauthier arrayed in his Sunday best went to Marguery's Restaurant, on the Boulevard Bonne Nouvelle, and there ordered a good dinner, selecting roast quail to follow the soup, and asked the sommelier first for a half-bottle of old Macon, and then a half-bottle of Moët et Chandon's finest Champagne. As he had an intense admiration for Ravachol and intended to avenge him, Leauthier felt it incumbent on him to imitate the martyr's habits. He was aware that Ravachol had always taken delight in the pleasures of the table and thus it was that he ordered a recherche repast, during which he glanced at those around him, with the object of selecting an appropriate victim. He was suitably armed for the crime which he contemplated, with his shoemaker's knife. When the bill was brought to him he fired up, remarking that he had dined because he was hungry, but did not intend to pay as he possessed no money. The waiter turned to Marguery, who came up exclaiming; You have no money, you say? In that case, a man does not drink Champagne! Léauthier retorted: But the bourgeois drink Champagne! Marguery replied: They pay for it. Léauthier's rejoinder was ~ With our money! The restaurateur did not argue the point. As he was a kind-hearted man, he contented himself with putting his hand on Léauthier's shoulder and leading him to the door of the restaurant, the young shoemaker simply went off.

The following evening Léauthier sallied forth again. On the Avenue de l'Opera, he ordered dinner at the Bouillon Duval and had almost finished it when seeing a well-dressed customer putting on his overcoat, he sprang up and stabbed him in the chest. Then he made off, but feeling certain he would be caught, he surrendered to the police, telling them that he had just had a good dinner and had stuck a bourgeois. On being asked why he had stabbed a complete stranger, he replied: Because he was well dressed and wore a decoration! On being told that his victim was Georgevic, the Servian Minister to France ~ An ambassador! Well, so much the better! At Léon-Jules Leauthier's trial, Public Prosecutor Bulot urged the jury to give such a verdict to entail capital punishment under the laws for the repression of Anarchism; but when Leauthier openly professed Anarchist principles in such a childish, incoherent fashion the jury regarded him as mentally unhinged. Thus extenuating circumstances were declared in his favour, and a sentence to penal servitude ensued. He did not live long however. Transported to Devil's Island, Leauthier was there in October 1894, when a revolt broke out among his fellow prisoners. One of them having been guillotined for a murder, considerable excitement arose, and then an Anarchist convict climbed a flagstaff and shouted Vive Anarchie! Repeatedly refusing to desist, a soldier fired on the man, who immediately fell to the ground dead. This provoked a rising of all 800 convicts. The outbreak was only quelled by repeated rifle volleys of the soldiers. Many prisoners were killed, one being Léauthier. E.A. Vizetelly

* * *

The next act of terror in Paris came when Auguste Vaillant threw a bomb into the Chamber of Deputies on 9th December. During his trial, Vaillant said ~ *At least I shall have the satisfaction of having wounded the existing society, this cursed society in which one may see a single man spending uselessly, enough to feed thousands of families, an infamous society that permits a few individuals to monopolize social wealth... Tired of leading this life of suffering, I carried this bomb to those who are primarily responsible for social misery.*

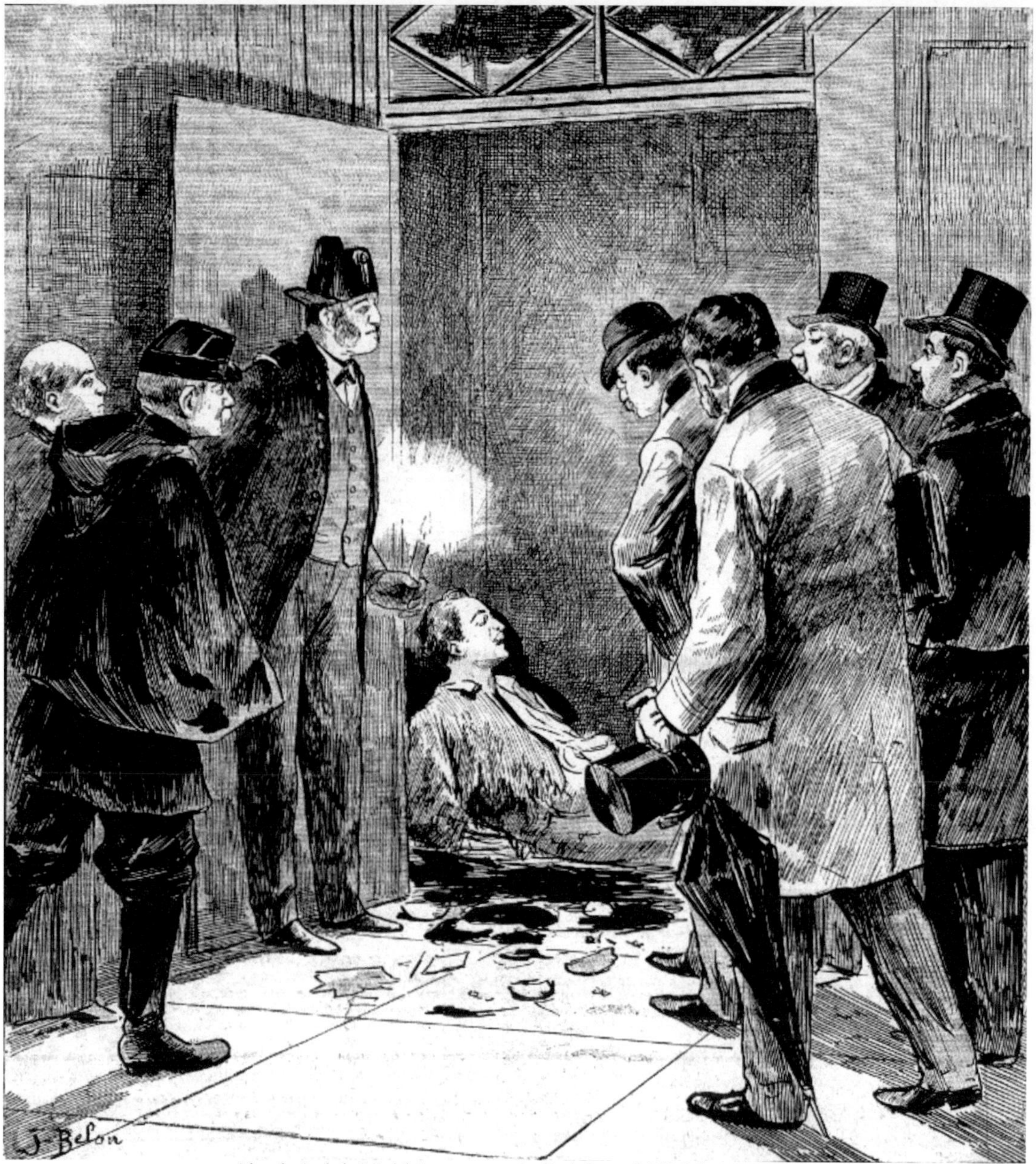

L'explosio de la Madeleine segons ~ Le Petit Journal 26th February 1894

On 19th February 1894, just two weeks after the execution of Auguste Vaillant two mysterious affairs occurred in two very different parts of Paris. One north of the Seine on the Faubourg St Martin, and another in the Latin Quarter south of the river on Rue St Jacques. In each district, the local police office received the suicide note of a man by the name of Étienne Rabardy who wrote of being so overwhelmed by misfortune that he had decided to take his life, and that nobody should be accused of an act for which he alone would be responsible. Two different addresses were given ~ two hotels.

It was a trap. The aim of the letters was to attract the police to the addresses, at each location there was a hotel, where Amédée Pauwels had taken a room, and in each room, he had left the entrance doors connected to bombs set to detonate and kill the responding police as they forced the door. The police naturally hastened to the two addresses. At the one on Rue Saint Jacques there was an explosion, which wounded three; one lady so severely that she soon afterwards died. Fortunately, on Faubourg Saint Martin, the bomb fell without exploding and was then carefully exploded by the police to prevent any further hurt or damage. Both bombings were traced to the Belgian anarchist Amédée Pauwells, who was active in the Saint-Denis anarchist group. Pauwells was never apprehended, because on 15th March 1895, as he entered the Church of Madelaine in Paris, the bomb he was carrying exploded, killing him instantly.
It was thus that the Anarchist Terror of 1892-94 sent shockwaves thru French society.

Much sprang from it, the consternation into which Parisians were plunged by repeated explosions, the loss of life, the destruction of property, and a succession of alarming occurrences in Paris, the enactment of exceptionally drastic laws, the revival of the *Cabinet Noir*, the arrests and sentences, and finally the assassination of President Sadi Carnot of France, and the Empress Elisabeth of Austria.

THE ASSASSINATION OF EMPRESS ELIZABETH OF AUSTRIA

The assassination of Empress Elisabeth of Austria ~ Queen Consort of Hungary, Bohemia and Croatia

In 1853, when the 15-year old Bavarian Duchess, Elisabeth von Wittelsbach accompanied her mother and sister Heléne to Bad Ischl to meet their 23-year-old cousin, the Emperor Franz Josef I, Heléne was the intended bride and Franz Josef wasn't supposed to see much of Elisabeth, but when Elisabeth burst into the room where Heléne was dining with Franz Josef, she realized her *faux pas*…

Franz Josef became enchanted and eight months later when they married in Vienna, Elisabeth became Empress of Austria and Queen of Bohemia. Franz Josef was unwavering in his love, and the Empress Elisabeth was much adored throughout Hungary as she took up the cause of the Hungarian people. Due to her influence, Franz Josef agreed to the Hungarian Compromise to re-establish the Hungarian Constitution, and to reform the Empire into the Dual Monarchy of Austria-Hungary.

* * *

On the morning of 30th January 1889, the bodies of the 30-year-old Crown Prince Rudolf of A-H, the only son of Emperor Franz Josef and Empress Elisabeth, and Rudolf's 17-year-old friend, Baroness Marie von Vetsera, were discovered in the Imperial hunting lodge at Mayerling in the Vienna Woods, 25km southwest of Vienna. The apparent murder-suicide death of Crown Prince Rudolf and Baroness Marie interrupted the immediate line of Hapsburg dynastic succession, as Rudolf had as yet, no son.

The line of succession passed to Franz Josef's brother, Karl Ludwig, and his son Franz Ferdinand. Rudolf, (like his mother Empress Elizabeth) was liberal, anti-aristocratic, and a supporter of Hungary. His death disrupted the growing reconciliation between the Austrian and the Hungarian constituents of the Empire, contributed to the drift toward war, and to the fall of the House of Hapsburg in 1918.

Following the death of her only son Crown Prince Rudolf, Empress Elisabeth was inconsolable. She wore black, and became so restless during in her travels under the name of *Countess Hohenembs* that she rarely stayed long anywhere, spending no more than a few weeks of each year in Vienna. During an overnight stopover in Geneva during 1898, a Geneva newspaper revealed that the elegant woman traveling under the pseudonym of *Countess Hohenembs* ~ was the Empress Elisabeth of A-H. On 10th September, as she walked from the Hotel Beau-Rivage on the shore of Lake Geneva to catch the steamship *Genéve* for Montreux, she was approached by the Italian anarchist, Luigi Lucheni who peered under her parasol, then stabbed her firmly through the left breast with a sharpened needle file of the type used to file thru the eyes of industrial needles, which he had affixed to a wooden handle.

Once the assassin's thin file poked a tiny hole in Elisabeth's heart, her blood leaked out so slowly that she was unaware of the full extent of her injury. Ghastly pale, she made it to the ship to collapse. Her companion at first assumed her Empress had merely fainted, and then loosening her bodice noticed a spot of blood and a hole in her camisole. She was rushed back to the hotel, but it was too late for the doctors to do anything. Her grief-stricken husband Emperor Franz Josef remarked to Prince Eduard of Liechtenstein, the couple's devoted equerry; *That a man could be found to attack such a woman, whose whole life was spent in doing much good, and who never injured any person, is to me incomprehensible. The world will never know how much we loved each other. I am a pechvogel* (German : bird of misfortune). Thereafter, Emperor Franz Josef who needing to secure his Monarchy and Empire against extremists, and who now doubly despised revolutionaries and assassins, naturally found such priorities high in his outlook and this would influence his decision in 1914 to take measures against the Serbian threat.

Crown Prince Rudolf of Austria

Baroness Marie Alexandrine von Vetsera

KING UMBERTO OF ITALY ~ FIRST ASSASSINATION ATTEMPT

King Umberto I of Italy

Queen Margherita & Prince Vittorio Emanuele

Giovanni Passannante

After the Italian House of Savoy led the Unification of Italy, Prince Umberto di Savoia became King Umberto I of Italy, and one of his first acts he undertook accompanied by Queen Margherita and Prime Minister Benedetto Cairoli, was a tour of the Italian Peninsula. During a parade in Naples on 17th November 1878, the anarchist Giovanni Passannante holding a dagger attacked Umberto. Umberto warded off the blow with his sabre, but PM Cairoli was severely wounded defending him…

The royal carriage was entering the Via San Giovanni when Passannante sprang forward. It was thought he wished to tender a petition, but in his right hand, he had a red flag ~ within the folds a dagger was concealed. Passannante reached the vehicle without impediment, raised his arm, and struck the King in the left shoulder. Umberto half rose from his seat ~ swung his sword round and hit his assailant on the head with the scabbard. But the blow was not effectual, and Passannante was about to repeat his attempt when Prime Minister Cairoli, hearing Queen Margherita, crying to him ~ Save the King! ~ sprang up and caught the assassin by the hair. At the same time, he received the blow intended for his sovereign, being stabbed in the thigh in a serious fashion, for the knife just missed the great crural artery. Still he did not relax his hold. A guard Telemaco Giamettine rushed up, and secured Passannante from behind. With the prisoner in custody, the royal carriage was driven in all haste to the palace, where Cairoli, whose wound had bled profusely, was put to bed and medically attended. The King's injury was less serious and he was the recipient of many telegrams congratulating him on his escape. Foremost was a message from the new Pope Leo XIII who prayed to God for his Majesty's health. E.A. Vizetelly

The failed assassin was condemned to death, but as the law only allowed the death penalty if the King had been killed, King Umberto commuted this to penal servitude for life with 18kg of chains. King Umberto quickly became known as *Umberto il Buono* or *Umberto the Good* by his people from the care and concern he showed them. When Venice and Verona were flooded in 1882, Umberto went to direct work to minimize the disaster, and gave generously from his own funds to displaced people. When an earthquake struck Ischia the following year, King Umberto again intervened to order rescue operations to continue five days longer than planned which resulted in many more lives being saved. He then went to provide a helping hand, even though the area remained quite dangerous. He showed similar compassion in 1884, when a cholera epidemic struck southern Italy. By these and other similar actions Umberto united the kingdom, and his reign was an early success. He oversaw Italian colonial expansion into the Horn of Africa, occupying Eritrea and then expanding into Somalia. However, the radical leftists and anarchists despised Umberto's conservatism and disapproved of Umberto's 1882 Triple Alliance with Germany and A-H, as the Austrians still occupied several areas claimed by Italy.

In 1889, King Umberto and his wife Queen Margherita di Savoia visited Naples for a holiday and asked the popular *pizzaiolo* Raffaele Esposito of *Pizzeria Brandi* to come to their southern palace so they could taste his finest pizzas. Esposito then made three: one with pork fat, provolone cheese and basil; one with garlic, olive oil, and tomatoes ~ and his specialita pizza with basil, mozzarella and tomatoes, in the colours of the Italian flag. Queen Margherita loved the specialita so much that she sent a note to thank Esposito Raffaele, who in turn dedicated this pizza to her, calling it ~ *Pizza Margherita.*

The Assassination of the King of Italy ~ and a characteristic image of His Majesty King Umberto I with Queen Margherita

The late 1890s saw an upsurge in radical activity in Italy. Rising food prices led many socialist agitators and anarchists to stir up strikes and anti-government protests, particularly in the areas of Bologna and Lombardy. By the spring of 1898 the strikes and protests began to spread southwards, gaining momentum in Tuscany and giving extra impetus to the influential anarchist presence there. As the strike movement spread across Italy, a state of siege was declared in early May across Tuscany. The increasingly insurrectionary nature of the strikes and protests in the region came to a head with a serious demonstration in Milan on 6th May 1898. Thousands of workers and their families marched towards the Savoia Royal Palace in the city ~ then under a very heavy police and military guard. Venting their anger at the high price of bread, workers raided bakeries, stealing bread and pizzas.

As the rioting approached the palace, the situation on the streets of Milan drifted out of control. With General Fiorenzo Bava-Beccaris prepared to order his command to fire upon the violent crowd; artillery pieces were fired at zero elevation and volleys of rifle fire shot into the crowds killing 100 people and wounding more. That put an immediate stop to the rioting. Many were shocked when the King decorated Bava-Beccaris ~ *You have rendered a great service to your King, and to your country.* Undoubtedly an unfortunate incident, yet the King recognized the General for taking decisive action to restore order and safeguard lives and property against an already violent crowd on Milan's streets.

Gaetano Bresci, an immigrant to the U.S. founded an Italian anarchist paper *La Questione Sociale* while working in the silk mills of Paterson, New Jersey. When he first heard of the massacre in the spring of 1900, Bresci surprised his comrades at the newspaper when he asked for the return of a loan he had used to help pay their printing costs. Refusing to explain, he quietly left to stalk King Umberto. Landing in La Havre, he eventually arrived in the small town of Castel San Pietro near Bologna. Staying at an inn owned by a relative, Bresci acquired a revolver and began target practice in the yard. Hearing of the King's plans to visit Monza, he quickly departed for Milan and from there to Monza, where he arrived on 26th July. He decided to act on 29th July, while the King was scheduled to distribute prizes to athletes after a sports competition. Arriving in the evening in an open-air carriage Umberto climbed the podium and distributed medals to the athletes. After words of congratulations, Umberto descended the platform and got back into his carriage. As he sat down, Bresci burst from the crowd holding a revolver and fired four times. Umberto died seconds later having been hit three times in the chest, with one shot going wide of its target. As he was tackled by the police, Bresci declared : *I have not shot Umberto. I have killed a King and a principle. I have avenged the people killed by Bava-Beccaris.*

Captain Dragutin Dimetrijevic ~ Apis

Major Voijaslav Tankosic

On 6th September 1901 in the small Kingdom of Serbia, a covert military clique formed to unite all the territories with a South Slav population ~ by violence if necessary ~ not ruled by the Kingdoms of Serbia or Montenegro in the manner of earlier unification processes, such as the Italian success of 1870. The founder of this group was Captain Dragutin Dimitrijevic, who as a teenager had attended the *lycée* in Belgrade, where he was a popular and brilliant student. He also displayed boundless energy and restless activity, for which his fellow students nicknamed him *Apis* (Latin : for Bee) on account of his beelike qualities ~ for he was four-winged in his patriotic élan, and he always knew the beeline from conception to execution ~ and he was certainly not without the industrious insect's sting. Captain *Apis* who considered King Alexander I an obstacle to the free progress of the Serbian people, now set forth on his mission to create a Greater Serbia. With his qualities that fascinated Serbian men, and reasoning which was always thorough and convincing; he understood how to make the worst deeds appear mere trifles, and the most dangerous schemes seem innocent and quite harmless ~ such as sending an agent into Albania in 1902 to assassinate the popular nationalist leader Haxhi Zeka.

For many years, the unattractive King Alexander I of Serbia tried to find himself a royal bride in the noble houses in Europe with which to form an alliance, to bring recognition and stability to Serbia. But, Draga Masin, the widow of an engineer (whom it was widely rumoured that she had poisoned) and ten years King Alexander's senior, notoriously tricked him into marriage with a false pregnancy. Soon his people scoffed that King Alexander's most important act had been to marry his concubine. The false pregnancy of Queen Draga swiftly spawned serious strife for King Alexander. In Belgrade, his army officers complained that the Queen's false pregnancy had diminished the international reputation of Serbia, and were furious from the constant temper tantrums thrown by her brother Nikola Lunjevica, (who once killed a policeman whilst drunk). Being childless, Queen Draga was soon suspected of intending to secure the succession to the Serbian throne for her most unworthy brother. Tsar Nicholas II then firmly declined to receive the Serbian royals on their planned visit to Russia. Following his repulsion by the Russian court, Alexander approached Emperor Franz Josef in Vienna during the autumn of 1902 to solve the question of his succession. Finally, in the beginning of 1903, Alexander was making ready to proclaim Queen Draga's now reviled brother as the heir to the throne, in spite of having agreed Prince Mirko of Montenegro would be proclaimed Crown Prince of Serbia if their union was childless. To prevent Queen Draga's brother from being decreed heir, and to replace Alexander ~ the secretive *Apis* (now in the pay of the Russians) prepared to change European history. John Goodwin of the *Evening News* reported: *The Black Hand party reached its peak when bottles crashed against the walls. Apis got up on the table shouting ~ That pair in the palace must die, so that Serbia can live!*

At 2 AM on 11th June 1903, Lieutenant Petar Zirkovic of the Royal Guard Cavalry Squadron unlocked the rear gate of the palace, to admit assassination squads who swarmed thru the corridors. Captain *Apis* and Palace Guard Zirkovic led them through the Belgrade Palace. The doors to the Royal bedroom were shattered with dynamite, along with every light fuse in the palace. With all power cut, the would-be assassins looking for Alexander and Draga, who were not in their bed, searched by candlelight for 90 minutes. Unknown to others, *Apis* had spotted a figure escaping down the stairs into the palace courtyard. Presuming it was the King, *Apis* ran after him. But, it was a Palace Guard. The guard turned his pistol to *Apis*, aimed and fired three bullets into his chest, and then scurried off. Hearing gunfire, feeling anxious from their failed search, with dawn on the horizon and *Apis* absent, (being sprawled wounded in the basement), the conspirators began to feel that their plot had failed.

The assassins re-entered the Royal bedchamber to search again for the hidden King and Queen. Behind a large mirror, was a small room which contained Queen Draga's huge wardrobe. This in turn, covered a hidden hatch in the floor ~ an entrance to a secret passage that led down to the next floor and thru to the Russian Embassy located opposite the palace and across the street ~ but the cautious pro-Austrian King Alexander, had recently sealed this only possible escape route. When the intruders finally found the barred secret passage under the Queen's wardrobe, they called on the royal couple to come out, threatening to bomb them if they did not. After the King and Queen emerged, the assassins raised their swords, then their revolvers and shot the royal couple. The U.S. diplomatic correspondent, C.L. Sulzberger wrote an account given to him by a friend of his who had participated in the action. *The assassination squad burst into the little palace. They found the King and Queen cowering in a large closet (both in silken nightgowns), they stabbed them and chucked them out the window onto garden manure heaps, hacking off Alexander's fingers when he clung desperately to the window sill.*

After the coup, life in Serbia continued as before. The Serbian senate elected as the successor to the throne of their Kingdom, Petar Karageorgevic the scion of an old banished house as King Petar I. The aggressive pro-Russian who newly reigned in Belgrade, began to exert a significant influence on international affairs. He realigned Serbia with Russia and France, and promoted the key conspirator, Cavalry Lieutenant Petar Zirkovic, to become his new CO of the Palace Guard. Captain Dragutin *Apis*, who had received three bullets which he would carry in his body for the rest of his days, was spirited away from the palace to recover from his wounds. The Serbian Parliament thanked *Apis* for his deed, called him the *Savior of the Fatherland* ~ and proclaimed his promotion to colonel. His enhanced reputation won him the dual influential positions of Professor of Tactics at the Military Academy, (where he taught more subversion and terror than military strategy) plus his appointment as new CO of the espionage department of the Royal Serbian Military Intelligence. According to C.L. Sulzberger; *Apis* and his clique of 57 officers who formed the secret Serbian society of *The Black Hand* in early 1902, as a rogue group, then faded into shadows until their next opportunity for nationalist intrigue arose. *Apis* and his secret society of scoundrels then looked forward to the day when Imperial Russia could help them create their vision of *Greater Serbia* ~ at the expense of Vienna and the Danubian Monarchy.

The Black Hand continued to cast its shadow over Serbia to become a secret state within a state. In Belgrade, it became dangerous to show any sign of sympathy for Austria-Hungary or to oppose Serbian nationalist organizations ~ especially any strongly represented amongst the Serbian military. Those bureaucrats and politicians not already members of this shady outfit, obeyed its express or perceived wishes out of terror, more imagined than exercised. How deep the influence, which *Apis* exercised upon the Serbian Government, was apparent from one particular meeting of *The Black Hand.* After *Apis* only half-jokingly suggested that one of the founders of their organization, the old ex-priest Bogdan Radenkovic, could perhaps be appointed as the next Archbishop of Serbia, the Government, to please *Apis,* at once accepted his suggestion and recommended Radenkovic as its own candidate for the high religious post, thereby provoking a storm of indignation through the ecclesiastical hierarchy. Of this and other schemes, British *Chargé d'Affaires* in Belgrade, Sir Dayrell Crackenthorpe wrote: *From the signal lack of religion noticeable among the large population of Servians, it seems that Servians look upon their Orthodox Church as a means of propagating not the Kingdom of Heaven, but the Kingdom of Servia.* For an English eyewitness account of how these events were viewed, we turn to Edith Durham…

It was in Cetinje that I first picked up a thread of the Balkan tangle, little thinking how deeply enmeshed I should later become, and still less how this tangle would ultimately affect the whole world. Chance, or the Fates, took me Near Eastward. Completely exhausted by constant attendance on an invalid relative, the future stretched before me as endless years of grey monotony, and escape seemed hopeless. The doctor who insisted upon my having two months holiday every year was kinder than he knew, he said ~ Take them in quite a new place. Get right away no matter where, so long as the change is complete.

I boarded an Austrian Lloyd steamer at Trieste, and with high hopes but weakened health, started for the ports of the Eastern Adriatic. Threading the maze of mauve islets set in that incomparably blue and dazzling sea; touching every day at ancient towns where strange tongues were spoken and yet stranger garments worn, I then began to feel that life after all might be worth living, and the fascination of the Near East took hold of me.

A British Consul, bound to Asia Minor, leaned over the bulwark and drew a long breath of satisfaction. We are in the East! He said, Can't you smell it? I feel I am going home. You are in the East as you cross Adria. He added; People don't understand. When you go back to England they say, how glad you must be to get home! They made me spend most of my leave on a houseboat on the Thames! I laughed. I did not care if I never saw England again. He asked whimsically; you won't ever go back again now, will you? Oh don't! Tell them you can't, and just wander about the East. Our voyage ended at Cattaro, and from there I would drive up to Cetinje, capital of Montenegro. It was there that I found the new place I sought. Thus Fate led me to the Balkans.

Cetinje then was a mere red-roofed village conspicuous on the mountain-ringed plain. Its cottages were one storied for the most part, and contained some 3000 inhabitants. One big building stood up on the left of the road as we entered. The driver said; that is the Austro-Hungarian Legation. Austria had started the great Legation building competition, which occupied the Great Powers for the next few years. Each Power strove to erect a mansion in proportion to the amount of influence it sought to obtain in this sphere. Russia at once followed.

Then came Italy, with France hard on her heels. England, it is interesting to note, started last; by way of economizing bought an old house to add to. In 1900 most Ambassadors parlour-boarded at the Grand Hotel, the focus of civilization, where they all dined at the Round Table of Cetinje, presided over by Monsieur Piguet, the Swiss tutor of the young Princes. As their number and variety caused me to ask: Why are so many Powers represented in such a hole of a place? The Italian architect who was designing the Russian Legation replied; Because Montenegro is the matchbox upon which the next European war will be lighted!

Early in 1903 it was hot at Cetinje. We were resting in one of the back bedrooms of the hotel on the afternoon of 11th June, when there came a knocking at the door and the voice of Ivan, the waiter ~ Telegramme! We jumped up at once, fearing bad news. Ivan cried excitedly as I opened the door ~ The King and Queen of Serbia are both dead! Later came full details. According to Belgrade papers a revolution had been planned for months, there were secret societies all over the country, and that the decision to slaughter both King and Queen had been taken by officers at Belgrade, and that the assassins gained access to the Palace at 2.00 AM and as the King refused to open the door of his bedroom, it was blown in with a large dynamite cartridge explosion. What followed was a shambles. The bodies of the victims, still breathing, but riddled with bullets, were pitched from the window. Draga, fortunately for herself, expired at once. But the luckless Alexander lingered till 4 AM.

According to current report, the assassins, drunk with wine and blood, fell on the bodies and defiled them most filthily, even cutting portions of Draga's skin, which they dried and preserved as trophies. An officer later showed a friend of mine a bit which he kept in his pocket book. Alexander was a degenerate. His removal may have been desirable. But not even in Dahomey could it have been accomplished with more repulsive savagery. The Russian Ambassador, whose house was opposite the Palace, calmly watched the events from his window. The assassins then rushed to also kill Draga's two brothers, one of whom was to be declared heir to the throne.

The afternoon of the 11th passed with excitement enough. Evening came and we all went in to dinner. Upon each table lay a black-edged sheet of paper informing us that the Serbian theatre company then playing in Cetinje ~ in consequence of the death of our beloved Sovereign King Alexander had closed the theatre till further notice. The tourist table was occupied solely by my sister and myself. The diplomatic one by Mr. Shipley, who was temporarily representing England, and Italian Ambassador Count Bollati. Dinner passed in complete silence. I was aching to have the opinion of the exalted persons at the other table on the startling news, but dared not broach so delicate a subject. However, when the servants withdrew, Count Bollati turned to me and said; Now, Mademoiselle, you know these countries. What do you think of the situation? I replied, Petar will be made King. Mr. Shipley said, I should advise you to say nothing about it here… you may very well get yourself into trouble. I have never seen them so excited, put in the wise Italian Count Bollati.

Both gentlemen commented on the grim telegrams. Count Bollati's blood was badly curdled by the fact that when he was in Belgrade he was well acquainted with Colonel Masin, the ill-fated Draga's own brother-in-law; who according to the telegrams, had finished her off with a hatchet. Bollati said disgustedly: Business carries on as usual during the alterations, and I have shaken hands with him! Mr. Shipley suggested; As you have first-hand information you have better write an article or two for the English papers. This I did at once. I had written my first Balkan book and hawked it unsuccessfully round the publishers, who told me that as nobody in England took the faintest interest in the Balkans, they could not take it, though they kindly added that as travels went it was not so bad. But the assassination of a King appealed at once to the great heart of the British people and I sold that book as an immediate result.

I came down early next morning to post the articles written overnight, and found a whole crowd of officers and intelligentsia (for in no land are these necessarily the same) around the hotel door. Vuko Vuletitch, the hotelier, in his green, red-embroidered coat, was haranguing them from the doorstep with the latest telegram in his hand. Loud and lively discussion filled the air. Vuko waved his hand as I approached ~ Petar will be King! The news had arrived.

More details came in. Murder is bound to be unlovely. This one was peculiarly so. One fact was prominent. And that was that although many persons expressed horror of the methods and condemned the treachery of officers who had sworn fealty, yet Cetinje as a whole regarded the affair as a blessing. Not only was the populace pleased, but also, with a childish ignorance of the Western point of view (and West Europe was very civilized), actually expected Europe to rejoice with them. It was a cleansing of the Temple; a casting out of abominations. And so ready was every one with a candidate for the throne that it was impossible not to suspect that there had been foreknowledge of the event. Subsequent enquiry through persons connected with the post office revealed to me the fact that a most unusual amount of cipher telegrams had been buzzing between Belgrade and Cetinje immediately before the bloody climax.

Vuko called to me; There is to be a gala performance at the theatre! I flew to the Zetski Dom. Not a seat was to be had. The ever-obliging Vuko said, if you don't mind a crowd, you can come into my box. And he hurried up dinner that we might all be in time for the Serbian Theatre Company, which had abruptly shed its mourning, to give a gala performance ~ in honour of the accession of our beloved King Petar. The theatre was packed from roof to floor. The performance opened with a tableau portrait of Petar I, bewreathed and beflagged. A speech was made. There were shouts of Zhivio! (Long life to him ~ an eminently suitable remark under the circumstances). The whole house cheered. I felt like an accessory after the act. We were a tight fit in our box. Gazivoda, head of the police at Podgoritza and brother-in-law to Vuko, was there. He, too, was assassinated a few years afterwards. And there was a crowd of Vuko's pretty daughters. The eldest being 15, still a pupil at the Russian Girls' School (Russia Institut) was shuddering with horror at the crime. She muttered at intervals; Poor Queen, poor Queen! She was still alive when they threw her from the window. If I had been there I would have wept on her grave. We shouted our last Zhivio! The play was over. Petar was King and the Near East entered upon a new path.

On the 14th June, I drove down to Cattaro with my sister to see her off by steamer. Cattaro, as usual in the summer, lay panting at the water's edge. No more news; any amount of gossip; the Petrovitches were tottering. The Austrian papers were full of digs. I arrived back at Cetinje on the evening of the 15th to find it beflagged and rows of tallow candles stuck along my bedroom window for the coming illuminations. I noted in my diary; Will the army, now that it has taken the bit between its teeth, be more than King Petar can manage?
In truth, no greater curse can befall a land than to be ruled by its own army.
A nation that chooses to be dictated to by its military has sunk low indeed.

At Cetinje, I gathered that though the populace still heartily approved of the murder of Alexander and were filled with anger and dismay at England's rupture of diplomatic relations, the mighty of the land had realized that in public at any rate, it was as well to moderate their image. King Nikola had been interviewed by several British and other journalists, had looked down his nose, lamented the wickedness of the Serbs, and assured his interviewers that the Montenegrins were a far more virtuous people. Montenegro posed as the good boy of the Serb race, and as the gentleman in question had not been present either at the thanksgiving in the church nor the gala performance at the Zetski Dom, they accepted the statement. Interviewing is, in fact, as yet the most efficient method by which journalism can spread erroneous reports. I returned to London shortly afterwards, and Cetinje showed signs of relapsing into dullness. Colonel Masin, who had himself helped kill his sister-in-law Draga, has in the interval been made head of the General Staff, and Colonel Damian Popovic, the apparent leader of the gang, has been given the command of the Belgrade-Danube Division of their army. Edith Durham

KING PETAR OF SERBIA ON HIS PONY AFTER HIS CORONATION

King Petar was crowned as King of Serbia in St Michaels Cathedral in Belgrade on 21st September 1904. The crown made for King Petar was made in Paris of bronze, gold, and precious stones and is now the only surviving crown of any Serbian monarch. No royal families ~ other than the Montenegrin Petrovic family came to the coronation ~ in response to the military coup d'etat, and Belgrade's false assumptions for a Greater Serbia.

Edith Durham ~ Balkan explorer, correspondent & aid organizer

Prince Ioann Constantinovic Romanov of Russia

The new king was welcomed in Belgrade with wild rejoicings, which King Edward VII described as, *in the worst possible taste.* At Edward's insistence, diplomatic relations with Serbia were broken off. Once Vienna and St. Petersburg had recognized the new régime, King Edward VII declared in June;

Russia and Austria are interested countries, and there is no need for England to recognize a Government consisting of assassins. It seems to me that these objectionable officers are the masters of the situation. Eastern views differ from those of Western nations. Till they know how to behave themselves, no British Minister should be sent to Servia. I certainly do not wish to be represented in any way.

With his decree, relations between London and Belgrade were terminated. When King Edward received a report in October 1904 that King Petar I had informed his confident, Crown Prince Danilo of Montenegro: *I knew about the plot, to murder my predecessor, and I am indifferent to British opinion.* King Edward VII scribbled on the report; *What a man King Petar must be ~ worthy of a sensational novel.* Petar then sold Draga's effects in London.

Christie's announce the sale of Queen Draga's jewels and costumes, including her wedding dress, made of white pleated satin, elaborately trimmed with fine old Brussels lace. A cabochon emerald and brilliant bracelet presented to Queen Draga on the occasion of her marriage by Tsar Nicholas II of Russia. A brilliant tiara, worn at Her Majesty's wedding, formed as a crest of ribbon and spray foliage, with two fine large brilliants in the centre, a Persian and Turkish order, and a gold pendant and a pair of earrings of Serbian design, set with pearls and diamonds which the Queen wore with the state costume. *The London Chronicle* ~ December 1904

On behalf of King Petar, his Chief Minister Nikola Pasic began to conduct a foreign policy viewed by Vienna as openly anti-Austrian. In 1904, Serbia and Bulgaria concluded a *Friendship Treaty* and established a customs union. This infuriated the Austrians. The Cabinet of Vienna who supposed that by the application of a few customhouse regulations, such as a Swine Tax, they could bankrupt the troublesome Kingdom of Serbia through the onslaught of economic pressure, so engaged in a *Pig War.* They only had to stop the pigs and oxen plodding into Hungary across the Danube, plus the plums wine, corn and leather that crossed the border at Semlin, the Hungarian town on the confluence of the Danube and the Save. So they established selective tariff barriers ~ allegedly for reasons of hygiene. Although this suited the Hungarians who also raised pigs and exported hams, it also resulted in Serbia gaining new markets in France and Belgium, and once German traders sought Serbian swine, the harsh tariff war failed to break the economy of Serbia. Worse ~ it now enflamed Serbian newspapers to open political and provocative rage against Vienna in their call for a *Greater Serbia,* which was hoped to include both Bosnia-Herzegovina and all of Macedonia.

To the Austrians, the rise of Serbian nationalism and aggression was a direct threat to the future of their Danubian Monarchy. By 1905, the disturbed youth of Bosnia and Serbia were even stirring to question the very future of the Austro-Hungarian Empire. Although pathetically impoverished and backward by the standards of the rest of Europe, education and culture begun to develop. Gymnasia, or schools whose graduates gained the equivalent of a junior-college education, had been established in Belgrade and Sarajevo, where, some students were captivated by the ideals of national liberation to become the spearhead of *Mlada Bosna ~ Young Bosnia,* a new generation of standard-bearers of revolutionary nationalism for Greater Serbia. Many Serbian revolutionaries gathered in Switzerland and came under the influence of Russians like Bakunin, Kropotkin, and Herzen, and the literature of other Russian Social Revolutionaries who had carried out a number of horrific assassinations, including that of Tsar Alexander II. The sole ambition of these young Bosnians was national liberation, and their chosen tactics were anarchic deeds. Whilst the seven assassination attempts on the life of Alexander II enthused and inspired violent South Slav extremists, it equally alarmed the threatened A-H Empire, whose people began to appreciate Emperor Franz Josef's limited options of response. Therefore, the Austrians began constructing an alternative rail route from Vienna to Salonika through the Sanjak of Novipazar for the *Berlin-Baghdad Railway* to bypass the turbulent Kingdom of Serbia. This railway came to be seen by Russia, as a symbol of Teutonic domination over the Balkans.

The Black Hand was responsible to none. Members joined the police force, and so secured their plans against police interference. By means of a paper called Pijemont, they spread violent chauvinism and advocated savage methods. Damian Popovic the head assassin, held an important post. Efforts on the part of politicians, who disapproved of its methods, to break up the society failed. The Black Hand brooked no interference. Unexplained deaths took place. *Edith Durham*

With the accession of King Petar, Servia began to make progress. First of all, the Russian influence that had been so cleverly undermined by Austrian agents and money, revived once more. The Russian Foreign Office had grasped at last the importance of keeping not only an attentive eye, but also a strong hand on the leading politicians of Belgrade. This was not a hard matter, as in general Russia was very much liked. In view of these developments it is not to be wondered that as soon as it was noticed that the old and cordial relations which had existed between the Petersburg and the Belgrade Cabinets seemed likely to be resumed, Pasic and his party, rallied to the new dynasty, whose accession discouraged Austrian emissaries and inspired sympathy in Russia.

It soon became evident that the new King meant to remain faithful to that Muscovite alliance which he firmly believed constituted the only chance that Servia still possessed of remaining an independent kingdom. He used certain influence which he had in St Petersburg as well as in Moscow to obtain the nomination at his Court of a Russian diplomat who had been in touch already with the East, and who could therefore form a dependable opinion as to the various Austrian intrigues, which at the time I am referring to had started with renewed vigour to undermine every manifestation of sympathy that was uttered in favour of the Tsar. He succeeded after some time in his design, and Nicholai Hartwig was appointed as Russian representative in Belgrade.

Hartwig thoroughly understood the Eastern character and the intricacies of Eastern politics. For a considerable number of years, he had been at the Embassy at Constantinople ~ at the time when it was administered by Count Ignatieff ~ and he had learned to appreciate at their proper value the protestations of friendship, or the declarations of enmity, which were liberally distributed all over the Balkan Peninsula to the people who happened, for one reason or another, to be prominent in that part of the world. He hated Austria, and he bitterly disliked Germany, whom he accused of fomenting all the troubles which were regularly taking place either at Sofia, Athens or some other Eastern capital. At the same time, he did not hold to the opinion that Russia, as some politicians suggested in that country itself, should slacken her interest in the affairs of the Balkans, and leave a free field to her enemies in that direction. He therefore applied himself to win the favour of King Petar, already very well disposed toward him, and he began negotiating the marriage of the pretty and attractive Princess Heléne of Servia, the King's only daughter, with a member of the Russian Imperial Family ~ perhaps Prince Ioann Constantinovic, the eldest son of the Grand Duke Constantine.

In this delicate affair Hartwig succeeded completely, and this gave him at once quite an exceptional position at the little Court of Belgrade, where he became a persona non-grata with the Royal Family, and also in Society. Among the embassies too, officials became very quickly aware that they had to deal with a strong personality, who would not allow himself to be circumvented either by Turkish or by Austrian intrigues. In fact, it was in regard to the latter that Hartwig occupied himself the most, trying to discover their importance and to find out their ultimate object and aim. He was fond of the Servians, whom he considered to have been very badly treated, and during the two Balkan wars Hartwig exerted himself in an active manner in favour of Servia. It was thanks to his efforts that the disputes existing between her and the new Kingdom of Bulgaria were at last settled to the satisfaction of both parties, and he laid with Pasic the foundations of a vast scheme which was to secure to the poor little kingdom on the Austrian frontier the protection of Russia against any possible aggression. He had always felt frightened at what Turkey might feel tempted to do, and had never looked with favourable eyes upon all that was going on at Constantinople, where his unerring instinct made him scent danger when no one else even suspected its nearness. He kept urging Pasic to declare himself openly in favour of Russian supremacy in the Balkan Peninsula, which he considered ought never to be shaken in the future. In a word, Hartwig wanted to bring back to life the glorious days when the power of Count Ignatieff at Stamboul was often pronounced to be superior to that of the Commander of the Faithful.

Hartwig found the ground very favourable to the active Russian propaganda he prosecuted from the very first day of his arrival in Belgradè. He was not only an exceedingly clever man, but also a sincere patriot. He was kept admirably well informed by his numerous agents as to the progress of Austrian intrigue, and it was said that he held in reserve several trump cards in the game, which he was playing against his Austrian colleague. He was exceedingly discreet, and handled his affairs alone. This fact gave him the opportunity to learn far more than a diplomat could do through the usual channels. He had friends everywhere and found servants who fully trusted him, and he held in his hand the thread of almost all the undercurrents, of which there exist so many in the East. But he did not allow his staff to learn who his agents were, and what they reported to him. Unfortunately, Hartwig died suddenly (a short time after the crime of Sarajevo) and carried with him to his grave many secrets that would have been of tremendous value to know today.

Princess Katarzyna Radziwillowa
aka Princess Catherine Radziwill
St Petersburg Courier

Meanwhile in the Far East, the intrigues of the new Russian régime were gathering momentum. Russia took complete advantage of the recent Japanese triumph over Qing Dynasty China and the subsequent signing of the Shimonoseki Peace Treaty on 17th April 1895, to end the Sino-Japanese War. This event then inaugurated a radical new epoch in the Far East. It brought Japan as a rising military and naval power before the attention of a surprised world, and within a decade into war with Russia.

Two events occurred as a consequence of the Sino-Japanese War, which proved to be new factors in the shaping of world policy and the relations of Empires to one another. These developments were :
1. Qing Dynasty China acknowledged the full and complete independence of Korea.
2. The Russian-French-German combination discarded the 1895 Treaty of Shimonoseki just six days after its signing to pursue their Triple Intervention of applied diplomatic pressure to force Japan to evacuate and to retrocede to China; the entire Province of Feng-Tien, in addition to the southern Liáodong Peninsula with the strategic deep-water Port Arthur (Japan's main war prizes from China) in exchange for a 15% larger indemnity. This was pursued on the pretext that the Japanese possession of this peninsula would menace the capital of China and render Korea's independence nominal. Then came the consequential aftermath of the leased acquisition by Germany of China's Kiao-chau Bay. As Imperial Russia had no intention of leaving the grand fortress of Port Arthur in Chinese hands, their new Minister for Foreign Affairs Count Mikhail Mouravieff deliberately mislead Great Britain concerning St Petersburg's intentions for Port Arthur and Port Talienwan in China. He first advised the British Foreign Office ~ *These will become open ports.* Subsequently, he modified his pledge.

Two weeks later on 27th March 1898, Russia took possession of a forced 99-year lease on the strategic deep-water Port Arthur at the southern tip off the Liáodong Peninsula that these same three European Empires had recently prevented Japan from keeping. Russia then strengthened her advantage by advancing money to enable China to pay the increased Japanese war indemnity to secure the paramount influence of Russia in Pekin. Russia also received further concessions in Manchuria, including the right to construct a railway from Harbin-Port Arthur down the Liáodong Peninsula to connect their Trans-Siberian Railroad to their newly acquired warm-water port to permit Russia to consolidate its presence and further expand into Asia and the Pacific. Port Arthur if it remained in Japanese hands would have ruined Russia's hopes for a warm-water port in the Far East. So, much to Japan's astonishment and alarm, Russian armed forces at once moved south to occupy the entire Liáodong Peninsula, and to further fortify the defensive positions at Port Arthur.

France got her share too. Two weeks later on 11th April, she exacted a lease from China of the Bay of Kwang-Chow-Wan. The meaning of this scramble was manifold. It was the first apparent attempt between the concerted Russian-French-German troika for the partition of China. Under the surface of this secret history of the Triple Intervention lurked a deep-laid, double scheme. The three Powers collectively had a common aim : an anti-British policy that was put to a test in the Far East by taking a foothold there. In addition to this collective aim, each of the three pursued a scheme of their own.

Russia sought expansion in the Far East with an ultimate aim of reaching a warm water port ; her ambition for 200 years, and incidentally to swallow Manchuria and Korea. Russia had induced China in 1895 to enter into a secret treaty with her, granting her certain rights and privileges in Manchuria, France, as an ally of Russia further followed her Asian interests. Germany, true to one of Bismarck's pet schemes, encouraged France in her colonial expansion, hoping that French energy and enterprise might be diverted from Alsace-Lorraine. Germany also encouraged Imperial Russia to expand into the Far East for the following reasons of geopolitical intrigue :
1. The Berlin militarists hoped that British and Russian forces might come into collision there.
2. Russia, as France's ally, should according to German aims, be always busily absorbed elsewhere than in Europe, therefore weakening the power of the Franco-Russian Alliance or Double Éntente.
3. Germany, with her policy of *re-insurance* (a heritage of Bismarck), always found it difficult to reconcile, harmonize, and satisfy both Russian and Austro-Hungarian aspirations in the Balkans.

From this point of view, more than from any other, it was imperative for Germany to divert Russian attention from the Balkans, in which she admirably succeeded until Russian forces were so decisively defeated by Japan in 1905. Russian foreign policy then returned its focus to the storm centre of the Balkan Peninsula with dire consequences, for as we shall see, the dreaded collision between A-H and Imperial Russia, so long warded off, would simmer anew to its fateful conclusion in 1914.

Great Britain was not slow to discern these moves, and pressed China to lease her Wei-hai-wei in 1898 on the terms and for a period equal to the Russian leases of Port Arthur and Port Talienwan. These events led to what Lord Salisbury so admirably called the *Battle of Concessions* amongst the powers in China. France and Germany soon saw Russia relentlessly pursuing her own policies, such as their striving for the Amur River region for reasons of navigation. France to her undoing in 1914, never opposed Russia's moves in the East. But Germany began to distrust Russia who she herself helped to plant there in the Triple Intervention. Britain had many old reasons of distrusting Russia, yet King Edward VII and Sir Edward Grey, with their eyes closed, would eventually sign an alliance treaty with Russia in 1907, which in essence became one of the main causes of the Great War.

In 1900, when news of the Boxer Rebellion reached St Petersburg, Minister of War Kuropatkin, beaming with joy said to Count Witte: *For my part, I am very glad. This will give us an excuse for seizing Manchuria. We will turn Manchuria into a second Bokhara.* The Boxer Rebellion, the devilish uprising that was the provocative work of Count Mikhail Mouravieff whose secret agents acting in Manchuria and Pekin had connived the agitation, saw Great Britain and Germany, prompted by their mutual distrust of continuous Russian intrigues come together in October to sign an agreement with three aspects;
1. To uphold and maintain an open-door policy in China.
2. Not to make use of complications following the Boxer Rebellion for territorial advantages.
3. To take common steps for protection if any other Power would do so. This was directed against Russia; who regardless of the above, poured troops into Manchuria and proclaimed all Feng-Tien a Russian protectorate on 11th November, and then occupied Niu-Chwang in Manchuria, promising that these occupations should be temporary ~ a promise she never kept. When the troubles came to an end, Russia then found fresh reasons for retaining the ground she had gained. About the same time Russian speculators obtained concessions along the Yalu River, and it became increasingly probable that Russia might soon attain at Seoul the same dominant position that she had already won at Pekin.

It is interesting to note, that Russian Foreign Minister Count Mikhail Mouravieff suddenly died on 21st June 1900, after a stormy interview with Sergei Witte and Tsar Nicholas, who laid considerable blame on Mouravieff for developing confrontational issues with Japan over China, as it had been Mouravieff who insisted on seizing Port Arthur against Witte's advice. With a prominent wound on his left temple when he died that stirred a rumor that he had committed suicide, officials announced: *After rising quite late, he had merely slipped in his study and grazed his temple on the sharp side of a bureau. The Spectator* 23rd June 1900: *A patriot in his own way, the Count had a strong dislike of British diplomatists, and had a fixed belief that the interests of the two countries were inherently opposed. As regards China, he never abandoned the double hope of benefiting Russia, which was his duty, and tripping up England, which was not.*

Japan, still deeply humiliated by the Triple Intervention of 1895, now looked with growing anxiety on Russia's expansion thru Korea and attempted to arrive at an understanding with Russia. However in 1901, when the Marquis Hirobumi Ito as Japan's foremost diplomat went to St. Petersburg hoping to achieve what he called the *Man-Kan Kokan* : the exchange of a free hand for Imperial Russia in Manchuria for a free hand for Japan in Korea : but the Russians now planning to retain their Korean ports for their navy, were in no mood to compromise as Foreign Minister Lamsdorff announced: *I think time is on the side of Russia, because of the Trans-Siberian Railway and there is no need to make concessions to Japan.* Russia therefore deflected the Trans-Siberian Railway from its original Vladivostok route to traverse Manchuria and the northern provinces of Qing China where wheat instead of rice was the staple food of the people, and for these rich agricultural areas to be developed into a Russian sphere of influence. The Russian refusal to Japan showed that Russia aimed at permanency in Manchuria and of all Korea. Since the Japanese also desired to be the masters of Korea (and much of China), they resolved to fight.

Britain's masterful response to this continued Russian expansion in China was to use Japan's Korean predicament to establish an Anglo-Japanese Alliance on 30th January 1902. After the signing of the Anglo-Japanese Treaty, Japan once more invited Russia to discuss the position of the Far East. These discussions went on, but it became clear to Japan in 1903 that their respective viewpoints were antagonistic. Tsar Nicholas expected trouble in China, and so transferred troops into Manchuria to protect his *Chinese Eastern Railway* and other projects. After this, there was every possibility of Russia and Japan coming to blows ; especially as Tokyo wished to precipitate a quarrel, and in early 1904, with Japan wary of Russian ambition, the Japanese further escalated the tension.

Imperial Russian Cossacks guard the Chinese Eastern Railway bridge over the Sungari River in Harbin ~ Le Patriote Illustré 1904

The two countries broke off diplomatic relations on 6th February 1904, and two days before a state of war was officially declared against Russia, Admiral Heihachiro Togo launched a surprise attack against the Russian Pacific Fleet in Port Arthur with ten destroyers, which launched 16 torpedoes. Three warheads hit Russian vessels, whilst the other 13 either veered off course or failed to explode. The loss of the *Petropavlovsk*, with her dynamic Vice-Admiral Stephan Makaroff, sounded the death knell of Russian sea power in the Far East ~ had he lived, the war might have run a different course.

Asserting their command of the seas off Port Arthur, in June the Imperial Japanese Navy lost two first rate battleships, the *Hatsuse* and *Yashima,* which constituted a third of their battleship strength. Although the loss of these battleships was a heavy blow, the Japanese held to their plans undaunted. The stouthearted Admiral Togo, knew that his surviving naval forces remained the foundation on which his sound strategy was based for an effective blockade of Port Arthur. For the remainder of 1904, Togo relied on the Japanese Army to capture Port Arthur to deprive the Russian fleet of its base and force it to fight at a disadvantage. After a series of engagements, Port Arthur fell at the end of the year, after which Russia tasked their 42-ship Baltic Fleet to make an arduous seven-month voyage from Kronstadt to the Korean Straits without undergoing any restorative repairs to optimize for battle.

The Russian vessels had to steam around the Cape of Good Hope, as Great Britain was alliance bound with Japan to deny them passage through the Suez Canal. Therefore, the Russian warships arrived in need of much repair. Their vessels were thickly overgrown with weeds below the waterline, and their boilers required cleaning, and thus could not develop their required optimal battle speeds. With the advantage of speed with the Japanese, they acquired the privilege of the attacking party. With a uniform speed of 17 knots, the Japanese battleships and armoured cruisers were able to steam ahead of the Russian ships averaging only 10 knots, and cut across their course heading N.E.

When the opposing fleets came in sight of each other in the Straits of Tsu-shima, it was soon apparent the Japanese not only had the advantage of speed, but also that of naval gunnery. The Japanese vessels carried a new type of shell containing *shimose* (melinite or lyddite) that delivered four times the explosive force of the now outdated Russian shells. The Japanese naval shells also contained picric acid that released clouds of poisonous gas on contact. Japanese gunners having trained to hit their targets accurately from 12,000 yards ~ twice the distance anticipated by the Russian sailors for the maximum effective range for naval combat ~ fired at almost double the rate of fire of their rivals. The Battle of Tsu-shima was decided by the Japanese Navy having developed their gunnery to a higher standard to score 6% of hits from 5,000 rounds fired, whereas the Russian gunners achieved scarcely 1% of hits from 3,000 rounds fired.

It was said by the Russians that the Japanese had trained and aimed their guns for a fixed range. Having done this, they did not alter it, but opened a rapid fire with all their guns on the leading Russian ship, their vessels all the time coming nearer. Then at the calculated moment, when an appropriate distance separating the Japanese warships from their Russian rivals was reached, a decisive number of Japanese shells would hit.

At noon on 27th May 1905, when the naval clash began, the Japanese sank the Russian battleships *Alexander III* and *Borodino.* By 7 PM, when darkness fell, and the seven-hour naval battle had ended, the Russian vessels had been scattered to the four winds by their IJN opponents. From the 42 vessels of the Russian Baltic Fleet that had steamed into Tsu-shima Straits that day, 19 were on the seafloor. A further seven were captured and six were interned in the neutral ports of either Shanghai or Manila. One escaped to Madagascar, leaving just one cruiser, the *Zhemtshug,* two destroyers, and a few supply ships, to steam into Vladivostok with the grim news that Imperial Russia had again lost to Japan. Although Japan had lost 3 torpedo boats with 117 of their sailors perished, the Imperial Russian Navy had suffered the loss of 4,830 sailors with 5,917 captured.

The defeat in the Straits of Tsu-shima forced Russia to the peace table, and from this came the Treaty of Portsmouth signed on 29th August 1905, to conclude the war. Russia agreed to hand over the disputed Liáodong Peninsula, including the fortified Port Arthur and the Island of Sakhalin to Japan, as well as an official Russian recognition that Korea and much of Manchuria were henceforth within the Japanese sphere of influence. Although the Battle of Tsu-shima Straits had established Japan as a major power in the Far East, the savage war had nevertheless near bankrupted Japan.

On the night of 11th September 1905, the flagship of the Imperial Japanese Navy moored in Sasebo harbour erupted in an explosion that caused almost 600 casualties among her crew. Admiral Togo himself had only gone ashore hours earlier, and the cause of this explosion that destroyed the Japanese flagship was never established ~ although Russian agents were always the prime suspects. In the same mode, when Prince Hirobumi Ito arrived at Harbin Railway Station in China to negotiate with the Russian Minister of Finance Mikhail Kokovstov some aspects of their *Chinese Eastern Railway*, a lurking assassin fired six shots into him from behind ~ Prince Hirobumi Ito was dead in minutes.

For Russia, the aftermath of the war with Japan, combined with lamentable social and political conditions at home, produced such internal disorders, which added to the immense losses of men and materiel in the Far East, rendered an effective Russian foreign policy impossible for years to come. With the defeat of Russia, and her being eliminated from international calculations for several years, a significant balancing force had been removed from the equation of world events, and with it a constant source of anxiety for Germany had disappeared ~ at least for a time. The Kaiser's hands were now freed, and what seemed attainable to Berlin within a calculable time was the mastery of the Ottoman Empire and all the lands that lay between it and the frontier of Austria-Hungary, and at this they aimed well. Once masters in these territories, they began to throw open vast regions to German enterprise and capital to even more exclude the influence of Russia from the Mediterranean.

Russia would undoubtedly resist and be aided by France, but Berlin was not deterred by France and Russia combined. In fact, Russian expansion in itself only weakened Russia as an ally of France, and further financially weakened France as an ally of Russia. Any observer could see these new moves ~ all except France, whose leaders seemed unaware of their fiscal reality that by so blindly supporting Russian foreign policy in the Far East, Paris was playing the game for Berlin. This was brought home to her during the Moroccan Crisis of 1905, spurred on partly as a result of the Russian demise at the hands of the Japanese Navy that same year. Furthermore, the millions of French savings lent to Russia for military purposes in Europe, were essentially squandered on further wild schemes in the Far East. (When the challenge of 1914 surfaced, France would fight simply because Russia had made Serbia's case her own : support of an attempted establishment of a virtual Russian protectorate in the Balkans. The deeply disappointed French found their northern ally ill-prepared, and themselves bleeding to death for a cause not their own). We now depart from the Far East to follow the converging threads of other scenes, with Kaiser Wilhelm II on the waters of N.W. Europe.

CIA Menu Collection

CIA Menu Collection

Midday Menu ~ Hohenzollern ~ 1905
Cold Jellied Boulllon served in Cream
Roasted leg of Veal with Roasted Vegetables
Roasted Young Turkey ~ Fruit Salad
Poached Peaches with Blackcurrant Liqueur
Butter and Cheese

An invitation to sail with Kaiser Wilhelm II

Midday Menu ~ Hohenzollern ~ 1907
Russische Suppe
Gestruffelter Rhinlaufb.
Lammoucban, ganmirt
Auflauf nou Geinplaber in Champagner
Tungen Puten, Fruchtc, Salat.

THE 1905 MEETING OF KAISER WILHELM AND TSAR NICHOLAS

Imperial Russian yacht Stellar Polaris

Nicholas Romanov was only 13, when his grandfather the Tsar Alexander II was assassinated, and as his father Alexander III ascended the throne of Russia, Nicholas became heir apparent. Nicholas spoke five languages fluently, had travelled the world, and he knew his European history. Therefore, when he came to the throne in 1894, and realized that Europe was becoming alarmingly militaristic, he made a personal effort toward the prevention of future wars by seeking to substitute legal methods for brute force in the settlement of international disputes. Pursuant to this, Nicholas invited foreign governments to his First Hague Peace Conference to discuss a viable disarmament. Unfortunately, nothing more redeeming came of Tsar Nicholas' proposition than the signing of two conventions for the peaceful settlement of international disputes in *The Definition of the Rules of War* that defined a prohibition of gas warfare and provisions for the humane treatment of prisoners of war. Tsar Nicholas II was an extremely worried world leader who was receptive to ideals of world peace.
What is going to happen to me and all of Russia? I am not prepared to be a Tsar. I never wanted to become one. I know nothing of the business of ruling. I will preserve the principle of Autocracy as firmly as my late father.

During 1905, Kaiser Wilhelm's summer cruise was not as customary among the fjords of Norway, but along the Baltic coast. Shortly before sunset on Sunday 23rd July 1905, the Kaiser's long white and golden yacht *Hohenzollern* steamed through a fresh breeze and calm waters toward Björkö Island in the Gulf of Finland. The *Hohenzollern* flying the Imperial pennant with the motto *Gott mit Uns* dropped anchor in the waters of Koivisto Sound alongside Tsar Nicholas' own moored *Stella Polaris* (*Polar Star*). The meeting had been arranged just four days before and had been kept secret by both. Wilhelm had telegraphed ahead to Nicholas in English, the language in which they both had the least difficulty.
Not a soul has the slightest idea of meeting. All my guests think we're bound for Gotland. My guest's faces will be worth seeing when they suddenly behold your yacht. Tableaux. What sort of dress for our meeting? ~ Willy

Nicholas boarded Wilhelm's *Hohenzollern* for dinner and remained until 3 AM in the morning, during which the Tsar declared that he would never enter into any form of éntente with England, most specifically one directed against Germany. At 9 AM, Wilhelm boarded the *Stella Polaris* where he and Nicholas retired to the Tsar's cabin to affix their signatures to their proposed Björkö Treaty. After an exalted Wilhelm wrote to Chancellor von Bülow, explaining how the two agreed to their treaty.

Visby, July 25, 1905

My Dear Bülow,

By my telegrams, you have already learned that the work of rapprochement has been crowned and the game won. And now that it is done, one is surprised and says ~ How is such a thing possible? For me, the answer is very clear! God has ordained and willed it thus; in spite of all man's wit, in scorn of all man's intrigues, He has brought together what belonged together! I have done so much thinking in the last days that my head has throbbed to be sure that I am acting right, always to keep in mind the interests of my country no less than those of the Monarchical Ideal in general.

Finally, I raised my hands to the Lord above us all and committed myself to Him and prayed Him to lead and guide me as He wished; I was only the tool in His hands and I would do whatsoever He would inspire me to do, though the task be ever so hard. And finally I also uttered the wish that if He did not wish to help me, He should at least not help the other side. Then I felt myself wonderfully strengthened, and the will and purpose became ever firmer and clearer within me ~ You will put it through no matter what the cost!
So ~ I looked forward to the interview, full of confidence.

And what did I find? A warm, amiable, enthusiastic reception, such as one receives only from a friend who loves one heartily and sincerely. The Tsar threw his arms around me and pressed me to him as though I were his own brother, and he looked at me again and again with eyes that revealed his gratitude and joy. The Tsar said he was burning to have a thorough going discussion. We lighted our cigarettes and were soon, in medias res. He heartily approved my hope, that it forms a lasting understanding, perhaps even an agreement with France might blossom forth. When I pointed out that in spite of egging on by England, France had down-right refused to take up our challenge and therefore no longer wanted to fight for Alsace-Lorraine, he said quickly ~ Yes, that I saw. Our talk then turned on England, and it soon appeared that the Tsar feels a deep anger at England and the King. He called Edward VII the greatest mischief-maker and the most dangerous and deceptive intriguer in all the world. I could only agree with him, adding that I especially had had to suffer from his intrigues in recent years, whereupon the Tsar interrupted me, striking the table with his fist ~ Well, I can only say he shall not get one from me, and never in my life against Germany or you, my word of honor upon it!

*I soon observed how deeply injured the Tsar felt by the attitude of France when he said;
The French behaved like scoundrels to me. By order of England, my Ally left me in the lurch.
And now look at how they fraternize with the English ~ what shall I do in this disagreeable situation?
(Now I felt the moment has come!) How would it be, if we too, should make a little agreement?
Last winter we talked about it... Oh yes, I remember well, but I forget the contents of it.
What a pity I haven't got it here.
Then I said ~ I have a copy, which I happen to have quite by chance in my pocket.*

The Tsar took me by the arm and he drew me out of the dining room into his father's cabin and immediately shut all the doors himself ~ Show it to me, please ~ His dreamy eyes sparkled. I drew the envelope out of my pocket and unfolded the paper on Alexander III's writing desk in front of the portrait of the Tsar's mother. He read once, twice and a third time, the text, which has already been sent you. I prayed God that He would be with us now and incline the young ruler. It was still as death. There was no sound, but that of the sea. The sun seemed gay and cheerful in the cozy cabin. Right before me, glistening white lay the Hohenzollern, and aloft in the morning breeze, fluttered the Imperial flag. On its black cross I was reading the letters, Gott mit Uns, when the Tsar's voice near me said ~ That is quite excellent. I quite agree!

My heart beats so loudly that I can hear it. I pull myself together and say ~ Should you like to sign it? It would be a very nice souvenir of our interview. He scanned the paper again, and said ~ Yes, I will.

I opened the inkwell and gave him the pen, and he wrote with a firm hand. He then handed the pen to me, and I signed. When I arose he clasped me into his arms deeply moved and he said ~ I thank God and I thank you. It will be of the most beneficial consequences for my country and yours; you are Russia's only real friend in the whole world. I have felt that, and I know it. Tears of joy in my eyes. Thus has the morning of 24th July 1905 at Björkö become a turning point in the history of Europe. Thanks to the grace of God, and a great relief for my dear Fatherland, which at last will be freed from the frightful Franco-Russian pincers. ~ Wilhelm II

Wilhelm then wrote to share his news with President Theodore Roosevelt of the United States.
The Tsar and I have concluded an agreement to lend each other mutual help in case any European power should attack one of us, and France is to be cosignatory to it. In fact, Germany enters the dual-alliance of Russia and France ~ originally concluded against it ~ as third party. Instead of glaring at each other for no purpose, the peace of Europe is guaranteed. This is the fruit of our understanding, upon which you sent me so kind compliments. I am sure, that this is leading to a general détente, will be of great use in enabling you to fulfill the great mission of peace, which Providence has entrusted to your hands for the good of the world. ~ Wilhelm II

On his return voyage to St Petersburg, Nicholas found the Björkö note that he and Wilhelm had signed, and which Russian Ambassador to Copenhagen Count Alexander von Benckendorff had also countersigned, was officially considered a *dead letter* because of Russia's commitment to France. Russian Foreign Minister Count Vladimir Lamsdorff and Count Sergei Witte, who were not present on the *Stellar Polaris* nor consulted beforehand, insisted that the proposed Björkö Treaty could never come into effect, unless it was approved by France. Nicholas succumbed to their reasoned opposition.
I did not consider the treaty from your point of view. When I signed it I did not think that my agreement with the Kaiser could be directed against France. Quite on the contrary, I always had in view to associate France.

Similarly, when the Kaiser returned to Kiel, his prayerful optimism and emotional fervor were soon given a dash of cold water by Imperial Chancellor Prince von Bülow, who threatened to resign. His pretext was that the Kaiser had ventured on his own responsibility without consultation with either government. Kaiser Wilhelm urgently appealed to Nicholas to stand by his written agreement.
Our treaty is a very good base to build upon. We have joined hands and signed before God, who heard our vows! I therefore think that the treaty can well come into existence. What is signed is signed! And God is our testator!
Whilst King Edward VII had earlier diplomatic success, these rulers were rebuked by their officials.

German and Austrian plans for an extension of their Teutonic influence and commerce eastward through Asia Minor and into the Middle East via the new *Berlin to Baghdad Railway* increasingly antagonized the opposing Byzantine dreams of Imperial Russians, who still intrigued for their own advance to Constantinople through a Greater Serbia and then down to the Adriatic Sea. These grand geopolitical lines of ambition lay across one another and remained forever mutually exclusive. Increasingly ambitious German militarists began to express an active interest toward Western Europe.

The Kaiser has a dramatic, but not a political instinct. He considers the momentary effect but not the consequences, and is actually for the most part unpleasantly surprised by them. Germany is on the brink of a period of danger and degradation and is perilously close to it. All provocations are either conceived by the Kaiser or are conceived to please him. The Kaiser must be made to realize that his prestige will suffer if he follows every impulse. This applies above all to the personal remarks of His Majesty. Baron Friedrich von Holstein
German Foreign Office

While Belgian neutrality was instituted after the defeat of Napoleon to act as a buffer between France and Germany, when King Leopold of the Belgians visited Berlin in January 1904 to attend the Kaiser's birthday celebrations, Wilhelm II would seize the opportunity to bind Belgium more closely to Germany ~ as Wilhelm indeed confided, to his Imperial Chancellor Prince Bernhard von Bülow. *We must remind King Leopold of the splendour and magnificence of Burgundy on the old days. If we give him the prospect of rising to similar heights through an alliance with us, Leopold will be ready for anything.*

The Kaiser looked King Leopold II squarely in the face, his left hand resting on his sword-hilt and asked him bluntly; *What attitude do you, as the King of Belgium, intend to adopt in case an armed conflict should break out between Germany and France ~ or Germany and England? Give me a written declaration now, in time of peace to the effect that in case of conflict, Belgium will take her stand on our side, and guarantee to us the use of Belgian railways and fortified places. If you do not do so ~ I will not be able to give a guarantee for either your territory or dynasty.*

The French want war. Good, they shall have it. In the gigantic struggle that is to come, Germany is sure of victory, but this you must decide. You must be with us or against us… If you are with us, I will give you back the Flemish provinces that France wrongly took away from you. I will restore the Duchy of Burgundy for you. Think about my offer and what awaits you.

The Kaiser complained to his Chancellor. *The contemptibility of my fellow monarch. I spoke to the Belgian King of the historic Burgundian Empire and told him that he could, if he wanted, stretch his realm over French Flanders, Artois, and the Ardennes. But Leopold smirked and replied ~ Neither my ministers nor the Belgian parliament will have anything to do with such ambitious plans.* Then the Kaiser lost all his patience. His eyes of light blue, now hard and stern, flashed like steel.

I told the King that I could not respect a monarch who felt answerable to deputies and ministers instead of to our Lord God in Heaven alone. I also told him I was not to be trifled with. Whoever was not with me, in the event of a European war, was against me. As a soldier, I belonged to the school of Frederick the Great, the school of Napoleon I. Just as the former had begun the Seven Years War with the invasion of Saxony and the later had always forestalled his enemies at lightning speed, so I would allow myself, if Belgium were not with me, to be guided only by strategic considerations. We would then, if the case arose, immediately invade Belgium, and you will have to suffer all the harmful consequences.

After this dramatic exchange between the Kaiser and the King, Chancellor von Bülow observed; *King Leopold looked extremely upset on his departure. In agitation, he put his helmet on the wrong way around, with the eagle at the back!* Count von Bülow expressed his misgivings too late, drawing the Kaiser's attention to the fateful risk he had run in exposing German plans thru such a challenge to Leopold. *Everything depends on the Belgians not suspecting in advance that if it comes to it, we intend to confront them with an either-or choice. Otherwise they can use their wealth to build fortifications against us, and tip off the French so that they too can adapt their plans accordingly. The Kaiser became, more nervous and vehement than was usually his wont with me, and had finally declared ~ If that is what you think, I shall have to look around for another Reich Chancellor in the event of war.*

As soon as King Leopold returned to Brussels, Kaiser Wilhelm's threats of war and occupation were recorded in Belgian documents, exactly as the Kaiser and von Bülow had spoken, and once King Leopold told his Cabinet of the Kaiser's ultimatum, criticisms of France, and of threats to himself, reports of this swiftly found their way to London and Paris, which served to confirm that the German Army was by this time developing a war plan ~ at least against France, and most likely thru Belgium.

The King grasped what was being said and informed his ministers of it, he saw it as another reason to take action against the anti-militarist trend in his country, and to strengthen ties of friendship with Great Britain, whose importance in view of the undisputed supremacy at sea, he foresaw. Count Louis de Lichtervelde

In spite of forty years having elapsed since the events led to France being seriously defeated, and suffering painful loss both of territory, power, and prestige ~ in spite of the country having achieved renewed prosperity and having increased its strength by the acquisition of extensive valuable colonial possessions, and having long since recovered her position of a Great Power as the result of wise policy, the wound of 1871 would not heal… The French still demanded satisfaction for the injury to their national pride. Baron Wilhelm von Schoen
German Ambassador to France

During the provocation of 1870, and the comparatively short struggle of the Franco-Prussian War, the forces of the North German Confederation with twice as many troops as the French, had achieved a complete victory for Germany. Emperor Louis-Napoleon III ignored General de Wimpffen's request that he ride out at the head of the last breakout attempt. He instead raised the white flag and dispatched his adjutant with a letter for the German HQ who asked Crown Prince Friedrich to read it; *Having failed to die amongst my troops, there is nothing left for me to do, but to place my sword in the hands of your Majesty.* Bismarck drafted his reply: *Regretting the circumstances you find yourself in, I accept the sword of Your Majesty and appoint General Moltke to negotiate the capitulation of the army that has fought so bravely under your orders.* Offered a flask of brandy by his nephew, von Bismarck toasted all present ~ *Here's to the unification of Germany!* ~ and drained the contents.

After the fall of Paris, the Treaty of Frankfurt concluded the Franco-Prussian War with the French forced to cede the two eastern provinces, Alsace and Lorraine to Prussia. The French regarded this as a severe crime ~ *the brutal dismemberment of a nation, the tearing of children from their mother.* Not just was Alsace-Lorraine taken by Prussia, but France was also compelled to pay an excessive indemnity of five billion francs to get the occupying German Army out of French territory. From this, the French acquired the very accurate assessment that the Prussians never blundered into war unwittingly, but conducted war with deliberate purpose, prepared long in advance, and carried through to victory with the intent to increase their own power and territorial sovereignty. Although the new Germany had achieved the aim of *Einheit* or unity, it had come at the price of their people's *Freiheit* or freedom, becoming as Karl Marx said ~ *A military despotism cloaked in parliamentary reforms, and guarded by police.*

The greed of Germany in depriving France of Alsace-Lorraine had done a great wrong in 1871, and the national honour of France demanded its reparation, and perhaps more. There were some military men who seldom spoke, but whose voices revealed the prevailing views at the Quai d'Orsay ; hopes that aimed not only at recovering the *Lost Provinces* but also at pushing forward the French frontier to the Rhine. Armand Mollard of French GHQ wrote; *The Rhine, is not a German river at all, it is a frontier. It divides Western Europe into two great districts, the French, extending from the ocean to the Rhine, and the German, extending from the Rhine to the Elbe. We have been, are, and shall remain, arch enemies.*

The people of France, aware of their inability to recover their lost provinces by their own strength, conceived exaggerated ideas of the power and helpfulness of Imperial Russia. And yet, the French people who subscribed enourmously to Russian loans, now knew that France was allied to Russia, and if Germany ever made war on Russia, then France could certainly resist the aggressors once again. The spirit in which the French welcomed the news was revealed in cheers of *Vive l'Alsace-Lorraine!*

However, not all cheered. Paris was the only European capital where there were frequent street demonstrations against militarism. French craftsmen and peasants did not want to go to war for Russian dreams of Constantinople or wish to waste French blood for the Balkans. But often it seemed, that Russian agitators were the passionate protagonists, and the French merely sympathetic seconds.

Having been vanquished by Germany in 1871, France turned her focus of foreign policy to acquiring a colonial empire ~ particularly to the colonization of North Africa. German foreign policy encouraged French colonial ventures to divert her strategic awareness from Alsace-Lorraine. France already had Algiers in North Africa, and in 1882 proclaimed a protectorate over Tunis, and Mzab too. In 1883, a French expedition claimed Senegal and Nigeria. In 1884, French settlers colonized the Congo and claimed Djibouti on the east coast's Horn of Africa. In 1885, they next acquired Tonking and Annam. Even so, France and Britain were still colonial rivals at this time, and several collisions between the two did occur. From 1893, France openly attacked the S.E. Asian Kingdom of Siam, which the British considered within their sphere of influence, however agreement was soon achieved on the Siamese Question and also in 1895, when English and French colonial forces met at Lake Tchad.

LA TACHE NOIR ~ THE BLACK STAIN

In Albert Bettanier's 1887 painting *The Black Stain*, a geography teacher shows his pupils the area of the two lost provinces of Alsace-Lorraine that he hopes might be regained by the next generation. Various details reveal the military spirit of the time. Next to the black stain, there looms a sinister blackboard representing the Teutonic threat in the East, accentuated by a cadet's drum in the corner.

The martial spirit of France is certainly no whit inferior to that of its neighbours. Frenchmen have assured us that the recovery of the lost provinces is not a leading motive in the minds of Frenchmen these days. Not to make aggressive war for that purpose has for years been the accepted policy of successive French Governments. But there is never a question that, if for any cause war may come about, this object will immediately come to life and impulse the fighting spirit. J.A. Spender ~ *Westminster Gazette*

Some in France might not have shrunk from bringing war out of the blue for their lost provinces, as their theft forever imperiled the atmosphere of Western Europe from clearing. From the French point of view, Germany was imperialistic by nature, and economic expansion was for Berlin only one means of conquering the world, and as Imperial Germany forged its *Weltpolitik* aspirations, some on Wilhelmstrasse envied the prosperous colonies of France and Great Britain. In particular, after the defeat of Russia by Japan in the Far East, many German military strategists supposed that France, now deprived of the military deterrence of her northern ally, could be intimidated with impunity.

Germany had no reason to be afraid, but she fortified herself with armaments and the Triple Alliance. France naturally was afraid after 1870, and she made her preparations and the Dual Alliance with Russia. Britain with a very small army and a very large Empire became first uncomfortable, and then when Germany began her big-fleet program wary of isolation. She made the Anglo-Japanese Alliance, made up her quarrels with France and Russia, and entered into the Éntente Cordiale. Sir Edward Grey

I was well aware of the antagonism between France and Germany and how that antagonism had been kept alive by the tearing from the side of France of two of her best-beloved provinces. In the year 1904, on the same day when the Anglo-French Éntente was announced, I arrived at Dalmeny for a two-day visit to Lord Rosebery. His first greeting to me was: Well, I suppose you are just as pleased as the rest of them with this French agreement? I assured him that I was delighted that our snarling and scratching relations with France had come to an end at last. He replied: You are all wrong. It means war with Germany in the end! David Lloyd George

French secret services of the era consisted of *Le Cabinet Noir du Quai d'Orsay*, the *Sûreté Générale* attached to the Interior Ministry ~ and the skillful operatives within the *Bureau de Poste* who reached such a high standard of efficiency that they were able to open diplomatic pouches and envelopes, copy the contents, restore the broken wax seals, and have the originals reposted within two hours. At other instances, their peeking at an activist's mail was done so to appear obvious.

Victor Hugo, whose opposition to Louis Napoleon forced him into 15 years exile on the island of Guernsey, became so exasperated with the tampering with his letters that he had the article of the penal code forbidding such acts printed on his envelopes, or would write on the back flap ~ family affairs of no interest. When this also failed to deter, at least he was able to economize by placing several letters in a single envelope with instructions to the Cabinet Noir to forward them to the addresses indicated, which they invariably did. Douglas Porch

The *Sûreté* maintained agents in Berlin, Geneva and elsewhere in Europe, plus in St Petersburg. There was often frequent collaboration between French and Russian operatives, who for example, often exchanged files on dangerous subversives. In 1887, the Russian Government formally asked France to stop expelling Russian Nihilists who then went to Switzerland or England where the police were less interested in tracking them, especially after a failed assassination plot on Tsar Alexander III in 1890 when Russia requested a crackdown by the French police on Russian Nihilist exiles in France. Such co-operation was an advantage used in France's courtship of a Russian Alliance, as it proved to St Petersburg that France too had a strong government capable of dealing with internal enemies that earned the Tsar's deep gratitude and helped secure the Franco-Russian Alliance of 1894.

The *Okhrana* was given two rooms in a side wing of the Imperial Embassy at 97 rue de Grenelle, with a separate entrance from the courtyard. The Russians installed an additional door with a lock in the hall and heavy bars in the windows. From there, the *Okhrana* ran an extensive network of French and Russian informers led by Paris *Okhrana* CO Leonid Rataev who bribed the French police to turn a blind eye to his various activities. While the *Okhrana* took care of their administrative aspects such as funds and communications, relations with case officers was a matter kept strictly under cover, not known to the *Sûreté* or Britain's Scotland Yard, despite the close liaison maintained with these groups. Grateful for their liaison, Russia gave France the Japanese diplomatic code.

Lieutenant-Colonel Jean Sandherr, who commanded the French Counter Intelligence Department (disguised under the name *Section de Statistique* and housed in a corner of the War Ministry on Paris' rue Saint-Dominique) ran agents throughout Europe ~ especially in Alsace-Lorraine and Germany, who specialized in the surveillance and the theft of documents ~ and Sandherr's principal target was the German Embassy in Paris once they knew new military attaché Colonel Max von Schwartzkoppen was funding a network of spies, and was in constant correspondence with the War Office in Berlin. Although the French had been warned by Spanish military attaché the Marquis de Valcarlos, that Colonel Max von Schwartzkoppen and Italian military attaché Colonel Alessandro Panizzardi had agreed to exchange the results of whatever discoveries they might make ~ what they did not know, was that Major Ferdinand Esterhazy who thru his occasional contact with French GHQ could collect classified information. In July 1894, Esterhazy offered his services to Schwartzkoppen to provide French mobilization documents, which he did in August 1894, and was very well paid for his services.

One agent employed by Sandherr was Madame Marie Bastien (the wife of a Republican Guard) who was hired as a cleaner by the German Embassy in 1889. Described as a woman of about 40 years of age, of Alsatian descent and who spoke fluent German, Marie Bastien as *Agent Auguste* was asked to collect all scraps of paper, torn or partially burnt that she found in the waste-paper baskets or fireplace of Schwartzkoppen's office. She took them in a paper bag to the Church of Sainte-Clothilde or the Church of Saint-François-Xavier for Major Hubert Henry of the *Section de Statistique*. The pieces were then carefully fitted together and gummed. From this, it was ascertained that since 1892, certain secrets concerning the national defense had leaked out. So poor was the security of the German intelligence attaché that he regularly discarded compromising personal and secret documents into his bin and continued to hire Marie Bastien until 1897, three full years after Dreyfus' wrongful conviction based on evidence taken from his office ~ a note signed by an initial *D* recovered by Marie Bastien during April 1894 between Colonel Max Schwartzkoppen and military attaché Panizzardi that read ; *Attached are 12 master plans of Nice which that scoundrel D gave me in the hope of restoring relations.*

Few diplomatic codes around the turn of the 19th century could resist the skills of the French code breakers, and in this, the *Cabinet Noir* superbly supported by French espionage would reconstruct : the Italian code from 1896 (after a French agent had copied part of the Italian diplomatic code book in the office of the Italian consul general at Marseille) plus the British diplomatic code through the copies of discarded telegrams of British Ambassador to France Lord Lytton (which were furnished by his valet) in addition to several German codes sourced from the double-agent, Guilliame de Balasy-Belvata.

Foreign intelligence gathered by the *Cabinet Noir* thru mail intercepts targeted the correspondence of people like British PM Gladstone and King Leopold II of Belgium during their frequent and lengthy vacations to the Riviera. Leopold, who suspected his mail was bang read, once summoned the postal inspector of Cap Ferrat on the Côte d'Azur to demand that his employees at least put the pages of his correspondence back in order once they had been read. Naturally, this now known activity lead to openings for misinformation and designs to deceive the French by foreign diplomats, such as during the second Moroccan Crisis when German Foreign Minister Alfred von Kiderlen-Wächter wrote to his mistress at Chamonix promising ~ *severe consequences if the French Government refuse to make concessions.* On another occasion, an aristocratic French lady wrote in a letter to her husband, which she expected to be intercepted ~ *Here are three violets which I have just picked from my garden. They are the first of spring.* When the letter arrived, it actually contained three violets ~ the *Cabinet Noir,* fearful that they had lost the flowers, had actually placed three violets into the envelope that had contained none.

The greatest pre-war success of the *Cabinet Noir* was their acquisition of a German plan to attack France via Belgium ~ as revealed in January 1904 when King Leopold of Belgium heard so from the Kaiser during his birthday celebrations. Two months after, a Prussian officer presented himself and offered to sell top-secret war plans. Calling himself *Le Vengeur ~ The Avenger* met his handlers with his face entirely wrapped in bandages. Maurice Paléologue noted in his diary entry for 25th April 1904; *A chief of the general staff reported to me on how a German officer, believed to be a general on their HQ staff, had contacted our Service de Renseignements in the War Ministry to offer the German plan of attack on France. An intelligence officer contacted him three times in hotels in Liège, Paris, and Nice to find a man whom he was unable to identify because only a Prussian moustache protruded from a mass of bandages that swathed his head, as if he had just undergone an operation. Although he claimed that revenge was his motive and even signed his letters, Le Vengeur, this did not prevent him from placing a price tag of 60,000 francs on his documents.*

The *Deuxiéme Bureau* were convinced *Le Vengeur's* papers were worth the asking price and handed over 60,000 francs in a parcel. *Le Vengeur* then handed over German deployment plans, together with charts that mapped where the Germans would concentrate and attack, and how the Germans planned to ignore Belgian neutrality to invade France thru the Low Countries. In his early war plans Schlieffen had indeed preferred a flanking movement through Luxembourg and Belgium, and as early as 1899, had added the manoeuvre to German war plans as a possibility. This possibility of a German thrust through Belgium prompted French intelligence and military attachés to study the supportive evidence of German railway construction along the Belgian frontier, the known theories of Friederich von Bernhardi and how observers at German military manoeuvres held each autumn had begun to note the virtues of flanking movements then being stressed over those of frontal assault. France responded in 1906 with *Plan XV bis,* based on the assumption that German forces would come via Belgium, rather than launching a frontal attack on France's defences on the Lorraine frontier, but the *Deuxiéme Bureau* underestimated the numbers of troops, and the western axis of the German plan.

General Henri Navarre confirmed that together with *Le Vengeur's* documents that French agents in 1908 had acquired a mobilization plan drawn up by Ludendorff, as well as a second plan showing that the German Army planned to march on Paris via Belgium, with the axis of advance on the Oise. Navarre wrote; *The German war plan had been known for a long time, and on the eve of war our general staff possessed the Plan of Concentration of the German Army in the west. The Generals refused to give intelligence warnings any credence because they clashed with their preconceived ideas of what the enemy was likely to do.*

Because of the bizarre circumstances of its sudden acquisition, many officers refused to believe either *Le Vengeur* or his documents were real ~ despite other intelligence which clearly corroborated the data *Le Vengeur* had given ~ thus General Joffre rejected the previous defence priority of France's Belgian frontier and formulated *Plan XVII* to primarily attack N.E. to recover Alsace-Lorraine.

Once France offered Great Britain a free hand in Egypt for support of French dominance in Morocco, Berlin had a legitimate cue for another sensational performance by the Kaiser, however this time Wilhelm would not be a willing actor. The German Colonial Society agitated, and Berlin listened. Although their commercial interests in Morocco were negligible and Germany's yearly trade there did not amount to that of an ordinary department store, nevertheless Berlin sought to confirm the status of Moroccan independence and German influence in the wider world. In the realm of international affairs, Chancellor Bülow and Holstein had a double motive : for all to see that Germany could not be ignored when it came to addressing colonial issues, and to disrupt the Anglo-French Éntente Cordiale. While Kaiser Wilhelm had no desire to overly antagonize France and needed some convincing, he was as quick to assert equal German economic rights in the disputed Sultanate, as von Bülow was quick to appreciate the strategic importance of Morocco's Atlantic coast, and its more northern shores facing the Straits of Gibraltar ~ therefore, he encouraged the Moroccans to confront the French.

German intervention when it came was dramatic in the extreme. On the last day of March 1905, Wilhelm interrupted his Mediterranean cruise to steam through gale-force winds to North Africa. In rough seas, the Kaiser disembarked at Tangier, stepping into a launch rolling heavily in the waves. Arriving ashore, he rode a white steed, with a military band pounding out Teutonic tunes all the way to the small German Legation in central Tangier. Once there, amid the welcoming din of rifle fire from whirling Arab horsemen, Wilhelm addressed the crowd.

I am pleased to make the acquaintance of the pioneers of Germany in Morocco and to be able to tell them that they have done their duty. Germany has great commercial interests here. I shall advance and protect our commerce, which shows a satisfying increase, and for that reason shall insist upon equal rights with all powers, which is only possible through the sovereignty of the Sultan and the independence of the country. For Germany both of these must be unquestioned, and I am, therefore, ready to intervene for them at all times. I hope that my visit in Tangier declares this plainly and emphatically and that it will call forth the conviction that what Germany undertakes in Morocco will be negotiated exclusively with the sovereign Sultan.

As Kaiser Wilhelm returned to his vessel to resume his cruise, he met and told the local French agent, Count Alexandré du Chérisay that he wanted free trade for all German commerce in Morocco, and then promptly dismissed him before he could reply. However, Kaiser Wilhelm enjoyed none of it. *I cannot see the point of the whole thing. I landed because the Chancellor wanted me to in the interests of the Fatherland, mounted a strange horse, and came within a moment of the horse costing me my life. I had to ride between Spanish anarchists because he wished it, and his policy was to profit from it.* After Tangier, the Kaiser continued on his voyage to Constantinople. Wilhelm also called at the Black Sea port of Odessa, where the mutiny by Russian Navy *Potemkin* sailors two months later, erupted in revolt.

Berlin's next step was to suggest that the Sultan of Morocco hold a European conference on the Moroccan question, and German diplomats wisely waited until the expected catastrophe befell Russia to emphasize her new weakness. Three days after the Russian Baltic Fleet was sunk by the Japanese Navy at Tsu-shima, the Sultan rejected the reforms presented by envoy Georges Saint Taillandier from Paris and issued invitations for an international conference to advise the Sultan, who Berlin said they now stood behind with all their power ~ and that Germany intended to call the French to account.

When defiant French Foreign Minister Théophile Déclassé said; *There is no need for a conference!* Chancellor von Bülow threatened war over the issue ~ it being necessary for his policy in Western Europe to throw down a gauntlet to challenge the irate French; *We need to see whether they will mobilize!* The crisis peaked on 15th June, when the French cancelled all military leave, but wisely refused to risk war over the issue. France and other nations came to attend a conference at the Algeçiras Town Hall on the Spanish coast overlooking the Strait of Gibraltar and when the conference concluded, Paris was pressured to postpone their proposed protectorate, and to recognize Moroccan independence ~ *The sovereignty and independence of his Majesty the Sultan, the integrity of his dominions, and economic liberty without inequality.* But the policied posturing of Bernhard von Bülow and von Holstein in their imposition on the Kaiser to visit Tangier, and their continuance of pressure thru veiled threats now resulted in the effect exactly the opposite of what these two clever ones had intended. Their sharp scheme so unforeseen, blundered to bring Britain and France together and gave the *Éntente Cordiale* all the weight and significance in European high politics it would have for many years.

After Kaiser Wilhelm, with the very best of personal intentions tried to make the most of his close connections with the British royal family to compensate for the hardening of public opinion on both sides of the North Sea, his sincere assurances of friendship were rarely taken at face value in London, and the Kaiser became disappointed by the failure of his attempts at a *rapprochement* with England. Therefore, he decided to give an interview to a British newspaper, but in a fit of frankness with flashes of impatient temper, he delivered a notorious critique on British foreign policy. On 28th October 1908, the German Ambassador in London, Count Paul von Wolff-Metternich, with shaking hands put down his morning copy of the *Daily Telegraph* and said to his staff ~ *Now, we might as well shut up shop!*

You English are as mad, mad, mad as March hares. What has come over you that you are so completely given over to suspicions quite unworthy of a great nation? What more can I do than I have done? I declared with all the emphasis at my command, that my heart is set upon peace, and that it is one of my dearest wishes to live on the best of terms with England. Have I ever been false to my word? Falsehood and prevarication are alien to my nature. My actions ought to speak for themselves, but you listen not to them but to those who misinterpret and distort them. That is a personal insult, which I feel and resent. To be forever misjudged, to have my repeated offers of friendship weighed and scrutinized with jealous, mistrustful eyes, taxes my patience severely. I have said time after time that I am a friend of England, and your press bids the people of England to refuse my proffered hand and insinuates that the other holds a dagger. How can I convince a nation against its will?

I repeat, that I am a friend of England, but you make things difficult for me. My task is not of the easiest. The prevailing sentiment among large sections of the middle and lower classes of my own people is not friendly to England. I am, therefore so to speak, in a minority in my own land, but it is a minority of the best elements as it is in England with respect to Germany. That is another reason why I resent your refusal to accept my pledged word that I am the friend of England. I strive without ceasing to improve relations, and you retort that I am your archenemy. You make it hard for me. Why is it?

There are mischief-makers, in both countries. I will not attempt to weigh their relative capacity for misinterpretation. But the facts are as I have stated. There has been nothing in Germany's recent actions, which runs contrary to the explicit declaration of my love of peace. I have done all that a sovereign can do to proclaim my goodwill. But as actions speak louder than words, let me also refer to my acts. It is commonly believed that throughout the South African War, Germany was hostile to her. German opinion undoubtedly was hostile ~ bitterly hostile. Let my critics ask themselves what brought to a sudden stop, and indeed, to absolute collapse, the official tour of the Boer delegates, who were striving to obtain European intervention? They were fêted in Holland. France gave them a rapturous welcome. They wished to come to Berlin, where the German people would have crowned them with flowers. But when they asked me to receive them ~ I refused. The agitation immediately died away, and the delegation returned empty-handed. Was that, I ask the action of a secret enemy?

Again, when the struggle was at its height, the German Government was invited by the governments of France and Russia to join with them in calling upon England to put an end to the war. The moment had come, they said ~ not only to save the Boer Republics ~ but also to humiliate England to the dust. What was my reply? I said that so far from Germany joining in any concerted European action to put pressure upon England and bring about her downfall, Germany would always keep aloof from politics that could bring her into complications with a sea power like England. Posterity will one day read the exact terms of the telegram ~ now in the archives of Windsor Castle ~ in which I informed the sovereign of England of the answer I had returned to the Powers which then sought to compass her fall. Englishmen who now insult me by doubting my word should know what were my actions in the hour of their adversity.

But, you will say, what of the German Navy? Surely, that is a menace to England! Against whom but England are my squadrons being prepared? My answer is clear. Germany is a young and growing empire. She has a worldwide commerce, which is rapidly expanding. Germany must have a powerful fleet to protect that commerce and her manifold interests in even the most distant seas. She expects those interests to go on growing. She must be prepared for any eventualities in the Far East. Who can foresee what may take place in the Pacific in the days to come? Look at the accomplished rise of Japan. Think of the possible national awakening of China, and then judge of the vast problems of the Pacific. Only those Powers which have great navies will be listened to with respect, when the future of the Pacific comes to be solved; and for that reason only, Germany must have a powerful fleet. It may even be that England herself will be glad that Germany has a fleet when they speak together on the same side in the great debates of the future. ~ *Wilhelm II*

During this entire period, the Ottoman Turkish Empire had shown no desire for social reforms. Sultan Abd-Ul-Hamid II had become increasingly adept in playing off the ambitions of one foreign power against those of its rival. His advisers with characteristic oriental fatalism were content with an administrative inefficiency that offered them ample opportunities for corruption. The people at large, who openly engaged in racial prejudice and religious fanaticism, were counted upon to assure the continuance of the Sultan's autocratic power and to prevent the spread of incendiary liberal ideas. Hence, secure in their archaic authority, the governing class resisted every opening for reform.

Then a small, but influential group of progressive Turks prepared to overthrow the old régime. These men, many of them officers in the army who had been educated in France or Germany or in foreign schools in Turkey, had some knowledge of the liberal governments of Western Europe. Inspired with genuine patriotic indignation at the inefficiency of Turkish administration and the corruption of Turkish officials, they were ashamed of the indignities and humiliations implied in the constant interference of foreign powers in Turkey's internal affairs.

The first serious stirring came in 1889, when a conspiracy in the military medical college spread to other Constantinople institutions. When their plot was discovered, its leaders departed overseas to Paris, Geneva, and Cairo, where they prepared a comprehensive critique of the Hamidian régime. *The Young Turkish Party* then established a central organization in Paris from where they published enormous quantities of seditious literature to smuggle into Ottoman Turkey to enlighten their people. They hoped for a new constitution.

In 1908, an opportunity presented itself for a revolution. With the Turkish Government focused on events in Arabia, where Turkish troops were unpaid and mutinous, and in Albania where the people were motivated to the point of rebellion by new taxes and recent restrictions on their few schools and printing presses, revolt was furthered hastened by the confirmation of Vienna in February that a new A-H railway would be made through Novibazar to link the Bosnian and Salonikan lines to bring the west of the Balkan Peninsula under the ever-increasing influence of Vienna.

The Young Turkish Party acted with promptness and passion. They transferred their HQ from Paris to Salonika, formed a Committee of Union and Progress to organize and direct a revolution, and worked with energy during the spring and early summer of 1908 to gain adherents to their cause. They selected the constitution which Abd-Ul-Hamid had granted to Turkey in 1876, however had withdrawn (with his suspension of the Ottoman Parliament in 1878) as their main demand. By July, the Young Turks had succeeded in pledging the Albanian chieftains and much of the Turkish Army in Europe to their constitutional movement. They all agreed to the principle of political and cultural self-determination, while the envoys of national minorities declared that they would be content to receive autonomy within the framework of the Ottoman Empire. Their early methods of struggle were ; the refusal to take the oath of allegiance in the army, provoking disturbances among the population, strikes of officials and police aimed at disorganizing the machinery of state, refusal to pay taxes, resistance to the authorities, and to plan for an armed uprising with the date fixed for October of 1908.

The proposed uprising came earlier, on 3rd July 1908, when an army Major of Albanian origin, Ahmed Niyazi Bey at Fort Rezna in Macedonia with his garrison force of 200 soldiers broke into their military store to appropriate money, 75 rifles and 15 cases of ammunition that they then distributed to volunteers to initiate a partisan uprising. After some success, they retreated to the mountains to join the forces of Ismail Enver, Mustafa Kemal, and Djemal Pasha. A week later, as Monastir commander General Osman Hidayet warned his troops to obey the Sultan's orders, one of his officers shot him. The revolutionary detachments that occupied Monastir, then threatened to march on Constantinople. On 23rd July, Ismail Enver speaking as the President of the Young Turks Committee of Union and Progress, proclaimed the new constitution at Salonika, and announced that they would march against Constantinople, if the Sultan did not yield.

Then the onward march to Constantinople, of the Army of Liberty. Volunteers were called for and Armenians, Jews, Greeks, Bulgarians, Kurds, and Lazes, all flocked to the Turkish standard. For the first time in the history of the world, Christian stood shoulder to shoulder with Muslim in a triumphant onslaught for the recovery of liberty and the reinstatement of the Constitution. Constantinople was besieged. Sultan Hamid's own guard and the officerless Constantinople Army Corps he had bribed were then conquered. Hamid was dethroned.

Robert Chambers

The following day, on 24th July 1908, Sultan Abd-Ul-Hamid, thinking that his troops in the capital and in Asia Minor had also sided with the Young Turks, yielded to the demands of the Young Turks. After determining that the demand for reforms did not violate Islamic law, Abd-Ul-Hamid called for elections so the Parliament could meet again. All were to be brothers. The Constantinople crowds that included Turks, Greeks, Armenians, Bulgarians, and Jews ~ all chanted in favor of the Constitution. On 26th July, Abd-Ul-Hamid pardoned and released all political prisoners and extended an amnesty to Armenians and Muslims in exile. He then issued decrees instituting the freedom of speech, the press with an end to censorship, and the right to free assembly. Imperial decrees abolished the secret police, ordered the new courts to be independent, and for no one to be imprisoned without cause.

The Young Turk Revolution was victorious, and a new era seemed to have dawned. However, theirs was only a partial victory because, as The Young Turks tried to come to an understanding with the Sultan's former ruling class ~ instead of forming a new cabinet, the revolutionaries allowed power to remain in the hands of Sultan Hamid and his cabinet, from which only the most compromised members were removed. The Turkish Revolution of 1908 suddenly disordered the balance of power. Although all the chancelleries of Europe had been aware of The Young Turk's schemes ~ none had anticipated their early success. The new face of constitutional Turkey startled the European powers. The sick man of Europe was rejuvenated. Instead of waiting for the break-up of the empire, the expectant nations might be forced to witness a new and efficient Turkey consolidate its dominions and reassert its sovereignty and rights of administration over all of its vast Ottoman Empire. Should this possibility become actuality, the new Turkey would imminently demand that Austria-Hungary withdraw their administrative officials from the Turkish provinces of Bosnia and Herzegovina.

Previously, the 1878 Congress of Berlin had handed over the two Turkish provinces of Bosnia and Herzegovina on the N.E. Adriatic coast to the administration of A-H. During the 30-year interval, a series of wise and considerate governors had overseen excellent constructive and administrative work throughout these provinces that had been more justly and more satisfactorily governed than during any preceding years under Turkish domination. However, the mountainous Balkan provinces of Bosnia and Herzegovina were still a part of the Ottoman Empire, and as such still owed their allegiance to Sultan Abd-Ul-Hamid. The actual legal rights of A-H, were merely administrative.

From the Turkish Revolution of 1908, came a long chain of consequences that led directly to war in 1914. The first link of this chain was forged with dramatic suddenness by A-H in early October, when the strategists of Vienna began planning in a determined manner to forestall any such demands by Turkey by annexing the Turkish provinces of Bosnia and Herzegovina at once, for inclusion into the lands of their own A-H Empire.

Serbia as a land-locked nation, also desired these two provinces, to give her an outlet to the sea. Belgrade as always, forever looked forward to a Greater Serbia inclusive of large parts of Macedonia (claimed by both Bulgaria and Greece), and all of Bosnia and Herzegovina and portions of Albania. The population in these districts was closely allied to Serbians in blood, religion, and in customs. Many enjoyed frequenting cafés to talk politics over tiny syrupy cups of Turkish coffee or glasses of fiery Serbian slivovitz taken to wash down mouthfuls of even more fiery Serbian onions and slices of sun-dried raw beef. Such rustic Serbs carried on an active secret propaganda to arouse sentiment for a Greater Serbia. Turkey had not surrendered her sovereignty ~ and never expected to surrender it.

On 5th October 1908, Prince Ferdinand of Bulgaria acted in agreement with the Cabinet of Vienna, and proclaimed the independence of his Kingdom, and in doing so thereafter became Tsar Ferdinand. The following day, the people of the Mediterranean island of Crete, from where the soldiers of the Ottoman Caliphate had disembarked in July, revolted in enthusiasm to join their Hellenic motherland, and were forthwith reunited with the Greek union. On that same day, the leaders of Austria-Hungary prepared for action, to formally acquire the fertile provinces of Bosnia-Herzegovina.

Soon the centre of European intrigue was no longer in St Petersburg, London or even in Berlin, but in Vienna, where the new A-H Minister for Foreign Affairs, Count Alois Lexa von Aehrenthal decided to bring to a head the animosity between the A-H Empire and the Serbian Government who had just purchased a significant consignment of war equipment from France, including artillery and ammunition; but not as expected from the A-H Skoda works in Pilsen. Aehrenthal's intention was to revive the grand foreign policy of Vienna and assume a powerful posture in Europe. As such he viewed the provinces of Bosnia and Herzegovina along the N.E. Adriatic under nominal Ottoman sovereignty with great interest. Aehrenthal instigated his plan to annex the region ~ before the new Turkish régime attempted to regain control over them, and he plotted to do so with the full consent of Russia. So with candour and duplicity, a diplomatic house party was arranged in Bohemia at Buchlau Castle, the residence of Count Leopold von Berchtold that became a milestone on the road to war, when Aehrenthal and his Russian counterpart, Alexander Izvolsky met there on 15th September 1908.

Russia's Foreign Minister Alexander Izvolsky for his part, planned to secure a new treaty to open the Straits to Russia's Black Sea Fleet ; conferring on them a new capacity to cruise through to the Mediterranean. To further his plan, Izvolsky first needed the full approval of Vienna's Ballhausplatz. To achieve this, Izvolsky agreed to von Aehrenthal's proposal to secretly meet him at Buchlau Castle for some preliminaries before calling a conference of the European Empires concerning such a move, as A-H, France, Great Britain, and Imperial Germany, all had strong interests in such a development.

Aehrenthal agreed in principle, saying that Vienna would not oppose an effort by St. Petersburg to open the Dardanelles to the Russian warships of the Black Sea Fleet ~ providing the same was allowed for other Black Sea nations, and that Constantinople would never be brought under the guns of Russia ~ if Izvolsky on Russia's behalf, agreed not to protest against the A-H annexation of Bosnia Herzegovina. This Alexander Izvolsky agreed to, provided that their meeting was never made public, as he had no desire for France or Britain to learn that Russia was prepared to turn a blind eye to the annexation of Bosnia and Herzegovina. To this, Izvolsky gave his written affirmation and the private diplomatic meeting was concluded. Count Alois von Aehrenthal was particularly well pleased with the outcome, as was Alexander Izvolsky ~ who remained unaware at von Aehrenthal's convenience, he had been diplomatically duped by the A-H Foreign Minister's careful wording in the agreement. While von Aehrenthal had given Alexander Izvolsky nothing substantive in the least, the Russian had just given Austria-Hungary *carte blanche* to annex the two Balkan provinces immediately.

Izvolsky then casually travelled from European capital to capital, having a few little holidays, and slowly seeking to arrange a meeting with the other concerned nations. Then on 6th October 1908, and less than three weeks after their secret meeting in Bohemia, the Austrians annexed the provinces of Bosnia Herzegovina without any new treaty for Imperial Russian naval access through the Straits. Britain promptly vetoed Izvolsky's Straits proposal, as did France. Only then did Alexander Izvolsky realize that if Russia protested, von Aehrenthal could wave the paper in his possession with his own signature on it ~ or Russia must remain silent. In a masterstroke of deceptive diplomacy, the Russian Foreign Minister had been duped by the façade of Austrian support for a notion that would never have made it to the conference table. To secure his winning position, Aehrenthal released his secret paper on which the inept Izvolsky had written ~ *Austria-Hungary has a free hand in Bosnia-Herzegovina.*

Aehrenthal then simply announced, that he wanted to give the provinces of Bosnia-Herzegovina as a 60th Jubilee present to his Emperor Franz Josef I. The reaction throughout Europe was explosive. Kaiser Wilhelm and King Edward were astonished that they had not been informed prior to the event. France was deeply indignant. Aehrenthal did not expect any serious difficulties with Russia, and said; *The Bear will growl and snarl, but won't bite or strike out with its claws.* Izvolsky was so outraged by Aehrenthal's trickery at Buchlau, he dishonestly denied the agreement and raised the threat of war, personally claiming solidarity with the Serbians. With wrath and consternation rising fast in Russia, war was avoided because Tsar Nicholas was not prepared to fight ~ and so he could do nothing other than concede and back down. The young Tsar had suffered another blow to his power and prestige, and to the Russians this condition was very humiliating ~ therefore from this time forward it became improbable that Tsar Nicholas would ever again submit peacefully to any similar stroke from either Germany or Austria. This revelation from Russia regarding their lack of expected support caused deep resentment in Serbia, a nation that had always insisted that Bosnia had belonged to them.

So began the Bosnian Crisis that would have enormous repercussions on the future of Europe. Kaiser Wilhelm II had not learned of Austria-Hungary's intentions until the very day of annexation. He was profoundly piqued, not only that he had been kept so long in ignorance, but at the Austrian action itself. The duplicity of the Viennese policy of deception severely embarrassed and discredited Germany, by the abrupt injury done by Austria to Turkey (both being German allies). This action also alienated Russia and so assumed a prominent place amongst the proximate causes of the Great War. Kaiser Wilhelm thought this unjustifiable offence to Turkey would damage German influence in Constantinople, and threaten the Bagdad Railway. *Vienna will be charged with duplicity and not unjustly. She has duped us in a most unheard-of fashion. My personal feelings as an ally have been most seriously wounded.* Such were some of Wilhelm's comments. He feared that this beginning of the partition of the Ottoman Empire could lead to war. *If the Sultan in his necessity declares war, and hoists in Constantinople the green flag of the Holy War, I should not blame him. With a policy of this kind Austria will drive us into a dangerous opposition to Russia. I only regret that Aehrenthal's fearful stupidity has brought us into the dilemma of being unable to protect and stand by the Turks who are our friends, seeing that my own ally has injured them. Thus ~ my Turkish policy, so carefully built up for 20 years is thrown away!*

Aehrenthal and the Ballplatz, by their deliberate infraction of the originally defensive character of the A-H German Treaty of Berlin had annexed Bosnia-Herzegovina, relying on the Dual Alliance clause that bound Germany to come to Austria's defense in the event of any Russian attack. Therefore, to head off any possible activation of this clause, Chancellor Bernhard von Bülow, simply demanded the Russian acceptance of the *fiat accompli* and that they cease their opposition to the annexation. Alexander Izvolsky duly notified Germany, that Russia had accepted the masterful A-H annexation. Russia thereafter became more and more accustomed to regarding Germany as the principle obstacle to the realization of its ambitions for exclusive control of the Balkan Peninsula and of Constantinople.

In the aftermath of Imperial Russia's humiliation, Alexander Izvolsky's diplomatic career and reputation were irreversibly damaged. He was dismissed from the high office of Foreign Minister and reappointed to France. In Paris, Premier Georges Clemenceau at a dazzling diplomatic reception, greeted the startled new ambassador with a loud, *Eh bien! Avez-vous bientôt fini de mettre le feu aux quatre coins de Europe? : Well! How soon are you going to be finished setting fire to the four corners of Europe?* British Ambassador Sir Francis Bertie tore the map from his wall, and holding it in his hand he said: *I feel Izvolsky is not being completely honest about the context whereby these understandings have come to be reached thru his ~ diplomacy. I have the impression, I may say the conviction that he did not tell me the truth…* Izvolsky's rage knew no bounds, and he secretly aimed to draw the Triple Éntente into open conflict against Austria-Hungary in revenge for being made Count Alois von Aehrenthal's dupe. Aehrenthal's assertive annexation in pursuit of foreign policy also led to a rapid cooling of relations with Berlin. The astonished Germans could not forget that Vienna had risked war, without consulting the view of their faithful ally in Berlin ~ and one year after the Bosnian Crisis, the new Imperial Chancellor, Theobald von Bethmann Hollweg with acuity and clarity commented: *Germany is not a Balkan Power. During the past year, for reasons of higher policy, we threw our influence into the scales in favour of Austria. In my opinion we should do well to prevent, as far as possible, a repetition of this procedure. For the future, we ought to preserve a free hand for ourselves, and allow ourselves to be drawn as little as possible into Balkan questions, so that we shall be able at the psychological moment to choose our policy freely as possible.*

The acquiescence, which Germany gave to A-H emanated from Berlin's desire to preserve the solidarity of their Dual Alliance, but it had the effect of emboldening Vienna's policy in the Balkans, which veered all toward to the chasm of 1914. Kaiser Wilhelm's irritation remained, for he had the right political instinct to realize that through his alliance with A-H, he had assumed a risky liability and had just created a dangerous precedent. From the day Wilhelm chose to give unswerving German support to the Hapsburg acquisitions on the Balkan Peninsula, he began to falter in his coronation promise as a guardian of peace. For the first 20 years of his reign, Wilhelm had kept his promise that he had made to the German people in his first speech from the throne ~ *I will have a peaceful reign.* Whenever the Kaiser reveled in the military magnificence of Imperial Germany, he preferred posturing on the periphery of a precipice, to plunging down it. This provincial Balkan episode thus became the catalytic point, and the immediate cause for the series of events connected with the affairs of the Balkan Peninsula, which filled the years of 1908-14 the drift to war.

* * *

MIKLOS HORTHY ~ AIDE-DE-CAMP TO EMPEROR FRANZ JOSEF

Korvettkapitán Miklós Horthy de Nagybánya

Vienna Court Opera House

We could see that the country was seething beneath the surface. A few weeks after my arrival in 1908, the revolution of the Young Turks broke out. On 23rd July 1908 at Salonika, Enver Bey proclaimed afresh the 1876 constitution that Abd-Ul-Hamid abolished soon after his accession to the throne. Elections were proclaimed, which put Vienna in a difficult predicament. To prevent elections from being held in Bosnia-Herzegovina, which though actually occupied was still nominally part of the Ottoman Empire, the Dual Monarchy proclaimed on 5th October 1908, that Bosnia and Herzegovina would be added to Austria-Hungary as a corpus separatum. The annexation of Bosnia-Herzegovina was diplomatically ill prepared. Serbia and Montenegro threatened war. The Turks began a general boycott of our exports, which caused much hardship, as Austria-Hungary was the main supplier and also the main customer of Turkey. Our merchant steamers were left with unloaded cargoes; they were no longer bunkered; and the Austrian firms in Constantinople were either picketed or ostracized.

It was with a heavy heart that I departed from Constantinople, where I had spent over two and a half years. I had come to know the Turks in diverse circumstances, and had grown to like them. They are a people of strong character, a noble, reliable race ~ and excellent soldiers. Nor was parting from our Ambassador Pallavicini easy. I respected him as a high-minded, wise, and clear-sighted diplomat to whom the other Ambassadors, in those difficult days of revolution bringing in their train many ambassadorial conferences, were only too glad to give the leading role. After the death of Aehrenthal, Pallavicini was offered the Ministry of Foreign Affairs, but he made conditions, which did not prove acceptable to Vienna. Had it been he who was resident at the Ballhausplatz at the time, it would not, in my opinion, have come to war. As it was, instead of the Marquis János Pallavicini, Count Leopold von Berchtold was successor to Aehrenthal. My appointment as aide-de-camp to Emperor Franz Josef came thru. I said goodbye to the sea and exchanged sea breezes for the air of the court and the big city.

His Majesty had four aides-de-camp to represent the branches of the armed services. During my years of service, Emperor Franz Josef resided at Schönbrunn. Twice a week, accompanied by an aide-de-camp, he drove to the Imperial Palace in Vienna, the Hofburg, arriving there at seven o'clock in the morning, to grant audiences. He usually returned at half-past four in the afternoon. The aide-de-camp on duty was dismissed each day at six o'clock by His Majesty personally. His Majesty retired about eleven o'clock and was at his desk the next morning at his customary hour. Franz Josef spent every working day, beginning at five o'clock in the morning, at his desk going through the documents submitted to him by his Ministers. Never idle or wasting time.

It was arranged, soon after I had reported to Count Paar in Vienna, that I should be received in audience by His Majesty ~ an interview I anticipated with tension and excitement. I was feeling extremely uncertain when, on the evening of 30th November, a guardsman came to inform me at what hour His Majesty would rise the next morning. He reported : The hour is four. We naval officers were in a more favourable position than our military colleagues, for on board ship we had been used to a four hours' watch at night and had become used to sleeping beforehand. Even on that first night I slept well, and rose feeling fresh and energetic as I hastened down to the aides-de-camp room on the first floor, separated from the Emperor's study by a baroque reception hall.

On his desk the monarch found the documents sent from the Chancelleries of the Cabinet and the Ministry of War, and his aide-de-camp was rarely in demand while he studied them. With the approach of eight, the two Adjutants-Generals arrived. On the stroke of nine, the aide-de-camp announced Count Paar and Baron Bolfras. They might be followed by archdukes, cabinet ministers, chiefs of the general staff and other high dignitaries with important communications or reports to make. These audiences lasted until lunchtime. Then His Majesty, usually alone, went for a stroll in the conservatory, after which he resumed work, going on until dinnertime at half-past five, dinner usually being served to him at his desk. At six o'clock, he dismissed his aide-de-camp.

My colleagues were right when they told me that in the main the rules consisted of tact and common sense. The Emperor was not a talkative man and preferred concise answers. Once, an aide-de-camp who was on duty for the first clicked his heels loudly every time he made a statement. The Emperor called for Count Paar and ordered him to have him replaced by someone else. That sort of thing was not done at the Viennese Court.

General audiences granted at the Imperial Palace began at ten o'clock and the list often ran to fifty names. The order of precedence had to be worked out by the aide-de-camp on duty. Those who had been called gathered in a hall where an official would conduct them one by one, in accordance with the aide-de-camp's list, to the aide-de-camp whose duty it was to announce them. Two officers of the Guards, one Austrian, and one Hungarian, stood with drawn swords at the door of the audience chamber. Each of the fifty individual audiences, usually to express thanks for an appointment or a distinction, or to make a personal request, was bound to be very short.

In the afternoon, more visitors were received until 4.30 PM, when His Majesty drove back to Schönbrunn. In spite of his age, His Majesty adhered to the traditional representative duties of the sovereign, among which were the gala dinners on the occasion of visits made by high-ranking guests, and in honour of the Diplomatic Corps in general. So large was the number of Ambassadors and Envoys that they had to be invited in groups.

Before audiences began, the aides-de-camp room was usually a hive of activity. Everyone present was drawn willy-nilly into discussions on a number of often-delicate problems. I learned to view the nationalities problem, from a new angle. It filled me with anxious forebodings to observe in the course of discussions that foreign influences were at work and to note how, even indirectly, the theories of irredentism and separatism were infiltrating. At times, socialist ideas were also mentioned. The people who voiced these were plainly unaware how well off they were. They wanted to see the country governed on the basis of abstract theory and failed to allow for the immutable laws of nature. Their gaze went as far as the destruction of what was in existence.

What the new state they were striving after would be like or what it would turn into, of that they had only the vaguest notion. Therefore we, that is to say those of us who lived in close contact with His Majesty, were not wholly without cares. The Bosnian crisis of 1908 had brought the dangers threatening the monarchy clearly before all eyes, we were obviously far from the spirit of defeatism that may have prevailed elsewhere. Strength of tradition and mutual interest, ensuring the stability of the Habsburg Empire was shown. An aide-de-camp was not in a position to make political comments or give political advice. Only among his friends could he discuss the advisability of taking stronger measures to counteract Slav propaganda, Serbian activities or Italian irredentism. The question whether such measures could have diverted historical development must remain unanswered.

In 1910, King Petar of Serbia had had an enquiry made to ascertain whether or not a visit from him would be welcome. Negotiations in this matter were protracted, as Franz Josef had no liking for the Karadjordjevich dynasty, which had acceded to the throne as a result of the murder of King Alexander in 1903. Since that time, Serbia had moved more into the Russian orbit. In the end, however, the Serbian envoy was informed that His Majesty would receive King Petar in Budapest. Shortly before the date fixed, the reception was cancelled on the plea of indisposition. In my opinion, that was to be regretted. The visit might have improved relations, whereas the cancellation could only aggravate existing tensions.

Miklos Horthy

MAD GEORGE OF SERVIA

WHOSE INFLAMMATORY SPEECHES HAVE HELPED TO DISTURB EUROPE

THE CROWN PRINCE OF SERVIA, WHO HAS BEEN ADVOCATING WAR WITH AUSTRIA

THE SPEECHES AND ACTIONS OF THE CROWN PRINCE OF SERVIA, WHO HAS DEFINITELY ESTRANGED HIMSELF FROM HIS FATHER, KING PETAR, HAVE CALLED FORTH A LITTLE FATHERLY ADVICE FROM THE SERVIAN FOREIGN MINISTER MILOVAN MILANOVIC, WHO HAS ADVISED THE IRREPRESSIBLE AND IRRESPONSIBLE YOUTH TO BEHAVE AS BECOMES THE HEIR TO A EUROPEAN THRONE.

THE FIREBRAND OF THE NEAR EAST.

Photograph by Jovanovitch.

THE ILLUSTRATED LONDON NEWS, 26TH DECEMBER 1908

PRINCE GEORGE, THE SPEECH-MAKER, HEIR-APPARENT TO THE THRONE OF SERVIA.

THE CROWN PRINCE OF SERVIA HAS DONE ALL THAT IS IN HIM TO RENDER THE CRISIS IN THE NEAR EAST AS DANGEROUS TO THE PEACE OF THE WORLD AS POSSIBLE. HE HAS MADE FIERY SPEECHES BY THE SCORE, AND ACCORDING TO REPORT, HE HAS ANNOUNCED THAT ~ IF NECESSARY ~ HE WILL EVEN TAKE UP ARMS AGAINST HIS OWN FATHER, THE KING. PRINCE GEORGE IS THE SECOND CHILD OF KING PETAR OF SERVIA AND WAS BORN ON 27TH AUGUST 1887. HIS BROTHER, PRINCE ALEXANDER, WAS BORN IN DECEMBER 1888, AND HIS SISTER, PRINCESS HELÉNE, IN OCTOBER OF 1884.

King Edward VII's efforts to maintain peace in Europe had inspired the popular music-hall song; *There'll be no war, as long as there's a King like good King Edward.* King Edward VII died on 6th May 1910, having helped create the Triple Éntente Alliance of Britain-France-Russia. British First Sea Lord Admiral *Jacky* Fisher wrote; *He wasn't clever, but he always did the right thing, which is better than brains.* The following year at Balmoral, his son and heir King George V remarked to King Gustav of Sweden.

I am quite convinced that the Kaiser, who professes to the desire to be the guardian and protector of peace, honestly and seriously means it, and that he did everything in his power to calm things down and prevent war. But will the Kaiser be in a position to guarantee the peace permanently? He has been strongly attacked in his own country and publicly called a coward. Will he put up with it? Nobody likes to be called a coward.

King George's view of the Kaiser as a protector of peace was confirmed in 1913, when Wilhelm sensibly insisted on a German veto of Austria-Hungary's contemplated intervention against Serbia. Kaiser Wilhelm and his ministers continually urged those in Vienna to come to some arrangement with Serbia and warned A-H against using force. Although, Count Berchtold and General Conrad were men of stronger will and purpose than the Kaiser and Chancellor Bethmann Hollweg, on the whole German influence routinely remained in favour of moderation and peace rather than the contrary, and this is evidenced in the contents of a telegram Wilhelm sent to Bethmann *a propos* A-H.

The Alliance with Austria-Hungary compels us to take action should Austria-Hungary be attacked by Russia. In that case France would also be involved, and in those circumstances England would not long remain quiescent. The present prevailing questions of dispute cannot be compared with that danger. It cannot be the intention of the Alliance that we, the life interest of our ally not being endangered, should enter upon a life-and-death conflict for a caprice of that ally.

Germany frequently found herself discomfited by the actions of Austrian intrigues in the Balkans, against Berlin's better judgment or without her approval. The acquiescence or rare encouragement, which Berlin gave to Vienna, emanated from by their desire to preserve the solidarity of the group. Bethmann's instructions to the German Ambassador in Vienna, Count Heinrich von Tschirschky, were repeatedly to restrain A-H from taking action against Serbia or from otherwise antagonizing Russia.

The 1908 Bosnian annexation course of action of Count von Aehrenthal, which can be qualified as one of the underlying causes of the Great War, had other unfortunate effects on the Dual Monarchy. The ease with which the triumph had been won, led the hawks of Vienna and Budapest to imagine that such high-handed methods would always meet with success. They fancied that the pacifist Tsar from fear of seeing the two Germanic Empires ranged against his domain, would not dare to cross Vienna's path, if one day Austria set forth to chastise Serbia. Aehrenthal's rash policy to restore the Danubian Monarchy to an active role on the European stage stirred the animosity of the Serbian people and revived the turbulent Eastern Question; so dreaded by generations of European diplomats that still smouldered beneath the ashes of the Balkan Peninsula. Thus far their manoeuvre succeeded, and to show their satisfaction, Count Leopold von Berchtold had a commemorative plate fixed upon a wall at Buchlau to remember the date.

~ 15th September 1908 ~
Conversations occurred here of major importance to Austria-Hungary.
When Isvolsky gave his consent to the annexation of Bosnia and Herzegovina.

Sometimes Vienna heeded Berlin's advice, sometimes she did not. But to represent Germany as exercising complete control over her ally, as so many writers have done, is altogether arid nonsense. This fact is evident in how Serbia felt no particular animus against Germany during the pre-war years, which was due in part to increased trade between the two countries during the ill-conceived Pig War. The Serbians at this time successfully defied the Danubian Monarchy and remembered very well how Berlin had used her influence to eventually restrain Vienna from their overbearing Balkan policies. Often, it was the Kaiser Wilhelm who justly complained of Count von Berchtold's *policy of surprises,* which Berlin was unable to prevent ~ and how Vienna abruptly presented her allies with *faits accompli* that they were compelled to honour from secured treaty signatures. *If Austria and Italy are unreasonable, we are not to blame! We do not have to join in every folly which they perpetrate. We have already taken over much at our expense for our allies. If they just go on making their situation worse in relation to the Éntente, we can warn them, but we cannot prevent them. But we do not need to join with them.* Wilhelm II

* * *

ADDITIONAL BLACK HAND INTRIGUES

King Nikola I of Montenegro

King George I of Greece

Ferdinand of Saxe-Coburg, Tsar of Bulgaria

Although President Nikola Pasic of Serbia was willing to await an auspicious moment, the secretive Colonel *Apis* was not; so his most effective Serbian nationalist group, *The Black Hand* stepped up their activities within Serbia, and within the new A-H provinces of Bosnia and Herzegovina. Colonel Dragutin *Apis* Dimitrievic, the veteran of the slaughter of his own king and queen, and several other political murder plots, was according to the new Russian military attaché in Belgrade Colonel Viktor Artamonov ~ *He is quite possibly the foremost European expert in regicide of his time.* Colonel *Apis* confirmed that he was responsible for the failed 1907 plot to assassinate King Nikola I of Montenegro, and that he had sent someone from Corfu to assassinate the King George I of Greece. In addition to these two failures, bombs were tossed and concealed pistols were discharged in several attempts against Tsar Ferdinand of Bulgaria. How much of this was Balkan delusion and how much reality is still open to debate, but *Apis* and his fanatical cult of assassination, certainly relished death, and believed in its transforming and catalyzing powers to realize their ambition of a Greater Serbia.

I have little sympathy with the Servians, for they are, I understand, treating the Montenegrins very badly. Do you remember the Bomb Affair of 1907? That ghastly outrage against the King of Montenegro was proved by the depositions of the witnesses, and by technical experts, that the plot had been hatched at Belgrade, and the perpetrators, who went to Cetinge to assassinate King Nikola and his entire family, were some malcontent students at the University of Belgrade, and that the bombs they carried were made at Kragujeva Servian Army munitions factory. The plot against King Nikola was an old plot ~ as old as the secret society of the Black Hand, whose programme includes the murder of many other rulers besides the failed assassination of King Nikola of Montenegro, and whose aim is the extension of Servian rule.

Colonel Sir John Roper Parkington
Consul-General for Montenegro in London

Cetinje, the capital of Montenegro, was no Constantinople or Vienna. The only two-storey stone buildings in the town were King Nikola's *Palais Royale* and the Russian legation. There were few public facilities, and most houses were thatch roof hovels. Throughout the spring of 1910, as Nikola tempered his bad feelings, Russia gave a large sum of money to his impoverished household, as well as a military subsidy of weapons and instructors. The money was happily received, but the Russians had not bought Nikola's good will. He still refused a rapprochement with the scheming Serbs, believing that his two daughter's marriages, to two Russian Grand Dukes (brothers of Tsar Nicholas) were more than sufficient to keep the power and prestige of Imperial Russia on his side.

Although the A-H Ambassador to Montenegro for 1910-13, Baron Wladimir Giesl von Gieslingen was familiar with the cunning of wily King Nikola and the oriental despotism of his *Palais Royale,* only during Giesl's frequent visits to Belgrade were his eyes truly opened to the hostility of the Serbian people toward A-H. Reports from the current Ambassador, Count Stephan von Ábránfalva and merely from reading the afternoon newspapers informed Baron Giesl aplenty about the secretive Serbian régime and the *Black Hand.* While Giesl became a heartened opponent of Serbdom in little time, he did not forget his duty, and maintained an outwardly accommodating and friendly disposition. He quickly came to like Serbian PM Nikola Pasic, who was cunning yet approachable. Soon Giesl felt more composed in the busy city of Belgrade, than he ever had in Cetinje.

L'ATTENTAT DE LISBONNE
Le roi de Portugal et le prince héritier périssent assassinés

THE LISBON REGICIDE

Assassination of King Carlos I of Portugal and his heir-apparent Prince Royal Luis Filipe by assassins of republican interests aided by the Portuguese Carbonaria, disenchanted politicians and anti-monarchists.

On 1st February 1908, as the Portuguese royal family returned to their Vila Vicosa Palace, their open carriage passed through the Terreiro do Paco fronting on the river. While crossing the square, shots were fired from the crowd by a pair of republican activists, Alfredo Costa and Manuel Buica. Buica, a former army sniper fired five shots from a rifle hidden under his long overcoat. King Carlos I died immediately. His heir Luís Filipe was mortally wounded, and Prince Manuel was hit in the arm. The Queen alone escaped injury. Royal bodyguards and police killed the two assassins on the spot. Days later, the young Prince Manuel was proclaimed King of Portugal, however the killing of King Carlos and Prince Royal Luis Filipe was the effective end of a constitutional monarchy in Portugal, which was later confirmed by the October Revolution of 1910 that deposed the Portuguese Monarch. The terrorist capacity to catalyze the downfall of a European dynasty alarmed the Hapsburg family.

Following a revolt in Morocco in January of 1911, Lieutenant Marchand and his French troops were all massacred by Moor rebels banded together under a pretender to the throne, who then leading from the front, drove back the forces of the Caliph of Marrakech and Fez, Sultan Abd-El-Hafid to besiege him within his palace in Fez (then under the command of Sir Harry Aubrey de MacLean). As the Moroccans considered their own Sultan the legitimate Caliph of Islam, they did not recognize the Ottoman Turkish Sultan as the supreme religious authority. While claiming their own Sultan was the true Caliph, they did acknowledge the Turkish Sultan as head of the most powerful Islamic state.

A French military expedition was mobilized to protect the people in the besieged Moroccan cities, but before the French force departed for its mission (and a fortnight before they would reach Fez), an approach was made on 7th May by the shady French Finance Minister Joseph Caillaux to the German Embassy in Paris to imply that Imperial German recognition of French interests in Morocco would be rewarded with compensation elsewhere. When the new German Foreign Minister in Berlin Alfred von Kiderlen-Wächter heard of this, he felt that he would be doing less than his duty if he did not somehow turn the suggested situation to an even greater advantage for his own country Germany. He looked at a map of French colonial interests in Africa, and spoke a single word ~ *Kongo!*

The occupation of Fez, would pave the way for the absorption of Morocco by France. We should gain nothing by protesting and it would mean a moral defeat hard to bear. We must therefore look for an objective for the ensuing negotiations, which shall induce the French to compensate us. If the French, out of anxiety for their compatriots, settle themselves at Fez, it is our right too, to protect our compatriots in danger. We have large firms at Mogador and Agadir. German warships could go to those ports to protect the firms. They could remain anchored there quite peacefully ~ merely with the idea of preventing other Powers from intruding.

Baron Alfred von Kiderlen-Wächter

While those in Paris and Berlin waited for the other to make the next move, and with no sign that France would sanction territorial compensation, time was running out for Kiderlen, By 26th June, his patience had evaporated. He now realized that the only way of obtaining compensation was by the application of dramatic threat. Together with Chancellor Theobald von Bethmann Hollweg, the two travelled to Kiel to meet the Kaiser on the *Hohenzollern.* When they met Kaiser Wilhelm during an interval of yachting at Kiel to obtain his reluctant consent on an assurance that the result was certain, and the risk negligible, Kaiser Wilhelm said ~ *If the French choose to involve themselves deeper in the affairs of that turbulent and ungovernable country, so much the worse for them. Germany now has better things to do.* With the use of a threat decided upon, they requested authorization for the dispatch of a warship to Agadir the key port of Morocco to force the issue. Neither the Kaiser nor many other Germans were much interested in Morocco ~ only to threaten Agadir ~ for a richer swath of territory in the Congo.

German Naval GHQ was informed, and the decision was taken to send the gunboat *Panther.* With a cruising speed of 12 knots, it would take her until 1st July to arrive off Agadir, drop anchor, and intimidate the locals and France with her two new 105mm guns. The Kaiser gave the following instructions to the captain of *Panther* before he sailed, which he was then asked to repeat back to him.

On no account, does it matter what official commands you receive, are you to use force when the Panther goes to Agadir. No matter what stress is brought to bear upon you, no matter what affront may be done to your code of naval honour, you are, under no circumstances to use any force against France or England.

So the *Panther* steamed toward Agadir, to arrive some distance offshore and to drop anchor. Kiderlen had sprung his trap for the now Prime Minister of France, Joseph Caillaux. The French saw themselves once more, as in 1905, given the choice between humiliation and war, whereas the British focused on the challenge to themselves of a German warship seemingly pegging out a claim to this Atlantic seaport. Caillaux saw the danger of a German naval station on the Atlantic coast at Agadir on the flank of their important Atlantic trade route. What further worried the Éntente was the potential for a German port on the Mediterranean coast of Morocco. The hawkish British Ambassador in Paris, Sir Francis Bertie was convinced that the Germans were sure that British support for the French would not be forthcoming. *I feel equally confident, that the German Government, if we join with the French in refusing to agree to the establishment of a German port, call it what you may, on the Atlantic coast of Morocco, may bluster, but there will be nothing else. If the Germans increase their naval force at Agadir and land men I think we ought to do so also with the French, and also the Spaniards whom I believe we can drag along with us.*

SMS *Panther* Armament: 2 x 105mm guns, range 12,000m

In the separate Franco-German discussions held in Berlin between Kiderlen and Jules Cambon (the brother of French Ambassador Paul Cambon in London) the German Foreign Secretary was not concerned about details. He wanted the whole Gaboan District, and the portion of the French Congo between the Atlantic coast and N'Goko Sanga River up to Lobay, where the tributary Oubanghi meets the River Congo. After Cambon asked what Germany wanted, Kiderlen in his big stick and burly manner replied: *See what you can give us in the Congo.* When the Kaiser was told of Cambon's reaction, *Cambon was aghast, he nearly fell over backwards.* Wilhelm was unconvinced ~ *He is a very good actor.* Kiderlen: *In order to reach a favourable result, we shall have to take a very strong line!* The Kaiser noted: *Then I must return home at once. For I can't let my Government take that sort of action without being right on the spot so as to keep a careful watch on the consequences and have them well in hand! Anything else would be unpardonable and too parliamentary ~ Le Roi's amuse! And meantime we are steering straight for mobilization! That must not happen WITHOUT ME! Besides, our Allies must first be informed about this! For it may draw them in, in sympathy!* The Kaiser clearly contemplated a mobilization of his forces as a result of this. When the second German warship, the cruiser SMS *Berlin* arrived off Agadir, the question sharpened. But as the Russian régime remained passive, Berlin knew ~ that their bluff, would not be called.

The German terms, as presented on 15th July, demanded from France the whole of the French Congo from the Sangha River to the sea. Franco-German negotiations on the Treaty of Fez led to a convention under which Germany accepted France's position in Morocco in return for territory in the French African colony of Middle Congo, and over 100,000 square miles of this prosperous territory, became German Kamerun and gave Germany an outlet on the Congo River. Kamerun proved to be a rich country with long valleys between mountains where palm oil abounded, and in the wild jungles, then full of elephants, was found a wealth of ebony and other valuable timbers. The Agadir Crisis remained Berlin's finest example of gunboat diplomacy and the *Panthersprung ~ Panther's pounce* became a popular figure of speech for any demonstration of power. France established a protectorate over Morocco in 1912, ending what little remained of the ancient Sultanate's former independence.

The French belief in *élan* was reinforced, and the capable approach of Cambon rejected. Yet the statecraft of von Kiderlen-Wächter and Jules Cambon was recognized when they sent each other signed photographs at the end of the *Agadir Crisis*. Cambon wrote on his to von Kiderlen-Wächter; *To my dear friend and terrible enemy* ~ while Kiderlen wrote ~ *To my terrible friend and dear enemy.*

The Moroccan imbroglio has brought the name of Daisy the Princess of Pless, a former English beauty, prominently to the front. The Princess is said to be in the diplomatic employ of the Kaiser and a strong factor in keeping the diplomatic corps of Germany informed as to the moves contemplated by Great Britain. The rumor has gained credence through the favouritism shown her by German Royalty and the elaborate entertainments she gives from the English Embassy, where she and her German husband, Prince Hans of Pless, mingle freely with both the German and English members of the diplomatic corps. The Prince and Princess are high in favor of the King and Queen of England. Both King Edward and Kaiser Wilhelm stood godfather to their son and heir two years ago. Fürstenstein is a pleasant meeting ground for British and German Royal society. The Daily Telegraph

The full gravity of the Agidir incident, though apparent to other antions, was never realised by the people of the British Empire. The crisis that arose suddenly in July 1911 subsided six weeks later, but it was not until the autumn that its meanings were grasped by a comparatively small section of the public interested in defence and diplomatic affairs. However, from October onwards more began to consider the possibility that war had been avoided, and to consider seriously ~ many for the first time in their lives : what would have happened if Britain had become involved in a European conflict? War, however had been avoided in 1911. Such was the position of affairs at the close of July 1911. Between that date and July 1914, when war broke out, various changes took place in the European situation, the general effect being adverse to Britain and the Éntente Cordialle.

In 1911, German estimates provided for considerable increases in artillery and also machine-guns. In the following year 1912, further German additions were made and two new army corps were formed out of existing units ~ one for the Polish, the other for the French frontier. As before, artillery and machine-guns were very greatly increased in the estimates of that year and again in those of 1913. In addition, Germany at the same time added a squadron to her fleet in the North Sea, announced the purchase of 27,000 additional horses plus the spending of millions more on fortifications and aircraft.

Berlin's proposals were timed to take effect the same autumn; so that by the following Midsummer of 1914, the military stregnth of Germany would reap the benefit anticipated from these enormous additions. Even so, Germany took time by the essentials, and began to carry through the contemplated programme before discussing the terms of the Army Bill to legislators in the Reichstag. Consequently the intentions of the German militarists were known in a general way to every Intelligence department in Europe, long before they were actually announced.

Thus the French Government took the bold step of asking their legistlature to sanction a lengthening of the period of active military service from two years to three, and an extension of the age limit of the reserves from 47 to 49. Power was also taken to summon in an emergency, the annual contingent of recruits of a year before their due time. Increases in artillery, engineers, railways, barrack accommodation, and subsidiary services were asked for and obtained. Apart from these material preparations, Germany would add 200,000 men to her striking force, and the intentions of France were similar. But in the case of Germany, the increases in stregnth would be operative by Midsummer of 1914, while France's would not impact until two years later. One British officer wrote: *N.B. ~ Most important : The German Bill takes immediate effect. The French only takes effect in 1916 because (1) The French Army are not going to retain the class which finishes its services this year with the colours; (2) There has been great delay in the French Parliament... A year from now will be the most critical time. Germany will have the full benefit from her Bill, whereas France will have a mass of young recruits still under her instruction. The strain on officers will be tremendous in order to knock this mass of raw recruits into shape.*

Russia meanwhile proceeded with her programme of military extension and reorganization which had been decided on after the Japanese war, with a great part of her expenditure being directed to the improvement of her exceedingly defective system of railways and communications, and to the fortification of the Gulf of Finland. The British Ambassador to Imperial Germany Sir Edward Goschen wrote in his diary ~ *Sometimes one cannot help thinking that people are right who say ~ War must come!*

In the British Army, during this critical period there had been no increases, but quite the reverse. The Regular Forces had been reduced in 1906 by nine battalions and the Territorials were 47,000 short. The Army Reserve was doomed to automatic shrinkage owing to the reductions in the Regular forces, from which the reservists were drawn at the expiry of their terms of service. The relative weakness of the British military position compared to the stupendous German Bill had undergone an even greater change for the worse, as Great Britain continued to lag far behind in aircraft. British Secretary of State Viscount Richard Haldane sought to justify earlier reductions when he attempted to explain ;

The infantry was in excess, while the artillery was deficient. I do not think that Compulsory Training will be adopted in this country until after England has been invaded once or twice. London, 1st December 1911

The British, however have the best reasons for feeling secure : they were a nation of splendid fighters. They were never ready, but they fought the better, the less ready they were... Glasgow, 6th January 1912

The policy of Berlin, in 1912 and 1913, to preserve peace, and her efforts ~ equally successful ~ in the following year to support A-H into a supposed limited war against Serbia, were probably due to one and the same cause. Two dates from Berlin's perspective were of importance, *the summer of 1914,* when the new German military preparations would be complete and when the Kiel Canal having been widened and deepened would be accessible for the transit of Dreadnoughts ~ *the summer of 1916,* by which date the French Army increases were due to take effect, and the Russian reorganization would have been carried through. And, if Italy (always a doubtful participator) could be tempted by her own reasons to make common cause with her allies in the summer of 1913, the certainty of her adherence (and the Italian Army of 700,000 troops) would turn the scales in favour of the earlier date. From the perspective of Berlin and Vienna, a limited war could be waged to greatest advantage as soon as the first of these two dates was reached.

On June 23rd 1914, Kaiser Wilhelm opened the new lock at the North Sea end of the Kiel Canal. On the following day he performed the same function at the Baltic end. The Times correspondent remarks that the Kaiser's passage through the Canal on this occasion was of symbolical rather than practical significance, as on the one hand German Dreadnoughts had already used the passage experimentally. The extension works, which were begun in 1907, are of vast importance to the German Navy. The Canal has been made two metres deeper, and has been doubled in breadth. The places at which large ships can pass one another have been increased in number, and at four of them Dreadnoughts can be turned. There are now four, instead of two at each end, which means a great saving of time in getting a fleet through. Above all, the distance between Kiel and Wilhelmshaven for battleship purposes is reduced from more than 500 to only 80 nautical miles. *The Times,* 25th June 1914

As to the the precise form of co-operation between the members of the Triple Éntente in case of war, there could be less mystery. It was obvious to those who paid attention to what had transpired during the summer and autumn of 1911, that in the event of Germany attacking France over the Agadir dispute, that London had let it be understood and expected, that Britain could send their swift Expeditionary Force across the Channel to co-operate with the French Army on the N.E. frontier; and from official statements, and from discussions which had taken place in Parliament, it was understood that the British Expeditionary Force consisted of six infantry divisions plus a full cavalry division, being in all about 160,000 men with the necessary national resources allocated to this force for it to be kept up to full strength for a period of at least six months. Nevertheless, when the British War Office and the Admiralty were not agreed about the transport of the Expeditionary Force, Winston Churchill went to the Admiralty with a mandate to co-ordinate their plans. For the Army and Navy alike, Agadir was a grand rehearsal of plans which were to take effect three years later.

Up to the very outbreak of the Great War it was said very frequently that; *The mere power and opportunity to make an outrageous attack are nothing without the will to do so. Every barber who holds his client by the nose could cut his throat as easily as shave his chin. Every horse could kick his groom, who rubs him down into the next world if he chose to do so. What sense could there be in allowing our minds to be disturbed by base suspicions of an enterprising and cultured neighbour? What proof was there that Germany nourished aggressive thoughts or was brooding on visions of conquest?* So ran the mood of the whole press. Warnings were not heeded. They came only from unofficial quarters, and therefore lacked authority. Those who expressed a different opinion were either mad or malefactors.

Nevertheless, proof had already been published to the world ~ and much more was known to the British Government ~ to show that Imperial Germany had both the will and intention to provoke war, if a favourable opportunity for doing so should present itself. When Winston Churchill stated bluntly in 1912, that the German High Seas Fleet was intended as a challenge to the British Empire, he was only repeating in shorter form, the boasts of a plethora of articles, speeches, lectures, and recent books in which the Germans had threatened Britain and France with attack at their own chosen moment.

It is significant that in early 1906, and a few months after Sir Henry Campbell-Bannerman's appointment as PM in December 1905, he found himself faced with the Morroco Crisis and the prospect of a European war that was partly averted when London made it clear to Berlin that in such an event, Britain would range itself on the side of France. *It was the general belief at this time that nothing but the support of the British Government gave to France averted a catastrophe in the early part of 1906, or induced the Germans to accept the Algeciras conference as a way out of a dangerous situation.* J.A. Spender

In sequential occurance, this was the *first* warning.

The *second* warning came in the form of subterranean rumblings, inaudible to the general public, but clearly heard by ministerial ears. In July of 1909, while the Imperial Conference on Defence was in session, Herr von Bethmann Hollweg succeeded Prince Bülow as German Chancellor. Up to that time there had been the menace of the German militarist's mailed fist, rattling sabre, and shining armour. Henceforward there came the additional menace of Wilhelmstrasse diplomacy playing for time, with a careless and unconcealed contempt for the British Empire. Berlin had clearly formed the opinion that British ministers were growing more aprehensive of asking their country and cabinet to support increased naval spending, and that it was only necessary for Germany to go on alternatively dangling and withdrawing illusory proposals for a naval understanding agreement, in order to steal ahead of Britain in the naval race. Here, as in many other instances, those in Berlin had observed, considered, but drawn the wrong inference from the facts.

The *third* warning was sounded from the gong of the German Chancellery during July of 1911. The Agadir incident remains one of the strangest to have occurred in history, in that its full gravity was not realized outside a very narrow circle at the time of its occurance. Three weeks after the *Panther* appeared, Lloyd George spoke at Mansion House. *Britain will not tolerate this encroachment.* This statement created a deep impression throughout the British Empire and a still deeper impression in other countries, as it was capable of only one interpretation ~ if Germany had persisted in her encroachment, Britain would have gone to war in August or September of 1911, in support of France. His words had no other meaning, and most high-ranking army and naval officers on becoming aware of this harsh fact, made appropriate preparations in his own sphere as their circumstances required. Except ~ for the enigmatic Lord Kitchener of Khartoum who told the Committee for Imperial Defence, *I expect the Germans to walk through the French like partridges.* And, when invited to participate in any decision that the CID might be contemplating, Lord K simply refused as Viscount Esher noted:
If they imagine I am going to command the Army in France, I will see them damned first.

The *fourth* warning of the series was delivered a few months later in February of the following year. It differed from its predeccessors as it was intended only for the ears of the British Government. Neither the Opposition or the people of Britain heard anything of it until more than two years later; until the battles of Charleroi and Mons had been lost, until the battle of the Marne had been won, until the British Army was moving north to take up a position in Flanders. Then the people of Great Britain learned that when Lord Haldane had visited Berlin in February of 1912, he had done so at the special request of the Kaiser, *to consider how Anglo-German misunderstandings might be removed.* Yet in Berlin, for two days before the appointed date for his visit, proposals for large increases of the German Army and Navy were laid before the Reichstag. Haldanes mission was to abate competition in armaments, and here was an unexpected beginning. Was it an attempt to overawe a civil emissary ~ or was it an effort to steal a march in negotiations by confronting Haldane on his arrival with a *fait accompli?* Possibly both, but from London it appeared exceedingly clumsy ~ and significant.

To the mission ~ Germany was willing in a vague mode to *retard* ~ though not abandon or reduce her naval programme, providing the British Government agree to remain neutral in any war which Germany might choose to wage. On no other terms would Kaiser Wilhelm consent to a *rapprochement* with Great Britain or allow the words *retardation of the naval programme* to be officially quoted. Such an undertaking was beyond the assurances of non-aggressive intent which Lord Haldane, on behalf of his own government was prepared to give. It was made clear that Germany desired a free hand to establish a position of supremacy astride Europe. So Lord Richard Haldane returned profitless, and the British Government was at its wits' end how to placate the implacable. To even have accepted the invitation was to fall into a German trap. Berlin had long delighted in such extra-official enterprises, undertaken behind the backs of accredited representatives. Confidences could be exchanged, *explanations offered in the frankest spirit,* even understandings arrived at, but as far as Imperial Germany was concerned, nothing was binding unless Berlin's interests made it desirable that it should be so.

Haldane explained German conduct as sudden change of spirit, such as once befell a collie dog which owned him as master, and which after a blameless early career, was possessed by a fit of depravity in middle life, and took to worrying sheep. Thus in metaphor, he extenuates the German offence and excuses his own blindness.
Frederick Scott Oliver

Lord Roberts addressed many meetings during the years which followed his return from South Africa, but his profound warning delivered in the Free Trade Hall at Manchester on 22nd October 1912 carried far beyond the walls of the meeting place, and caused a sensation in political circles. This was mainly due to his prescient words of clear sanity on the consequences of military unpreparedness, which the First Balkan War then unfolding before an astonished Europe provided a warning to those concerned, still puzzling their heads over the true meaning of Agadir, who had become impressed with the need for arriving at a clear decision on the adequacy of Britain's defences.

In the year 1912, our Imperial German friends, I am well aware do not ~ at least in sensible circles ~ assert dogmatically that a war with Great Britain will take place this year or next. But in their heart of hearts they know, every man of them, that ~ just as in 1866 and just as in 1870 ~ war will take place the instant the German forces by land and sea, by their superiority at every point, are certain of victory as anything in human calculation can be made certain ~ Germany strikes when Germany's hour has struck. That is the time-honoured policy of her Foreign Office. That was the policy relentlessly pursued by Bismarck and Moltke in 1866 and 1870. It has been her policy decade by decade since that date. It is her policy at the present hour.

It is an excellent policy. It is or should be the policy of every nation prepared to play a great part in history. Under that policy Germany has, within the last ten years, sprung, as at a bound, from one of the weakest naval powers to the greatest naval power, save one, upon this globe.

Such, gentlemen, is the origin, and such the considerations which have fostered in me the growth of this conviction ~ the conviction that some form of National Service is the only salvation of this Nation and Empire. I seem to see the gleam in the near distance of the weapons and accoutrements of this army of the future ~ the Citizen Army, the warder of these islands, and the pledge of peace and the continued greatness of this Empire.

The whole speech made it abundantly clear that Lord Roberts' sole concern for the British Empire with regard to Germany was, how to defend their possessions and be secure against German schemes of aggression. Following the Haldane Mission, the British Government knew for certain what few else did.[1] They knew of the aims of Imperial Germany, and knew that Lord Roberts had spoken the truth. The British people were told nothing, knew nothing, and therefore thought nothing about the matter, for that was a Govenrment secret. After Haldane had returned from Germany, concealing under a smiling countenance and insouciant manner, a great burden of care in his heart. He had learned the mind of Germany and it was not a light lesson. He imparted his secret to his Cabinet colleagues thus: *The rulers of Germany, to put it quite plainly, had asked us for a free hand to overbear and dominate the European world, whenever they deem the opportunity favourable, at a time when Germany was enormously increasing both her aggressive and defensive resources, and especially upon the sea. To such a demand but one answer was possible, and this the British Government had promptly given in clear ringing tones of scorn.* Consequently, Lord Haldane's dreadful secret kept by the concealment of facts by the Government was not in accordance with the principle of trusting the people. Lord Roberts had trusted the people, while the democratically elected Government could not bring itself to do so. The British Cabinet, which knew the full measure of the foreign danger, concealed it out of a mistaken notion of policy. The spirit of British Foreign Policy had been simply ~ security, yet failed to achieve security ; as policy at the last resort depends on armaments, and for armaments to be effective, their purpose must correspond with policy. Political leaders of all British parties up to the outbreak of the Great War failed to trust the people with a full knowledge of the dangers which threatened their security, and of the means by which alone these dangers could be withstood.

One of our leading statesmen had even gone the legnth of suggesting an alliance, and had been denounced immediately by the whole German press, although it was understood at the time that he had spoken with the august encouragement of the Kaiser and his Chancellor. It was Germany herself, deprived of the guidance of Bismarck, who by blustering at her various neighbours, and threatening them in turn, had aroused their suspicions and achieved her own isolation. Frederick Scott Oliver

1. With the known exception of King Albert of Belgium, six ministers of Belgium and France, and the Paris-based *Cabinet Noir.*

Sophia Schliemann ~ co-discoverer of Troy Wearing treasures recovered from Hisarlik

Baron Max von Oppenheim

Wooley and T.E Lawrence at Carchemish

Construction on the *Berlin-Baghdad Railway* from Constantinople eastwards began in 1903, and by 1912 linked Constantinople to Ankara plus Adnar near the Mediterranean coast. Then the railway ran in to unexpected difficulties in the mountain ranges of Cilicia and fell considerably behind schedule. While most of the line in Anatolia was complete, long lengths of branch rail lines remained under construction in Syria and Persia. On 1st June 1914, the first train pulled out of Baghdad Station on a line that tracked to the north for 38 miles through the desert to Sumaykha ~ a virtual train to nowhere. Work continued, and four months later the Bagdad line reached Samarra 74-miles to the north.

As German engineers toiled on the *Berlin-Baghdad Railway* along the harsh Turkish-Syrian border, Baron Max von Oppenheim travelled widely across the region as a scholar and an oriental adventurer. Living in Cairo from 1892, and throughout his travels in the East, Oppenheim's dedication and identification to Imperial Germany was such that he insisted he won the confidence of the Bedouins, not as other explorers had done, by adopting the disguise of a Muslim Arab ~ but by presenting himself as nothing but the German aristocrat he was. Well known as an attentive and sympathetic observer analyst of the customs and social structure of the desert Bedouins, his study of Arabian tribes *Die Beduinen* still remains to this day, the standard reference.

The Kaiser who desperately needed capable diplomats with local knowledge to further his Eastern plan therefore appointed Oppenheim as an attaché to the German consulate in Cairo. From there, Oppenheim became a hero of German archaeology comparable with three other amateurs, Heinrich and Sophia Schliemann who discovered Troy, and Carl Humann who found the Pergamon Altar to Zeus. Oppenheim then travelled through northern Mesopotamia on behalf of Deutsche Bank planning routes for the *Berlin-Baghdad Railway* and its associated branch lines. Stopping for a rest at a Kurdish camp he was told a tale of stone idols buried beneath the sand and ventured forth to discover a trove of ancient sculptures and artifacts at Tell Halaf ~ some of which dated from around 3,000 B.C.

I was being entertained by the tribal leader Ibrahim Pasha, head of the nomadic Kurdish troops in Ottoman service with whom I had become friendly, when I learned of remarkable statues said to have been found on a hill by the small village of Ras el Ain. The village was inhabited by Mohammedan Chechens from the Caucasus who after the Russian conquest of their home had fled as religious refugees to an Islamic land. The Chechens, I was told, had wanted some years before to bury one of their dead on the hill, and while doing this had come upon stone statues of animals with human heads. Filled with superstitious dread, they filled in the hole and buried the body at another place. In that same year, the neighbourhood was visited by drought, locust swarms and cholera. This was attributed by the Chechens to the evil spirits that they believed to have been in the ancient statues, and now set free as the vengeance of the dead disturbed. As a result, they carefully avoided speaking of the mysterious statues, fearing that others might dig out the fabulous beings and thereby bring ill upon them once more. Immediately I set off for Ras el Ain ~ to the house of the Mukhtar, the village headman of the Chechens.

Baron Max von Oppenheim in the tent of Ibrahim Pasha

It was not until we had partaken of the feast that I started to speak with great care of the remarkable statues. As I had foreseen, they denied everything. However, I did not give up. I described the stones and promised the Chechens a good reward if I could get guides to the place of the finds. It was all in vain. Then I appealed to the laws of hospitality, and demanded as the guest, not to be told what was untrue, but to have my request granted. Thereupon the Mukhtar and the village elders swore on the Koran that they had not lied. I now played my last trump. I hurled a curse at my hosts for having sworn falsely on the Koran. There was then a dramatic scene. All the Chechens stood up. Some drew their long narrow daggers. My soldiers ran to my side; the situation grew threatening. I shouted to the Chechens to cap their false oath with the murder of their guest in their own house. At the very last moment, three old Chechens and the guides sent with me by Ibrahim Pasha came between us. A sudden silence fell on them. The Mukhtar acknowledged his wrong and asked me to stay, and on this the Chechens solemnly promised to take me up the hill on which the statues were said to have been found.

The excavation was begun with a gang of ten men made up of my own servants and of Arabs. Supplies and flour for their bread had to be brought in at our own cost from the villages. In the work of excavation, the men with the pickaxes first loosen up the ground, whether the object is to dig trial trenches or to lay bare a layer. Then come the mattock men and take it into the baskets of the women and older boys, who carry them away under the arm, on the shoulder, or on the head, and empty them where they are told. It was the overseer's duty to see that the bearers do not get too little soil put in their baskets through politeness shown to the ladies.

* * *

Wilhelm II : *During my vacations I busied myself with archaeology and was active in excavation work. I kept in view one basic idea : to discover the roots from which ancient Greek art developed and to find a bridge in the endeavour to establish the cultural influence of the East on the West. It appeared to me that Assyriology was important, since from it might be expected an elucidation and vitalizing of the Old Testament, and hence of the Holy Scriptures. Therefore, I accepted with pleasure the offer of the Presidency of the German Orient Society and devoted myself to the study of its work, which I promoted to the best of my ability, never missing one of its public lectures on the results of its explorations. I had much to do with those at the head of it, and requested detailed reports to be made to me of the excavations at Nineveh, Assur, and Babylon, in Egypt and in Syria, for the protection and facilitation of which I often personally brought influence to bear on the Turkish Government.*

Oppenheim's passing through Constantinople on an enigmatical journey to the Khabur River and the Urfa ostensibly was to exhume the statue of a Hittite goddess on the banks of the Khabur. According to Syrian advice, the real object of the Baron's visit was to purchase the support of Arab and Kurdish tribal chiefs in the Khabur region for the Baghdad Railway Company. John Logie Baird, Cairo correspondent ~ *New York Times*

The purpose of state-funded excavations was to both highlight the presence of Imperial Germany in the Middle East and also to acquire spectacular items for the recently founded museums in Berlin. At a group of mounds on the banks of the Khabur River von Oppenheim stumbled across the ruined palace of an Aramaean king buried in the desert within the Aramaic city of Goza (cited in the Bible). Tell Halaf proved to be a remarkable and plentiful site. They discovered large statues of sphinxes, scorpion birdmen, lions, idols, plus reliefs of the Western Palace built for the Aramaean King Kapara. These were identified as 3,000 B.C. Hittite and attributed on the basis of inscriptions on some of them. Oppenheim held that many of the relief orthostats and sculptures had been part of an earlier palace, and that as King Kapara cleared the site for the construction of his palace, he reused them for his own building as often happened in ancient times. Kapara simply put his own name on the old sculptures.

Several other archaeologists working in the Middle East at the time also served intermittently as agents of their Empires such as T.E. Lawrence and his eccentric teachers, the Oxford archaeologists D.G. Hogarth and Leonard Woolley who both worked at Carchemish in Northern Syria. Nevertheless, these political agents in disguise were keen sincere scholars, genuinely interested in both the objects of their investigations, and in the present-day inhabitants of the ancient sites. Max von Oppenheim once visited the Oxford Group while his team was excavating Tell Halaf. In the words of Lawrence : *Oppenheim came about 5 PM stayed till 11, then went off to eat & sleep: came back at 4 AM and stopped till 10. He then invited me over to his place by his relay of post-horses. I found him such a horrible person that I never took him up on the invitation.* Later, when von Oppenheim tried to spirit away his best finds out of Syria to Berlin and his smuggling efforts were discovered, T.E. Lawrence wrote: *He was an ass to have his things taken so, and not very virtuous to take all those things from his excavations.* Lawrence and Woolley while searching for the *Wilderness of Zin* (also referred to in the Bible) conducted a parallel secret military survey of the Negev Desert to map features of strategic relevance such as sources of water ~ of vital importance to the British military, as they knew that if war came, the Negev would need to be crossed by Ottoman Turkish forces, if they were to attack British interests within Egypt.

Apart from his scholarly archeological pursuits, von Oppenheim who was openly hostile to the British Empire became the prophet of *Islampolitik* when he advocated using German friendship with the Moslem world as an effective weapon against the British and French. As Oppenheim acquired the confidence of the Kaiser ~ *The Kaiser's Spy* as his British contemporaries referred him to in Cairo, he continued to keep an eye on the activities of the British Empire. Each summer he returned to Germany to meet with Kaiser Wilhelm II, and to fully brief him on the condition of the Moslem world.

Freiherr von Oppenheim is attached to the Consulate-General in Cairo as an Oriental specialist attaché to keep an eye on the supposedly powerful ferment in the Islamic world in order to warn Europe in good time of any imminent outbreak. To this end, he develops good relations with the native-born people and joins their caravans in order to get a sense, in the great markets of the interior, of the prevailing mood in the Islamic world. Oppenheim has made many caravan trips, sent in reports of his conversations with the locals, devoted himself to serving distinguished German travelers as a guide, and keeps in addition, an open house for guests, Oppenheim will be coming to Berlin to advertise the value of his local knowledge for his Majesty. Friedrich von Holstein

Once in Berlin, Baron Max von Oppenheim predicted: *In the future Islam will play a much larger role. The striking power and strength of Islamic lands will one day have a great significance for European Empires.* Strategic planners in Berlin foresaw the potential to deploy Islam with more than 100 million Moslems living in British India, around the Persian Gulf and in Egypt as an effective measure against the British should the need arise. If Berlin could enthuse the religious aspect of Egypt's 13 million Moslems in addition to the millions of Moslems living in British and French colonies throughout Asia and Africa to weaken the Éntente Powers from within ~ combined with a declaration of *Jihad* or Holy War from the Caliph ~ in time, the British and French hold in some of their colonies could become untenable.

While Berlin's planners called such concepts *Islampolitik*, some secular Young Turks initially dismissed such hopes as *Jihad made in Germany*. However, Enver Pasha had personally appreciated the power of Islam when he fought in 1911 against the Italians in Libya and came to believe that religious fanaticism could be deployed against the Éntente. Arriving in Tripoli, he came to view the war in terms of *Jihad* : *The Libyan volunteers are fanatical Moslems who see death before the enemy as a gift from God,* and duly noted their devotion to him as the son-in-law of the Caliph. Kemal also viewed Islam as a bond between Turks and Arabs and supposing a religious war could further strengthen such ties, he was convinced that a *Jihad* ~ such a powerful weapon in the early period of Islam ~ could be revived to serve as a source of strength in any confrontation with an opposing European Empire ~ he wrote: *The majority of Arabs would not hesitate to make any sacrifice in war for the liberation of the Moslem Caliphate.*

I have been a friend of the charming Baron Max von Oppenheim whose learned talk I much enjoyed in Cairo, and yet Oppenheim's scientific work was merely the cover for his work in the secret service of the Kaiser. Oppenheim was a man of real scholarship who had early on become a common spy. As early as 1886, for instance, he was in Morocco making an attempt to detach the Moroccans from their connection with France, and at the same time he made his first attempt to gain for Germany the aid of the powerful Sanussi Fraternity. In 1893, a long journey from Beirut through the Syrian Desert to Bagdad and the Persian Gulf was not unconnected with the Bagdad railroad. In 1894, he made a journey down to the Tchad Sea in anti-English interests to gain the assistance of Mohammedan tribes for his work in Egypt against the English occupation. More recently, during his stint in Cairo, he went to considerable lengths, seducing the Khedive's European mistress in order to obtain and have photographs made of important papers in the Khedive's possession. There is much more that I could say about the learned Baron von Oppenheim. Richard J.H. Gottheil
Professor of Languages
Columbia University

While the German adventurer and master spy founded the magazine *El Jihad* in 1914 to incite the Arabs to prepare to wage some Holy War against the British and French occupiers in the Middle East, his adversary T.E. Lawrence would find future fame playing a finer role to guide a real Arab uprising. John Buchan's *Greenmantle* captures the essence of Europe's apprehension of Islam:

Islam is a fighting creed, and the mullah still stands in the pulpit with the Qur'an in one hand and the sword drawn in the other ~ there will be hell let loose in those parts soon. I have reports from agents everywhere, peddlers in South Russia, Afghanistan horse dealers, Turcoman merchants, pilgrims on the road to Mecca, Sharifs in North Africa, sailors on Black Sea coasters, sheep-skinned Mongols, Hindu fakirs, Greek traders in the Gulf, as well as decent consuls who use ciphers. They all tell the same story. The East is waiting for a revelation.

Serge Sazonoff
Russian Foreign Minister 1910-16

General Franz Conrad Graf von Hötzendorf
Army Chief of Staff of Austria-Hungary

King George V as the 26 year-old
Prince of Wales ~ signed George 1891

Prince Bernhard von Bülow & Isvolsky
Swinemünde 1907

Count Alois Lexa von Aerenthal
A-H Ambassador to Russia 1899-1906
A-H Minister of Foreign Affairs 1906-12

Count Theobald von Bethmann-Hollweg
Chancellor of Imperial Germany 1909-17

Baron Nikolai Hartwig 1912
Russian Ambassador to Persia 1906-08
Russian Ambassador to Serbia 1909-14

Baron Alfred von Kiderlen-Wächter
Imperial German Foreign Secretary 1910-12

Mary Edith Durham
British Balkan authority and aid organizer

Mata Hari
German Secret Agent H-13 / H21

Oberstleutnant Walter Nicolai
Head of German Secret Service ~ Section IIIb
Recruiter of Mata Hari in 1913

Margaretha Zelle ~ Mata Hari

Count Heinrich von Tschirschky
German Ambassador to Vienna 1907-16

Princess Daisy of Pless

Jules-Martin Cambon
French Ambassador to Germany 1907-14
Close friend of Mata Hari

Baron Wladimir Giesl von Gieslingen
A-H Ambassador to Montenegro 1910-13
A-H Ambassador to Serbia 1913-14

Count Artur Giesl von Gieslingen
Commander of A-H VIII Corps in Prague

Pierre-Paul Cambon
French Ambassador to Britain 1898-1920
Brother of Jules-Martin Cambon (above)

SERGE SAZONOFF'S LETTER TO COUNT SERGEI WITTE

Count Sergei Witte
Finance Minister of Russia 1893-1903

Grigori Rasputin
Mystic faith healer & Romanov advisor

Pyotr Stolypin
Minister of Internal Affairs 1906-11

Following the bombing derailment of Tsar Alexander II's freight train, and the 1888 appointment of Count Sergei Witte to the Rail Ministry, he was endorsed in 1893 as Finance Minister of Russia. Count Witte was thereafter in great favour with both Tsars Alexander III and Nicholas II who highly praised him for his considered frankness, absence of servility and independence of judgments. In his new role of Finance Minister, Witte prepared plans for Russia's modernization efforts. He also noted :

Several weeks before Stolypin's assassination when I was at that time at Biarritz I received a remarkable letter signed by S. Sazonoff, whom I had known for a number of years. The career of this man is worth mention. In his youth Sazonoff is said to have been intimate with Zhelyabov, a plotter of the assassination of Alexander II. At one time, he wrote for the radical press, but when the revolution came, he found it profitable to join the extreme reactionaries. He joined hands with Professor Migulin and later was befriended by such influential clergymen as Archbishop Hermogenes, and Staretz Rasputin. He became especially intimate with the latter. When visiting St. Petersburg, Rasputin stayed with Sazonoff, who gradually assumed the role of a circus sideshow manager demonstrating an outlandish prodigy to an avid public. Highborn ladies who were among Rasputin's clientele would come to see him at Sazonoff's house. Naturally enough, Sazonoff became a person of importance himself, for Rasputin wielded an enormous influence at the Court. Sazonoff succeeded in obtaining from the Minister of Finance Vladimir Kokovstov, direct and indirect subsidies for his weekly, The Economist. Then Kokovstov granted to Sazonoff and Professor Migulin a license to open a banking institution, which license the worthy pair sold for 250,000 rubles. With a portion of this sum ~ the directors of the bank at Kokovstov's suggestion added 100,000 rubles ~ Sazonoff founded a newspaper, where blackmailing by means of the printed word was practiced under the guise of a fairly progressive tendency.

In his letter, Sazonoff informed me that Stolypin was done for, and that the Tsar had formed a firm resolution to get rid of him immediately after the Kiev solemnities. His Majesty, Sazonoff wrote, had chosen as Minister of the Interior, Alexei Khvostov Governor of Nizhni-Novgorod province. Sazonoff further wrote that he and Rasputin were now going to Nizhni for a final conference with Khvostov. They were quite certain that he would be an admirable Minister of the Interior, but they had some doubts as to whether, on account of his youth, Khvostov would be a fitting substitute for Stolypin in his capacity of President of the Council of Ministers. Therefore, Sazonoff wondered whether I would accept the post of President of the Council in order to lend prestige to the new Cabinet.

Khvostov, I may remark, is one of the worst specimens of officialdom as it had existed under Pyotr Stolypin. In his contempt for the law he actually outdid all the other provincial governors. Shortly before, he had submitted to His Majesty a memorandum in which he asserted that Russia was in a state of latent unrest, that the revolution had been driven underground by Stolypin and might break out again, should the government fail to take proper measures. On his part ~ he suggested that all suspects should be killed off in one way or another. My answer to Sazonoff was to the effect that I was in receipt of his letter and that their proposal made me wonder whether they themselves were out of their minds or whether they imagined me stark mad. Sergei Witte

In September 1907, Russia and Great Britain concluded a treaty relating to Persia, Afghanistan and Tibet. The agreement inaugurated the policy of philandering with England. Since we did not give up our traditional flirting with Germany, the situation became rather ambiguous. At present, we are trying to adjust ourselves to it by assuring Germany that we love her best and that we are flirting with England merely for appearance's sake, while to England we say the reverse. I believe we shall soon have to pay for this duplicity.

The rapprochement with England, the ally of France, who is our own ally, has resulted in the formation of a Triple Éntente, as opposed to the Triple Alliance of Germany, Austria-Hungary, and Italy. The history of the Éntente is as follows : On my way from Portsmouth I stopped in Paris and met there, among other people, Kozell-Poklevski, first secretary to our Embassy in London. He brought me an invitation from King Edward VII to pay him a visit, but I could not accept it without my Monarch's express permission, which I failed to obtain. At the same time our Ambassador in Paris, Izvolsky, submitted to me a project of an agreement with Great Britain, substantially identical with the one which was later actually concluded. I asked Kozell-Poklevski to inform the King that should I on my return to Russia assume the governmental power I would use all my influence to establish friendly relations with Great Britain. I added, however, that I was decidedly opposed to the idea of concluding the treaty sketched to me by Izvolsky, for the reason that it was best for us not to tie ourselves down by treaties. I feared that an agreement with Great Britain would arouse the jealousy of Germany. As a result, we would perhaps be forced into making an agreement with that country too, and be cheated in the end.
It was owing to my opposition that the agreement was not concluded before 1907.

The agreement was a triumph of British diplomacy. It dealt chiefly with Persia. The North which includes its most fertile and thickly populated sections, had from times immemorial been within our sphere of influence. With the conquest of the Southern parts of the Caucasus, formerly provinces of Persia and Turkey, the Northern part of Persia was naturally destined, so to speak, to become a part of the Russian Empire. To prepare for that eventuality we sacrificed a great deal of our blood and treasure. The agreement set all these sacrifices at naught. According to it, Southern Persia was to be under the economic influence of Great Britain, while the North was left to us. As for Persia's central Government, it was to be controlled by Russia and Great Britain acting jointly. Since Teheran, the seat of central Government is situated in the North; this meant British influence in the North as well as in the South.

Russia has no annexationist designs upon Afghanistan. We are merely interested in preserving its status quo as a buffer State between Russia and British India. True, the agreement provided for the preservation of this status quo, but stipulated that the country should be under the exclusive influence and protection of Great Britain, so that we were not even allowed to have our diplomatic representative there. This meant that all our negotiations with the Government of Afghanistan were now to be conducted through the British authorities. Under these circumstances the buffer became something in the nature of a loaded gun pointed at us.
In Tibet, the contracting parties obligated themselves not to introduce any missions or troops.
We also renounced all claims to the Southern Persian ports.

The agreement was concluded without regard to the claims of the other Powers upon Persia. For that reason the division of Persia was rather futile. No sooner was the treaty published than Germany began to seek to safeguard its economic interests in Persia. As early as 1904 the German Government in the person of von Bülow complained to me that we were hindering the freedom of importing German goods to Persia. In 1911, we concluded an agreement with Germany, agreeing to connect the railroads of Northern Persia with the German Berlin-Bagdad line and also to give her a free hand in Northern Persia with her imports. In sum, what have we achieved? By signing the agreement with Great Britain, we made it impossible for us to annex Persia politically, and by entering into an agreement with Germany we lost Persia economically, for economic competition with Germany under equal conditions means certain defeat for us. In a word, Persia has now slipped out of our hands. At present, we can play there merely the part of a policeman, until the native Government grows strong enough to restore order. Count Sergei Witte

Persia remained a neutral state, but one still dominated by Russian and British spheres of interest. Its significance lay in its location between the vast Russian, Ottoman Turkish, and British Empires. While the Anglo-Russian Agreement of 1907 stabilized relations between their Empires after decades of antagonism over influence in Persia and the northern approaches to India, and whilst it appeared that Great Britain did not seek territorial gain (as it was the security of India that was paramount) the British Royal Navy had commenced their conversion from coal to oil-fired warship engines, and were aware of a British oil exploration concession in Persia, which would soon be of strategic importance.

Jacques de Morgan

William Knox D'Arcy

Kerosene was first distilled in a clear form from petroleum that could be used for lighting by the Canadian geologist Abraham Gesner in 1846. By 1860, kerosene was one of the wonders of the age, having largely replaced whale oil and tallow for most lighting in North America and Western Europe. Petroleum or oil still had few uses, but with the development of the internal combustion engine to power vehicles, new uses were found for petroleum. As well as providing fuel for vehicles petroleum could also be used to fuel oil- fired boilers to run the latest machines ~ and importantly at a time when almost any coal-fired engine that heated water to form steam could be replaced by a new version to operate on the new fuel ~ British naval engineers began designing new oil-fired engines for warships.

The first reports of oil in Persia's Qasre Shirin region were the articles of famed French geologist and archeologist Jacques de Morgan (who discovered the pyramids of Amenemhet II and Sesostris III) in the 1892 *Annales des Mines* in which he confirmed the abundance of rich oil seeps in S.W. Persia where oil leaked thru fractures to the surface, and the possibility of establishing a petroleum industry. Indeed the local desert people who had known of these oil seeps and bitumen deposits for centuries had already extracted, traded and used these materials for traditional purposes such as fireworks, fire-hurling weapons, oil lamps, and waterproof glue. The archeologist Jacques de Morgan also knew that the Zoroastrian fire temples of ancient Persia had all been constructed near such oil seeps.

At the 1900 Paris Exposition, the Persian representative Kitabchi Khan approached British and French investors to talk about Jacques de Morgan's papers and prospects of Persian oil. He also talked to Sir Henry Drummond Wolff (who was recovering from a mental breakdown which had recently ended his career as the British Ambassador to Persia), and it was he, who introduced Khan to D'Arcy. Sir Henry explained: *Lately I have met with some high-ranking Persians who believe Persia possesses vast, untapped oil reserves. Persia is unquestionably petroliferous territory. All the Persians need is a financial backer. Muzzaffar-al-Din Qajar, the Shah of Persia who all but owns the country is a kindly, but decadent individual. Like his father, who was assassinated in 1896, Muzzaffar is completely out of touch with his people. He spends fortunes on his grand tours of Europe while the Persians suffer food shortages, unemployment, and inflation.*

For William Knox D'Arcy, Shah Muzzaffar of Persia was the perfect person to strike a deal with. D'Arcy who some years before had acquired a mining lease on Ironstone Mountain in Queensland, had found his claim so enriched of gold, he became one of the wealthiest men in the British Empire. He purchased two country houses in England, apartments in Paris and Bruxelles and two in London. He also commissioned the Holy Grail tapestries from William Morris and Edward Burne-Jones, thus financing one of the great achievements of the Arts and Crafts movement. *Morris & Co.* was also commissioned to furnish his grand residences throughout. He held shooting parties on his 19,000 acre Norfolk estate where Melba and Caruso were among the stars who performed at his private concerts.

W.K. D'Arcy whilst living luxuriously in London was looking for new commercial ventures so he swiftly dispatched the geologist H.T. Burls to verify the oil seeps that Jacques de Morgan had so conveniently well mapped. Burls' report was optimistic ~ *The territory as a whole is one of rich promise.* In March 1901, D'Arcy sent his agent Alfred Marriott together with Kitabchi Khan to Tehran to negotiate an oil concession from the Persian Government. Sir Arthur Hardinge as British Ambassador to Persia helped D'Arcy's representatives meet with the Persian PM Amin al-Sultan, whereupon Alfred Marriott was reported to have paid £10,000 to Persian officials to secure a 60-year concession from the cash-strapped Shah Muzzaffar of Persia for oil exploration and exploitation rights in Persia. This granted D'Arcy exclusive rights to explore, drill, obtain, market, and export petroleum, asphalt, and natural gas in Persia for 60 years. However, D'Arcy was required to form a company within two years for this purpose and because of Russian opposition, the five northern provinces were excluded from the deal. In return, the Persian Government was to receive £20,000 in cash, £20,000 in shares from the company, and 16% of all profits made by the first or any other company formed by this concessionaire. W.K. D'Arcy swiftly formed the First Exploration Company in May 1902 with a capital of £600,000, half of which belonged to him, and the search for marketable oil in Persia began.

For field operations, D'Arcy hired George Bernard Reynolds, an Englishman who had graduated from the Royal Indian Engineering College, served in India and thereafter as a self-taught geologist had drilled in the Dutch oilfields of Sumatra. Working conditions in Persia as Reynolds soon learned, were harsh ; the climate hot, the drill sites remote, and the local tribesmen suspicious of foreigners. Reynolds put together a technical team of Polish and Canadian drillers, plus an Indian doctor, and proceeded to hire workers and security guards, and to pay the local tribal chiefs for their cooperation.

In November of 1902, drilling started at the seepage area of Chiah Surkh in the far west of Persia (which had been one of the key areas mapped by Jacques de Morgan). In the summer of 1903 a slight show of gas and oil was encountered at 1500 feet. In January 1904, a second well found oil at a similar depth, however D'Arcy's excitement was temporary when the oil flow fell from 180 barrels a day to a mere 25 b/d. After a few months, the well ran dry. At a time when a deeply disappointed D'Arcy appeared willing to sell his arid Persian concession to a German company keenly interested in a deal, D'Arcy wired to George Reynolds ~ *Good news from Persia would be very acceptable now.*

Admiral Sir John Fisher, who became First Lord of the Admiralty in 1904, accelerated British plans to convert the Royal Navy from coal to oil fueling, and needed to act fast to prevent the D'Arcy concession from slipping out of British hands, so Fisher met with W.K. D'Arcy and discussed how the Royal Navy could become a market for Persian oil that provided many advantages for naval vessels : they could remain on operations longer since they could be refueled at sea, were faster and got up to speed faster than coal-powered ships, and thus were more maneuverable and required fewer hands.

On Admiral Sir John Fisher's initiative, the new Concession Syndicate Ltd was formed in 1905 to support D'Arcy's operations in Persia. With new finance from the Scottish Burmah Oil Company, D'Arcy remained director of the new company that decided to abandon the Chiah Surkh drilling site. Also in 1905, (and two years before signing the Anglo-Russian Persian Agreement) a dossier based on information from an informant working on the *Berlin-Baghdad Railway* was received by the British F.O.

SECRET ~ From H.M. Ambassador in Constantinople.
Report on the Petroliferous Districts of Baghdad, Mosul, and Bitlis.

William Knox D'Arcy then sent the geologist, W.H. Dalton, to report on oil prospects in the southern Khuzistan province. Dalton advised: *Shardin 110 miles N.E. of Ahavaz City is by far the best.* Based on this report, Reynolds drilled two test wells at Shardin. One reached a depth of 2170 feet and another in 1907 to 1940 feet deep. Both explorative wells were dry. Reynolds decided to try a third; Dalton's suggestion N.E. of Ahavaz. This area was called Maidan-i-Naftun, literally; *Plain of Petroleum.* It was also home to the ruins of an ancient Zoroastrian fire temple, and for this reason local people (mistakenly) called it Masjid-i-Suleiman *Mosque of Solomon*. This place, so favourably mentioned in the geological reports of de Morgan, Burls and Dalton, was particularly attractive for Reynolds as on a recent trip to Kuwait, he again heard about the oil seeps of Maidan-i-Naftun from the British official Louis Dane who had travelled to this desolate place in 1904, and once again in 1906 with Lord Curzon, to see these legendary sedimentary rocks, which were indeed constantly saturated from seeping oil.

In 1907, after British oil exploration efforts moved to Maidan-i-Naftun and W.K. D'Arcy had risked his entire fortune on oil exploration in Persia, he was soon on the verge of losing everything; both of his grand country houses, and even his London mansion. With his personal fortune almost spent, William D'Arcy was close to despair. In Persia, staff were dismissed. Even the Burmah Oil Company, whose investment had saved his expedition during 1904 (on the delusion of an imminent discovery) were tiring of finding nothing of consequence in Persia. Reynolds received orders from London for his final well ~ *Drill to 1,600-ft and then give up.*

Reynolds was enthusiastic about this well. When he received the telegram from company HQ to stop the drilling, he cabled back; *The drilling should go on.* Giving up was not part of Reynold's nature. It had taken his crew 10 days just to get to Shardin, eight months to start drilling, and six years of toiling in the desert to find little of significance. Torrential rains had washed away nearly four months of roadwork on a link to Masjid-i-Suleiman, where two weeks before a drill bit had fallen off in one of two last-chance wells and had taken more than a week to retrieve. In a narrow gorge with sheer cliffs of gypsum overlaid with gravel, the Canadian drillers and their Persian workers continued to hammer through layers of rock with simple percussion rigs in search of the elusive oil…

On 16th May 1908, a gas smell wafting through the rocks at Masjid-i-Suleiman was first noted. With a scent of sulphur hovering in the air on 25th May; a very good sign for an experienced oil hand, it was a smell that George Reynolds could see in the rising vapours through the last evening sunlight. He instructed his men to keep drilling through the night. At 4 AM, when the drill reached 1180 feet below the desert it struck oil and the gusher sprayed a black fountain 60 feet above the rig to the sky. Reynolds, in high spirit called for a camel courier to take a message to the telegraph office in Baghdad. *I have the honour to report that this morning at 4 AM oil was struck in the No.1 hole at a depth of 1180 feet.* The site was so remote that it took five days for Reynold's telegram to reach London. D'Arcy replied: *If this is true all our troubles are over. I am telling no one about this until I have the news confirmed.*

Lieutenant Arnold Talbot Wilson, who commanded a detachment of the 18th Bengal Lancers to guard the British consulate in Ahvaz, was deployed to safeguard the British oil exploration site. Lt. Arnold Wilson who formed an instant friendship with Reynolds, described him as; *Solid British oak, dignified in negotiations, quick in action and completely single-minded in his determination to find oil.* Lieutenant Wilson wrote of the oil find: *I was sleeping outside my tent and ran to the spot as soon as I heard the rumbling noise and shouts of jubilation from the Persians. The oil rose 60 feet or so above the top of the rig, smothering the drillers and their devoted Persian staff who were nearly suffocated by the accompanying gas.*

Wilson quickly swallowed some yoghurt and some flat bread, mounted his stallion and galloped to the Persian telegraph office in Shustar 30 miles away to report to HQ at Bushire on the Gulf coast. However, when he arrived, he realized that he didn't have the current secret codebook with him. Instead he took out his Bible that he always had, and consulted the Old Testament for the right verses.
As I had no Telegraph Code I wired; See Psalm 104 Verse 15 third sentence and Psalm 114 V 8 second sentence.
Who turned the flintstone into a springing well. That he may bring out of the earth oil to make him cheerful.
This told them the important news, and in the circumstances, was quite as effective as a cipher.

The following day, No. 1 hole was tested to flow at 297 b/d. Two weeks later on 5th June, No. 2 hole at Masjid-i-Suleiman tapped oil at 1010 feet ~ another gusher. These two wells struck near the Persian Gulf changed the geopolitical dynamics of the Middle East. British Government involvement in the oil concession connected to D'Arcy's company through intricate financial arrangements and political maneuvers saw D'Arcy's original concession become the new Anglo-Persian Oil Company. Winston Churchill described the development as ~ *A prize from fairyland beyond our wildest dreams.*

On 14th April, when D'Arcy's Persian concession, consolidated with some additional investment from Burmah Oil went public as the Anglo-Persian Oil Company (later British Petroleum) its initial public offering of Anglo-Persian stock sold out in 30 minutes in London. In Glasgow, investors stood five deep around the cashier's cages to buy APOC shares, and from that day, William Knox D'Arcy who had almost lost it all, was richer than he ever was, and the race for oil accelerated throughout the Middle-East ~ additionally W.K. D'Arcy now had a long-term a contract with the British Admiralty for APOC to ensure a steady supply of fuel oil to the Royal Navy (at substantially reduced prices).

Admiral Sir John Fisher and Home Secretary Winston Churchill at work ~ the First Sea Lord in his room at the Admiralty

In 1910, Winston Churchill as Home Secretary had opposed higher naval expenditure in favour of an Anglo-German naval agreement, but by July 1911 having witnessed provocative German naval aggression in Morocco, he converted to the case for modernization of the Royal Navy for Britain to maintain its mastery of the high seas at a time when they were in a serious naval race with Germany: both in the size of the fleets and in technological advancement, so a fuel that was much more efficient and cheaper than coal had immense military benefits. Already Great Britain was building destroyers and submarines that were oil-fired, but her battleships still burned coal. To hasten this conversion, Admiral Fisher led the *Royal Commission on Supply of Liquid Fuel and its Application to Warship Engines.*

With Britain now fully committed to a strategic involvement at the head of the Persian Gulf, a 222km pipeline was constructed from the Persian oilfields to an APOC refinery on Abadan Island on the Shatt-al-Arab waterway near the confluence of the Tigris and Euphrates; and as the waterway formed the disputed border between Persia and Ottoman Mesopotamia and was within a day's march of the nearby Turkish troops in Basra, Great Britain signified her oil interests by stationing a flotilla of gunboats in the Shatt-al-Arab Waterway, and supported the nearby protected Sheikdom of Kuwait by dispatching the 18th Indian Brigade for a show of strength in the area.

After the Abadan Island oil refinery opened in 1912 with an initial annual capacity of 120,000 tons it rapidly expanded to become the largest in the world, and within a few short years, Anglo-Persian were supplying much of the Royal Navy's fuel oil requirements. At this time, there were five principal suppliers and refiners of crude oil or petroleum in the world: Rockefeller's Standard Oil Company that first discovered oil in Pennsylvania in 1859; Branobel, the Nobel Brothers Petroleum Company operating at Baku on the Caspian Sea in Russia; the Rothschild family with concessions at Baku, operating in association with the Shell Oil Company founded by London's Samuel brothers who revolutionized the transport of oil by tanker; the Royal Dutch Company that discovered oil in Sumatra and merged with Shell in 1907; and now the British Anglo-Persian Oil Company oilfields of Persia.

THE ITALO~TURKISH WAR OF 1911-12

Giovanni Giolitti ~ PM of Italy five times between 1892 & 1921

Encouraged by British success in Persia and the prosperity of European Empires from their many colonial acquisitions around the globe, Italians decided to seize a swathe of territory in North Africa. The Italo-Turkish War, fought between the Kingdom of Italy and the Ottoman Turkish Empire from September 1911 to October 1912 became a significant precursor to the Great War, as after observing how easily Italian forces could defeat the defenders of the Ottoman Caliphate, such success stirred the provocateurs of nationalism in the Balkans to agree to the Russian plan to form the Balkan League, and strike while the Ottoman Army were occupied in their losing struggle against the forces of Italy. Another reason for the formation of the Balkan League resides in the sharp reaction of Russia, whose trade was damaged by the closing of the Straits, thus St Petersburg established a (supposed defensive) Balkan Alliance under Russian leadership to acquire Constantinople as Protector of the Slav peoples.

In the summer of 1911, while secretive preparations for the invasion of Ottoman-occupied Libya were carried out, Italian Prime Minister Giovanni Giolitti began to probe the other European leaders to gauge their reactions to a possible Italian invasion of Libya. The Italian fleet appeared off Tripoli in the evening of 28th September. War was declared at 3 PM the following day, and the Italian Navy bombarded the port on 3rd October. After 1,500 Italian sailors conquered Tripoli, the Turks hastily organized the local Arabs and Bedouins as an improvised defensive force against the Italian invaders.

The actual disembarkation of 20,000 Italian Army troops came on 10th October, and they easily conquered Tobruk, Derna and Khoms, but not Benghazi. The first setback for the Italians occurred on 23rd October, when their troops near Tripoli were encircled and attacked by Arab horsemen and Turkish regulars who practically wiped out the Italian formation. The Italian force was then enlarged to 100,000 men to face an estimated 20,000 Arabs and 8,000 Turks. The Italo-Turkish War was notable for the first use in warfare of fixed-wing air power. Carlo Piazza flew the first recon flight on 23rd October 1911, and a week later the first aerial bombing of an enemy position came when *Sottotenenté* Giulio Gavotti piloting a German *Taube* dropped four grenades on Turkish positions near Ain Zara.

The Arabs and Turks continued to make frequent attacks by day and night on the strongly entrenched Italian garrison around the southern suburbs of Benghazi. Four Italian infantry regiments acting on the defensive (and backed by the cruisers *San Marco*, *Agordat*, and the mighty *Regina Elena*) were attacked during the night of 15th December. The Turks attacked, but were repulsed by naval fire from the Italian warships. At the Battle of Tobruk in December, Turkish troops and Arab volunteers organized by Captain Mustafa Kemal achieved a victory for the competent command of Kemal. Following this achievement, Mustafa Kemal was assigned to Turkish Army HQ at Derna to coordinate future operations. On 3rd March 1912, 1,500 Libyan volunteers attacked Italian positions near Derna. The Italians were outnumbered, but having superior weapons, held the line until further Italian reinforcements stabilized the situation. The battle ended in the afternoon with an Italian victory, and the Ottomans withdrew from Libya. As a result, Italy acquired the Ottoman Provinces of Tripolitania, Cyrenaica, and Tripoli. Italian forces also steamed in to occupy the Dodecanese Islands in the Ægean.

Le Petit Journal

Le Petit Journal
CHAQUE JOUR — 6 PAGES — 5 CENTIMES
Administration : 61, rue Lafayette
Les manuscrits ne sont pas rendus

5 CENTIMES SUPPLÉMENT ILLUSTRÉ 5 CENTIMES

Le Petit Journal agricole, 5 cent. ~~ La Mode du Petit Journal, 10 cent.
Le Petit Journal illustré de la Jeunesse, 10 cent.
On s'abonne sans frais dans tous les bureaux de poste

ABONNEMENTS

	SIX MOIS	UN AN
SEINE et SEINE-ET-OISE..	2 fr.	3 fr. 50
DÉPARTEMENTS..........	2 fr.	4 fr. »
ÉTRANGER....	2 50	5 fr. »

Dix-neuvième Année — DIMANCHE 18 OCTOBRE 1908 — Numéro 935

LE REVEIL DE LA QUESTION D'ORIENT

La Bulgarie proclame son indépendance. — L'Autriche prend la Bosnie et l'Herzégovine

The Bosnian Crisis of 1908 altered the balance of power in the Balkans and precipitated events for the formation of the Balkan League. Bulgaria declared its independence and Prince Ferdinand with much initiative further entitled himself as its Tsar. Austria-Hungary in the person of Emperor Franz Josef annexes Bosnia and Herzegovina, while Ottoman Turkish Caliph Sultan Abd-Ul-Hamid looks on helplessly.

THE RUSSIAN FORMATION OF THE BALKAN LEAGUE

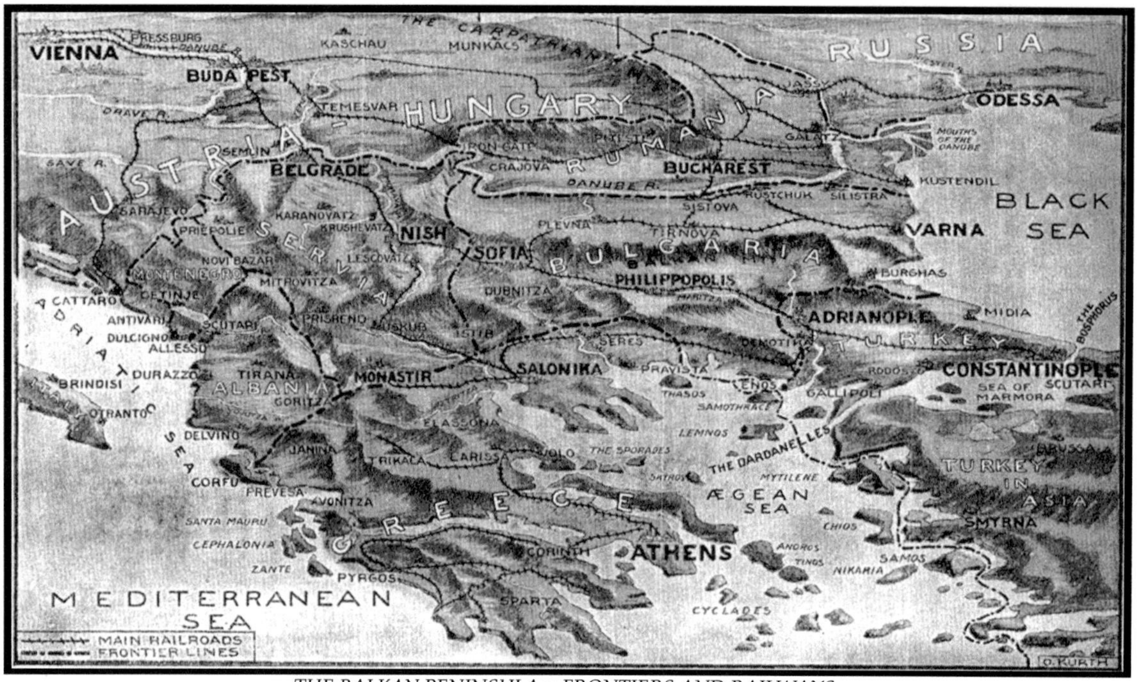

THE BALKAN PENINSULA ~ FRONTIERS AND RAILWAYS

Russia, whose trade had been severely damaged by the closing of the Dardanelles by the Turks in the Italian War, was more than ever determined on the opening of the Straits ~ for a Balkan Alliance under Russian leadership to take possession of Constantinople as *Protector of the Slav peoples.*

After the annexation of Bosnia Herzegovina, Russian diplomacy, more concerned than ever to establish a barrier against the further advance by Austria to the southeast, seized firmly upon the idea of a Balkan League. In this, Hartwig, the buttress and hope of Russo-Slavic opportunist policy in the Balkans, gave Pasic some formal assurances regarding support. Russia's idea was to create a defensive alliance against Austria-Hungary. However, the Balkan allies formed an offensive alliance against Turkey, and war began. Leon Trotsky

For decades, Austria-Hungary and Imperial Russia had been keen rivals on the Balkan Peninsula. Russia, swayed by her sentimental and territorial considerations, sought to exercise increasing control over the Southern Slav countries of the Balkans based upon the racial affinity of the Slavic nations that owed their existence to the wars waged by Imperial Russia against the Ottoman Turkish Empire. More and more through her Balkan affairs, Russia infringed upon the commerce of Austria-Hungary, and this rivalry between Russia and Austria-Hungary sparked perpetual diplomatic strife. This was all aggravated by the underground activities of Russian consuls, reinforced by their unofficial agents.

The next blow aimed at the Dual Monarchy by Russia in her persistent attempts to exclude A-H influence from the Balkans came when Russian foreign policy focused on forging a Serbian-Bulgarian alliance. Russian Foreign Minister Serge Sazonoff, instructed Russian Ambassadors Nikolai Hartwig in Belgrade and Anatoli Neklyudov in Sofia to encourage Serbia and Bulgaria to establish closer diplomatic relations to counter Austrian influence in the Balkans. Sazonoff specifically mobilized the powerful apparatus of Russian paradiplomatic operations in the Balkans when he wrote to Hartwig; *I have the impression that Austria-Hungary, despite the German brotherhood in arms, is on her last legs. Serbia's promised land lies within the orbit of present-day Austria-Hungary. Under these circumstances it is a matter of vital importance for Serbia to put herself by hard and patient work in the position of readiness necessary to face the inevitable outbreak of future war.*

The Russian Ambassador to Serbia Nikolai Hartwig was the grandson of a German doctor who had immigrated to Russia. After completing his studies at the University of St. Petersburg he entered the Asiatic Department of the Foreign Ministry where he worked with Foreign Minister Vladimir Lamsdorff from 1901-06 and was rewarded with the position of Head of the Asiatic Department. While there, Hartwig attacked the official Russian policy in the Near East, as too weak and compliant. An appointment as the next Russian Ambassador to Belgrade in 1909, eventually enabled him to carry out a policy more in line with his own convictions, and immediately Hartwig took an active role to reverse the situation ~ to make the small Kingdom of Serbia the center of Russian Eastern Policy.

Convinced of the necessity of close Russian involvement in Slavic causes, Hartwig's stand soon led to hostility with Vienna. He enjoyed a major influence not only over Serbia's foreign but also its domestic politics, and seemed second in influence only to the Radical Party leader, Pasic. Hartwig misled foreign representatives, Serbian politicians, and his own government as to his precise position. In Belgrade, he cultivated the impression that his word represented official policy and soon Belgrade and Vienna became convinced that he was carrying out official Russian foreign policy in the Balkans. The A-H diplomat, Count Stephan von Ugron zu Ábránfalva compared Hartwig to a *malicious demon.* Bulgarian Ambassador to Serbia Dimitar Rizov wrote; *Hartwig enjoys such an unlimited influence in the Serbian Government and their society that his diplomatic colleagues privately call him The Regent, for in reality; he fulfills the functions of the Serbian King.* British diplomat G.H. Barclay: *Thanks to Hartwig, Serbia is practically a Russian province. I have never known the Serbian Government to act against the directions of him.* Nicholson: *He is a faithful and active supporter of Russian policy; and his Serbian policy is as far as Russian interests are concerned, distinctly a success.* Nikolai Hartwig, like many Russian emissaries, may have run ahead of his instructions, but there is no doubt that St Petersburg was up to the neck in this business.

By the spring of 1912, Ambassadors Nikolai Hartwig and Anatoli Nedlyudov had formed the defensive Balkan League alliance of Serbia, Bulgaria, Montenegro, and Greece to supposedly put the diplomacy of these Balkan States under the closer control of the Russian Foreign Office. Although Hartwig and his military attaché Colonel Viktor Artamonov expected to see the Balkan League focus on confronting Austria, the leaders of Serbia and Bulgaria were motivated by different desires. Tsar Ferdinand of Bulgaria moved to expel Turkey from Macedonia and Thrace, and the Serbians sought sufficient coastline for a port on the Adriatic Sea. Urged on by Hartwig's unofficial Russian diplomacy that represented itself as genuine actual diplomacy, the Serbs surprised him when they went forward into conflict, not as expected against A-H, but to wage war against the Ottoman Empire. *To rectify the historical anachronism of life under Turkish barbarism on European soil in the 20th century and the liberation of our fellow Southern Slav nationals from terror by their unification.* Among the militarists and royal courts of Greece, Bulgaria and Serbia there was no compromise. Tsar Ferdinand of Bulgaria concluded a Treaty with Greece, proclaimed her adhesion to the cause, and the Balkan Alliance was complete by March 1912. The Russians, who had plotted and planned to lead this alliance to their own advantage, now became exceedingly uneasy of its consequences. Russia was still unprepared for war, and when they heard that Tsar Ferdinand of Bulgaria aspired to the possession of Constantinople they could do no more than hope to defer his offensive with a few diplomatic discussions.

Nikolai Hartwig was a friend of social events; he loved good dinners and willingly made apparent his admiration for Vienna, where he acquired his entire wardrobe. He often said to me ~ One can live only in Vienna. He was a true and faithful friend of the Serbians; he was just as sincere and ardent a Russian patriot; but his vehement and domineering character, his intolerance of any opinion differing from his own, prevented him from forming an equitable judgment on men and matters, and from seeing the terrible dangers accumulating on the horizon.

Baron Wladimir Giesl von Gieslingen
A-H Ambassador to Montenegro 1912

As it became necessary for the Russian Foreign Office to ascertain : if in the future, a war between Serbia and Austria broke out ; if Germany should help Austria, and if Russia should help Serbia : whether Russia could count on the support of both France and Great Britain? Russia received this assurance (in secret) when Premier Raymond Poincaré of France visited St. Petersburg in August 1912. Before this, Raymond Poincaré protested to Sazonoff when he read the recently signed Balkan Treaty, *This convention in no way corresponds to the account of it that was given me. It is a convention for war. Moreover, the treaty contains the germ not only of a war against Turkey, but also of a war against Austria.* After this disturbance was settled, a secret naval agreement between France and Russia was made, whereby France agreed to concentrate her naval forces in the Eastern Mediterranean to support the Russian Navy in the Black Sea. This agreement was confirmed when the French IIIrd Naval Squadron was permanently transferred from the Atlantic to the Mediterranean Sea. Poincaré told Isvolsky; *This decision has been made in agreement with England, and forms the further development and completion of the arrangement already made previously between the French and British Staffs.* (Referring to the conference of Asquith, Churchill, and Kitchener at Malta one month before, where the new naval dispositions of the British and French fleets were there decided).

No. 542.

T. Russell to Sir Edward Grey

F.O. 2880/2467/12/3
(No. 8) Very Confidential — Vienna, D. January 18 1912
R. January 22 1912
Sir,

With reference to my telegram No. 7 of yesterday, I have the honour to report that Count Aehrenthal's numerous enemies are closing in around him and that his resignation of office now only depends on the already enfeebled determination of the Emperor to retain his services. It is being reported that His Imperial Majesty will be unable much longer to resist the importunities of the Heir Apparent who declared at the time of the dismissal of General Conrad von Hotzendorf that he would give his uncle no rest as long as Count Aehrenthal controlled the foreign policy of the Dual Monarchy. The Emperor who, a short time ago, resented any sort of suggestions on the part of his nephew, has now ~ I learn privately ~ reluctantly admitted that as his own end could not be far distant, it was perhaps right that the future occupant of the Throne should be permitted to assume a larger share in the administration of the State. My informant told me that at the time of the dismissal of General Schonaich, the Minister of War, last summer the venerable Monarch, in reply to a protest from one of his advisers, remarked that it was easier to get rid of a War Minister than it was of an Heir Apparent. The crimes of Count Aehrenthal in the eyes of the Archduke Franz Ferdinand are two-fold : firstly, a lack of cordiality towards Germany, and secondly, an excess of cordiality towards Italy. With regard to the first I hear very confidentially that the Emperor Wilhelm II wrote to His Imperial Highness in December to say that though Imperial Germany had not always had reason to be thoroughly satisfied with the attitude of Count Aehrenthal, he felt bound to admit that His Excellency was at the present moment the only possible Austro-Hungarian Minister for Foreign Affairs. On my remarking that the Archduke's admiration for the German Emperor would have led one to suppose that, this letter would have acted in a favourable manner for Count Aehrenthal, I was assured that his character was not one to brook the slightest interference even on the part of his exalted and much admired friend ~ nay more, that this interference tended to intensify his desire to see a prompt change at the Ministry for Foreign Affairs.

With regard to Count Aehrenthal's second delinquency; namely a correct attitude towards Italy in conformity with the letter and spirit of the alliance existing between the two countries, there is now but little doubt that this attitude has thwarted an ardent desire on the part of the military party, led by the Archduke to push matters to a head with Italy while a large portion of her army was shut up in Tripoli. Though you, Sir, are aware and it has often been reported from this Embassy ~ that very unfriendly feelings exist between Austrians and Italians, it is only since the Italian seizure of Tripoli that a belief has been generally and openly expressed that actual hostilities are inevitable sooner or later, and that Albania will provide a fitting scene for the struggle.

Count Aehrenthal, whose policy it has been to improve relations with the junior member in the hope of thus diminishing Austria-Hungary's dependence on the predominant member of the Triple Alliance, can naturally not co-operate with War Ministers and other leading military men who share the views of the Heir Apparent. He may moreover be reluctant to see the worries of the aged Emperor increased on his account. For these reasons, he will probably in the near future place his resignation in the hands of his Imperial Master, pleading the state of his health as an excuse for laying down the burdens of office. Should this prove to be the case, the history of the future will, I think, decide that the Heir-Apparent has rendered his country a great disservice in ridding her of a Minister who, with all the mistakes he has committed, was the one strong man capable of raising the House of Hapsburg from a state of thraldom to a position of comparative independence. *I have, &c.* THEO RUSSELL

MINUTES. *Extremely interesting. We may assume from this that Count Aehrenthal's early resignation is not merely a probability, but a certainty; and that his successor will be anti-Italian and pro-German. From the British point of view this is deplorable; but in any case, on the Emperor's death, which cannot be long deferred, the anti-Italian and pro-German sentiments of the Archduke Franz Ferdinand would have found expression. In view of the objection of the Archduke to any interference with his plans, it is not by any means certain that Austria on his succession will become submissive to German policy. In fact, the Archduke's character and the methods of German diplomacy are not unlikely to clash, and to lead to friction.*

No. 545

C. Barclay to Sir Edward Grey

F.O.5027/5027/12/39.
(No. 6) Belgrade, D. February 1 1912
Sir, R. February 5 1912

In his despatch No. 61 of September 27 last, Sir R. Paget reported upon the violent attacks of the press against the Crown and the Government, and the feeling of unrest attributed to conspirator intrigues.

It would appear that the younger officers who took part in the murder of King Alexander and Queen Draga have for some time past been discontented with their treatment at the hands of the Court and also with the weak policy of the Radical Government. They realize that the Karageorgevitchs are no better than the Obrenovitchs, that neither the King nor the Crown Prince take any interest in military affairs, that the army does not progress like that for instance of Bulgaria in spite of the millions spent on it, that the country is ruled by men whose only object is to enrich themselves at the country's expense. They reproach the King and the Government with weakness and their object is to turn the Radicals out of power and to replace them by a strong Government, which would pursue a policy more in accordance with Pan-Servian aspirations. With that object in view, they have formed a secret society under the name of ~ Union or Death ~ better known as the Black Hand.

This society has of late been rapidly gaining ground until it now numbers among its members not only officers of all ranks, but also civilians of every stations of life. High officials are members and Minister of War, General Stepan Stepanovitch, is supposed to be at the head of the movement, with practically the whole of the Belgrade garrison as well as the Police at his back.

Such is the influence of the society that some of those most directly aimed at, such as the King, the Crown Prince and the Prime Minister, have it is rumoured, thought it expedient to join the Black Hand, whose overt intentions are the pursuit of an active policy to unite the scattered members of the Serb race under one Flag. The present attitude of the King would seem confirm this rumour.

To complicate matters, another society has been formed, named the Servian Union. Its promoter, General Sreckovic, an officer on the retired list, issued last week a manifesto to the Servian people, calling their attention to the rotten state of the army and urging all Reserve officers to become members of the Union, with the object of opposing the process of disorganization now ruling in the army, and of furthering the Pan-Serb movement.

Although I have tried several well-informed sources, it is impossible to obtain satisfactory information as to the actual scope of the movements. But one thing is certain and that is that at least one powerful organisation headed by the Military does exist, and that its influence will have to be reckoned with as an important factor in the attitude of Servia in case of complications arising in the Balkans.

All the Foreign Representatives, with whom I have spoken on the subject, are of opinion that the movement is, or will be, serious. *I have, &c.*

COLVILLE BARCLAY

Whilst clear that originally the Black Hand was on excellent terms with the Serbian Government, which used it as a secret instrument for the promotion of national ambitions, this harmony, would not be of much longer duration.

At the demand of Hartwig, PM Pasic was ready to accept unconditionally the arbitration of the Tsar. The Black Hand had received information from St. Petersburg that the Tsar would decide in favour of Bulgaria, and they were not strong enough to force the resignation of Pasic. Hartwig prevailed upon the King not to accept the resignation, and the premier was about to announce in the Skupstina that Serbia would agree to Russian arbitration when the news arrived of the Bulgarian attack on the Serbian Army. This prevented an open rupture between Pasic's Radical Party and the Black Hand, but henceforth Hartwig is a bitter enemy of the secret society which he perceives is not ready to obey Russia. As Hartwig's influence is enormous with both the King and the Crown Prince, the Radical leader was able to reconstitute his Cabinet and make no concessions. King Petar soon after abdicated, alleging the state of his health, and now Crown Prince Alexander has assumed the Regency. The government, sure of the support of the Crown, dissolved parliament on 24th June. Baron Giesl the A-H Ambassador expressed the wish that Pasic should triumph over his adversaries, for this will facilitate the pending negotiations on the Oriental Railway. Ugron to Berchtold, 24th June 1913

No. 556

Sir C. Barclay to Sir Edward Grey

F.O. 9375/9375/12/39.
(No. 11.) Belgrade, D. February 29 1912
Sir, R. March 4 1912

The death of Count Alois d'Aehrenthal elicited considerable comment in the Belgrade Press; but considering its low level it behaved with surprising correctness, only two of the numerous rags inveighing against the late statesman in abusive language. The general view adopted of political consequences of the death of His Excellency is that it will change nothing in Austria's attitude towards Servia. Occasion is also taken to comment on Austrian and Russian policy as affecting this country, and in view of the almost coincident improvement in the relations between the two Empires, it may be of interest to quote some of the remarks made in this connection.

The Pravda ~ organ of the Progressists, says : Most of the years spent abroad by Count d'Aehrenthal were spent in Russia and it was the thorough knowledge which he there acquired of Russia's worth that enabled him to proclaim the annexation of Bosnia. We Serbs, during the crisis that followed, consoled ourselves for the loss of Bosnia with the idea that Russia, owing to the damages caused by the Japanese war, was not ready to take action against Austria, otherwise that she would not have allowed the annexation to take place. However, this is not true, for it becomes more and more evident every day that Russia would never have drawn the sword for Bosnia even had she been prepared. Russia had already ceded by special secret treaties those two Serb provinces to Austria. Today again Russia is too much engaged in Asia and she will take good care not to come up against any Power in Europe. On the contrary, she is endeavouring to come to terms with Austria to have her hands free. The only reason why Russia was dissatisfied with the annexation of the provinces was simply that she herself obtained nothing in return. On Count Berchtold's appointment, it is not chance that makes the Emperor choose a man whose knowledge of Russia and popularity in St. Petersburg are greater even than those of his predecessor, it is because Austria in all her endeavours to advance in the Balkans finds herself confronted by Russia, and it is therefore essential that the man responsible for her policy should have a thorough knowledge of Russian methods. For us Serbs, Count Berchtold is as dangerous as Count d'Aehrenthal. He is even more dangerous, for he is younger and abler, and has a larger circle of influential friends. Berchtold is a rejuvenated Aehrenthal and the consequence will be a more active policy on the part of Austria in the Balkans.

The Srpscka Zastava ~ organ of the Liberal party says : Count Berchtold will endeavour to bring about an improvement in Austrian and Russian relations, a necessity which Count d'Aehrenthal had already seen. But this desire will not change the foreign policy of Austria which will remain imbued with imperialistic tendencies. What Count Berchtold will do will be to give more scope to that policy by coming to terms with Russia and thus having his hands freer. Serbs cannot therefore hope for any good to come out of an Austro-Russian rapprochement. Russia is ever ready to come to terms with Austria. In Vienna, they know that full well and that is the reason they always laugh at us.

The Politika ~ endeavouring to console Servian public opinion says : Not much harm will come to Serb races out of an Austro-Russian rapprochement. According to its usual practice, the Vienna press have suddenly changed their attitude towards Russia, they think they will beguile Russia, but in this they are mistaken for Russia cannot forget everything. The first thing- that she can never forget is that Austria can never be trusted. That is a sufficient reason why events will not be altogether as Austria wishes in the Balkans. These remarks of the Politika are worth noting, for that paper, besides being the semi-official organ is also in the pay of the Russian Legation. They are evidently intended to convey that Russia will not throw over Servia.

Confidential. – *I have endeavoured to ascertain the feeling in official circles and I find that Pasic is optimistic, being confident that Russia will not abandon Servia, seeming to hint that they have assurances on the point. But what the scope of these assurances may be, or whether they come merely from Hartwig or St. Petersburg itself, and in this there is a considerable difference, no one seems able to tell. Spalaikovitch, the Servian Ambassador in Sofia was at the beck and call of Hartwig when he was here as Secretary General at the Ministry of Foreign Affairs, and is still in constant touch with him. He has been going backward and forward frequently during the past few months between Sofia and Belgrade and every time he has had long conferences at the Russian Legation, when Hartwig no doubt pressed his views, which he has for a long time been advocating, as to the desirability of a rapprochement. The facts are probably that the Bulgarian Government do not discourage the advances of the Servian Government, although they have no intention of coming to any arrangement, and on the other hand that Spalaikovitch, who though an able man is an enthusiast rather apt to be carried away by his convictions, may be inclined to adopt too sanguine a view and to take wishes for facts.* COLVILLE BARCLAY

MINUTE. *It is a curious estimate which makes Count Berchtold abler than his predecessor.*

No. 577.

Sir C. Barclay to Sir Edward Grey.

F.O. 26739/26739/12/7.
(No. 58.) Very Confidential — Sofia, D. June 18 1912
Sir, — R. June 24 1912

Daneff has returned from his special Mission to Livadia and subsequent visits to Moscow and St. Petersburg where he had the occasion to meet a number of leading political men in Russia. Knowing that the Servian Minister was on intimate terms with Daneff and that he had seen him yesterday, I took an opportunity of asking Spalaikovitch this morning what impressions His Excellency had brought back from his visit to Russia.

It appears that Daneff said that he had found great determination on the part of all political men with whom he had come into contact to work for the settlement of the Balkan Question. He had gathered that Russia was better prepared now than ever before, though there was still room for improvement, but that she realised that the moment was not propitious for any definite action. It was not due, however, to inadequateness of military preparations that Russia held back, but to the yet incomplete formation of the European constellation such as she desired it to be.

Daneff said that from the Russian standpoint England was far too preoccupied in keeping on good terms with the Turks with a view to gaining the pre-eminence in Constantinople : France had too many financial interests at stake in Turkey to desire the disturbance of the status quo ; Italy was too unreliable and, they feared, undetachable from the Triple Alliance ; with Germany it remained to be ascertained whether she would always be prepared to support Austria as she had done in 1908 ; as to Austria, Russia felt that any attempt at coming to terms with her would be hopeless.

It would seem from Daneff's statements that the sole aim of Russia is directed at present to bringing about a state of affairs which will enable her to impose her will, when she is of opinion that the time has come to put to Europe the question of the solution of the Balkan Problem.

With this object in view, Spalaikovitch said; Russia has successfully induced Bulgaria and Servia into contracting an alliance, to be joined ultimately by Greece and Montenegro, as a barrier to an Austrian advance. It would only be required for Russia to persuade Great Britain and France to share her views as to the desirability of driving the Turks out of Europe, leaving them with Constantinople.

To my remarking that I failed to grasp what advantages Russia and especially England and France would reap from such a policy, which would probably cause a European war and spell ruin to the vast interests that both the latter Powers have at stake in the Ottoman Dominions, Spalaikovitch replied that it would only mean a change of names on the map and that those interests would subsequently benefit thereby.

Besides, he went on, victory would mean the crushing of Germany, the recovery of Alsace Loraine to France, the saving of millions a year in shipbuilding to Great Britain, in fact, the beginning of an era of peace to Europe! But, Spalaikovitch added, a European war was not a necessity; Russia's real desire was to 'keep the ring' for the Balkan States to fight out their own battle with Turkey without any interference from outside, and her aim was directed to being in a position to achieve the object with the support of the Western Powers. Much also would depend on the extent of Germany's support of Austria, which Russia seemed to believe would not be unlimited.

Russia was increasing her armaments : not from fear of aggression, for she knew she would not be attacked, but with a view to pursuing an active policy.

Of course, I am not in a position to judge how far the above views affect Russia's policy in the Near East. But I am reporting this conversation as indicative of the strong sentiments which animate these small States to the extent of deluding them into believing that the Great Powers would deliberately run the risk of a general conflagration merely to bring about the realisation of Balkan aspirations.

I have, &c.

COLVILLE BARCLAY

MINUTE. *Balkan statesmen are rarely able to see anything except the Balkans.*

In September 1912, Poincaré gave the further assurance to Foreign Minister Serge Sazonoff, that if Germany supported Austria-Hungary in any struggle in the Balkans and if Russia were drawn in; *France would not hesitate for a moment to fulfill its obligations towards Russia.* That same month, Isvolsky had an interview with the British Foreign Minister Sir Edward Grey, in which Sir Edward was said to assure him of the fullest British co-operation in the same event. Isvolsky reported to St. Petersburg. *Grey, upon his own initiative, corroborated what I already knew from Poincaré ~ the existence of an agreement between France and Great Britain, according to which England undertook, in case of a war with Germany, not only to come to the immediate assistance of France on the sea, but also on the Continent, by landing troops.* These understandings between Poincaré and Sazonoff, Isvolsky and Sir Edward Grey were top secret. And so assured, Sazonoff urged the ambitious Serbs ~ *Play a waiting game, to deliver a blow at Austria.* Then on 8th October 1912, King Nikola of Montenegro declared war and attacked the Turks.

It is doubtful whether the Balkan League could have been formed except for the very active part taken by Hartwig and Neklyudov at Belgrade and Sofia. The Serbians were greatly elated by their conquests from the Balkan Wars, which doubled their territory and seemed to foreshadow the possibility of the early realization of their Greater Serbia ambitions at Austria's expense. They were actively encouraged by Hartwig at Belgrade. So extreme was Hartwig in the promotion of Greater Serbia that Sazonoff was obliged to distance himself by noting that ~ Hartwig has great sympathy for the Slav cause, is of a passionate character, and lets himself be carried away by his Slavophil sympathies.

Sidney Bradshaw Fay
Professor of History
Harvard & Yale

In November 1912, Serbian forces overran Kosovo, and the Serbian IIIrd Army seized the Adriatic port of Durazzo giving the Serbs an outlet to the sea and establishing a presence in central Albania. The Italians objected to the Serbian presence on the Straits of Otranto and the Serbian appearance on the Adriatic sorely antagonized Vienna who foresaw an Adriatic port becoming a Russian naval base. To prevent this, Vienna moved to create an Albanian state. *The territorial boundaries of this new state will be whatever is proper to insure the interests of its vitality and stability and should include if possible all the Albanians living on Turkish soil within its natural borders.* In particular, the Viennese leadership intended the new Albania to include all areas of Serbian and Montenegrin aspiration, such as the strategic towns of Scutari, Prizren, and Djakova, to prevent a Serbian presence on the Adriatic. In pursuing this policy, the Austrians were prepared to use force against the rogue Slavic states, and all A-H forces in Bosnia, Herzegovina, and on the Dalmatia Coast were placed on a war footing. A warning telegram was sent to Belgrade, which resulted in the evacuation of Scutari and other strategic parts of Albania.

The French Ambassador in St Petersburg Georges Louis reported to Paris a remark of Hartwig's on the Balkan victories. *The affair of Turkey is settled. Now it is the turn of Austria.* Hartwig continued to play a leading role in promoting conflict between A-H and Serbia and act as the link between Russia and the Greater Serbia group. On 12th November, he told Romanian Foreign Minister George Filaliti. *Russia counts on making Serbia, enlarged by the Balkan provinces of A-H the new vanguard of Pan-Slavism.* With Hartwig's active encouragement for Serbia to confront Austria-Hungary, he extended the protection of Russia and therefore, and when Nikola Pasic later visited Russia to suggest a marriage between the Serbian Crown Prince and Tsar Nicholas' eldest daughter, Pasic also requested;
120,000 rifles and ammunition, and some cannon: especially howitzers.

The A-H Army again took military measures in November of 1912 in Galicia and in Bosnia Herzegovina because of another Serbian incursion into Albania. Foreign Minister Count Berchtold explained to his subordinate; *I could easily provoke a war in twenty-four hours, but I do not want to do that.* Vienna considered going to war on both occasions, but her military threats succeeded in compelling the Balkan states to yield and the Serbian Army withdrew from all territory then assigned to Albania. The A-H Ambassador to Montenegro, Baron Giesl von Gieslingen was involved in both actions. In December 1912, Giesl received special instructions from his friend, the Archduke Franz Ferdinand to uphold the integrity of Albania, and was sent to the London Conference of Ambassadors to assist the A-H Ambassador to Great Britain Count Albert von Mensdorff. Beginning on 17th December 1912, the ambassadors met with Sir Edward Grey to oversee the Balkan peace process. They decided on the disposition of the Ægean islands and the determination of a new Thracian border between Bulgaria and the Ottoman Turkish Empire, while A-H directed the independence of Albania. Sir Edward Grey advised Russia; *The question of a Serbian port on the Adriatic Sea is not worth a European war.*

From the Albanian coastal fortress of Scutari overlooking the port of the same name on the Aegean Sea.

During the First Balkan War, Scutari, the most important town in the area fell into the hands of the Montenegrins, who at first absolutely declined to leave it. The policy taken up by King Nikola was in entire opposition to the outlook of Vienna and for a time it even seemed destined to lead to war. Then Germany and Britain joined cordially in the part of honest brokers to address certain fixed points. Austria was determined that neither Serbia nor Montenegro should set foot within the boundaries of the Turkish province of Albania.

However, the King of Montenegro's persistence in occupying Scutari raised the question of how to get him out by any measures to which both Austria and Russia could consent. A blend of threat of coercion and the offer of compensation settled the matter to the satisfaction of A-H and to the satisfaction of the King of Montenegro, and this danger to European peace was laid to rest. The current of European affairs was setting towards war. Austria and Russia were drifting with it, and dragging the other Powers in the same fatal direction. In agreeing to a Conference in London during 1912-13, it was as if we put out anchors to prevent ourselves from being swept away. Then the current seemed to slacken and the anchors were pulled up. The Conference was dissolved. We seemed to be safe. In reality it was not so; the set of the current was the same, and in a year's time we were all swept into the cataract of war.

At this time, Germany received a warning from King George V himself. In December of 1912, when the Kaiser's brother, Prince Henry of Prussia visited London, Prince Henry asked, *In case of a war between Germany and Austria-Hungary on one side and Russia and France on the other, would Britain come to the assistance of the two latter Powers.* King George V replied; *Undoubtedly yes, under certain circumstances.* Prince Henry, without asking what the circumstances would be, expressed his surprise and regret. The King wrote to Grey, *Of course, Germany must know that we could not allow either of our friends to be crippled.* Grey found this declaration fortunate. *While Great Britain is not committed in the event of war, and public opinion is very averse to a war arising out of a quarrel about Serbia, nevertheless, in the event that Austria-Hungary were to attack Serbia aggressively, or Russia in assisting Serbia were to be attacked by Germany, or France were finally to be drawn into the conflict, it might become necessary for England to fight for the defense of her position in Europe and for the protection of her own future and security.* To Berlin's view, such consolidation of Balkan policy in the hands of the Russian Foreign Office signified their increased responsibility to A-H in case of a quarrel, as it now had to confront an entire coalition of the Balkan States, instead of just a single one. Imperial Russia, which had been previously humiliated because of its inability to support Serbia during both the Bosnian Crisis and the First Balkan War, announced a forthcoming reconstruction of its military forces.

Uniforms of the Ottoman Turkish Army *Uniforms of the Balkan Federation: Greece, Serbia, Bulgaria & Montenegro*

As organized by Hartwig and Isvolsky, the Balkan Federation made its début in S.E. Europe at the gates of the A-H Empire, and with a trail of gunpowder then laid from Belgrade to St. Petersburg, any spark in the Balkans could cause an explosive flash thru the entire length of Europe. As Isvolsky wrote to Sazonoff on 23rd October 1912. *Although a victory of the Balkan states over the Turks was unlikely, it would bring forward in full historical development the question of the struggle of Slavdom not only with Islam but also with Germanism. In this event, one must prepare for a great and decisive general European war.*

When Baron Wladimir von Giesl returned to Montenegro, he was widely blamed for outcome of the London Conference that had denied all Montenegrin demands; particularly that A-H had established an independent Albania on the Adriatic coast, thus cutting Serbia off from the Adriatic and robbing Montenegro of her cherished hope of securing Scutari. His position became so unviable that he could no longer remain at his post, so Baron Giesl was reappointed to Belgrade to face the less intense Serbian resentment over the October ultimatum. Giesl: *The officer corps represents the stronghold of the Greater Serbian, Austrophobe direction. No political party here supports the Hapsburg Empire. The authoritative factor in Serbia, the army, is filled with South Slavic chauvinism and hatred against A-H and will force upon the policy of any regime a national chauvinistic and Austrophobe direction.*

On 29th June, conflict renewed in the Second Balkan War when Bulgaria fought a loose coalition of Greece, Montenegro, Serbia, Romania, and the Ottoman Empire. Five weeks later, Bulgaria capitulated, and treaties were signed in Bucharest and Constantinople in September 1913 to conclude the conflict. The Balkan Wars of 1912–13 were the first concerted effort by the Balkan peoples to emulate the Italian and German patriots to establish their own analogous and large nationalist states. The Great Powers of Europe intervened to resolve the Balkan Wars and reconfigure the borders of the Balkan Peninsula. From previous Turkish lands, the new Kingdom of Bulgaria obtained the Rhodope Mountains and western Thrace. Greece and Serbia divided most of Macedonia. Greece also gained most of Epius and was also left in possession of three Ægean islands seized from Turkey : Chios and Mytilene dominating the entrance to Smyrna on the Turkish mainland, and Lemnos near the Straits of the Dardanelles. Serbia divided the Sandjak of Novi Pazar with Montenegro, and acquired Kosovo. An independent Albania emerged, and the Ottoman Empire all but disappeared from Europe.

Edith Durham wrote; *December 27th 1912 ~ The Serb Ambassador in St Petersburg telegraphs: The Minister for Foreign Affairs replied that in view of our successes he had confidence in our strength, and believes we can give a shock to Austria. For that reason, we should feel satisfied with what we are to receive and consider it as a temporary halting place. Small wonder, that in May 1913 that the Montenegrins boasted to me: We, the Serb nation, are a danger to Europe. We have the Russian Army with us, and shall take what we choose.*

Russian public opinion demanded that Russian diplomacy intervene actively in the interests of Serbia. But Russian diplomacy did not heed this call, and gave no support to Serbia's claims on the Adriatic coast. It might seem that in the light of this undoubted fact, the Serbian régime had the right to develop its own policy. This Nikola Pasic did, seeking to rebuild the diplomatic bridge across the Danube. But then Russia voiced its dissatisfaction in Belgrade, and protested constantly in a most determined way, against any concessions being made to Berchtold. To be sure, our government and Russian diplomacy made a big mistake there. Leon Trotsky

The failure of the Balkan League followed this inconsistent attitude of the Russian Government. The Russian representatives in Belgrade and Sofia continued to give conflicting advice, and as a result, St Petersburg lost much of their influence in Bulgaria when pervasive Austro-Hungarian and Italian interest in the security of the Adriatic Sea motivated them to create an independent Albanian state to thwart all Serbian and Russian ambitions of Adriatic access. The Russians then found themselves still involved in the less strategic position with Serbia, and followed Belgrade into the catastrophe of 1914.

The antagonism between Vienna and Belgrade grew steadily more acute, because each had vital interests at stake: Austria to preserve her very existence ~ Serbia to satisfy her ideals of national unity. Dangerous and reckless territorial ambitions took a stronger hold of most Serbians. Pasic, an obsessed politician with the skill of a horse trader, once reflected in a remark to the Greek envoy Nik Politis at the 10th August Bucharest Peace Conference as they divided up the spoils of the Second Balkan War. *The first round is won; now we must prepare the second against Austria.* Serbia having more than doubled in size, now threatened A-H both politically, and also militarily. Sazonoff wrote to Nikolai Hartwig: *Serbia has only gone through the first stage of her historic road and for the attainment of her goal must still endure a terrible struggle in which her whole existence may be at stake.* Russian FM Sazonoff re-confirmed: *Serbian expansion into A-H, Serbia will have to wage the inevitable future struggle*

From this moment, Serbia, assured of the protection of Imperial Russia, which had been put to the test during two Balkan wars, now abandoned every reserve and openly plunged into a campaign of defiant provocation against the neighbouring Austro-Hungarian Empire for the creation of a Greater Serbia upon the ruins of a disintegrated A-H. This concerned even the most optimistic minds in Vienna. The gravest repercussion of the Balkan conflicts was indirect. Germany, anxious of further Balkan attacks on the Ottoman Empire that might lead to its final downfall, drew close to Austria and dispatched General Liman von Sanders to Constantinople to re-organize the Ottoman Turkish Army. This move infuriated Russia. If German military power supported the Turkish régime and established itself astride the Straits to restrict Russian ambitions, it assumed the potential to fracture her security. Sazonoff to Russian War Minister Soukhomlinoff on 12th November 1913 : *Conquering Constantinople would give Russia a global position, the natural crown of her efforts and sacrifices over two centuries of history. Such a triumph, might also bring to our internal life ~ government and society; achievements and enthusiasm which could unite them in the service of a matter of indisputable pan-national importance.*

On 4th April 1913, Belgian Ambassador to Berlin Baron Napoléon Eugéne Beyens reported that the arrogance and contempt with which the Serbians received Vienna's protests over various issues could only be explained by the belief that St Petersburg would support them. Baron Beyens observed;

The Serbian Government would not have persisted in its course for the last six months in the face of the Austrian opposition unless it had not been backed up and incited by Hartwig, the Russian diplomatist of the Isvolsky school. On the Kaiser and his diplomats, Baron Beyens said; *There can be no doubt that the Kaiser, the Chancellor, and the Secretary of Foreign Affairs are ardent admirers of peace. Whatever may be the plans of Kiderlen-Wächter, it is certain that they firmly intend to avoid a European conflagration.*

In 1912-13, the current of European affairs was setting towards war. Austria and Russia were now drifting with it and dragging the other Powers in the same fateful direction. If the crisis passed in 1913, it was because Germany restrained Austria, while Russia kept Serbia in check, and the London Conference offered the means of stopping the drift to war. Confusion and agitation continued in Vienna, with different cliques of policy fighting with each other within the Ballplatz, and reducing Foreign Minister von Berchtold to a state of indecision and distraction. Berchtold wavered between the two themes of Conrad; for peace one moment and for war the next ~ Slavophil in the morning, and breathing fire and slaughter in the evening. It was intimated to Vienna that the Germans would not fight to keep the Serbs out of Durazzo, and to Belgrade that the Russians would not fight to put them there. But when the Serbs flouted the London decisions of the Powers by again crossing into Albania, burning Albanian villages, and massacring Albanians, Count Leopold von Berchtold issued a very sharp note and brought upon Serbia a warning from all the powers, that she must respect their decisions. J.A. Spender

CAPTAIN VOJISLAV TANKOSIC

Captain Vojislav Tankosic, as a member of *Young Bosnia* and co-founder of the Serbian *Black Hand,* had helped to write the constitution of the *Black Hand* whose aim was the unification of all Serbdom. Before the outbreak of the First Balkan War in March 1912, he was tasked with training the volunteers who arrived from all Serb-inhabited regions. Tankosic was very strict in his selection, and out of 2,000 candidates, he chose only 245 volunteers. In the First Balkan War, he commanded the Laplje Chetnik detachment that began operations in the Turkish rear two days before the actual outbreak of the war, at the border post at Merdar, where Chetniks and Albanians skirmished in the first battle of the war.

Tankosic's Chetniks fought for three days by themselves, until regular Serbian Army forces arrived to secure victory, which then enabled Tankosic's Chetniks become the first Serbian forces to enter Pristina. As Tankosic's initiative had opened the door for the Serbian Army to liberate all of Kosovo, he was awarded the Order of the Star of Karageorge with Swords and promoted to Major. When Belgrade tried to retire Tankosic and *Apis,* King Petar I disagreed. Conflict between the civilian and military authorities simmered thru 1913, and escalated with open threats to particular ministers.

During the winter of 1913-14 I gathered funds for Albania, and the American missionaries worked hard at feeding the refugees of Gostivar and Dibra. General Phillips, in command at Scutari, did all his funds would allow for the refugees there, but reported that the Serbs' victims were dying of hunger in the Gashi mountains at the rate of twenty a day. But the Mansion House refused to start a fund. Mr. Willard Howard took cinema photographs of the starving people in their burnt ruins, hoping to rouse public feeling against the Serbs and stop their further war plans. At the Foreign Office I begged protection for the Balkan Moslems, who were being barbarously exterminated, and stated that until it was seen that the Balkan conquerors were capable of just rule, the Capitulations should remain in force. Those with whom I spoke admitted that the consular reports from Uskub and Monastir were very bad, but that it was not advisable to publish them. In truth, we were hopelessly tied to Russia and could say nothing about her pet lambs, even though the truth of the accusations had been proved up to the hilt by the Carnegie Report.

Edith Durham

Italian FM, the Marquis Antonio di San Giuliano conducted a policy of friendship toward France, while remaining faithful to Italy's commitments to A-H and Imperial Germany. He served as Italian Foreign Minister in 1905–06 and during 1910–14. He resisted the expansion of A-H in the Balkans, while supporting the Italian economic penetration of Montenegro, and a fully independent Albania. He advanced Italian national interests by balancing his efforts between the two competing alliances.

Italian Foreign Minister Marquis Antonino di San Giuliano & A-H Foreign Minister Count Leopold von Berchtold

A-H Foreign Minister Count Berchtold focused almost exclusively on Balkan foreign policy aims to preserve peace and maintain the principle of non-intervention, but the Balkan Wars of 1912-13 shattered any hopes for such a peace-based policy. From the outset of the Balkan Wars, Berchtold pursued a hardline policy, reviewed all options for war against Serbia, but declined from the brink. Although he prevented Serbia from securing an outlet to the Adriatic Sea by supporting the creation of an independent Albania, the Balkan Wars failed to contain a rising Russian influence in the Balkans.

LONDON CONFERENCE OF AMBASSADORS : AUGUST 1913

Sir Edward Grey (GB), Count Benkendorff (Russia)

Paul Cambon (France), Prince Lichnowsky (Germany), Count Mensdorff (A-H), Marquis Impériale (Italy)

Fair criticism can be made that British diplomacy lost touch with the A-H Empire when the London Conference of Ambassadors ended in August 1913. The British Foreign Office found itself sharing the common opinion of those days, that Imperial Germany was so much the predominant partner in the Triple Alliance that it was sufficient to address Vienna through Berlin. After all, everyone at the conference had seen Austria following the German lead and all had assumed that Austria was under German control. Therefore, the urge towards war from the Austrian camp, and the growing pressure being applied to Germany by Austrian statesmen and soldiers went on unknown to the rest of the world (and even now this is scarcely realized by most English writers on this period). More accurate intelligence sourcing at this time might have enabled British statesmen to apply more effective pressure against Vienna. But it became the fatality of this time, that the leaders of Vienna laid their perilous plans in almost complete ignorance of immediate British policy, whilst British statesmen continued to conceive of Viennese policy as being both indistinguishable and subordinate to German. Eventually, the reversal of roles has become more apparent.

For example; few knew in the summer of 1913, that Austria-Hungary planned to chastise Serbia. Vienna notified her allies, Germany and Italy of her imminent intention. Vienna was dissuaded by protests from Berlin and Rome, and this influence exerted, prevented Vienna from taking action that might have begun a wider war. The great friendship between Rome and Berlin could not prevent the Cabinet of the Quirinal from remaining neutral. Vienna alluded to this Italian *Walzertour ~ waltz turn, tour de valse* as a step taken without regard to the consequences or a light-hearted escapade. After Italy's first breach of her Alliance contract, Rome would not scruple to resume her freedom of action.

On 9th August, I received from the Marquis di San Giuliano this telegram: Austria has communicated to us and to Germany her intention to act against Serbia and defines such action as defensive hoping to apply the casus foederis of the Triple Alliance, which I consider inapplicable. I intend to join forces with Germany to prevent any such action by Austria but it will be necessary to say clearly that we do not consider such eventual action as defensive and therefore do not believe that the casus foederis exists. I replied: If Austria intervenes against Serbia, casus foederis does not arise. It is an action that she undertakes on her own account since there is no question of defense, as no one thinks of attacking her. It is necessary to make a formal declaration to Austria, and for some German action to dissuade Austria from her most perilous adventure. Giovanni Giolitti

THE SCANDAL OF COLONEL ALFRED REDL

Imperial Russian Okhrana

Black Hand ~ Union or Death

A-H Evidenzbureau

Evidence of espionage prior to the outbreak of war is fragmentary. Most was never written down, much was deliberately destroyed, and some details dwell deep in secret archives. Scrap by scrap, information has accumulated over the last century. One particular incident of extraordinary espionage and deception in the dark ops duel between the *Okhrana* and the A-H *Evidenzbureau* requires attention. It is the scandal of one of the most infamous traitors in history, and his betrayal of Vienna's secrets that contributed to the downfall of the old Austrian Hapsburg Monarchy that ruled A-H for 600 years. In any investigation of Russian intrigues directed against Vienna, one in particular hastened the Russian decision to mobilize ~ the Redl Scandal. The same month that the savage Balkan Wars ended, Vienna suffered the revelation that Colonel Alfred Redl of the Austrian General Staff had been selling critical military secrets to the *Okhrana* ~ the Russian Secret Guards Department. At the time, for two or three days Europe was interested. The affair was a brief sensation ~ then the scandal was forgotten. The Great War had to come and go, the A-H Empire had to fall, and extraordinary militarist societies had to crumble before; one by one, here and there, enough documents have been found and analyzed. Only now can the full story be told.

Alfred Victor Redl, the son of a poor railroad clerk in Lemberg in the eastern Austrian province of Galicia, (now Lvov in Ukraine) at the age of 15, enrolled in a cadet school, and five years later was commissioned a lieutenant in the A-H Army. Ruthlessly ambitious, Redl rose rapidly in the army. At 28 he won a coveted appointment to the War College. Promoted to captain, he was sent to Russia in 1889 to improve his Russian and serve as a spy in the realm regarded as A-H's most potent foe. On his return in 1900, when Baron Artur von Giesl was CO of the A-H secret-service Information Department (*Kundschafts Stelle* or the K.S.) Giesl assigned Redl to the Russian Group of the *Evidenzbureau* (espionage & counter-espionage). Once established, Redl accelerated the age of modern espionage, being first to record conversations on gramophone discs, and to use concealed cameras. Every visiting suspect was photographed in profile and full face without ever realizing it through two paintings on the interview room wall, thru tiny openings for camera lenses, all operated from an adjoining room.

With similar subterfuge, fingerprints were lifted from every visitor to the Viennese *Evidenzbureau*, (without his or her being aware of their contribution). While apparently taking a break to make a telephone call, the interviewer with one hand would point to a closed polished silver cigarette case on the table and say, *Do have a cigarette.* The visitor would take the case and help himself to a cigarette and perhaps a treat from a varnished chocolate box both of which had been treated with minium, and retained the fingerprints (invisible till developed) of those who handled it. The lighter and ashtray, which any smoker was required to make use of, were also prepared in this way. The telephone talk was *faux conversation* of course, for the officer had *phoned himself* by pressing a knob under the table.

If the visitor declined both cigarettes and chocolates, then the *Evidenzbureau* officer would suddenly recollect that a file, lying at the visitor's end of the table, should have been taken to Room #. He would snatch it up, and with an ~ *excuse me for a couple of minutes,* would run out of the room. Another file was then exposed. It was marked *SECRET ~ Restricted Inspection.* If the suspect were inclined to espionage, and few visitors to the *Evidenzbureau* were able to resist a peep into a secret file, they would take in hand the file on the table; the boards of it were also treated with minium. If the visitor, watched from the next room, resisted that temptation, others were tried till one was successful.

At the age of 36, Captain Redl was a soldier with a very bright future ~ but one with a dark secret. There was a private villa on the outskirts of Vienna, where gay gentlemen of means and cross-dressers gathered. Redl often attended such unmentionable activities, and posed for cross-dresser photographs (which he kept under lock and key). Sybaritic living, expensive gifts to friends, and his foible for luxury forced him to live beyond his means ~ soon he was way over his head in debt. He managed to hide his secret vices from his colleagues and superiors in Vienna, but Colonel Nikolai Gumiljow of the *Okhrana* detachment in Poland, searching for useful information on the private lives and personal eccentricities of the A-H General Staff GHQ was rewarded in 1902 when one of his agents discovered Redl's secret. The Russians continued to collect evidence on Redl and eventually assessed him as; *More sly and false than intelligent and talented ~ a cynic who enjoys dissipation*. The *Okhrana* detachment based in Warsaw and next under the energetic and successful leadership of Colonel Nikolai Batjushin, quietly waited and accumulated more information to blackmail Redl into spying for Russia.

In late 1903, a Russian agent provided Redl with a proposition, and an ultimatum. Colonel Nikolai Batjushin wanted particular secret documents from the classified files of the A-H General Staff GHQ and he was willing to pay well. If Redl refused, convincing proof of his perversions would be sent to the A-H Army GHQ. Forced to choose between treason and certain ruin, Redl chose treason. After the *Okhrana* succeeded in enlisting Redl from 1903 to 1913, he became Russia's principal spy. Once Redl provided the *Okhrana* with *Plan III* : the complete A-H Army invasion plan for Serbia, Russia's Military Attaché in Belgrade Colonel Viktor Artamonov forwarded a copy of *Plan III* to Serbian Army HQ, and thenceforth *Apis* and the *Black Hand* became aware of the threat of *Plan III.*

Redl also provided reports on army deployments, pertinent data on the great Austrian fortress of Przemysl in Galicia, plus blueprints and evaluation reports of new weapons. He not only gave away A-H's secrets and plans; he also supplied incorrect estimations of Russian military strength to Vienna. It was a simple task for an officer in his position to extract classified documents from the files, photograph them, and then return them to GHQ files. The photos, sealed in innocuous envelopes, were mailed to various spy drops in Paris, Zurich, and Oslo. By return mail, Redl received regular remittances of money, from which he purchased a large apartment and an expensive new Daimler. Redl's sudden affluence came from his faux-inheritance, from a fictional female relative in Geneva.

Nevertheless, Redl was aware that he was living on the edge of a precipice, and the moment his usefulness was over, Colonel Nikolai Batjushin would not hesitate to expose him and turn him over to an A-H Army court-martial. So, casting around for a means to avert his fate, he had an interesting thought. *Surely this Russian spymaster has some secret weakness of vice that he is hiding from his superiors!* Redl assigned agents to investigate Batjushin, and discovered that the *Okhrana* Colonel was living well beyond his visible means. Thousands of rubles vouchered as pay for foreign agents actually went straight into the Russian's pocket. Incognito, Redl arrived in Warsaw to confront Colonel Batjushin with what he had exposed, and to exhibit the proof of embezzlement that his agents had uncovered.
To this, the Russian had to yield. *I suppose this puts an end to our convenient agreement?!*
Redl quietly replied; *Not at all. Except that from now on we deal with each other as equals.*
Patriotism, honour, and loyalty are all empty concepts to us.

As each of the scoundrels still wanted to enjoy life after their own fashion, and to advance their own careers, they sought a mutually agreeable and profitable outcome as duplicitous double agents. From time to time Batjushin would send him the names of Austrians on the payroll of the Russians, and in exchange Redl would send Austrian agents into Russia, only to betray them to the *Okhrana.* They shook hands on this new twist in cynical treachery. Redl knew of an Austrian agent within the Russian Imperial Staff, so he sold him out to be hanged. He also betrayed every Russian officer who had contacted the Austrian *Evidenzbureau*. During the next 12 months, Redl acting on information supplied by Batjushin was able to track down and apprehend 12 spies who worked for the Russians. His superiors were astounded. During the previous five years, only three spies had been caught. Redl's accomplice (an ethnic Serb from Bosnia) *Oberleutnant* Cedomil Jandric, met with Russian military attaché Colonel Mikhail Zankievic in a Russian Orthodox Church in Vienna, and while apparently at devotion, handed him secret documents about the aviation corps and an artillery report. Redl continued to receive regular large sums of money selling secrets to the Russians. Redl was so trusted by Colonel August von Urbanski, that he appointed him Deputy Director of the *Evidenzbureau.*

Alfred Redl ~ 1890

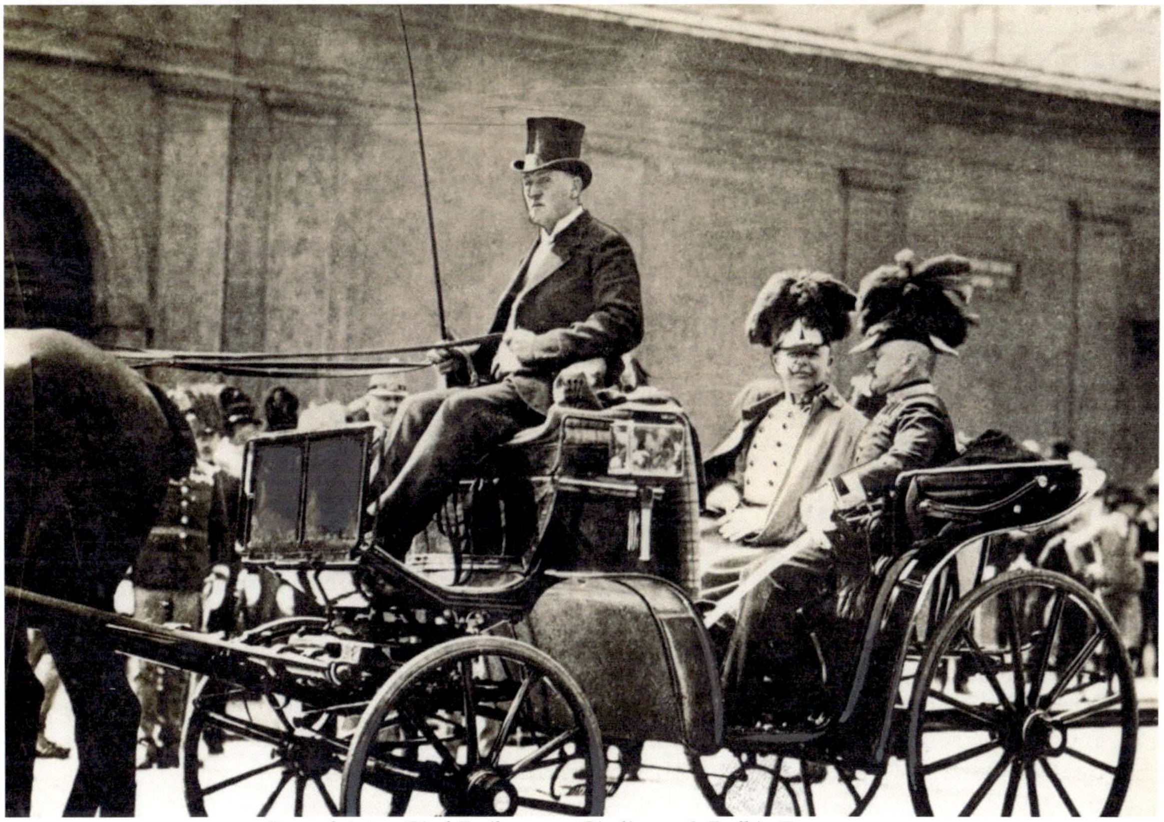

General Artur Giesl Freiherr von Gieslingen & Redl in Prague

So great was the efficiency of the K.S. Evidenzbureau when Major Redl was promoted to the rank of Colonel and eventually became General von Giesl's chief-of-staff, that his very name was an incentive to his successor, Captain Maximilian Ronge and to the whole K.S. staff. Remember Redl! What would Redl think? Were indeed mottoes in the Evidenzbureau, and the department was kept in a high state of efficiency. The legacy which Redl left to the Evidenzbureau was, strangely enough, to be the cause of his undoing eight years later. The new chief of the Evidenzbureau kept up the tradition of success, and was always on the lookout for methods and plans whereby he could out-Redl Redl.

In 1908, when the annexation of Bosnia and Herzegovina by A-H made the European situation uneasy, Captain Ronge and his chief, General August Urbanski von Ostromiecz, (who was General von Giesl's successor of the whole Secret Service) had to be more than ever on the alert, and one of the things Ronge did was to establish a secret postal censorship. Only three persons ~ Urbanski, Ronge, and the officer put in at the head of the Black Bureau, as the censorship department was called ~ knew the real reason for the step of monitoring suspect mail via a secret locked room. The staff, sworn to secrecy, was told that it was mainly for the detection of customs swindles. Thus, the staff paid special attention to letters coming from frontier places! It was by means of this censorship that during the Annexation Period, the Russian military attaché in Vienna, was found to be engaged in espionage. He could not be arrested of course, but at a ball, the Emperor cut him in pointed fashion. The attaché knew from that, that he had been found out, and within a week he had engineered his own recall. An early success for the Black Bureau. *Oberstleutnant* Walter Nikolai, CO *Sektion IIIb* Berlin

So well did Redl appear to do his work, that when Giesl was appointed to the Prague command, (one of the most important in the Dual Monarchy) he insisted on having Redl as his chief-of-staff. Once Redl had relocated to Prague as Intelligence Chief of the VIIIth Army Corps, his clandestine pay was still sent in bulky envelopes, mailed at fixed dates to a postal-box address in Vienna, from the same small East Prussian town of Eydtkuhnen on the Russo-German frontier to provide the packages with a German postmark. Major Max Ronge received a communiqué from *Oberstleutnant* Walter Nikolai of *Sektion IIIb* Prussian Intelligence in Berlin who had just sent him a suspicious intercepted letter. Instead of a recipient's name, there was a cryptic designation; *Opera Ball 12 ~ Herr Nikon Nicetas ~ Care of General Delivery ~ Vienna Post Office.* The letter remained unclaimed in the Vienna Post Office for weeks and was returned to the postmarked city of Berlin.

Early in 1913 a letter, with the poste restante address of Nicetas Vienna, was returned to Berlin, as it had not been called for. It was opened and found to contain Russian money with two addresses in Paris and Geneva. From the contents, it was obvious that this was for purposes of treason. Austrian General Staff was informed, and it arranged for the particular P.O. in Vienna to be kept under observation. *Oberstleutnant* Walter Nikolai

Major Max Ronge set a spy trap in the Vienna Post Office, and waited. (In the interval, Walter Nikolai visited Cologne to interview Mata Hari, who portrayed herself as a dancing Javanese Princess, and recruit her as agent H-13. As a flirtatious courtesan, Mati Hari continuously toured Europe and being free with her favours, many foreign leaders whispered indiscreet secrets to her over the pillow).

In 2nd March 1913, Ronge's diligence paid off. Two more letters addressed to *Herr Nicetas* arrived. One contained 6,000 Austrian kronen and the other 8,000 Austrian kronen. Neither contained any covering letter, so it was natural that they should arouse suspicion. If the money were owing to anyone as the result of an honest business transaction, why had it not been sent in a normal manner? At the time, most operational spies of Europe knew of the Prussian border station of Eydtkuhnen. Little wonder that the *Evidenzbureau* became interested and determined to discover whom the person was for whom the letters were intended.

Ronge had the letters replaced, and assigned two detectives to monitor the Vienna Post Office to follow whoever came to claim it. Electricians installed a push button under the ledge over which the postal clerk would hand the letters to the person who should call for *Opera Ball 13 & 14.* From the button, a wire led across the street to a police station in the *Fleisch-markt.* The clerk was given orders. *The moment the letter is called for, push this button, and be as slow as you can in handing over the letter.* In the police station, two detectives were assigned to be ready day and night to rush out the instant the bell rang to arrest the person receiving the letters. March and April went by and the letters still lay there. Then on the 83rd day of waiting, Saturday 24th May : *B-r-r-r . . . b-r-r-r . . . b-r-r-r.*

The bell was ringing. As the mysterious *Nicetas* had arrived in civilian clothes to claim his mail, the post office clerk stepped on the buzzer to summon the detectives, but as bad luck would have it, one of the detectives on duty was out of the room while the other was washing his hands. Still, not too many seconds later, they both dashed across the *Postgasse.* By the time they arrived at the scene, their target had taken his mail. The clerk said ~ *Oh, you have been long in coming. He has just gone to the left.*

The detectives rushed out onto the street, to see a taxicab at the corner move off.
As he went off in the taxi, the watching detectives could only note the number of the vehicle.
While the detectives stalled, wondering what to do, an incredible stroke of luck came their way.
The same taxi that Colonel Redl had taken 20 minutes before was now driving back towards them. They whistled, yelled, and ran. The taxi stopped. It was empty. The detectives took the taxi, and asked to be driven to the address of his previous client ~ *Oh, to the Hotel Klomser. To the Hotel Klomser then. En route,* the detectives thoroughly examined the inside of the cab and found a pocketknife sheath in bright grey cloth. That was all. Arriving at the Klomser, the detectives asked the portier if anybody had driven up in a taxi within the last hour. Portier: *A quarter of an hour ago Colonel Redl came in. Dressed in civvies he was. He has his usual room, No. 1.* One of the stunned detectives asked: *Colonel Redl? Of Prague,* added the portier. One of the detectives showed the hotel portier the pocketknife sheath. *You may as well take it and ask your guests, as you get the chance, if any of them has lost it.* The portier took the sheath, and just then a fellow in fashionable civilian clothes came down the stairs and gave up the key of No. 1. Portier: *Pardon me, but has the Herr Colonel, by any chance lost the sheath of his pocketknife?* As he held out the sheath to the Colonel. Redl: *Thank you. Yes of course that's mine. Where did I use -----*

Redl stopped, and went deadly white. He glanced quickly at the portier, now hanging up the key. Another man was near, apparently engrossed in a newspaper. For a moment, the colonel stood still. Did it flash through his mind that he was cornered ~ caught after ten years of treachery? He pocketed the sheath, looked slowly around him again, went to the door, and walked out of the Klomser.

The man with the newspaper then rushed to a telephone-box to call 12-348, the secret number of the *Evidenzbureau* to report Redl. The methodical Major sent an aide to pick up all recently signed post office receipts from the *Poste Restante* counter, where persons who asked for letters had to fill in a form defining the nature of a packet, the address on the packet, and to say (if possible) where it was from. He was given the form that had been filled in by the person who had taken the two *Opera Ball* letters, and drove back to *Evidenzbureau* HQ.

What were Redl's thoughts? How did the portier get hold of his sheath? He had used the knife when he was in the taxi to get the money out of the envelopes. He had been careless and may have asked; *Am I being followed? If so, have I been betrayed or have I betrayed myself.* In that case, his experience at the Klomser confirmed that his pursuers now knew his identity. His immediate task then was to throw the men off his trail for a while. Redl went off at a good pace down the Tiefengraben.

He could not make a dash for it, so drew some papers from his pocket, tore them in pieces, and threw them away. A detective trailing Redl from some distance spotted him tearing up and throwing away the pieces of paper. The fragments were all collected, and then taken by the detective to Ronge. When the bits were put together, it was seen that the papers that Redl had torn up were :
~ A receipt for the dispatch of money to an officer of Uhlans, *Leutnant* Hovora.
~ Three receipts for registered letters sent to Bruxelles, Warsaw and Lausanne. The Bruxelles address was known to be the joint office of the Russian and French secret services in Belgium. The Warsaw address was for the *Okhrana.* The Lausanne address was the foreign HQ of the Italian Secret Service.

Meanwhile, Ronge also compared the writing on the Post Office delivery applications with a recent specimen of Redl's handwriting. From a shelf behind his chair he pulled down a slim volume *Advice on Espionage Detection.* It was a 40-page manuscript by Colonel Redl, and the last document he had written at the *Evidenzbureau.* Max Ronge laid the tiny form on one of the pages. Yes, there was no doubt about it ~ both were in Redl's handwriting. The evidence confirmed that Redl had been in contact with foreign agents. With these details in his possession, Major Max Ronge went to Conrad. Redl was not immediately arrested, but kept under observation.

When Colonel Redl returned to the Klomser, he was greeted in the hall by Dr. Viktor Pollack, Redl's colleague and collaborator in court proceedings in connection with important espionage cases. They were great friends. Dr. Pollack said; *We dine at the Riedhof,* and the colonel excused himself for a little while in order to get into evening clothes. One of the detectives overheard what had been said; he reported to HQ and was given his instructions. At the Riedhof he saw the manager, explained his objective, and later, another detective who was disguised as a waiter and acting the part perfectly, received the two guests and served an excellent dinner in a private room. Pollack was in a jovial mood, and Redl at first, tried to cast the haunting fears out of his mind ~ for his last dinner.

Elsewhere General Conrad von Hötzendorf was enjoying his dinner at the Grand Hotel when Major Max Ronge interrupted him with the news. As soon as General Conrad recovered from the initial shock, his visage turned livid. *He must be caught! He must be made to reveal the extent of his treason. Then he must die! And under no circumstances must the manner or reason of his death be made known! Good God, if Plan III should be known. Everything must be taken care of tonight! Major Ronge, you will see to it. Then report back to me at once! Redl. Of the VIIIth Corps too. Just the point where treason may be so deadly. My God! If Plan III has gone!*

Then there was silence. Conrad's face showed he was in agony. Then he pulled himself together and said in a steady, clear tone. *Call together four officers. Report direct to me.* That night, Major Ronge's desire to contain the scandal took precedence to determining what Redl had betrayed to the Russians. Conrad knowing the safe of the VIIIth Corps in Prague contained *Plan III* for use against Serbia had given clear and specific orders to Ronge and his aides : *He must be made to reveal the extent of his treason.*

At 11.30 PM, Redl, having returned to the Klomser, took his key and went slowly to Room No. 1. Around midnight four officers arrived at the Klomser. There was a sharp knock on Redl's door. Without invitation, the officers in full uniform entered. Colonel Redl, white-faced, rose as they entered. *I know why you are here. I am guilty. I want only to judge myself. I have written everything in these letters. Any accomplices? None. Your activities, how long and to what extent? In my apartment in Prague you will find all the proofs. Do you have a revolver, Herr Redl? No. You may request a firearm. May I borrow yours Major?* Ronge put his loaded revolver on Redl's table. *Thank you. Good night gentlemen!* The officers without a word filed out to the street below. One posted himself across the street from the hotel, while the others went to the Central Café to wait, taking turns going for black coffee.

5 AM. An *Evidenzbureau* detective who had been shadowing Colonel Redl was then summoned to the Central Café. *Go to Colonel Redl's room at the Klomser with this letter. If his door is locked use this key. No matter what has happened in his room, you are to raise no alarm. You report back here: Regrettable Suicide.* The *Evidenzbureau* man went to the hotel and showed the night porter his Imperial badge of office. *Colonel Redl expects me to deliver this letter in person.* But there was no reply to his knock on Redl's door.

The detective found it unlocked ~ the lights were on. Entering, he closed the door behind himself. Redl now lay dead at the foot of a tall mirror. The detective left the hotel quietly and reported to his superiors in the café. Minutes later the telephone bell on the reception desk of the Hotel Klomser rang.

The night porter answered it. So the corpse would be found before sunrise, another detective said: *Request Colonel Redl to come to the phone you are now using. He is sleeping. Who are you? Do as you're told!* The porter felt the voice of authority and obeyed. Minutes later the night porter called the nearest police station to report a suicide on the premises of the Hotel Klomser. So far, the Viennese and wider Austrian populace were none the wiser, but a bizarre set of concurrences soon exposed their cover-up.

The *Evidenzbureau* team who knew of the real situation was next closeted with General Conrad von Hötzendorf who pledged them to silence. *GENTLEMEN, I shall require of every one of you to take a special vow of absolute silence in the matter of Colonel Redl's treason! Not even our Emperor must know of it!*

Vienna's leading newspapers the next day all carried a discreet note of Redl's death. *Lamenting the loss of the highly-gifted officer, who was on the verge of a great career. His death is attributed to mental overexertion resulting from severe neurasthenia following a long period of melancholia. Colonel Alfred Redl, serving with the VIII*th *Army Corps in Prague in a moment of depression, committed suicide last night.*

Egon Kisch ~ August 1914

Redl reaches for Ronge's pistol…

Major Maximilian Ronge

Redl picked up the pistol, and took most of his secrets to the grave.
As the shot rang out, officers returned to find a note near the body.
Passion and levity have destroyed me. I pay with my life for my sins
I will die now. Please do not permit a post-mortem examination.
I am going to die now. Pray for me. ~ *ALFRED. 1.15 AM*

Denied a chance to interrogate the traitor, Conrad and Ronge were understandably curious about Redl's treason, so they took the first express train to Prague, the 6.15 where they met *Evidenzbureau* CO August von Urbansky and Military Commander of Prague, General Count Artur Giesl von Gieslingen (who had received a letter from his brother, A-H Ambassador to Belgrade Baron Wladimir Giesl, advising that Serbia regarded war as inevitable. Both brothers were in constant correspondence with each other as *Plan III* tasked VIIIth Corps to advance across the River Sava into Serbia). In civilian clothes the four went to secure the contents of Redl's apartment. A door with specially constructed locks halted them. It was Sunday and all the locksmiths in Prague were closed. Major Ronge went in search of a locksmith and found Hans Wagner. *Come with me!* He ordered. *But I don't work on Sundays. And I'm fullback on my football team this morning!* Drafted for temporary military service, the young man had no choice. He released the locks on Redl's doors and cracked the code on his private safe. As the military men pounced on the contents, Wagner caught glimpses of maps, sketches, photographs, blue prints, and reports of all kinds. He heard one officer exclaim, *How was it possible!* After Count von Giesl asked him to open the desk, the locksmith overheard an officer remark ~ *Russia!* In other rooms the décor was acutely tasteless. Furniture in various shades of red, photographs framed in snakeskin, embroidered underwear, silk stockings, cosmetics, and other such objects of a gay cross-dresser's life. A further inventory of Redl's effects gives a measure of his motivation. At his Vienna quarters were, 160 cases of champagne, 100 dress shirts, a score of uniforms, and a dozen fur-lined greatcoats.
A locked closet held ball gowns, lingerie, cosmetics, and other cross-dresser accoutrements.

If only the *Evidenzbureau* investigators had picked a different locksmith. Hans Wagner missed an important soccer match as a result of his unexpected duty. His team *SK Union*, one of Prague's leading clubs, lost their match against the German team *GBC Storm*. In the Prague newspaper *Prager Tagblatt* the following day 26th May 1913, appeared the report of the football match that Wagner had missed;

FOOTBALL IN PRAGUE

Storm v. Union Score : 5-7 (Half-time : 3-3).
Storm was the weaker team, suffering from the absence of Wagner.
Atja alone was not strong enough to withstand the opposition he had to face.

Certainly it was a sharp report, for it was written in annoyance by his defeated team's captain, Egon Kisch ~ who was also the sporting sub-editor of the *Prager Tageblatt* ~ yet, from these few lines of small print the story would emerge of the most sensational espionage scandal of the era. Had Wagner played, Storm might have won, and the tale would in all probability have been hushed up forever.

Only ten people in all of Austria knew the full story: the commander-in-chief of the A-H Army General Conrad von Hötzendorf, the command group of the *Evidenzbureau,* some in Vienna's War Ministry, and the commander of the Vienna police. Great precautions had been taken to keep the scandal secret. Each of the ten persons had taken a special oath not to breathe a word of it ~ even Emperor Franz Josef and the Archduke Franz Ferdinand were to be kept in the dark ~ but all their precautions failed, because a full-back was absent from his place on a playing-field of Prague.

The captain of *GBC Storm,* Egon Kisch went on Monday afternoon to look for his football friend, Hans Wagner to ask why he had not appeared at the match the previous day. Wagner apologized. *I really couldn't make it. The military sent for me.* Kisch: *And what did they want?* Wagner: *Oh, they made me break open some locks in an officer's apartment, and then open up all the drawers, wardrobes, and desks with locks on them. There was a commission from Vienna there looking for Russian papers, and photographs of maps. When they examined the large number of papers and photographs found, the officers were all amazed saying ~ How is it possible? ~ Who would have imagined it? A good deal of the money was found. The officer must have been very rich, and his house was gorgeously furnished. Some papers were in Russian; there were lots of plans.* Kisch: *Really? Whose apartment was it?* Wagner: *Some high-ranking officer I think. He died yesterday in Vienna. It all looked so strange ~ as if it were a woman's apartment, with manicure sets and perfumed letters. I saw one letter which said; Under no circumstances tell her anything about me! Women interfere in everything, and the one thing they understand is what they are not supposed to understand!* Wagner told Egon Kisch all he knew ~ *Looking for the will, I think. Some of the locks were very difficult to break.*
Oh yes, I'll be at next Sunday's match. A thing like that doesn't happen every Sunday.

The football captain Egon Kisch, now no less amazed than the officers, rushed back to his office to match the locksmith's tale with the official release from the War Ministry which he had dealt with in the course of his sub-editorial work during the previous evening concerning Redl's suicide and took it to his editorial chief ~ Russian documents, photos and plans, and a suicide ~ there was no mistake. The man whose supposed genius was everywhere recognized and who might soon have been in command of the A-H Army must have been a Russian spy. The captain of *Storm* had a superb story, but he could not use it as such. Were the news to appear in the *Prager Tagblatt,* the police would swoop down and confiscate every copy, and send to prison as many members of the staff as possible. So he consulted his editor, and this is what appeared in the *Prager Tageblatt* on the Tuesday morning :

We are asked by a high authority to contradict rumors which have been spread, particularly in army circles about the Chief of Staff of the Prague Army Corps, Colonel A. Redl who as already reported, committed suicide in Vienna on Sunday morning. The rumors are to the effect that the colonel had been guilty of betraying military secrets to a foreign power ~ now believed to be Russia. As a matter of fact, the commission of high officers who came to Prague to carry out a search in the dead colonel's house were investigating quite another matter.

That was all that could be safely printed in a Prague paper ~ but the people knew what it meant, for newspaper readers in Bohemia had long been forced by a stringent Imperial censorship to read not only between the lines but also ~ *behind the paragraph.* If they read, *Colonel Redl is not a traitor,* they knew that meant, *Colonel Redl is a traitor.* Thus Austria-Hungary learned the main fact of the scandal. However, on Wednesday the rest of Europe read a much more definite story, for the captain of *Storm* Egon Kisch, was also the Prague correspondent of a Berlin paper.

The War Ministry confirmed the basic details of Kisch's story. That Redl was a gay cross-dresser, that he had accumulated debts, and that he had sold military secrets to agents of a foreign Power. Viennese society was outraged when they read there had been an officer selling official secrets. While telegraph wires and cables burned as the story spread around the globe, the handling of the aftermath led to more complications. The Archduke Franz Ferdinand turned his anger against the *Evidenzbureau. I cannot comprehend why an office of military intelligence even exists when an officer owns two automobiles, and holds orgies without attracting the least attention!* The authorities seized Redl's estate to be auctioned without it being properly examined. A student bought a camera containing a roll of undeveloped film. When he and his teacher processed the film, it revealed the A-H mobilization orders.

Wien Kriegsministerium (Vienna War Ministry) Richard Moser 1913

Franz Ferdinand was similarly outraged by the discovery that Redl had been allowed to blow his brains out before revealing everything he knew about Russia's *Okhrana* operations. When Conrad had ordered a pistol for Redl, the last thing on his mind had been the fate of the traitor's soul, however the ardent Catholic successor to the throne, Archduke Franz Ferdinand was perturbed that the Chief of the General Staff had condemned an officer to perdition by opting him to commit suicide.

It is un-Christian to aid in a suicide. Suicide in itself is un-Christian, but when one extends his hand in order to facilitate it, it is outright barbarity. No other name for it! Even if he was a swine ten times over! Anyone who gets hanged receives the blessing of religion on the gallows, and that swine should have been sent to the gallows! I would have gladly let him swing, but to order a person to kill himself is un-Christian.

Conrad wrote; *To have sacrificed forty years to a profession, and then have to endure what I have had to endure in the past year is more than I can bear. This beast has destroyed everything I have worked so hard for.* From the time Redl betrayed *Plan III* to the *Okhrana*, it became a dormant cause of future military action between A-H and Russia. Imperial Russia had sought this gathered information, and now perfectly forearmed with the details of potential A-H Army deployments, such specifics ensured early victories to both Serbia's and Russia's armies. In the coming year, Russia would be first to mobilize.

During his years as an *Okhrana* secret agent for Russia, Alfred Redl disclosed many details of the A-H Army and *additionally* blind-sided the A-H General Staff; and to a degree German military GHQ, to the existence of at least 75 Russian divisions. In the months before the outbreak of the Great War, Redl's treason gave the A-H Empire's enemies every military and political advantage, therefore with good reason, Austrians still refer to the traitor Redl as ~ *the hangman of the Austro-Hungarian Empire.*

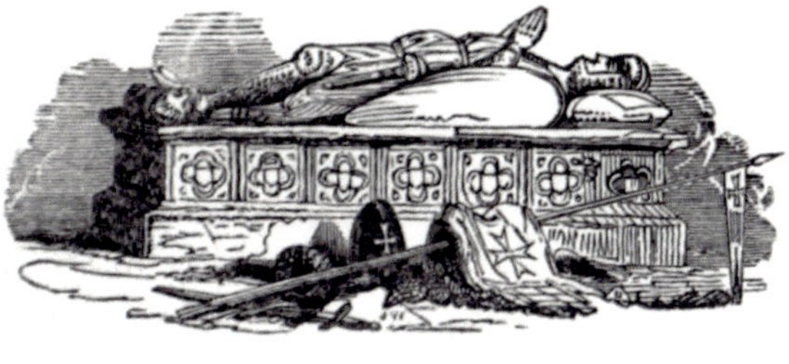

The French often commented on how they saw Prussia impose its will, as well as its methods of thought and life on the rest of Germany, and how it was comparable to the old warrior state Sparta. More than 200 years ago, the Marquis de Mirabeau, at the beginning of the French Revolution said; *War is the national industry of Prussia.* Napoleon proclaimed that *Prussia was hatched from a cannon ball.* When Napoleon engaged in war with Prussia in 1806, the swiftest of his military triumphs followed. In two months the Prussians had surrendered all their fortresses, and a period of humiliation followed for some years as the people lived under the dominance of France. Deliverance came when Napoleon, stretching his power too far was defeated by Europe in arms. The liberation which Prussia alone could not have accomplished, was preceded by a regeneration in which her military system was reformed. But what remained as the principal heritage from these years was the abiding sense that Prussia had suffered from being weak, and that only through military strength could there be safety in the future.

The expansion and greatness of a free Prussia left unfulfilled their old idea of a united Germany. From the middle ages, Germany had remained disunited and weak because of it. The smallest Teutonic states had disappeared, but there were still larger ones grouped under Austria in her vague and shadowy empire. Thus the history of old Germany in the half century following the downfall of Napoleon was a record of yearning and striving on the part of the German people filled with aspiration after strength and union which had come to their neighbors and yet, had not come to them. Soon it was seen that Austria would not or could not achieve this aim, and that Prussia could do it.

The Kingdom of Prussia decided to attack Denmark in 1864 to seize the Duchies of Schleswig, Holstein, and Saxe-Lauenberg. Austria and Bavaria were then defeated and a confederation formed. The Prussians chose not to acquire Austrian territory for itself, making it possible for Prussia and Austria to ally in the future, since Vienna felt more threatened by Italian and Pan-Slavic irredentism than by Prussia. The war left Prussia dominant in German politics and German nationalism began to encourage the remaining independent states to ally with Prussia in the Franco-Prussian War in 1870. Shortly before the Franco-Prussian War, the French military attaché in Berlin had reported to Paris, *Other countries possess an army, but in Prussia, the army possesses the country.* Finally, the German group struck France down to seize her eastern provinces, and Imperial Germany was established in 1871.

This era is important not only for what it ended and began, but also for the methods employed. So sharp were the intrigues of Bismarck, that had his methods been different he may have stood greater in history. Bismarck was no worse than most of the diplomats who had preceded or would follow him, yet too often, his instruments were cunning, force, and fraud. With him, means were justified by end. Ordinary justice and moral considerations had with him, as with Napoleon, small place. There was pity for the weak and mercy to the fallen only as such things were politic. He left behind him imperial glamour, but he left also fear and hatred, and desire for revenge. As by force and violence he had accomplished what he had done, so he knew that only by force could his work be maintained. Bismarck isolated his enemies, then struck them down. From Austria he had taken little, and so he was able to create friendship based on identity of interests, but from France he had taken too much. Before France recovered from her wounds, Bismarck sought readjustment by a new grouping of nations about Germany. Austria, who soon came to be dependent upon German support, was firmly attached in alliance. Meanwhile Bismarck seized upon old suspicion and colonial rivalry, played upon Italy's fear of France, and urged her into a Triple Alliance with A-H and Germany.

Friendships cooled, some enemies became friends, and newcomers entered the opposing camps. Of this there were many causes, but most of them may be traced ultimately to the prodigious growth of the German Empire, which was the most striking phenomenon of Europe during those decades. A nation, which had been mainly agricultural in 1870, by 1910 was second only to the U.S.A. in manufactures and second only to Britain in shipping. So fast had their national wealth increased that Germany was the wealthiest nation, having in two generations outstripped both England and France. It was filled with exuberant strength and aggressive energy, the basis of which was organization, method and efficiency, but it was also their increasing population that had grown amazingly and continued to enlarge by a million a year. Germany with expanding industry and increasing wealth, where the birthrate had been surging and for decades had showed scant prospect of diminution had seen German statisticians gloat over their old enemy, and look forward to the day when there would be in Germany twice as many people as in France, and as many as in England and France together.

The strength of Imperial Germany was never in doubt. Those who passed down the Rhine or went to Berlin saw everywhere colossal power. Rising cities, mounting numbers, and vast material creations of a new empire with a vitality which one saw not so much in England, and little in France. In Germany, the very purpose of their cultured civilization seemed different. The army was the most impressive in the world; nowhere did numbers and equipment go so far to make a nation invincible; and no army could be so swiftly assembled and hurled upon its enemy with such speed and precision. As an inheritance from previous wars, German soldiers believed that their army was unconquerable. The navy, a newer creation, had grown greatly and was now feared. Finally, the power of an energetic and intelligent people was wielded by an autocratic government, in character capable and efficient.

Imperial Germany like a giant conscious of destiny waxed ever more ambitious and aggressive, and the pressure of numbers, first felt within her own bounds, was soon felt by all her neighbours, and at last by every nation in the world. For where intelligence and efficiency are nearly equal, numbers will usually be decisive. It was the greatness of her population which so long gave to France the leadership of Europe; and now behind commendable organization and thoroughness of method, it was increase of people which gave Imperial Germany an impregnable position across the continent, and let its rulers dream of the hegemony of the world. A-H and Imperial Germany acted as one in most aspects of Pan-Germanism, which in the beginning meant the union of all Germans in Europe and those scattered in lands beyond, but in time, came to mean the rising ambition of the ruling elite in Berlin and Vienna to obtain the mastery of the sea and the hegemony of Europe, by building an empire from the Baltic Sea to the Persian Gulf.

Geo-political success would probably come to whichever power controlled the Balkan Peninsula. If the Tsar's regiments guarded the Bosporus and the Ægean, then Russia would have her window upon the world, and from her (more frequently) proposed new capital at Constantinople, Russia could control all Danube trade, and with her finger thus upon the artery of Vienna, would come to an end the dream of the Germanic powers of stretching their dominion down to the Persian Gulf. Conversely, if ever Germany and A-H won the Bosporus and the Hellespont, then Russia, thrust back from the south and east, would lie like a giant bound among its enemies, a vast but an inferior power. Therefore, the Balkans remained the danger-zone of Europe, and the destinies of cultured and wealthy nations were there in the sights of plotters and peasants.

A peculiar cause of the quarrel is said to have been the changing character of the German people, in that the German psyche had been much altered by two significant successes. The unparalleled triumphs of the Franco-Prussian War had given German leaders supreme confidence, and the soldiers believed that they were invincible. This feeling, which might have become noble and disciplined, tended to haughty arrogance, because of the completeness with which it possessed people's minds, and because of a materialism based upon immense material prosperity. Kindliness and homely virtue while still being the character of many Germans, was now less evident than before. The ruling caste and hosts of followers now admired Imperial might, force and power. Antique, pernicious doctrines that *might was right*, and the end makes good the means, reappeared in a modern, dangerous guise. In other nations arose distrust and fear that Germany would strike without pity and without remorse, and in Germany such boasts were fierce and frequent, this was brusquely expressed in the writings of Prussian General Friedrich von Bernhardi;

Germans must think no more of poisonous peace doctrines. They must become convinced that war is not only a necessary factor in civilization, but the highest expression, among civilized people, of power and life. War is necessary and ennobling, that no consideration outweighs necessity, that France must be struck down so that never again can she be an opponent, that the reckoning with England must follow, and that superior German culture must be spread by the sword. Much that our rivals regard as arrogance and aggression is always the expansion and increase and prosperity of Germany. Merely to put up a vigorous defensive fight if attacked is not the whole duty of a nation. That duty includes the carrying on of offensive warfare for the best interests of the Nation's greatness. Strong, healthy, and growing nations, need a widening of their boundaries. They need new lands, on which their overflow population may settle. But since the Earth is occupied almost in its entirety, new lands can only be obtained at the cost of those already occupying them ~ through conquest. Thus, the right of conquest becomes a law through necessity. German power is not an end in itself, but must be justified by its employment for the highest human good.

In practice, the noble caste of Prussia dominated the German Army and Navy, and this secured the high ranks for professional officers of the noble class ~ those with the prefix *von* before their name. Within this otherwise cultured caste in Prussia were a number who believed in war for its own sake. The noble descendants of the Teutonic Knights, who had conquered the Slav population of Prussia, and bound the civil population to their will. Visitors to Germany increasingly returned to comment on how the Prussian officers often behaved in a harsh and arrogant manner toward the civil population.

Time and again, in Germany while waiting in line at some ticket office, an officer shoved himself ahead of all others without even a protest from those waiting. On one occasion, I went to the races in Berlin with my brother-in-law and bought a box. While we were out looking at the horses between the races, a Prussian officer and his wife seated themselves in our box. I called the attention of one of the ushers to this, but the usher said that he did not dare ask a Prussian officer to leave, and it was only after sending for the head usher and showing him my jockey club badge ~ and my current pass as U.S. Ambassador, that I was able to secure possession of my own box. U.S. Ambassador to Imperial Germany ~ James W. Gerard

The worst example of this was the notorious Saverne Affair of December 1913. While marching through the Alsace garrison town of Saverne, 19-year-old *Leutnant* Baron Günter von Forstner had a clash with Karl Blank, a shoemaker who laughed at his uniform. When the other townsfolk joined in, von Forstner lost control and cut down the jeering Blank with his sabre, causing severe head injuries. This brutal act caused an outrage throughout Germany. Forstner was tried by a court martial for wounding an unarmed civilian and sentenced to one year's imprisonment ~ then he was acquitted by the German High Court *as having acted in supposed self-defence.* With such sinister and threatening decisions on display, Germany appeared to outsiders as a militarist and fierce-minded nation.

The Saverne Affair sparked a debate in the Reichstag. Chancellor Bethmann Hollweg tried to downplay the military's actions, but seemed visibly shaken and nervous by observers at the time. War Minister General Erich von Falkenhayn spoke for the first time before the Reichstag to defend the officers, and in a sharp commanding voice told all present. *This military officer has only done his duty, and I will not be swerved by press agents or hysterical individuals. I support von Forstner as a young officer who has been harshly punished. This kind of courageous young officer is just the type that Germany needs.* After von Falkenhayn's speech, and having stirred all present to the highest pitch of bitterness, it was moved that such attitudes did not meet with the approval of the representatives of the people. Falkenhayn went on to criticize the press and their handling of the matter claiming they had inflated the issue with propagandist methods. When the SPD demanded Bethmann's resignation he refused. A vote of censure directed against the government was debated in the Reichstag, and voted down.

Demonstrations demanding change spread across Imperial Germany, as people watched in horror at the way the military handled the situation. Protests in 17 German cities demonstrated against the despotic rule of the military and demanded the resignation of Bethmann and Falkenhayn. Finally, Bethmann announced that the Kaiser had decreed that military authorities should be kept within legal bounds, all guilty persons would be punished, in addition to the removal of the military units from Saverne as a temporary solution. While the court publicly apologized for the actions of the soldiers, it blamed the civil authorities for not maintaining order. This all had consequences...

The Saverne Affair and the changing attitude of the whole nation greatly alarmed the German militarists. It was perhaps the final factor, which decided the advocates of the old military system of Germany in favour of a European war. Usually in past years when the Reichstag in adjournments had risen and cheered the name of Kaiser Wilhelm II, the Social Democrats absented themselves from the Chamber, but when the Reichstag adjourned on 20th May 1914, these members remained in the Chamber and refused to rise or to cheer the Kaiser. The President of the Reichstag immediately called attention to this breach of respect to the Kaiser, upon which the Socialists shouted ~ That is our affair! ~ and tried to drown the cheers with hoots and hisses at which the other parties applauded tumultuously. This occurrence I know greatly incensed the Kaiser and did much to win his consent to the war. Their evident and growing dislike of militarism indicated a new spirit of resistance to autocracy; and autocracy saw that if it was to keep its hold upon Germany it must lead the nation into a short and successful war. This is no new trick of a ruling and aristocratic class. Autocracies have always turned to war as the best antidote against the spirit of democracy. U.S. Ambassador James W. Gerard

* * *

In October of 1913, Kaiser Wilhelm II with an entourage of his closest advisors visited Archduke Franz Ferdinand at Konopište Chateau, some 50km S.E. of Prague, and then travelled on to Vienna. At the German Embassy, Count Leopold von Berchtold explained how; *The inextinguishable animosity of Serbia and their nationalist dreaming makes any proposal impossible.* The German leadership encouraged Vienna to prepare for future action in the Balkans, in the expectation that an A-H punitive campaign would remain localized. The Kaiser suggested: *Austria need not ask much ~ only that Serbia should place her army at the disposal of Austria, and make it subordinate to A-H forces in return for perpetual protection.* When Berchtold asked ~ *What if Serbia refuses?* The Kaiser advised :

Force would have to be employed, for if His Majesty Franz Josef demands something, the Serbian Government must yield, and if she does not, then Belgrade will be bombarded and occupied until the will of His Majesty is fulfilled. You may rest assured that I stand behind you and am ready to draw the sword whenever your initiative makes it necessary. (With these words, Wilhelm motioned his hand towards his sword) *We find ourselves in the margin between the military and the political ~ treacherous and obscure ground on which the diplomat goes astray. I, as a military man do not entertain the smallest doubt that Russia is now systematically preparing for war against us, and govern accordingly.*

The Kaiser knew the situation unmistakably when he spoke of the margin between the military and the political. It was precisely on this ground that most governments now moved, with many in the German and Austrian camps using their influence to distort the Kaiser's judgment and draw him on to a collision course. With every month, the same reports came ever more persistently from Vienna, that Germany's position was dangerous and likely to become even more dangerous in the future. Count Conrad von Hötzendorf, the A-H Chief of Staff, who during previous crises had called for war, lamented that Austria had not attacked Serbia in 1908, and argued vehemently for a final reckoning. All that was required by the Viennese hawks was for a pretext to appear, as an instrument of justice. German Military Attaché Count Karl von Kageneck reported to Conrad. *The Kaiser is not for war, but the German Staff are convinced that war is inevitable, and are counting on it, and on a war with Russia as well.*

In any future war, if it is undertaken without a pressing reason, not only the crown of the Hohenzollerns but the future of Germany will be at stake. Of course, our policy must be a bold one. But to rattle the sabre in every diplomatic complication when the security of and the future of Germany are not threatened, is not only foolhardy but criminal. Kaiser Wilhelm II

Militarily, the character of the alliance between the two central empires was now transformed. The decision taken by the German General Staff to support Austria-Hungary if she invaded Serbia, (even if Russia intervened) now inevitably moved to recrystallize German strategy into a fatal pattern. Russia now had her ally of France, who might join her in declaring war on the Danubian Monarchy. In that case, where would Germany launch her first counterblow? When General von Hötzendorf raised the question in 1909, with his Berlin counterpart General Helmuth von Moltke. Moltke replied ; *To hurl the main body of the German forces first against France.* Therefore, the defensive alliance created by Bismarck had turned into an offensive alliance. Worse ~ the decision to strike the first blow at France depleted the prospect of localizing an eventual conflict in S.E. Europe, to the point that any Balkan war involving A-H and Russia could become a wider European war. And since a blow against a modern military power like France could only succeed if it were delivered with stunning speed ; strategy implied the minimum delay in mobilizing and concentrating forces, and then launching them headlong across the French frontiers. Once set in motion, the German Army could not be halted temporarily without compromising the outcome. The minimum time lag between mobilization and total war for the German Army, massed at many railheads and marching to battle with their horse artillery echoing through the cobblestone streets of German border villages, offered only an extremely fleeting logistical interval of opportunity between a severe diplomatic crisis, and an imminent military clash for any last-minute efforts of peacemakers.

The German Army is vital, not merely to the existence of the German Empire, but to the very life and independence of the nation itself, surrounded as Germany is, by other nations, each of which possesses armies as powerful as her own. Germany herself has nothing like that superiority over France alone, and she has of course, in addition, to reckon with Russia on her eastern frontier. Germany has, therefore, become alarmed by recent issues, and is spending huge sums of money on the expansion of her military resources. David Lloyd George
British Chancellor of the Exchequer

To win, we must endeavour to be the stronger of the two at the point of impact. Our only hope of this lies in making our own choice of operations, not in waiting passively for whatever the enemy chooses for us.

General Alfred *Graf* von Schlieffen

For decades, there had been a stirring sense of insecurity. All Europe felt restless and suspicious, but in Germany, nervous ambition led the Prussian militarists to plan a secret stroke to put them beyond reach of all rivalry. But, if their stroke were delayed, the growing wealth of Russia and the increasing military strength of France would soon suppress any opportunity for German dominance. The Schlieffen Plan was named after Count Alfred von Schlieffen, Chief of German General Staff who for 18 years had devoted much of his time to one problem : the danger of the Franco-Russian Éntente and the inherent potential of a two-front war in both the east and west. Assuming the Russians would move slowly, his answer was simple. To defeat the French, before the Russian Army entered the field, one German Army would join the A-H Army to contain the Russian threat while other German forces attempted to defeat France in 40 days, then turn east on rapid rail routes to vanquish Russia.

Count Alfred von Schlieffen's invasion strategy was to swing counter-clockwise across Belgium toward France like a huge swinging door, with Luxembourg as the hinge, and the last man on the right side of the swinging German door supposed to brush his sleeve on the English Channel. Over a million German troops would outflank the French, by marching west through Belgium by Brussels, swing south after the Belgians were disposed of, then thrust east to encircle the entire French Army. But the too clever Schlieffen Plan which called for the violation of Dutch and Belgian neutrality to bypass formidable French frontier fortifications, proposed that the German Army could achieve this by marching a grand circumference thru France, however they forgot to reason that the French Army would always be able to switch their own troops by rail across the chord of the circular enemy thrust. After von Schlieffen retired, he expressed his concern about the supposed role of the British in Europe. *The whole of Germany must throw itself on one enemy ~ the strongest, most powerful, most dangerous enemy, and that can only be the Anglo-French! I am in favour of the earliest possible clearing up with France at arms. No waiting ten or twenty years for a wider war, but so thorough a settlement that thereafter there should be no fear of a wider war. France should be provoked until she has no course left but to take up arms.*

When France grew alarmed at the disparity between her army of 450,000 men and that of the 900,000 which the laws of 1911-14 assured to Germany, the French voted for three years' service and a slight increase in associated armaments. Immediately the German Press denounced this as a provocation. Prussian Baron Wilhelm von Stumm, declared during a dinner at the Dutch Legation; *If France presumes to challenge Germany's right to be stronger than she is ~ it must be that she desires war.* Colonel Erich Ludendorff as CO of GHQ Operations Department working on strategic plans under the direction of Count Alfred von Schlieffen, when asked to compose a report on the methods to be pursued in arousing German national enthusiasm, and how to shift the perceived responsibilities for their proposed violation of Dutch-Belgian neutrality in a future war, wrote the following;

We must act with prudence to awaken no suspicion. In the next European war, the small nations must be forced to follow us or they must be crushed. Under certain conditions their armies and their fortresses can be rapidly reduced or neutralized ~ which would be the case with Belgium and Holland ~ so as to shut out our enemy in the West from territories which could be used as a base for operations against our flank. This will be a vital question for us. Our aim must always be to take the offensive with greatly superior forces from the start. From now on, the military leaders are not alone in the secret of this aggressive plan. The Governments of the German States are informed that France is to be attacked through Belgium. Germany cannot respect Belgian neutrality. The Chief of the General Staff has declared that even English neutrality would be too high a price to pay for respecting that of Belgium. For an offensive war against France is possible only through Belgium.

Schlieffen's successor, General Helmuth von Moltke was not a particularly daring commander. After he was heard confessing ~ *I lack the capacity for risking all on a single throw* ~ he modified the old Schlieffen Plan to create *Plan 1914.* After discarding the former planned advance through Holland and the associated violation of Dutch neutrality, he reduced the size of the sweeping German invasion. *Plan 1914* remained a contributory cause to the quarrel, in that while it offered Germany a chance of swift triumph in the west, in reality it meant that if Germany actuated its intention to violate Belgium, it then risked detonating a wider war with France and Russia, and further ensured that Great Britain joined her continental allies ranged against the Triple Alliance.

Lunéville April 1913 ~ two contemporary French postcards of the German Z4 airship incident

Knowing in advance that they may need to fight a war with France and Russia, German GHQ decided in advance which officers would lead the troops, and sent men like Paul von Hindenburg to study the geography of Poland, and men like Alexander von Kluck to study the geography of France. All too soon, the French noticed vast numbers of German tourists walking through their countryside, German artists making detailed sketches of French scenes, and others with cameras taking pictures. Some were generals planning their campaigns. Others were infantry officers who were going to lead their troops across that topography when the time came. These agents were selected with the utmost care so that they might send on to Berlin regular up-to-date reports of Germany's intended targets. In this manner, the German leaders knew what was going on elsewhere, whom they might meet, what was being done to offset their plans, and if possible, spies would steal military and naval secrets.

Similarly, from the air ~ on 3rd April 1913, Zeppelin Z4 ascended from Germany on a trial flight. At 9 AM as the Zeppelin crossed the French frontier, it turned to traverse the entire line of French forts facing the German frontier from south to north, causing a sensation through France and Germany. Trouble compounded for Z4, when almost out of fuel it was blown off course in obscuring mist and its flight came to an end at 2 PM being forced to descend among French cavalry on the parade ground of Lunéville in Meurthe-et-Moselle. French troops swiftly surrounded the Z4 to keep back sightseers, and that evening while the Mayor of Lunéville entertained the Germans at dinner; Z4 was thoroughly inspected and assessed by French technicians, who confirmed their suspicion that the German airship could indeed be equipped to carry bombs.

The French authorities, in a word, behaved in such a chivalrous manner that the German Government at first felt compelled to express its gratitude to them publicly. The dirigible had scarcely got back to its hangar, however, when the German Press opened an attack on the French officials and troops and on the inhabitants of Lunéville, who it was asserted, had been lacking in courtesy towards the German officers. The German Government accepted these calumnies as true, and Baron von Schoen, with the stormy weather expression on his features which he invariably assumed on such occasions, called at the Quai d'Orsay to make formal complaints. He complained of alleged difficulties that had been placed in the way of the captain and the pilot of the Zeppelin in connection with telegraphic communications, of measures that had been taken to keep the crew at a distance from their airship, and of insulting inscriptions that had been penciled on the car of the vessel by spectators.

Raymond Poincaré, President of the Republic of France

On 5th November 1913, King Albert I of Belgium made the first visit to Berlin by a Belgian king since his uncle Leopold II had done so in 1904. The Belgian royal family of German decent, from the House of Saxe-Coburg and Gotha (as are the British royal family) thus the two countries still enjoyed strong economic and cultural ties. Therefore, when King Albert of the Belgians was invited for a diplomatic dinner party where he was the guest of honor at Kaiser Wilhelm's *Neuen Palais* in Potsdam, he had expected a friendly encounter, limited to the usual aristocratic pastimes of horseback riding, dancing, champagne, and gossip. It was however, to be the scene of several alarming conversations.

Speaking with Albert at the pre-dinner ball, Wilhelm pointed to General Alexander von Kluck. *He is the man who will lead the march on Paris.* This astonishing statement was merely the *amuse-bouche* for a four-course meal of inebriated invective. Similarly, Belgian Ambassador Baron Beyens recalled: *The Kaiser discoursed at length on the political situation in Europe. He thinks it so bad, though the fault of France, that he regards war with the Republic of France as ~ inevitable et prochaine, imminent. Albert protested; It is a travesty of intentions of the French Government to interpret them in that sense, and to be so mislead by the ebullitions of a few irresponsible spirits or the intrigues of unscrupulous agitators. The Kaiser obstinately went on declaring that given the overwhelming superiority of her army, Germany is sure of victory.*

Wilhelm had no doubt that in these circumstances Albert would remember his family connections with the House of Hohenzollern (King Albert's mother was a Princess of Hohenzollern-Sigmaringen). Moltke confirmed that war with France appeared imminent and that this time they must finish the job. *Cette fois, il faut en finir. Your Majesty can surely imagine how the whole German people will be swept along in the irresistible enthusiasm on The Day. There will be no resisting the Furor Teutonicus once it is unleashed. Small countries, such as Belgium, would be well advised to rally to the side of the strong if they wished to retain their independence, as the consequences will be hard for those who are against us. The matter must be settled.*

Albert's military advisor Captain Émile-Joseph Galet noted; *This was more than intimidation; it was a shameless threat against the neutrality and independence of Belgium.* When the Belgian military attaché Major Charles de Melotte resolutely demurred ~ *Europe can surely look forward to a long period of peace!* Moltke retorted: *Do not have any illusions. War with France is inevitable, and much nearer than you think. We do not want it. We have nothing to gain in that quarter. But we have had enough of these alerts that retard our progress. And we shall win. We are sure of being victorious… we shall lose battles but shall win in the end.* Moltke again enquired: *What would Belgium do, if say, one of the Great Powers violated her neutrality. Would she actually fight ~ even if it were hopeless, or would she bow to the inevitable and lay down her arms?* Major de Melotte boldly replied: *Belgian honor would require her to fight any invader with all her strength.*

A few days after Albert's departure, Kaiser Wilhelm wrote in the margin of a report from Paris. *Just let them come! Then we will finally, with God's help, settle accounts with them. It is a question of our standing in the world, which is under attack from all sides! So head held high and hand on sword-hilt.*

When King Albert returned to Belgium, he wrote a report that would be entrusted to just six men in Paris. President Raymond Poincaré was one. King Albert had believed before, that as the Kaiser had so far exercised his influence in several critical situations for the maintenance of peace, he would do so in the future. But to Albert, this felt like an exact repeat of King Leopold's earlier visit to Berlin in 1904. *This time he was completely different. Wilhelm II is no longer the upholder of peace against the bellicose tendencies of certain circles in Germany. The Kaiser has become convinced that war with France is inevitable and that it must come sooner or later.* Baron Leon Lambert confirmed the nature of Moltke's militarism, when he recalled the Kaiser's words of restraint; *They had better take care in Paris ~ I shall not live forever!*

With King Albert's permission, in November 1913, Baron Eugéne Napoléon Beyens reported the incident to the French Ambassador in Berlin Jules Cambon, who said to President Raymond Poincaré. *If I may be allowed to draw a conclusion, I would submit that it would be well to take account of this new factor, namely, that the Kaiser is becoming used to an order of ideas which were formerly repugnant to him and to borrow from him a phrase which he likes to use ~ we must keep our powder dry.* Citing Cambon's report, Poincaré warned Quai d'Orsay principal Maurice Paléologue and Chief of General Staff Joseph Joffre, that war with Germany was coming. *France has only one aim, the maintenance of the concert of Europe and of peace.* Wilhelm replied: *We have heard this rubbish before, and it does not impress us at all! As soon as they have the opportunity to attack us they will do so without further ado.* Fear and suspicion now fed upon itself in a vicious cycle as Wilhelm said ~ *inevitable et imminent* ~ as a whirlpool dragging in the Empires.

Elsewhere, the strengthening link between Berlin and Constantinople soon provoked a crisis in European affairs and the traditional balance of power. Following the loss of the Ottoman European territory in the two recent Balkan wars, Grand Vizier Said Halim Pasha requested Kaiser Wilhelm to nominate an experienced general to lead a mission of German officers to reform the Ottoman Army, still significantly impaired by its reverses suffered during the Balkan Wars. The Kaiser nominated the Prussian *Generalleutnant* Otto Liman von Sanders, who accepted the commission with no hesitation and set off by train for Constantinople to improve the fighting capabilities of the Ottoman Army. Shortly after his arrival, von Sanders met the leading Young Turks and was impressed by Enver, Interior Minister Talat's *charm and attractive personality*, and remarked that Ahmed Cemal Pasha, the current Commander of the First Army Corps *combined great intelligence with a very determined attitude.*

St Petersburg opposed the appointment of the German military mission from the first report of it. Although it wasn't uncommon for European officers to train and command the troops of other lands, the appointment of Liman von Sanders to command the First Army Corps guarding Constantinople far exceeded the usual scope of such arrangements. St Petersburg's opposition developed into a crisis, the moment Cemal Pasha handed over command of the Ottoman First Army to Liman von Sanders, since its garrison was responsible for the security of Constantinople, in addition to the strategic Straits. This infuriated the Russians, as Germany had just acquired control of terrain in which St Petersburg had an interest, and foreclosed any chance for Russia to seize the Straits, and Constantinople as their proposed future capital of *Tsargrad* for themselves. Apart from threatening Russia's foreign trade ~ most of which flowed through the Straits ~ the German mission raised the potential for a German-led Ottoman assault on Russia's Black Sea ports and, as it imperiled Russia's secret plan for expansion in eastern Anatolia, Russia threatened to occupy the eastern Anatolian city of Erzurum to redress events.

On 10th November 1913, Russian Foreign Minister Sazonoff, instructed the Russian Ambassador in Berlin Sergei Sverbeev to advise: *The von Sanders mission is regarded by Russia as an openly hostile act.* While the von Sanders mission troubled Sazonoff, he understood that Germany couldn't simply back down for reasons of prestige and therefore sought a solution that would allow them to retire. On 18th November, Russian Premier Count Vladimir Kokovtsov while already visiting Berlin paid a visit to Chancellor Bethmann to suggest that Liman von Sanders be given a different assignment somewhere other than Constantinople. As Bethmann was only vaguely aware of the von Sanders mission, it being an initiative of the German Army that increasingly seemed conducts its own course of foreign policy, and even if Berlin was willing to reach some new adjustment ~ it wasn't solely their decision to make. The new Ottoman régime, fed up with years of European intrigues were in no mood to compromise.

On 25th November, Sazonoff sent formal requests to Paris and London for diplomatic support against Germany in the Sanders affair, with a warning that Berlin was trying to pry the Éntente apart. On 1st December, Sazonoff explained to the British *chargé d'affaires*: *This matter will be a test of the value of the Triple Éntente. I believe that if the three Powers show themselves to be determined, Germany will not persist in her intentions. British participation is particularly important, as Germany might disregard the remonstrances of France and Russia ~ if she has not also before her, fear of the British fleet.* Secretly however, Britain and France were in reality more determined to prevent any Russian action that would lead to a premature partition of the Ottoman Empire : Sazonoff in a 6th January 1914 memo to Tsar Nicholas : *If our War Minister and our Navy Minister believe it possible to risk complications, in the case France could support us with her forces, and England lent us assistance, we can engage in a confidential exchange of views on this question with these two Powers ; for Britain to land troops at Smyrna, France at Beirut, and us at Trabzon, ostensibly to pressure for the expulsion of the Liman mission, but in reality as a prelude to partition. We would remain until the fulfilment of our demands.*

When Sazonoff enlisted France to pressure Britain. French Ambassador in London Paul Cambon, urged British Foreign Minister Sir Edward Grey: *Join France and Russia in delivering a note warning the Turks that to entrust the Constantinople First Army Corps to a German General would mean virtually handing the keys to the Straits to that Power and upset the equilibrium of the Powers which is the guarantee of the existence of the Ottoman Empire.* On 2nd December, Grey sent the British Ambassador to Constantinople a telegram stating: *Control of the Straits is a matter of concern to every Power that is interested in Turkey.* Grey wisely limited himself to asking Constantinople to, *clarify the extent of Sanders' responsibility.*

Once again, it was Kaiser Wilhelm who came up with a peaceful resolution to a dramatic crisis, when he simply promoted von Sanders to a rank too elevated to command an Ottoman Army Corps. When *Generalleutnant* became Field Marshal von Sanders, command of Constantinople's First Army Corps reverted to an Ottoman officer, von Sanders stayed on despite more protest from St Petersburg, and Kaiser Wilhelm cruised through the crisis with well-deserved merit.

After the resolution of the Liman von Sanders crisis, Tsar Nicholas gave an audience to former French FM Théophile Declassé during which the Tsar's resolve to resist Germany became evident. *The Liman von Sanders affair represented a severe threat to Russia's vital interests, for the free passage of goods through the Dardanelles is essential for our commerce. I am convinced of the need for much closer co-operation between St Petersburg, Paris and London ~ and the prospect of the collision between German ambitions and Russian interest. We will not let them tread on our toes. This time, it will not be like the war in the Far East. The national mood will support us.*

The Young Turk régime saw Russia as the single greatest threat to the survival of their Empire. With Russian territorial ambition for eastern Anatolia, the Straits and their capital of Constantinople, they knew that the plotters of St Petersburg openly sought the demise of their Ottoman Empire. Therefore, the leaders of the Ottoman Empire entered 1914, in search of an effectual defensive alliance.

While previously the Ottomans could expect France or Great Britain to restrict Russian ambitions, these three Empires were now allied in their Triple Éntente ~ therefore neither London nor Paris could be counted on to come to the diplomatic aid of Constantinople in the future. A-H having seized Bosnia-Herzegovina precluded them, so the obvious contender for consideration remained Germany, and from Berlin's *Weltpolitik* perspective, the Ottoman Empire held a strategic position for Germany.

From the time of the Kaiser's first visit to Constantinople, Britain and Russia were engaged in an intense and intricate rivalry for predominance over Central Asia, in the classic era of the *Great Game.* Seen from Berlin, the Ottoman provinces of eastern Anatolia were a route to Persia and Central Asia for Germany to become a bold new player in the *Great Game* through an alliance with the Ottomans.

Nevertheless, the Ottoman leadership showed an openness to secure a defensive alliance with various Empires ~ and the Young Turk leaders had quite different opinions on potential alliances. Enver and Talat favoured an alliance with Germany, while Cemal considered that only an Éntente power could peacefully restrain the increasing threat that Russia showed towards Ottoman territory.

Cemal was appreciative that France was the Ottoman Empire's principal financial creditor since the conclusion of a $100 million public loan in early 1914. Cemal thought Britain an alternative, as for most of the 19th Century, Britain had preserved the territorial integrity of the Ottoman Empire, and more recently assisted with the restructuring of the Ottoman Navy through the Limpus mission and the construction of new ships for the Ottomans. Since becoming Minister of Marine, Cemal had worked closely with the British naval mission, and appreciated the manner of British professionalism.

When Cemal visited France in response to an invitation to attend French naval manouevres, he took advantage of his trip to Europe to meet with the Ottoman officers liaising with the British shipbuilders, who were putting the finishing touches to two new Ottoman vessels of the dreadnought class. The lead ship to be named *Reshadiyeh* was laid down by Vickers in 1911, launched in September 1913, and was scheduled for delivery in August 1914. The Turkish officers reported to Cemal;

The British are often in an interesting frame of mind. They seemed to always be explaining some new excuse for delaying the completion and delivery of our warships. Cemal replied; *Return and take delivery of the warships as soon as possible, leaving any final fittings for our Ottoman shipyards to complete.*

Following his attendance of the French fleet review in Toulon, Cemal returned to the offices of the French Foreign ministry in Paris where he suggested; *You must take us into your Éntente, and at the same time protect us from the terrible perils threatening us from Russia. In return, the Ottoman Turkish Empire will be a faithful ally in helping France and Britain forge an iron ring around the Central Powers.* One French diplomat responded cautiously; *The French Government could only enter into an Ottoman alliance with the approval of our other allies ~ which would seem very doubtful at the moment.* Cemal returned to record; *I understood perfectly that France was convinced that it was quite impossible for us to escape the iron claws of Russia, and that under no circumstances would she vouchsafe us her help.* As before, Kaiser Wilhelm II's offer of friendship remained the best avenue of possible alliance, and the Ottoman Turkish Empire reaffirmed to their special relationship with the Triple Alliance of Germany, A-H, and Italy.

In mid-1914, the British would requisition the two warships commissioned by the Ottomans, which would deeply disappoint Cemal, who as Minister of Marine recalled his discussions with the Ottoman officers in Paris and realized : *The repeated British delays had been nothing but pretexts which revealed the design England had long cherished of making these ships her own. Given that these ships were paid in full and mostly by public contributions. Schoolchildren had contributed their small-change paras and kurus to fundraising efforts at schools, and some women had sold their hair, and through many popular subscriptions.* The ships having become symbols of national pride, the British decision to requisition the ships was a humiliation for Turkey ~ and ruled out the option of any accord between London and Constantinople.

Kaiser Wilhelm writing letters at his villa on the island of Corfu ~ Achilleion

In March of 1914, the A-H Ambassador to Serbia Baron Wladimir von Giesl, was directed to hand to Pasic yet another military ultimatum that demanded the immediate withdrawal of Serbian troops from Albania that they had violated ostensibly to pursue bandits, but in reality, were attempting to destabilize the fragile new government. Pasic notified Giesl that the Serbian troops were withdrawing immediately, and there was therefore no need for Austria-Hungary to mobilize its imposing army. Peace came, but the Belgrade press roared with calls of vengeance and war against Austria.

For years, Kaiser Wilhelm had been of the opinion that Vienna was unnecessarily nervous about Serbia, and ought to try to come to some friendly understanding with Belgrade. In the spring of 1914, when Vienna was greatly alarmed by rumours that Serbia, instigated by Russia, might attempt some union with Montenegro, the Kaiser appeared to be pro-Serbian rather than pro-Austrian. He regarded Vienna's efforts during the Balkan Wars to exclude Serbia from access to the Adriatic Sea as *nonsense* and of A-H's new effort to thwart Serbia from reaching the Adriatic via a Montenegro union, declared: *Unbelievable! This union is absolutely not to be prevented. And if Vienna attempts it, she will commit a great stupidity, and stir up the danger of a war with the Slavs, which would leave us quite cold.*
One must realize that in the long run Serbia and Montenegro will come together anyway.

The Kaiser agreed with Hungarian PM Count Tisza, who calmly accepted the union as imminent, rather than with Count Berchtold and Emperor Franz Josef who both proclaimed it as unacceptable. On 5th April from Corfu, Kaiser Wilhelm telegraphed his views to Chancellor Bethmann Hollweg.

It is absolutely necessary that the people in Vienna should face the possibility of a union of Serbia and Montenegro seriously ~ and be clear in their minds whether under all circumstances they would stand by the position taken by the Emperor and Count Berchtold, or whether they adopt Count Tisza's view. The first would only be possible in case they were absolutely firmly determined to prevent the planned union by a force of arms. In any case Austria must not put her prestige at stake, and publicly declare unacceptable things which she will ultimately be willing to permit. If they will agree to the sensible views of Tisza, Austrian policy will without further ado be able to adapt itself to the changed conditions in the direction which we have been preaching for years. There must be found a modus vivendi with the Dual Monarchy which will be attractive to Serbia.

U.S. Ambassador to Germany James W. Gerard at work in his Berlin office.

The U.S. Embassy on Wilhelm Platz

Ballroom of the U.S. Embassy

Grand Salon in the American Embassy

Working in the Embassy Ballroom

In early 1914, when the immediate danger of conflict seemed to pass for a few months, a détente in Anglo-German relations assisted by an offer of British cooperation for the *Berlin-Bagdad Railway* appeared to provide an opening to end the friction of alliances ~ the U.S. moved to preserve the peace.

Colonel Edward House, who was never in the U.S. military, had acquired his honorific title of Colonel from Governor James S. Hogg of Texas in appreciation for his political services rendered. *Unwilling to offend the Governor, I accepted the designation, although I immediately presented the resplendent uniform that came with it to my old Negro butler, who proudly wore it to many lodge meetings and funerals.*

House became one of the most powerful men in American federal politics. He was close to the Morgan banking dynasty. He personally chose Woodrow Wilson and secured his nomination for President on the Democratic ticket in 1912, then convinced various people with influence to support Wilson, which empowered him to win the election and become the 28th President of the United States. The new Administration, in favour of the proposal of First Lord of the Admiralty Winston Churchill for a naval holiday in the construction of warships, asked House to proceed quietly to Berlin to secure a private audience with the Kaiser in the early summer, and explain that ~ *Britain has buried the hatchet, and there is now much new hope that Imperial Germany can come into this good feeling with an innovative disarmament agreement between the United States, Great Britain, and Germany herself.*

In January 1914, Colonel Edward House wrote to U.S. Ambassador James Gerard in Berlin to learn of the Kaiser's plans, and was advised; *Wilhelm II will be in Corfu in the late spring, that he will return to Potsdam, go on to Kiel for the races, then cruise in Norwegian waters, and later to his Rhine estates.* House, as the President's confidential emissary chose the earliest option and cabled Gerard to arrange an interview in June, after Wilhelm's return from Achilleion on the island of Corfu. Achilleion Palace, previously the summer residence of the assassinated Empress Elizabeth who had it built it in the Pompiean style themed on Achilles. Among the islands' fragrant wooded hills, it overlooked the sea to mainland Greece. Wilhelm II purchased Achilleion from Elizabeth's daughter, Archduchess Gisella.

In Berlin, Ambassador Gerard requested for House to have a private discussion with the Kaiser. The wits of Wilhelmstrasse protested ~ they were perfectly willing that the Colonel should receive the satisfaction of an interview, but they insisted that some member of the Government must be present. House was equally firm in his insistence that it must be a téte-a-téte or nothing. *It is bluff on my part, but I decline to see him unless it can be arranged that I can see him alone. If that is not possible, then please do not bother about it at all.* House wanted to ensure that an official atmosphere would not encumber the frankness of their conversation. Colonel House departed for Europe, and soon after his arrival, he wrote to President Woodrow Wilson of his first impressions in Germany.

I was fairly well informed as to the situation here when I reached Germany. Prince Münster von Derneburg and the Count von Moltke were fellow passengers, and I came to know von Moltke well. Münster is what we would call a reactionary, and I let him do all the talking... von Moltke on the contrary, is perhaps the only noble in Germany who has a detached point of view and sees the situation as we do. He gave me valuable information, which merely tended to confirm my opinion as to the nearly impossible chance of bettering conditions.

I have not seen the Kaiser, but have been invited to lunch at Potsdam on Monday. Just what opportunity there may be to talk with him is an uncertainty. I have had long talks with von Jagow, Minister for Foreign Affairs, and Admiral von Tirpitz. Jagow is a clever diplomat without much personality. Von Tirpitz is the father of the navy and is forceful and aggressive. I was told not to talk to von Tirpitz, because of his known opposition to views we hold, but finding that he is the most forceful man in Germany excepting the Kaiser, I concluded to. We had an extremely interesting hour together ~ and I believe I made some dent. The situation is extraordinary. It is militarism run stark mad. Unless someone acting for you can bring about a different understanding, there is some day to be an awful cataclysm. No one in Europe can do it. There is too much hatred, too many jealousies. Whenever England consents, France and Russia will close in on Germany and Austria-Hungary. England does not want Germany wholly crushed, for she would then have to reckon alone with her ancient enemy, Russia; but if Germany insists upon an ever-increasing navy, then England will have no choice. The best chance for peace is an understanding between England and Germany in regard to naval armaments. It is an absorbing problem, and one of tremendous consequence. I wish it might be solved. E.M. House

25th May 1914. I had insisted before coming that we should not be entertained. There are only a few people I desire to meet. However, the Gerard's did not literally follow our desires and we had several people every day. On Tuesday, they gave a dinner of twenty-four covers, at which were Minister of Marine Admiral von Tirpitz, Minister for Foreign Affairs von Jagow, and British Ambassador to Berlin Sir Edward Goschen. E.M. House

I arranged that the Colonel should meet von Tirpitz at dinner in our house. We did not guess then what a central figure in this war the great admiral was going to be. At that time and until his fall, he was Minister of Marine, which corresponds to our Secretary of the Navy Department, and what is called in German Reichsmarineamt. The Colonel also met Bethmann Hollweg, von Jagow, Zimmermann and others. J.W. Gerard

The bluff of Colonel House and the diplomacy of Ambassador J.W. Gerard accomplished its task, and Gerard finally received word, that if he and House would come to Potsdam on 1st June, an occasion would be made for Colonel House to talk alone with Kaiser Wilhelm II. On that day came the ceremonies of the *Schrippenfest*, a traditional presentation of devotional militarism in the grand Prussian style, as the Kaiser loved dearly. House later wrote: *I cannot state too strongly my appreciation of the part the Ambassador played in finessing with Wilhelmstrasse in order to bring about the desired result.*

On 31st May, we took Colonel and Mrs. House to the aviation field of Joachimsthal. Here the Dutch aviator Anton Fokker was flying, and after being introduced to us he did some stunts for our benefit. Fokker was employed by the German Army and later became a naturalized German. The machines designed by him, and named after him, for a long time held the mastery of the air on the Western Front. J.W. Gerard

At noon on 1st June 1914, Colonel Edward House prepared for an audience with Kaiser Wilhelm in the Shell Hall of the *Neuen Palais* at Potsdam ~ where all the walls were encrusted with seashells. The Kaiser, in military uniform was seated at a crescent-shaped luncheon table. Opposite him, in evening dress sat Colonel House and James Gerard, in the Kaiser's words ~ *Looking like two black crows.* It was the day of the *Schrippenfest* ~ literally the *White Roll Feast*, (*Schrippe* being a roll of white bread) which was held annually on Whit-Monday in Potsdam for the *Lehr-Infanterie-Battalion.* This particular unit was an instruction battalion drawn from every regiment of the Prussian Army, and this day was the one occasion of the year when the soldiers received white instead of black bread, and when they were also treated to such luxuries as roast beef courses, stewed prunes in rice pudding and wine. The feast was given by the Kaiser who invited foreign military and naval attachés, the Berlin ambassadors, and distinguished strangers. The Kaiserin Augusta, and most of the Imperial family also attended. The outstanding feature of the ceremony was the Kaiser's presence at the table, sitting in the midst of his loyal troops, eating their white rolls, and drinking from a glass already used by one of his soldiers. This time, Colonel House and Ambassador Gerard were the Kaiser's only foreign guests.

House was flanked on one side by the Prussian War Minister General Erich von Falkenhayn, who insisted on discussing with House, highly intricate issues of military strategy. House vainly protested that he was not an actual military man, but his efforts were wasted. *Most of my time at luncheon was used in explaining to my neighbours the kind of Colonel I was ~ not a real one in the standard European sense, but as we would say home in America, a geographical one. My explanation finally reached von Falkenhayn's consciousness, but my neighbour was hopelessly befuddled and continued till the last to discuss army technique.*

When the luncheon of *Schrippenfest* was finished, the Kaiser took Colonel House aside, and the two withdrew to the terrace of Potsdam that overlooks the Park of Sans Souci. For nearly half an hour the Kaiser and the Colonel walked up and down ~ the German Generals at a respectful distance, watching the proceedings, puzzled and curious as to what it was all about. Two or three of the highest ranking Generals and Admirals present were in on the secret, but the mass of officers watching the meeting little guessed that the purpose of House's visit was to persuade the Kaiser to abandon everything for which the *Schrippenfest* stood ~ to enter into an international compact with the United States and Great Britain for reducing armaments, to reach an agreement about trade and the treatment of the peoples of all colonies, and to form some permanent association for the preservation of peace. They talked for half an hour ~ half an hour that possibly could have changed the fate of our world, if their circumstances had been more auspicious. But before it was possible to work out a more definite plan for world co-operation, the shot of a Serbian assassin would wreak its havoc.

As the Kaiser became more absorbed in the subject under discussion, his entourage tried to break up the interview several times. With an abrupt gesture every time, the Kaiser waved them away. Out of hearing, stood Ambassador James Gerard and Gottlieb von Jagow in keen conversation.

Edward House: *President Wilson feels that Europe is divided into two armed camps, guarding huge stores of ammunition. He fears the spark that may set the world on fire.*

Kaiser Wilhelm: *We Germans keep our powder dry, but we do not play with fire. Germany wants no more war, she has nothing to gain from war. I have maintained the peace of Europe for 25 years. Before Germany was poor. She is now growing rich. A few more years and she will lead the Old World in commerce.*

House: *But why Sir, do you constantly increase your military and naval budgets? Having already the largest army in Europe, why do you insist upon rivaling Great Britain on the high seas?*

Wilhelm: *I challenge the statement that ours is the largest army. I admit that we must maintain our strength. We are menaced on every side. The bayonets of all Europe are directed at the heart of Germany. We are surrounded by jealous and unfriendly neighbours.*

House: *Your neighbour across the Channel is not unfriendly. My conversations with Sir Edward Grey and Sir William Tyrell convince me that England is willing to cooperate with Germany. She recognizes Germany's right to more colonies, and a place in the sun. It seems to me that an understanding between England and Germany would make it possible to reduce the frightful burden of armaments. There are new worlds to be peacefully conquered. Asia, Africa, South America offer vast fields for concerted action. England and Germany, with the aid of the United States, can explore and exploit the waste spaces of the world.*

Wilhelm: *Nothing would please me more. I have always loved England, and everyone knows that I was Queen Victoria's favourite grandson. Do you think that I could ever forget her love for me? I have been attacked in my own country because I love England too well. Germany and the English-speaking nations are kinsmen. Many years ago, when I began the culture exchange between the United States and Germany, I paved the way for closer cooperation between England, Germany, and the United States. My hopes were frustrated by England. How could English statesmen form an alliance with Latins and Slavs, who have no sympathy with the ideals and purposes of our people, and who are vacillating and unreliable allies? England, Germany, and the United States are the light-bearers of the world. They alone can evolve the ultimate ideal of a Christian civilization.*

House: *I believe that the culture and the history of France entitle her to participate on equal terms.*

Wilhelm: *Perhaps… I have pocketed my pride more than once to conciliate the French, but my friendly gestures have been misunderstood. Now France is tied to Russia, and England to France. Both will pay dearly for their Russian alliance. The English will learn by experience the impossibility of making permanent or satisfactory alliances with either Russia or France.*

House: *But Your Majesty, it is your ever-growing navy that together with your enormous army, frightens England into an unnatural alliance with her natural enemies. England cannot permit Germany, almost a next door neighbour, to possess the largest navy as well as the largest army.*

Wilhelm: *It is Russia, not I, that threatens Great Britain in India. As a soldier, I have no doubt that Russia is systematically preparing for war against us, and I therefore frame my policy on that assumption.*

House: *Nevertheless, the day may come when English statesmen will have to decide whether they run more danger from a successful German invasion than from the possibility of losing their Asiatic colonies to Russia.*

Wilhelm very surprised, interrupts: *We never dreamed of an invasion.*

House: *British statesmen include every contingency in their calculations. Great Britain has no desire to see Germany crushed.*

The Kaiser looks at House in incredulous amazement…

House: *Germany's downfall would leave her to reckon alone with her ancient enemy, Russia. But she would rather envisage that possibility than the spectre of German naval supremacy.*

Wilhelm: *I am not building my fleet to attack Great Britain. Germany, without coaling stations, without naval bases, must build a navy capable of protecting her commerce everywhere on the seven seas, a navy commensurate with her growing importance and power. We must be strong enough to protect ourselves against the combined attack of Russia and France. We do not seriously consider the possibility of a clash with our British cousins. What grounds for conflict exist anywhere between England and Germany? Like Germany, America and England have nothing to gain from war. They can only lose.*

House: *I recognize the community of interest between England, Germany, and America. If they will stand together they can maintain the peace of the world. But there can be no understanding between England and Germany so long as you, Sir, continue to increase your navy. The English statesmen to whom I have spoken wonder when Germany will reach the end of her naval program.*

Wilhelm: *That is no secret. The matter has been thrashed out in the Reichstag. We have formulated a definite policy. We shall make no attempt to expand beyond the point needful to our defence. I am a friend of England. My army and I render an incalculable service to her by holding the balance of power against Russia in Europe.*

House: *Your Majesty's point of view and that of British statesmen is so nearly identical that it should not be difficult to harmonize the policies of the two countries. Both President Wilson and I feel that an American may be better able to compose the difficulties which still remain and to bring about an understanding with a view to peace. There is too much distrust and dislike in Europe. We are in a better position to act as go-betweens because we have no selfish interest at stake.*

The Kaiser nods approvingly…

House: *This is the work I have undertaken. I have come to Germany before going anywhere else to win, first, the approval of Your Majesty. After leaving Germany it is my purpose to go directly to England, where I shall take up the negotiations with the British Government in the same informal manner in which I have talked to you. If the reaction is favourable, I shall return and report to Your Majesty.*

Wilhelm: *By all means keep in touch with me. Any communication will reach me through our friend Zimmermann here in the Foreign office.* (One of the Kaiser's sons appears on the terrace to remind the Kaiser that their Imperial train from Potsdam is waiting ~ he waves him away. Kaiserin Augusta also comes out, but the Kaiser continues to talk to House. Finally, the Grand Chamberlain approaches in an embarrassed manner ~ Wilhelm dismisses him curtly, and continues the conversation.

Please tell the President from me that nothing is nearer to my heart than to preserve world peace.

House: *May I ask Your Majesty why Germany refused to sign the Bryan Treaty for Arbitration providing a cooling-off period of a year, before hostilities can be inaugurated?*

Wilhelm: *You would not ask this question, if you have realized Germany's unfortunate geographical position. For centuries, Germany has been the cockpit of Europe. No natural borders protect her from her enemies. Our only advantage is that we are ready. We have not signed, and never will sign a treaty that would deprive us of this advantage. Our strength lies in being always prepared for war at a second's notice. We will not resign that advantage, and give our enemies time to prepare.*

House bids farewell to his host, Kaiser Wilhelm.

The two men clasp hands, never to meet again.

Two weeks after his talk with the Kaiser, House wrote to U.S. Secretary of State, William Bryan from London on 13th June. *I had a most interesting visit to Germany, and will tell you of it all when I return. I have never seen the war spirit so nurtured and so glorified as it is there. The situation is dangerous in the extreme and I am doing what I can in a quiet way to bring about better conditions.* In London, after the British agreed in principal, Colonel Edward House attempted to persuade Sir Edward Grey to visit the Kiel Regatta, to see the Kaiser and discuss the plan with him personally ~ but the British Government were apprehensive that such a visit would be too disturbing to France and Russia. Already Winston Churchill's proposal for a naval holiday had so wrought up the French that a hurried trip to Paris by Asquith had been necessary to calm them. The consternation, which would have been caused in Paris by the presence of Sir Edward Grey at Kiel, can only be imagined. Thus, unseen German opposition and French suspicion left the British unmotivated toward the U.S. proposal, which thus rendered House's mission a failure. Unknowing of this, House reported to the Kaiser, the situation as he saw it.

Sir Edward Grey was sympathetic to the last degree. England must necessarily move cautiously lest she offend the sensibilities of France and Russia. While this communication is, as Your Majesty knows, quite unofficial, yet it is written in sympathy with the well-known views of the President, and I am given to understand, with the hope from His Britannic Majesty's Government that it may bring a response from Your Majesty, which may permit another step forward. Permit me, Sir, to conclude by quoting a sentence from a letter which has come to me from the President ~ Your letter from Paris, written just after coming from Berlin, gives me a thrill of deep pleasure. You have, I hope and believe, begun a great thing and I rejoice with all my heart.

I have the honour to be, Sir, with great respect,
Your Majesty's very obedient servant,
Edward M. House

They were skating on the thinnest of ice, and the peace of Europe was at the mercy of a chapter of unforeseen and unforeseeable accidents. British Prime Minister H.H. Asquith

PROGRESS ON THE BERLIN-BAGHDAD RAILWAY

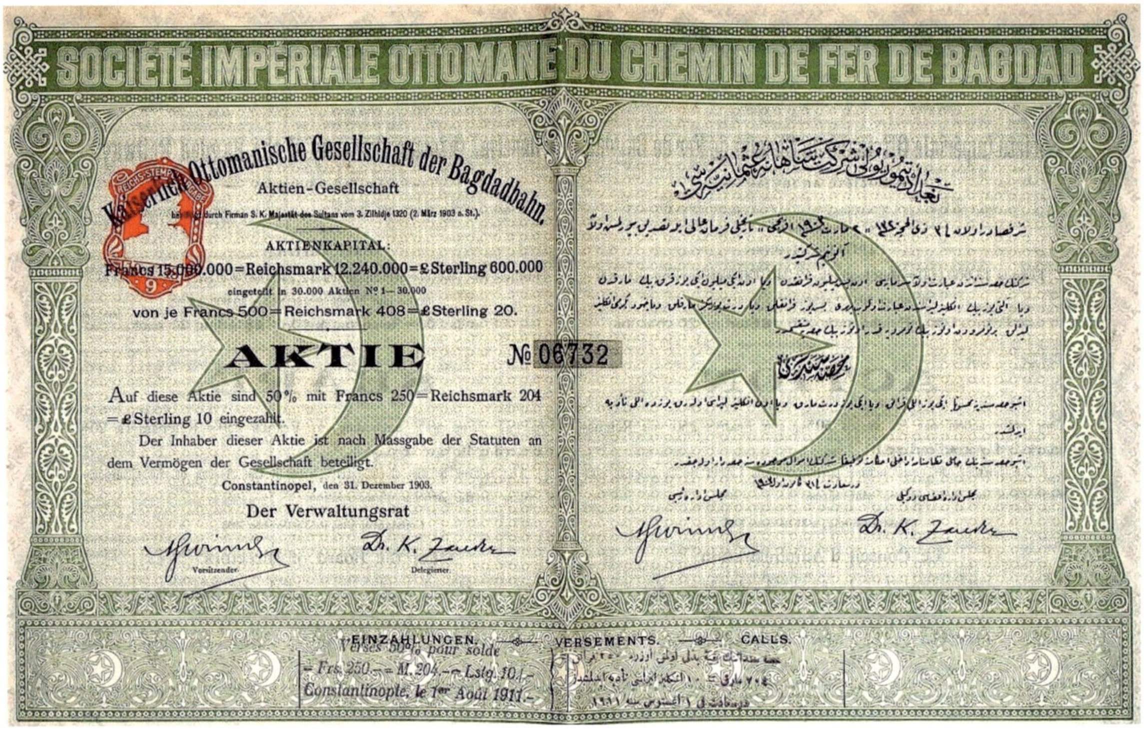

Share of the Baghdad Railway ~ issued 31st December 1903

The agreement for the Berlin-Baghdad Railway stipulates for the provision of a fortnightly express between Constantinople and the Persian Gulf. This train was to run at an average speed of about 28 miles per hour, including stops, for the first five years from the opening to traffic of the whole of the main line, with that speed subsequently to be increased to 37 miles per hour, including stops. This means, were the express train to run at its lower speed, that the journey from Constantinople to Baghdad could be accomplished in 55 hours, and from Constantinople to Basra in 66 hours.

The tourist desirous of travelling by the Anatolian and Baghdad railways is not provided with the comforts which would have led him to choose such a route in preference to one followed by a first-class ocean steamer. Leaving the Bridge at Constantinople at an early hour in the morning, one traverses the Bosphorus in a steamer run in connection with the train. From Haidar-Pacha to Eskisehir the first part of the journey took 10 ½ hours. Eskisehir is a tableland 2000 feet high; we had therefore a steady rise on the whole journey up the valley formed by the Kara Su River, which has its source in the neighbourhood of Eskishehr. On each side, rounded hills shut out the horizon, where here and there a tributary valley revealed, through steep-sided gorges, a distant view of purple ridges with snow-clad tops. It was night when I arrived at Eskishehr, and I groped my way to the Grand Hôtel d'Anatolie, kept by Greeks. As there were no night trains, one was compelled to sleep at the place.

Starting at 5 AM for Konia, I arrived after a journey of 15 hours. The dervishes are strong in Konia. Their founder is buried here, and his tomb is an object of pilgrimage. The chief feature of the mosque is its wonderful polished floor, where the dancing ceremonies take place. From Konia, starting at 6.30 AM the subsequent 190 miles to Karapunar is reached after 11 hours. The speed on all of these sections only averaged about 18 m.p.h. Practically nobody travels first class, and as the majority of natives, cannot afford anything better than third, the second-class passenger when he gets well into the interior is generally the sole occupant of the carriage. The trains are mixed passenger coaches and baggage cars, and therefore at all the larger stations one stops for a sufficient time to allow for the loading and unloading of goods. More indirect advantages are now accruing due to the fact that in railway districts the peasants, instead of only cultivating what they require for their own use and local markets, develop their land to full advantage as soon as they are able to easily send away their goods. The change in the increased prosperity of the people can be seen as the peasant's trade in gold instead of silver, with the result that a vast quantity of gold coin has disappeared into the interior.

C.H. Woods

The tragedy of the Great War has been ascribed to many different causes. It was of course, the natural outcome of foreign policies deliberately pursued, and behind the policy was the personality of those in whose hands were the destinies of Empires, especially those capricious personalities who guided the destinies of Russia and the Balkan Kingdoms. To thoroughly understand matters that might seem marginal but which were of themselves more important than the published treaties with which history deals, it is useful to assess the characters of the Russian Foreign Ministers concerned.

Eugene de Schelking of the Russian Diplomatic Service, held posts in Greece, France, Spain, and in Germany for fourteen years, where he was First Secretary of the Imperial Russian Embassy there until 1903, when he decided to leave diplomatic life for a career in journalism, as a correspondent in Paris for the Russian papers, *Rouss* and *Novoie Vremya.* After Paris, he wrote for *Birjivia Wiedemosti,* the largest paper in St Petersburg. While in this position he was officially asked to undertake missions for his government, and in consequence travelled a great deal in Europe ~ especially in the Balkans. He knew the Tsar's Foreign Ministers well, and noted some anecdotes relevant to our investigation.

The appointment of Count Mikhail Mouravieff was unexpected. The new FM had occupied as an official in the diplomatic service, many important posts. He had spent five years in Paris as First Secretary and had also been Councilor to the Embassy in Berlin for ten years, but he only held one post, as Ambassador to Denmark, in which no official was directly over him. This last post was of no great importance as it was out of the path of high politics and the more serious affairs of Russian diplomacy. Mouravieff had the reputation of being rather ignorant, having not graduated from a High School, or University. From the archives of the Embassy at Berlin, I was astonished to find that during his ten years as Councilor there he had only written three letters in his own handwriting and these, written in French, were full of grammatical and constructional errors. On the other hand he had a good deal of common sense, and extraordinary self-assurance, the last being his chief characteristic.

On his way to assume office in St Petersburg he had to pass through Berlin. I never remember having seen a man happier than he was at that time. He radiated sunshine. In his delight, he was eager to do everybody a kindness and sent for each of us secretaries in turn and asked us our ambitions, saying he would see that our wishes were granted. Our Ambassador gave a banquet for him, to which came the chiefs of the diplomatic world of Berlin with the Imperial Chancellor Prince von Bulow at their head. It so happened that Count Mouravieff's luggage had gone astray and he had to appear at dinner in his travelling suit. He was not in the least embarrassed and showed perfect confidence in himself and the impression he was going to make on our Ambassador's guests. In his conversation with the Princes Hohenlohe and Bulow he steered a very clever course, hiding his own ignorance of affairs and avoiding all dangerous subjects with consummate craft.

Diplomatic circles in Berlin were not displeased at his appointment. His good qualities and his weaknesses, which later Berlin proposed to exploit, were well known there. Kaiser Wilhelm however, was not at all pleased. Count Mouravieff, when Councilor in Berlin, always posed as a great admirer of the German Kaiser and the latter had shown him special favour. He had given him a large signed portrait of himself, and added some very flattering words over the signature. But Wilhelm learnt later that Mouravieff, while at Copenhagen ~ a Court much disliking the Kaiser, had made fun of him to please his éntourage. Ever after, he was hostile to Mouravieff.

As Minister of Foreign Affairs, Mouravieff was faithful to but one person and that was himself. He had little real personality, was very careless and remarkably trivial. He treated the problem of Kiao-chau recklessly. In many other matters, just as serious, he acted in the same way. His sole preoccupation was to cling to his office as long as possible. He knew how to win the sympathies of Tsarina Alexandra by means small in themselves, but large in results. He used to play at being a horse on his hands and knees with the Imperial children, which with other similar tricks pleased the Imperial Couple. As he fully appreciated the weak character of Nicholas II, he became a flatterer of the boldest kind to gain his own ends, and was successful in guessing his master's mind.

At this time, the personal relations between the Russian and German Monarchs were very strained. Instead of endeavouring to smooth over matters, Mouravieff added fuel to the flames and in consequence Russo-German relations became steadily worse and worse. Count Mouravieff had no sure policy, and no visible objective. His policy deteriorated into a series of vacillations. His enemies worked hard to bring about his downfall and had all but succeeded when he died very suddenly. His valet found him in his study lying on the floor with a gash in his temple, which had been caused by his head coming into contact with a sharp edge when he fell. It was rumoured he had committed suicide, being unable to bear the disgrace of dismissal which he knew was coming. Personally, I do not believe this. He used to drink a quart bottle of champagne every night before he went to bed. So ~ it was probably too great a satisfaction, rather than dissatisfaction with himself that carried him off.

Count Lamsdorff, who succeeded Mouravieff, was absolutely his opposite in temperament and character. Nicholas II the Unreliable, could always be relied upon to appoint ministers whose characters and opinions had not the slightest resemblance to their predecessors. Russian policies were continually chopping and changing.

Although far better educated than Mouravieff and with an intensive knowledge of affairs of state, Lamsdorff was very narrow minded and small of soul and lacked the saving grace of Count Mouravieff ~ commonsense. He was a religious bigot. Women had never played any part in his life and consequently rumour gave him the reputation of being a pervert. He lent colour to these rumours by showing great favouritisms to some of the men about him who were for the most part, strikingly handsome young men.

At that time, there was attached to the Chancellery a very handsome young man, Savinsky by name, who held the minor position of Third Secretary. Lamsdorff noticed him and chose him as his travelling companion in the Crimea when the Emperor went to Livadia. Savinsky returned from this short voyage a Gentleman of the Chamber and Second Secretary. A few months later he was made First Secretary. Lamsdorff wished to make him Master of Ceremonies at Court but was opposed by Count Hendrikoff, then Grand Master of Ceremonies. Lamsdorff insisted! He threatened to present his resignation to the Tsar in case his protegé was not appointed. As usual, Nicholas gave way and consented to Savinsky's appointment.

One day, at a gala performance in honour of the visit to Petersburg of the German Kaiser, Savinsky fulfilled the duties of Master of Ceremonies. During one of the entr'actes Tsar Nicholas, turning to Admiral Lomen, his aide-de-camp, who told me the story, said to him: Point out Countess Lamsdorff to me. He meant Savinsky!!! The Tsar showed that he felt something was wrong, but this did not prevent him appointing Savinsky his minister at Stockholm and, later, at Sofia, where he remained until the declaration of war. (See 130 189 294-95)

Prior to his appointment as Foreign Minister, Lamsdorff had never been beyond the confines of Russia, not even on a pleasure trip. He visited Central Europe for the first time when he accompanied the Tsar to Compiegne in France. Count Lamsdorff was popularly known as a rubber cushion such as invalids use in an inflated form. He was surrounded by nonentities, but he was sure of them and had known them for many years. In his policy, Lamsdorff tried to avoid complications. While carefully keeping up the friendly feeling and intercourse which existed between Russia and France, he also did his best to better our relations with Germany. The result of his policy was apparent during the Russo-Japanese War. The Berlin Cabinet surrounded us with favours, and the Kaiser sealed this friendship with us by a commercial that which was disastrous to our industrial interests.

Count Lamsdorff was frankly hostile to the war with Japan, and acted with great courage and honesty in this matter. He presented a long memorandum to the Tsar, and in this memorandum, he called serious attention to the grave dangers of any policy which would provoke or annoy Japan, as the well-known Yalu River Company with Monsieur Bezobrazoff, a new favourite of the Tsar at that time, at its head, was doing. Count Lamsdorff concluded this memorandum by stating his open opposition to an anti-Japanese policy, with almost brutal frankness and threatening to resign if it were continued any further. Nicholas II refused to accept his resignation and promised amendment, but the activities of the Yalu River Company in no way lessened, and the subsequent Russo-Japanese War was the result.

At first sight it seems strange that Count Lamsdorff gave proofs of such real statesmanship. The key to the puzzle is to be found in the reports which passed between Count Lamsdorff and Count Witte. The latter, was the originator of the policy of Lamsdorff, by whom he was constantly consulted. Lamsdorff had sufficient acumen to make himself Witte's mouthpiece and the Tsar knew perfectly well of the two men's relations and the credit with which the Foreign Minister thought he had covered himself, suffered severely in consequence.
Lamsdorff fell into disgrace and died soon after.

Count Witte, after the death of his first Imperial master, Tsar Alexander III preserved the greatest loyalty and pious affection for him. In speaking to me of the deceased sovereign he always referred to him as ~ My Tsar. But Alexander III could be severe when he deemed it necessary. If he discovered that the men whom he trusted were not worthy of his confidence and that his trust was misplaced, he rid himself of them forthwith. This happened in the case of Prince A.A. Lieven, Minister of Domains, and Governor Kryjanowsky of Orenburg. Both were deprived of their positions at a day's notice, Tsar Alexander III having personally found out that they were illegitimately trafficking in state lands.

The Tsar's choice of FM successor fell upon Monsieur Isvolsky.

Isvolsky graduated from the Imperial Alexandre Lyceum, with honours, receiving a gold medal ~ and even at that time his comrades prophesied he would eventually become Minister of Foreign Affairs. He had always dreamed of obtaining the post of Ambassador to Berlin and intrigued a great deal to secure it, but though he failed in this he achieved the higher office as had been prophesied. Incontestably very intelligent, with the very wide outlook and broad point of view of the real statesman, he had however, two great faults : limitless ambition and a snobbishness, which amounted to a disease. His acute snobbishness at least equaled his great ambitions. Of an unpleasing physical appearance, he nevertheless thought himself irresistible where women were concerned. In order to further his ambitions and his career, he would hesitate at nothing.

When he was First Secretary in Romania he fought a duel with a Romanian Officer, Lapteff by name. Prince Gortchakoff had once said ~ A diplomat who fights, is equal to a soldier who does not. This principle was always one of the rules of the Russian Foreign Office, but Isvolsky cleverly knew how to make use of the duel to further his interests. The cause of the duel had been a woman, but he let it be understood at Petersburg that, in reality, he had fought to defend the honour of his Imperial Master who (he claimed) had been insulted adversely. His version was believed and he was created a Gentleman of the Chamber, and as he could no longer remain in Bucharest because of this contretemps, he was transferred to Washington. Isvolsky only had a very small fortune of his own and he spent the whole of it in the United States in order to make a good impression in Washington, Newport, and in the Bar Harbour Society. All his friends were sure that he would marry an American heiress, but he transferred to Copenhagen, where he married Countess Toll, daughter of our Ambassador to Denmark, whose parents were very high in favour with Alexander III. This was really the beginning of his career.

After having been Minister of Foreign Affairs, he was made Ambassador to Paris and there had to receive the various members of a Republican Government, which to him, was torture. He preferred to associate with Dukes and Princes; but, of course, his position as Ambassador necessitated his throwing open his drawing rooms to powerful Republicans. Monsieur Briand, who was then Minister of Justice before being made President of the Council, knew Isvolsky's weakness in this matter very well. For that reason, he avoided all invitations to the Russian Embassy. When he became President of the Council, however, he could no longer refuse an invitation, which was personally sent to him on one occasion by the Russian Ambassador. He went and was received by Madame Isvolsky, who said to him: Monsieur Briand, it would seem that you have no pleasure in coming to us. Briand answered, Oh, Madame, surely not as much in coming, as you have in asking me!

On another occasion when he was giving a fête to which all classes were asked and, of course, all the members of the French Government, the Russian Ambassador was particular that the Duke of Vendôme, a member of the Royal Family of France, should be present. This was not easy to accomplish as the members of the French Royalty studiously avoid coming into any contact with the Republican Government representatives. Isvolsky therefore went to his British colleague, Sir Francis Bertie, and the following conversation took place; Isvolsky: I want very much to have the Duke of Vendome present at my reception. Sir Francis Bertie queried: Well, why don't you ask him? Isvolsky insisted: What would you do were you in my place? Ambassador Bertie: I wouldn't ask him at all. Isvolsky: But why then advise me to do so? British Ambassador Bertie replied: Because that sort of thing amuses you, whereas it bores me. Many other anecdotes of a similar character floated about the Paris salons and were most annoying to our Ambassador as they made him appear ridiculous. Isvolsky's snobbishness combined with his colossal self-assurance, made him follow policy dangerous for Russia. The annexation of Bosnia-Herzegovina was a striking example of his self-assurance.

In September 1908, Isvolsky was taking a holiday and was invited by Count Leopold von Berchtold, Ambassador of A-H at St Petersburg to come for some shooting to his castle at Buchlau, in Bohemia. Isvolsky accepted the invitation and there met Count Alois Aerenthal, Minister of Foreign Affairs of Austria-Hungary. Count Aerenthal had occupied for many years the post of Ambassador to Russia, and very naturally, knew our Minister of Foreign Affairs. Berchtold did things in a big way. Isvolsky was royally received at Buchlau Castle. Knowing the tastes of his guest, a whole galaxy of lovely women were present, and it was there that was discussed the question of the annexation of Bosnia-Herzegovina, which was to begin an era of complications in the Balkan Peninsula which had as their result the two Balkan Wars and ~ indirectly ~ began the Great War. His snobbishness and his incredible egotism had ruined him. In the salons of St Petersburg, he was given the soubriquet of ~ The Prince of the Bosphorus. In his conceit Isvolsky could not see that he was being mocked, accepting the nickname with pleased smiles. The foreign policy of the Russian Empire changed completely. We now swam openly in British waters, with the Anglo-Franco-Russo Éntente. Isvolsky was the real founder of Russia's new policy, which was continued by his successor, Monsieur Sazonoff.

Finding himself attacked on all sides by the press and by his enemies, and seeing the foundation of his power crumbling away under his feet, Isvolsky appointed Sazonoff Vice-Foreign-Minister in the hope that he would secure the support and influence of Stolypin. But his hope proved vain. A few months later Isvolsky was made Ambassador to Paris, and Sazonoff became Foreign Minister. Serge Sazonoff thus owed his career partly to Isvolsky, whose secretary he had been at the Holy See in Rome, and partly to the fact that he had married the sister of Mme. Stolypin, when Stolypin was President of the Ministerial Council and omnipotent in the Empire.

I knew Sazonoff intimately. He had been my colleague in the Foreign Office where our desks were opposite each other. We had both been appointed Imperial Chamberlains in the same year, and we made our début at the Court at the same time in March 1898, at Moscow during the Imperial Visit to the old capital. The new minister of Foreign Affairs ~ although not nearly so intelligent as his predecessor ~ he was by no means stupid. He was chiefly handicapped by lack of experience. His entire experience had been gained in London and at the Holy See, and included only one appointment as an Ambassador of Russia ~ as Resident Ambassador to the Papal Court.

As Chief of the Foreign Office at first he was modest, and showed himself eager to learn, allowing himself to be guided by his brother-in-law Stolypin. But he gradually became stubborn and insisted on displaying his own abilities, especially after the assassination of Stolypin. His successes in the Duma completely spoiled him, and the deputies credited him also with our understanding with Great Britain, and Italy. Each time that Sazonoff appeared in the Duma he was received with vociferous applause. This completely turned his head. In making decisions he refused to listen to any advice tendered by others, seeming to consider he was above criticism. Zinovieff, our ex-Ambassador to Constantinople and a diplomat well-known in all European Capitals for his intelligence, said to me ~ I never see Sazonoff anymore. He does not talk over affairs these days! He decrees!

But notwithstanding his self-assurance, Sazonoff was quite amenable to the influence of the Foreign Ambassadors accredited to Petersburg. Sazonoff naturally endeavoured to carry on Isvolsky's policies, especially where England was concerned. But at the same time, he was carefully watching Germany, and he concluded with the Berlin Cabinet an éntente regarding Persia, which was in my opinion very humiliating for Russia. He promised to link Teheran and Bagdad by rail at our expense! In this way, we lost our influence in Persia. In the south, our understanding with England gave the British a sphere of influence in the empire of the Shah, and in the north, where England had given us a similar sphere, this railway allowed the Germans to penetrate.

But his lack of experience and his unfortunate self-confidence were still more apparent in his Balkan policy. In the month of April 1912, Serbia and Bulgaria concluded an offensive and defensive alliance. It was obvious that this alliance had Turkey in view, and that war would ensue. I was then in the Balkans as special correspondent of my paper (Birjevia Wiedemosti ~ the largest daily in Petersburg) and had stayed for several weeks in Belgrade. From my conversations with Serbian Crown Prince Alexander, Prime Minister Pasic, and also with our Ambassador in Serbia Hartwig, quite probably the three best informed men on the Balkan situation I had gained the firm conviction that Serbia was at that moment ready for war, that this was quite understood and agreed to by Montenegro, and that Sofia was to give the signal which would precipitate war.

Later when I was in Bulgaria my conviction was confirmed. I was perfectly certain war would break out very shortly. More than that, on the 24th July 1912, I sent a telegram to my paper in which I predicted that hostilities would begin about the end of September. This telegram was quoted everywhere at the time and subsequent events proved I was right. On the 2nd of October 1912, the first shot was fired by Montenegro.

When the report was published, Sazonoff was visiting European capitals. He was absolutely certain that war would not ensue and expressed himself strongly in Paris and London. In speaking of the new Balkan Allies; They will never dare! When he reached Berlin, he learned that war had been declared and said to his intimates; The hounds have slipped my leash!

On his return to St Petersburg, I met him at the railway station of Tsarkoe-Seloe, the residence of the Tsar, where we were both staying. Serge Sazonoff asked me into his carriage to ride with him to St Petersburg.

He asked ~ Where the devil did you get your information? My answer astonished him ~ From exactly the same source as you could have had it just as easily. More especially of course, thanks to Nikolai Hartwig. You however, did not agree with Hartwig, and did not follow his line of reasoning, preferring to listen to the reports of Anatoli Neklyudov, our envoy in Sofia. Naturally as you did not want war, you only listened to those who were of the same opinion and desires.

It must be admitted that on the outbreak of hostilities, Sazonoff knew how to get out and keep out of trouble. Relying on our understanding with France, he soon succeeded in persuading Vienna to recognize two principles:
1 : *The dis-interestedness of the European Powers in the war, and*
2 : *The non-limitation of the theatre of war in the Balkans.*
Austria-Hungary evacuated the Sanjak of Novibazar, a Turkish Province she had militarily occupied since the annexation of Bosnia and Herzegovina, and thus permitted the Serbs to join with the Montenegrins. It is certain that Germany, and especially Austria-Hungary, were convinced the Slavic Balkan alliance would be crushed by Turkey, but when the victorious Serbs reached the shores of the Adriatic, Vienna received a severe shock.

Count Berchtold, at that time A-H Minister of Foreign Affairs, promptly proposed a conference of the Powers to discuss Balkan problems, and Sazonoff was weak enough to accept this proposal. The conference took place, and resulted in the evacuation by the Montenegrins of Scutari which they had occupied by force of arms; the creation of Albania as an independent state but naturally dependent on the Central Powers with a German Prince ~ Prince Wied, as Sovereign; and the retirement from Durazzo on the Adriatic of the unfortunate Serbs. The Serbians retired ~ nay, were forced into Macedonia. Bulgaria shrieked; Treachery! and a second Balkan war, this time between the Allies of the previous day, broke out in 1913.

On one of my visits to ex-Prime Minister of Russia I.L. Goremykin, he strongly criticized Sazonoff's policy. He said to me; Why did he go to London at all? Had I been Minister of Foreign Affairs and had in my pocket the treaty with A-H which did not limit the theatre of war, and had been asked to a conference, I should have simply replied; Gentlemen, let the Slavs and the Turks cut each other's throats to their hearts' content! and when they have had enough of this, they will make a treaty of peace, to hand around to the Chancelleries of Europe.

Count Witte too did not hesitate to criticize Sazonoff's work severely. Unfortunately, Goremykin, through indolence, and Witte, through impotence, did not bring their respective points of view to the Tsar's attention. Thus Sazonoff remained master of the situation. Afraid of a second Balkan War and desirous of preventing it, he persuaded the Tsar to intervene personally. This was a great mistake. The telegram of the Tsar to the Kings of Serbia and Bulgaria gave no practical result, and Russian prestige in the Balkans was seriously damaged.

The Second Balkan war was concluded by the Treaty of Bucharest in 1914. This treaty pleased no one. Serbia victorious was very dissatisfied at not having obtained an outlet on the Adriatic. Greece did not receive all she hoped for in Macedonia. Romania was not content with what she received in Dobrudja and felt she had been prevented by Russia from marching on Sofia. Bulgaria lost all round. The Bulgarians were furious with Russia because they believed Russia had forced Romania to intervene. The Treaty of Bucharest was chiefly the work of Sazonoff. It was obviously an expedient which sowed a harvest sure to be ripe for reaping one day. Yet when the harvest came in the Great War, as far as the Balkan States were concerned Sazonoff pursued his old policies, blind to their consequences. I am personally convinced that Sazonoff did not believe in the possibility of a European war which would develop into a world-wide war, and he impressed this belief on France and England. They considered their interests in the Balkans small beside those of Russia and naturally preferred to follow the lead of Sazonoff.

On the 1st June 1914, Tsar Nicholas II had an interview with King Carol I of Romania, at Constanza. I was there at the time, and learned through Romanian statesmen, and intimate friends of mine, that King Carol had made certain promises to Russia. For example a military convention that had been in force between A-H and Romania for more than 20 years would not be renewed. A-H in consequence, lost strong support in the Balkans. For my part, after I left Constanza I spent ten days in Bucharest, and then went to Budapest and Vienna.

Unfortunately, Russian diplomacy lacked initiative, and drew the Allies into the same mischievous situation. Dynastic considerations were allowed to affect diplomatic and military action, and this unhealthy state of affairs was ultimately the cause of the Serbian disaster. Greece in 1913 had concluded an offensive and defensive alliance with Serbia, which was primarily aimed at Bulgaria, but also had as an objective the possibility of a European war. It was stipulated in this treaty that if Serbia were attacked, Greece would come to her assistance with all her military and naval forces. In Paris, the true situation was realized. But as much cannot be said for the authorities either in London or in Petersburg. The King of Greece sent one of his brothers to England and another to Russia to plead his cause. Prince Christopher worked in London, and Prince Nicholas ~ married to a Grand Duchess of Russia, did his best in Petersburg. They both used every possible argument, and both Sir Edward Grey and Sazonoff seem to have been blind to the real situation.

Eugene de Schelking

Those directing Russia's Foreign policy neither trusted Vienna or Berlin, nor were they prepared to risk forgoing French support for the current all-important balance of power in Europe which involved thwarting any situation that led to an Imperial German military preponderance in Europe. Thus, in October of 1905, Count Lamsdorff denounced the attempted Björkö Treaty ;

...Kaiser Wilhelm's crude attempt to use the Tsar's signature on a compromising document to bring about disclosures in Paris and London which would be harmful for Russia, forcing her into isolation and dependence on Berlin. From many years' experience I have drawn the conviction that to be on genuinely good terms with Berlin, our alliance with France is necessary. Otherwise, we will lose our independence, and I know nothing more burdensome than the German yoke. If an isolated Russia ever found herself in an alliance with her more powerful German neighbour, the later would of course, not lose the opportunity to make her feel the whole difficulty and price of these bonds of iron.

On 14th May 1906, Russia's Ambassador in Berlin Count Nicholas von der Osten-Sacken advised: *German ambitions in Asia and the wider Moslem world, have for the first time made the German Empire a possible adversary of Russia's well-established interests in these areas.*

The Anglo-Russian Agreement of August 1907 that enabled Britain and Russia in a time of crisis, at first to diplomatically, and then to militarily confront any German attempt to dominate Europe, was made in an atmosphere of Russian Foreign Minister Isvolsky being astutely aware that Berlin would look with disdain on any agreement between London and St. Petersburg as he recognized by stating ; *Such an agreement will only benefit Russia if we do not arouse German resentment.* Striving to placate Berlin to ensure that German interests were unaffected by the Anglo-Russian agreement, he initiated negotiations with Berlin regarding the *Berlin-Baghdad Railway* and German interests in Central Persia. In a meeting in October, Isvolsky signed a secret agreement with Berlin to maintain the *status quo* in the Baltic, specifically drafted to clearly show Berlin : that Russia was not allied to a bloc opposed to German interests, but on the contrary, was willing on occasion to negotiate separate agreements with Berlin whenever their mutual advantage required appropriate joint action.

As the Viennese protested that London only viewed A-H policy thru the distorted prism of its suspicion of Germany ~ so Russian observers saw Berlin's obsession with its sparring with London leading it to misinterpret fundamental intentions. One of the psychological conditions out of which armed conflict between nations might arise was in the undeniable existence of what Viscount Haldane described as ~ *a set of colossal suspicions of each other by all the nations concerned.* With these suspicions, being industriously fostered in all countries, Isvolsky's foreign policy was seen as the precarious plot it was by Sir Arthur Nicholson, who wrote in January 1908 ; *Isvolsky's attempt to find balance between the Anglo-French and German-Austrian blocs, and to maintain good relations with all powers is a policy which will require some skill and adroitness to pursue for any length of time, and it is doubtful if it will be a feasible one.*

With trouble on the Balkan Peninsula expected at any time, FM Isvolsky warned in January 1908; *Events in the Balkans do not depend on us. Russia at any time can have a crisis imposed on it against its will.* Russian nerves were stressed when the A-H Foreign Minister Aehrenthal failed to warn St Petersburg in advance of Austrian railway construction in the Sandjak of Novibazar on the Serbian-Montenegrin border to link Sarajevo to Salonika ~ and with a complete lack of caution dually derived from a desire for a diplomatic triumph and an over-confidence in his own skill, Isvolsky foolishly confirmed a secret Austro-Russian agreement in writing for the Hapsburg Monarchy to annex Bosnia, Herzegovina and the Sandjak ~ if Russia could soon secure its historic interests in the Straits. After Isvolsky went to London, where he found the British opposed to his plans for the Straits, Aehrenthal declared that the Russian FM had given his unconditional consent to the annexation. When the Tsar heard of this, he knew further Russian protests would be fruitless, even dangerous, as the annexation was of two provinces already in practical rule under A-H administration. Nicholas II ~ *It sickens one's feelings...*

As a finale to the annexation crisis, in the last week of February 1909, Vienna delivered notes in Belgrade and St Petersburg demanding full recognition of their annexation; and making it abundantly clear that failure to comply would eventuate in an A-H invasion of Serbia. Russian attempts to seek mediation thru Berlin on 21st March only sparked a stern German response when Berlin stipulated their firm demand for an immediate, unequivocal, and unconditional acceptance of all A-H claims.

After the September 1910 appointment of Serge Sazonoff as subsequent Russian Foreign Minister, he travelled to Berlin for talks with Bethmann Hollweg to promote an atmosphere of enhanced trust between the two Empires, and to show that Russia was willing to grant concessions to Germany in order to secure good relations. Once Berlin agreed not to back any hostile A-H moves in the Balkans, not to construct any railways on areas bordering Russia's Caucasian provinces, and not to seek any concessions in Russia's northern zone of influence in Persia ~ Russia, much to the dismay of her Éntente partners; dropped all long-held opposition to *The Berlin-Baghdad Railway* and consented to link the *Berlin-Baghdad* to Russia's Persian Line by themselves constructing a link from Tehran-Khanikhin. In the two years following this agreement, Russo-German relations improved to become even warmer than those between Paris and London ~ as experienced in the Moroccan Crisis, when Russia's new Ambassador to France Isvolsky cautioned the French ~ *The Franco-Russian alliance is purely defensive and cannot be invoked automatically to cover an obstinate French stance towards German claims in Morocco.*

In September 1912, French PM Raymond Poincaré assured Isvolsky. *If an Austrian invasion of Serbia or any other Balkan crisis brings a war between Russia and the Central Powers, the French Government recognizes this in advance as a casus foederis and will not waver for one moment to fulfill the obligations lying upon it with respect to Russia. The French General Staff is confident of victory in the event of a European war.*

Still Sazonoff regarded French support as insufficient. *Russian action with the support of France is not especially dangerous for Germany. Both powers are scarcely capable of dealing Germany a mortal blow even in the event of success on the battlefield, which is always uncertain. But a struggle in which England took part might be fatal for Germany, with the danger of being reduced to total internal catastrophe given English action.*

British Ambassador to Russia, Sir George Buchanan cabled Sir Edward Grey on 15th May 1913: *Like Nicholas II and Kokovstov, Sazonoff believes that the Germans are so scared of Britain that a firmly-based defensive alliance between the three Éntente powers will effectively deter any German dreams of aggression, thereby forcing Berlin to treat with respect the interests of Europe's other powers.*

Of Russia's historic ambition for the Straits, Chief of Naval Staff, Prince A.A. Lieven advised ; *The secure possession of this region requires the domination of its hinterland ~ both in the Balkans and Asia Minor.* For the successor-state of Byzantium to witness the possession of Constantinople by a rival would have been difficult. With Russian economic fundamentals based on shipping access to the Straits through which 70% of Russian grain exports and 40% of other exports was shipped, and whose brief closure during the Balkan Wars had alarmed Russian society. Nicolas de Basily advised ; *Any attempt to ship these bulky exports via overland routes would have been extremely expensive, moreover these routes are controlled by the Central Powers.* Sazonoff held the opinion that ; *The Straits in the hands of the wrong state would signify the complete subordination of the economic development of Southern Russia to this state.*

Therefore, when the announcement came from Berlin that a German General would be appointed to command the garrison at Constantinople, fury raged thru St Petersburg. German influence ensured the existence of a pro-German Turkish régime to intrude upon Russia's aim of securing the Straits.

Growing economic power entails the expansion of German influence and conflicts with the established interests of other states. At the root of everything, I see the gigantic force of expansion of Germany which carries along with it its influence and inevitably its flag. This expansionist force in no way necessarily means that the Berlin Cabinet is deliberately waging an aggressive policy, but it entails countermeasures on the part of the other powers, which always creates the danger of conflict.

Count Alexander Benckendorff
Russian Ambassador to Great Britain

All that other nations attained in centuries of natural development ~ political union, colonial possessions, naval power, and international trade, was denied to our nation until quite recently. What we now wish to attain must be fought for, and won, against a superior force of hostile interests and powers. We must rouse in our people, the unanimous wish for power in this sense, together with the determination to sacrifice on the altar of patriotism, not only life and property, but also private views and preferences in the interests of common welfare. Then alone shall we discharge our great duties of the future, grow into a World Power, and stamp a great part of humanity with the impress of the German spirit.

Friedrich von Bernhardi
Germany and the Next War

REVELATIONS OF RUSSIAN FOREIGN POLICY

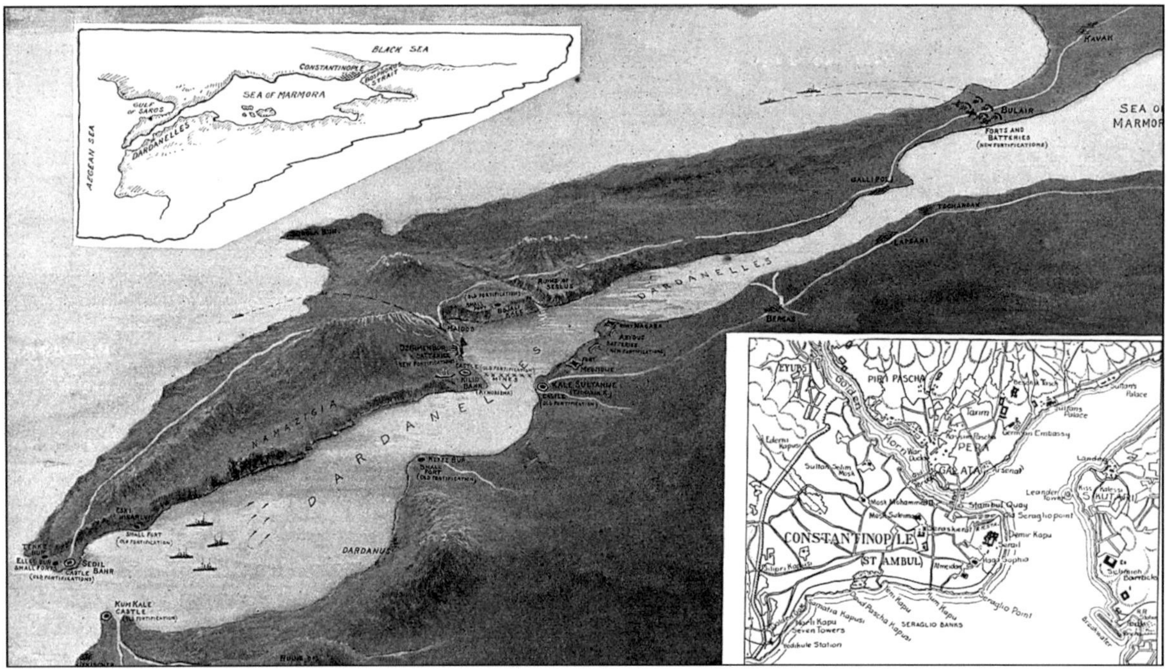

Russia of the Tsars, was in a state of continual internal ferment, and every Tsarist Government had to keep the possibility for a foreign conflict in ever readiness, in order to always be able to deflect attention from inner strife to foreign difficulties; to have a safety valve as an outlet for the passions that might lead to revolution at home. Russia still clamored for a maritime outlet thru to the Ægean, and there was the significant Austro-Russian focus of conflict in the Balkans that concerned Germany. From this perspective, several prescient items of alarming evidence appeared as early as 1913.

The Tsar himself told the Austrian Ambassador in the spring of 1913, that if Germany and Austria insisted on a war they would have one, but before it was ended, all three Emperors would be tottering on their thrones. There is no reason to suppose that he reconsidered this well-justified presentiment. John Alfred Spender

The military activity in Imperial Russia displayed from 1913, gained extraordinary intensity. The restored Russian Army on a peacetime footing of 1,400,000 was in the opinion of the St Petersburg correspondent of the London *Times* : *an effective numerical strength hitherto unprecedented.* From January, the Russian Government made immense purchases of war material, and in February, concluded a loan in Paris for the improvement of its strategic roads and railways towards the A-H and German frontier. Always with aspirations of Constantinople, the Imperial Crown Council of Ministers convened on 21st February 1913 in St Petersburg to conclude that only a European war would enable them to realize their dream of seizing Constantinople to gain control of the Straits, and had all decided ~ *To work out a programme of action, in order to secure for us a favourable solution ~ of the historical question of the Straits.* In his report of the Crown Council presentations, Sazonoff formally addressed the Tsar. *In the expected crisis, which will possibly very soon give us an opportunity for settling the Straits question ~ the success of operations will be bound up with international conditions. To prepare the ground for this is the task of the Foreign Ministry. The St. Petersburg Cabinet having calculated on opening the floodgates of war to steer the Russian Ship of State on a high tide of blood, it is now for Russia to settle this question during a European war.*

A Russian member of the Duma and a close friend of Sazonoff heard about this Crown Council meeting where Sazonoff had addressed the group, suggesting the possible capture of Constantinople. *The Triple Alliance will not acquiesce in it, as this would cause a war against Germany and Austria. Italy will break away from these two, and in the course of events, France is to be trusted absolutely, and England probably.* Nicholas agreed and gave orders to take the necessary preliminary steps. This same Russian member of the Duma would later write: *Two days after the outbreak of war I was invited by Sazonoff to breakfast. Sazonoff came up to me beaming with joy, rubbing his hands together ~ Come now, my dear Baron, you must admit that I have chosen the moment for war excellently, haven't I? When I, rather worried, asked him what stand England would take, Minister Sazonoff smote his pocket, and, with a sly wink, whispered ~ I have something in my pocket which, within the next few weeks, will bring joy to Russia and astound the entire world. I have received the English promise that England will go with Russia against Germany!*

In January 1914, after visiting Belgrade, the ex-Chief of the Russian Cabinet of Foreign Affairs, and current Minister Plenipotentiary to Bulgaria, Alexander Savinsky reported from his observations ;

On January 21st I took the Cannes express. After a short stop at Vienna, I arrived in Belgrade, where Sazonoff had asked me to talk with Pasic and Hartwig, our Ambassador to Serbia in order to agree upon the line of conduct we should follow in our corresponding activities. I did not see Pasic as he had just left for Russia with Prince Alexander. As to Hartwig, I found him in rather a nervous state. He had wanted to accompany Prince Alexander to Russia and had applied to the Ministry for the permission, but as yet no reply had been given. Knowing what Sazonoff thought of Hartwig's impetuous temperament, I suspected that the Minister did not wish to have him in St. Petersburg at the same time as Prince Alexander and Pasic. Subsequent talks I had with my Belgrade colleague confirmed my supposition. After two days at Belgrade, I reached Sofia on 27th January.

I have become convinced that Hartwig's thought is to set Serbia on to Austria. The information that Hartwig passes back to St Petersburg from Belgrade is on occasion highly selective. Moreover, one looks in vain in his published dispatches for an appreciation of the dangers posed by organizations such as the Black Hand to a power guaranteeing Serbia from Austrian invasion. In addition, Hartwig cannot always be relied upon to fulfill his instructions loyally, nor is the effect of the Russian Government's counsel of restraint in Belgrade likely to be of value, when its representative has not endeavoured to conceal the fact that he considers Sazonoff's policy despicable and misguided. This is especially true, given the open knowledge about this Ambassador's powerful connections in St Petersburg and Imperial approval signified by the award to Hartwig, of the Order of the White Eagle in the Spring of 1913. Alexander Savinsky

* * *

The reason why Pasic and his entourage were in St Petersburg (as mentioned above by Savinsky) is wholly relevant to our investigation. By early 1914, two aspects were becoming apparent to both King Petar of Serbia, and to the European Powers who surrounded his turbulent Balkan kingdom. Firstly ; the inevitability of a showdown between the *Greater Serbia* militarists led by Colonel Dragutin Dimitrievic or *Apis* and the comparatively moderate civilian régime led by Radical PM Pasic.
Secondly ; if any outside emergency support were needed ~ it could only come from Russia.

The focal point of this dependence came in January 1914, with a baptism in St Petersburg. The mother of the baby boy was Princess Heléne, King Petar's daughter who was married to Prince Ioann, the son of Grand Duke Constantine. King Petar had agreed to go to the ceremony with Crown Prince Alexander, and their PM Pasic. King Petar was aiming to bind his minor Slav dynasty of Karageorevic to the powerful Romanov dynasty by arranging the marriage of his heir to one of the Tsar's daughters. After the christening, Pasic and Crown Prince Alexander would remain in St Petersburg to hopefully secure the much-desired family link. In a report dated 20th January 1914, sent to Belgrade for King Petar, Pasic described his one-hour audience with Tsar Nicholas, during which he offered to the Tsar ; through a proposed marriage bond, a second empire for his daughter, part of it at the expense of the rival Hapsburg Monarchy.

There was much preliminary political talk of how best to counter the Austrian thrust, backed by Germany, into the Balkans, now that these two mighty allies are arming Bulgaria ~ with whom Serbia has remained on disturbed terms since the Second Balkan War ~ with large consignments of rifles, ammunition, and howitzers. Could Russia, who has been rearming so vigorously herself since the 1908 crisis, give similar help to Serbia? The Tsar has agreed to look benevolently at a list already handed over to his Minister of War, Soukhomlinoff.
Then, I came to the real point.

If only it could be granted to us to have as our queen, a daughter of the Tsar of Russia, she would not only have the feelings of the whole Serbian people behind her. She could become, if God and the situation permitted it, Tsarina of the whole South Slav and Serbo-Croatian peoples. Her influence and her glory would stretch over the entire Balkan Peninsula. Tsar Nicholas heard these words with visible joy. On the subject of the marriage he said neither yes nor no in advance, but indicated ~ It is up to the Crown Prince, who should stay on in St Petersburg to do his wooing for himself. I withdrew from his presence convinced of success in this matchmaking mission. The Tsar's final words to me were ~ Tell your King that ~ For Serbia, we will do everything. Nikola Pasic
PM of Serbia

SOUKHOMLINOFF'S BOAST

Colonel Viktor Artamonov
Russian Military Attaché to Serbia

Count Vladimir Kokovtsoff
Prime Minister of Russia 1911-14

General Vladimir Soukhomlinoff
Russian Minister of War

In March of 1914, a St Petersburg newspaper that served as the mouthpiece of the Minister of War Vladimir Soukhomlinoff, printed an article stating: *Russia's strategy will no longer be defensive but active. Time is coming when the crossing of the Austrian frontier by the Russian Army will be an unavoidable decision. It is important that Russian public opinion be conscious that the country is prepared for all possibilities.*

When, this bellicose article under the heading ~ WE ARE READY ~ announced to the world that we were ready not only for a defensive but also for an offensive war appeared in one of the leading Russian newspapers, and when it became known that this article of an interview with Minister of War General Soukhomlinoff had been dictated by the Minister himself to a representative of the influential Gazette : Birjhevye Viedomosti, it had the effect of bringing grist to the mill of the German militarists by enabling them to point out that the Russian menace was not a creature of their imagination, but now a serious reality. It is not easy to understand what could have prompted General Soukhomlinoff to publish this boast ~ for it could hardly have been intended as a bluff to intimidate a potential enemy, which would have been a silly, and deliberate provocation. One could only attribute it to that same irresponsible recklessness which in union with the wrong-headed incompetence of the Minister of Foreign Affairs, would at the critical moment, push the country over the brink of the precipice. General Soukhomlinoff's idle boast in his published interview about our readiness not only for a defensive but also for an offensive war, must have been extremely welcome to the German militarists. Baron Roman Rosen

At the end of March 1914, the Director of the St. Petersburg War Academy General Sherbatshev addressed his officers : *War with the powers forming the Triple Alliance has become unavoidable on account of Austria's anti-Russian Balkan policy. There exists the strongest sort of probability that it could break out as early as this same summer, and for Russia, it is a point of honor to assume the offensive immediately!*

In April 1914, trial mobilizations were begun in Russia that were thereafter continuous. Belgian Ambassador in Berlin Baron N.E. Beyens reported on a Japanese military mission that had arrived from St. Petersburg. *At the Russian regimental messes, the Japanese officers have heard quite open talk of an imminent war against Austria-Hungary and Germany. It was stated that the army was ready to take the field, and that the moment is as auspicious for the Russians as for their allies, the French.* German Marine-Attaché in Tokyo *Kapitän* Eduard von Knorr cabled to Berlin : *I was greatly struck by the certainty with which everyone here foresees a war of the Triple Éntente against Germany in the near future. There is now something in the air as if, so to speak, people are expressing their condolences over a death sentence not yet pronounced.*

An American, writing of his trip through the Russian Caucasus in the spring of 1914 wrote:
When I arrived in the Caucasus, at the beginning of May 1914, I saw on my way to Tiflis, long columns of troops of all arms with war equipment. I had feared that a revolt had broken out in the Caucasus. When I made inquiries of the authorities at Tiflis, while having my passport inspected, I received the news that the Caucasus was peaceful, that I might travel wheresoever I wished, and what I had seen, had to do with practice marching and maneuvers. At the close of my trip, I wished to embark at a Caucasian port, but all the vessels there were so filled with troops that only after much trouble could I manage to get a cabin for myself, and my wife. The Russian officers told us that they were to land at Odessa, and march from there to take part in great maneuvers.

In the spring of 1914, when Nicholas II was questioned by his Court Marshal of his spring and summer holiday plans, the Tsar replied ; *Je resterai chez moi cette année parceque nous aurons la guerre... I shall stay at home this year, for we are going to have a war.*

A memorandum written by Hungarian PM Count István Tisza to Emperor Franz Josef II of A-H contained a predictive new word ~ *Weltkreig* ~ World War. *It is my firm conviction that Germany's two neighbors Russia and France are carefully proceeding with military preparations, but they will not start the Weltkrieg so long as they have not attained a grouping of Balkan states against us that confronts the Monarchy with an attack from three sides and pins down the majority of our forces on our eastern and southern front.*

Then, came the reports of Tsar Nicholas and FM Sazonoff travelling abroad in June of 1914 ;
Whether official Russia, four weeks in advance, really counted on the assassination of the Archduke and the outbreak of a war ensuing remains an open question ~ but one thing is certain, that Russia within a measurable space of time had prepared for war as being inevitable and had endeavoured to secure Romania's co-operation. When the Tsar was at Constanza a month before Sarajevo, Foreign Minister Sazonoff paid a visit to Bucharest. When there, he and Bratianu went on a walking tour together to Transylvania. I shared Berchtold's surprise... By chance, I overheard a conversation between two Russians at Hotel Capsa, a known resort for anti-Austrians. They were sitting at the table next to mine in the restaurant and were speaking French quite freely and openly. They appeared to be on good terms with the Russian Ambassador and were discussing the impending visit of the Tsar to Constanza. They agreed that the Emperor Franz Josef could not live very much longer, and that when his death occurred and a new ruler came to the throne it would be a favourable moment for Russia to declare war on us. They were exponents of the "loyal" tendency that aimed at declaring war on us without a preceding murder; and I readily believe that the majority of the men in Petersburg who were eager for war held the same view. On 2nd June Sazonoff came to Bucharest after having accompanied Tsar Nicholas on his visit to King Carol of Romania at Constanza. I took hold of the opportunity to write him and imbued with fearing lest Sazonoff's desire for a rapprochement with Romania should lead him to lose sight of Bulgaria, I wrote to him ; While fully realizing how greatly important it would be for us to separate Romania from the Triple Alliance, I hold that Bulgaria's political orientation cannot be a matter of indifference for us, and this not solely because of our traditional relations with her. Bulgaria's attitude should be far from indifferent to us. How much stronger would our position be if, at the moment of complications, we could be sure of the support of Bulgaria and Serbia.

The season of 1914 was one of the most brilliant ~ as it was to be the last that St. Petersburg would see. Society was gaily dancing on an unsuspected volcano, quite unconscious of the approaching catastrophe ; nor could anyone even dream of the depth of misery and unspeakable horror to which a once magnificent capital, with its teeming population and thousands of happy homes, was to be reduced in so near a future. Among the memories of that fateful season, I recall an afternoon dance at the palace of the Grand Duchess Vladimir, to which she had invited the Emperor and Empress with their four daughters, the youngest two mere children. It makes my heart bleed when I see before my mind's eye the radiantly happy faces of these innocent young ones rapturously enjoying their first ball ~ which, alas was to be their last ~ blissfully unconscious of the unutterably awful fate which was in store for all of them... Baron Roman Rosen

The Russian military attaché to Belgrade, Colonel Viktor Artamonov saw *Apis* nearly every day as they were conducting a joint secret-service operation across the nearby Hungarian border with the help of Serbian customs inspectors and frontier guards, who had been recruited as operatives of *Apis*. The agents of Colonels Artamonov and *Apis* collected much military intelligence, and also engaged in subversive propaganda activities, and among other things encouraged extreme nationalist groups. The *Black Hand* was all subsidized by Russian Ambassador Hartwig in Belgrade as the *alter ego* of Russian Foreign Minister Sazonoff, and according to former collaborators of *Apis*, Victor Artamonov was fully informed about the subversive and intelligence aspects of all projects by Serbian operatives.

There were already four coded telegrams in the possession of the French Government on 8th July, sent by the Belgrade Government to their Ambassador Vesnic in Paris; further there was a photograph of a letter, in which Colonel Dimitrievic Apis, the chief of the Black Hand, had given detailed instructions for his delegate to Paris, the Serbian military attaché. (The secret Serbian code was already in the possession of the French Government). Pan-Serbian secret organizations are directed and financially supported by Ambassador Hartwig in Belgrade. Unity or Death is supported by him with 1500 Rubles monthly. On 19th June 1914, Hartwig submitted 3000 Rubles to Major Voija Tankosic for the purpose of the Attempt of Sarajevo. Hénri Pozzi, *Le Deuxiéme Bureau*

Kaiser Wilhelm II and the Archduke Franz Ferdinand

~ ENTR'ACTE ~

The Archduke Franz Ferdinand & below with A-H Emperor Franz Josef

THE ARCHDUKE FRANZ FERDINAND

Franz Ferdinand generally prevailed in internal debates over Vienna's response to Balkan issues, whenever he spoke out against Chief of Staff Conrad, who in 1913 alone had called for a pre-emptive war against Serbia no less than 25 times. When different factions vied for Emperor Franz Josef's decision, Franz Ferdinand's clarity of wisdom was the decisive factor when he frequently warned that any severe treatment of Serbia would bring A-H into direct conflict with Serbia's patron Russia, and as Franz Josef's views were more in tune with those of his nephew the Archduke, Conrad and Count Leopold von Berchtold's plans were repeatedly frustrated by the heir apparent to the A-H throne. Franz Ferdinand kept the military in their place and vetoed away all Balkan military proposals. *These would merely win cheap laurels and would run the risk of escalation, with a war on two or three fronts.* Having repeatedly chastised Conrad for urging Franz Josef to attack Serbia, the heir apparent was estranged with Conrad over this issue. In September 1913, Conrad suggested to Count von Berchtold; *Now would be the opportunity to put things in order down there. An ultimatum, and if Albania is not evacuated in 24 hours, then mobilization.* Berchtold: *I personally support military measures, but feel no confidence that authoritative quarters will stand firm.* Conrad: *On peace and war, the decision lies solely with the Emperor.* Franz Ferdinand, who called Conrad's militarist clique ~ *a witches' kitchen of war*, wrote to Berchtold;

Excellency! Don't let yourself be influenced by Conrad ~ ever! Not an iota of support for any of his yappings at the Emperor! Naturally he wants every kind of war, every kind of hooray! Rashness that will conquer Serbia and God knows what else... It's unforgivable and insane to start something that would pit us against Russia. Suppose we wage a separate war against Serbia? In no time at all, we will overthrow it, but what then, and what good will it do us? God save us from ever annexing Serbia; a country over its head in debts, brimming with regicides. And scoundrels! As it is, we cannot cope with Bosnia ~ and Serbia will be far worse! We can throw away billions there and still be faced with terrible irredenta. Irredentism in our country will cease immediately if our Slavs are given a comfortable, fair, and good life instead of being trampled on.

The Habsburg Empire genuinely tried to advance the level of civilization and higher education of the people of Bosnia-Herzegovina during the years before the war. After annexing the territories formally into the domain of the Danubian Monarchy in 1908 to obviate any Turkish or Serbian hope of regaining the territory, the Austrians built 4,000km of railway, numerous schools thru which more than half the population became literate by 1913 plus roads and public buildings throughout Sarajevo. More important however, from the point of view of our investigation into the causes of the Great War, was that Franz Ferdinand once tried to establish a three-part division of A-H which would entail setting up of a South Slav Kingdom (reigned over by a Habsburg monarch) which would include the Croats and the Serbs in a unified realm with the same autonomy as the Kingdom of Hungary.

Scholars known to the Archduke suggested the creation of *Vereinigte Staaten von Groß-Österreich* or a *United States of Greater Austria* a federation of autonomous units based on language and ethnicity. Franz Ferdinand thus hoped to resolve other ethnic issues and planned to reform A-H by remaking the Empire as a federal state to absorb Serbia peacefully, and advised Baron Albert von Margutti; *The introduction of the dual A-H system in 1867 has been disastrous, and when I ascend to the throne, I intend to re-establish strong central government ~ this objective I believe, can be attained only by the simultaneous granting of administrative autonomy to all the nationalities of the monarchy.* Berchtold said of the Archduke; *If he had succeeded to the throne, he would have tried to replace the dual system by a supranational federation.*

But as A-H was in such a state of latent crisis, its mosaic of nationalities could hardly be held together without radical constitutional reforms. Proposals for *U.S.G.A.* were opposed for different reasons by the Hungarian nobility in the east, by German nationalists in the west, and bitterly opposed by Serbian nationalists hoping to become the nucleus of a Greater Serbia who could never brook the concept of being incorporated as an appendage into a peaceful multi-ethnic state of Franz Ferdinand's new Danubian Monarchy in a futuristic form. Imperial Russia also dreaded any outcome of Franz Ferdinand ascending to the throne to see him unite the southern Slavs under the Hapsburg flag to foil Romanov expansion in the Balkans. With the expectation of Franz Ferdinand having to subdue ethnic resistance to his *United States Plan*, as well as implementing it, A-H would not be ready to join German military operations for years. With German High Command's fundamental belief that starting a war would be futile after 1917 due to Russian rearmament, and as Germany could not launch a war without her A-H allies, it is highly probable that had Franz Ferdinand endured the events of Sarajevo, the German militarists would have been compelled to abandon their war plans

Countess Johanna von Schönburg, Major Baker-Carr, Victoria Bentinck, the Duke of Portland, Duchess of Hohenberg, Winifred Portland, Archduke Franz Ferdinand, Count Alois von Schönburg

When the Archduke Franz Ferdinand and Duchess Sophie visited England in the autumn of 1913, they spent 17-21st November with King George V and Queen Mary at Windsor Castle. The A-H royals spent their time shooting, shopping, and meeting dignitaries such as Prime Minster Herbert Asquith. One of the principal reasons why Franz Ferdinand and his wife Sophie liked to travel abroad was that ever since their marriage ~ the Countess Zofie Chotek von Chotkova ~ her rank though high, did not meet the A-H level of protocol demanded of a future Empress. She was not accorded the title of Archduchess, nor could she appear as an equal to him in public within their own Hapsburg Empire. This naturally aggrieved Franz Ferdinand, and the way he found to publicly show his love for his wife was to travel abroad where his wife could accompany him in public; where she could ride thru foreign streets with him in an open carriage and be jointly greeted at more welcoming receptions.

On 22nd November, the Archduke and Duchess left King's Cross by special train to pay a visit to the Duke and Duchess of Portland at Welbeck Abbey in Nottinghamshire. On their arrival at Worksop the platform was covered with crimson carpet and the station garlanded with foliage and flowers. Waiting for them were the A-H Ambassador Count Albert von Mensdorff, the Duke and Duchess of Devonshire, Lord Curzon of Kedleston, Lord and Lady Salisbury, and ex-British PM Arthur Balfour. After church, the Archduke went game shooting with the Duke of Portland: *One of the loaders fell down. This caused both barrels of the gun he was carrying to be discharged, the shot passing within a few feet of the Archduke and myself. I have often wondered whether the Great War might have been averted, or at least postponed, had the Archduke met his death there and not in Sarajevo the following year.*

Welbeck Abbey

Franz Ferdinand (far right) shooting with the Duke of Portland

SHELTERING THE NEW YEAR

F.H. Townsend ~ *Punch* ~ December 1912

THE THREE PREMONITIONS OF FRANZ FERDINAND'S FATE

In the summer of 1911, Princess Zita of the Franco-Italian royal house of Bourbon-Parma had just become engaged to the Archduke Karl who was second-in-line to the A-H throne of his great-uncle Emperor Franz Josef who had reigned over A-H since 1848. The quiet Karl, and the temperamental Zita complemented each other well, and their engagement was to be celebrated on 13th June 1911. During the weeks before the wedding, Zita travelled to Rome with her mother for an audience with Pope Pius X who would bless them, and congratulate them on the engagement.

During the audience with Pius X and Cardinal Merry del Val who congratulated the young Princess on her upcoming nuptials, the Pope being a great supporter of the Habsburg Monarchy, began the conversation: *I am very happy with this marriage and I expect much from it for the future.... Karl is Heaven's reward to Austria for her faithfulness to the Church.* Pius X then appearing to wander in his thoughts when he referred to Zita's future husband; *Now you are going to marry the heir to the throne. So ~ I augur all of God's blessings upon you.* When Princess Zita pointed out that her future husband was not the direct heir to the throne, for first came his Uncle Franz Ferdinand. Pope Pius X firmly insisted; *No! Your fiancée will be Emperor, and when this event takes place, you both should work zealously for peace.* Princess Zita: *Franz Ferdinand will certainly not abdicate on account of the fact that he is in the prime of life.* Pope Pius X: *No, Karl will be the heir of Franz Josef. Whether there will be a renunciation I do not know.*

Franz Ferdinand had some reason to suppose he may be bound to die. A-H legend ~ not to be found in any academic history, but in folklore preserved in the oral tradition of Austria's huntsmen, records that in 1913 the Archduke had shot a rare white stag, and that it is widely believed by hunters; *Any hunter who kills a white stag ~ he or a member of his family shall die within a year.*

The Archduke Franz Ferdinand had other presentiments of his fate that his journey to Sarajevo would be a journey to his death ; for in the beginning of May 1914, he invited the couple next-in-line to the throne, his 27-year-old nephew the Archduke Karl and Princess Zita who were now married, to supper at the Belvedere. The future Empress, Princess Zita described what happened:

It was just a small family meal, with ourselves as the only guests. Everything passed off quite normally, indeed quite merrily ~ until after supper, when Sophie went to take the children up to bed. After his wife had left the room, the Archduke suddenly turned to my husband and said ~ I have something to say but I must say it quickly as I don't want your Aunt to hear anything of this. I know I shall soon be murdered. In this desk are papers which concern you. When it happens, take them, they are for you. Karl protested ~ *This must be a joke!* Their uncle replied; *I am in deadly earnest* ~ only having time to add ~ *Fortunately the special crypt being constructed at Artstetten is now finished.* When Sophie reappeared, their normal conservation resumed...

In 1914, the Archduke Franz Ferdinand as heir-presumptive to the Danubian Monarchy was appointed Inspector General of A-H armed forces ~ with the intent to restore military morale in the wake of the Redl scandal ~ and he was pressing ahead with his plans to attend the upcoming annual maneuvers in Bosnia in June, which were to be followed by a visit to the capital of Sarajevo. While the Archduke hoped to avoid war with Serbia and planned to conciliate the Empire's own Slavs, he knew that some sabre-rattling could help to sustain the peace. Archduke Franz Ferdinand shrugged off Count Leopold von Berchtold's warnings concerning any potential of Serbian threat during his visit ~ *all such Serb horror stories leave me cold!*
Later at a toast at a public event in Vienna ~ *To peace! What would we get out of a war with Serbia? We'd lose the lives of young men and we'd spend money better used elsewhere. And what would we gain, for heaven's sake? Some plum trees and goat pastures full of droppings, and a bunch of rebellious killers. Long live restraint!*

According to *The Assassination of the Archduke,* (a recent published biography of Franz Ferdinand and Sophie sponsored by their descendants) Franz Ferdinand was actively pushing to remove Conrad from his post and was prepared to expose the general's affair with Gina von Reininghaus to achieve it. When Conrad met with the Provincial Governor, General Oskar Potiorek, to make arrangements for Franz Ferdinand's visit to Sarajevo, the security provisions soon proved to be hopelessly inadequate. Word of the Archduke's impending visit spread and Serbian Ambassador to A-H Jovanovic wrote: *From the month of December 1913 onwards, the maneuvers in Bosnia were spoken of in Vienna. The Inspector General of the Austro-Hungarian Army was to take part ~ both as future Emperor, and Commander-in-Chief. It was to be, as was said towards the end of 1913, a lesson and warning to the Serbs of both Bosnia and Serbia.*

Few imagined any Serbs within Bosnia wanting to pay homage to the Hapsburg Monarchy who blocked their way to uniting all Serbs within a Greater Serbia. Even to contemplate a public visit by members of the ruling A-H family to a hostile provincial capital like Sarajevo, seething with intrigues, terrorist plots and dangers of all kinds was an act akin to madness. Many people foresaw a disaster. The Serbian Ambassador cautioned: *Some Serb participating in the A-H Army manœuvres might seize the opportunity to fire at the Archduke and Duchess.* Officials urged the visit be cancelled and police warned: *On the long route that is presently proposed ~ we cannot guarantee the safety of the Archduke and Duchess.* When the Archduke was warned by a friend that he might be killed in Sarajevo, he sighed: *I am sure your warning is justified, but I do not let myself be kept under a glass cover. Our life is constantly in danger. One has to rely upon God.*

On the evening of Tuesday 23rd June 1914, shortly before boarding the train that was to take him from Vienna to Trieste, the Archduke Franz Ferdinand dined with his wife Duchess Sophie in the place he described as *our Vienna apartment* ~ the Belvedere, a magnificent baroque Viennese palace. Although he was apprehensive about heading into a known hotbed of Serbian opposition to A-H rule, Franz Ferdinand had decided that for the anniversary of their marriage, and as the Inspector General of all A-H forces, he would review the annual A-H Army military manœuvres that month in Bosnia. So now their marriage certificate became also the entre'acte to their death warrant in Sarajevo, Bosnia. As they prepared to travel south, the couple would have discussed recent events and what could come once they steamed south through Slovenia and on to Sarajevo.

By far the cleverest thing I ever did in my life was to marry my Sophie. She is everything for me: my wife, my doctor, my advisor ~ in a word my whole happiness. And then our children! They are my whole pride and joy. I sit with them all day long in amazement that I can love them so much. And then the evenings at home when I smoke my cigar and read my papers. Sophie knits and the children tumble about, knocking everything off the tables. It's all so cozy and precious… Archduke Franz Ferdinand d'Este

A Great European War was far from inevitable. The same political elites capable of achieving peace prior to 1914 could have continued to do so in the years that followed. Continuing social changes, and in particular German economic growth, would have adjusted the geo-political future. History is shaped by actual events, the perceived context in which these events occur ~ and the response by leaders capable of reacting to those events. Sarajevo was such a pivotal historical event, and the response by struggling European leaders to it, enough to trigger the catastrophic Great War. It could have been very different, and our global culture completely changed ~ certainly for the better.

The Archduke Franz Ferdinand ~ if we are to believe those who knew him well ~ was not at all the kind of man he has been commonly represented. In fact, all through his life he kept the general public profoundly ignorant as to his real aims and ambitions. Franz Ferdinand, though heavy and dull in appearance, was gifted with more insight into human nature than he has been given credit for, and he had nothing of the warlike character which has been attributed to him. On the contrary, he showed himself on more than one occasion the restraining spirit among the advisers of the Emperor. In particular, he could never be brought to endorse the policy of Count Aerenthal, whom he profoundly disliked, and mistrusted even more than he detested. He had applied himself to the reorganization of the Austrian Army, not because he was anxious to bring about a war, but because he did not intend, in case his country should be drawn into one ~ an entirely different thing ~ that Austria should be compelled to play second fiddle to her Prussian allies.

At the time of the last Balkan crisis he had displayed extreme moderation, and is wrongly supposed to have urged the annexation of Bosnia and Herzegovina to the Austrian Empire. This measure, the forerunner of so much trouble, originated entirely in the fertile mind of Count Aerenthal, who had always nursed the ambition of becoming the Bismarck of his country and of earning for himself a name in her diplomatic annals by making her come out of her outwardly relative passivity to assume the leadership in the Balkan Peninsula. Moreover, he saw clearly that in that part of the world, with a little management, Austria could easily find the opportunity sought. The annexation of Bosnia-Herzegovina was hinted at in private conversation when the old Monarch declared that he thought it was a shame this measure had not been resorted to long ago, and finally took upon himself the responsibility of this bold step about which he would never have troubled but for the activity of Count Aerenthal.

The Archduke Franz Ferdinand was perfectly well aware of this state of things, and it made him dislike even more than he had done before his uncle's adviser and Minister. The Archduke, whatever has been related to the contrary, had no hostile intentions against anyone among the neighbours of Austria. Indeed, he was very averse to any complications likely to bring the Monarchy into trouble ~ complications for which public opinion might have made him answerable. His great ambition was to consolidate the position of his wife and children, and if possible, to arrange matters so that his son might succeed him, if not in Austria, at least in Hungary.

Franz Ferdinand, who was always supposed to be very rich, had in reality very little beyond his allowance as a member of the Imperial Family to live upon, together with some estates that he had inherited from his father, which were encumbered by the large dowry which his stepmother drew from them. When people in Vienna, who did not like either him or his wife, reproached them for what they called their meanness, they did not realize that he drew no revenue from the Modena estates, and far from having the millions attributed to him, the Archduke had to be very careful with his money ~ the more so in order that he might make provision for his children. The Countess Chotek also had no money of her own. The children, would have been left with a very modest fortune if the Archduchess Marie Therese, their father's stepmother, had not intervened on their behalf and persuaded the Emperor to give them an allowance of the same proportions as enjoyed by all the Archdukes and Archduchesses.

Considering these facts, of which no one abroad and but few people in Austria were aware, it will not be difficult to believe that Franz Ferdinand's immediate interests commanded him to try and smooth over any political difficulties likely to compromise the peace of the Austrian Monarchy which might have arisen. He was perfectly well aware that should his wife survive him before her position had been put on an unassailable footing, she would find herself in terrible difficulties, both from a moral and material point of view. The Imperial Family detested her, and the society of Vienna had not forgiven her for a marriage which had at once raised her to a pinnacle to which the daughters of the haughtiest princely family of the upper aristocracy could not have aspired.

As time went on the Duchess Sophie of Hohenberg became important, though she always kept in the background. Kaiser Wilhelm, however, was clever enough to seek her friendship. He saw that if he won her good graces she could persuade her husband to enter into the political views of Germany and espouse her quarrels. But she and Wilhelm II were at cross purposes; the consort of Franz Ferdinand merely wanted the Kaiser's help to consolidate her own position by obtaining official recognition at foreign Courts. The Kaiser, on the contrary, simply wished to make the Duchess a pawn in the complicated game he had already made up his mind to play sooner or later. But it became clear to Wilhelm II, that the Archduke Franz Ferdinand did not intend following him in the adventurous policy he was dreaming of, and that his thoughts were concentrated on family matters.

Princess Catherine Radziwill
St Petersburg Courier

COUNT OTTOKAR CZERNIN ON ARCHDUKE FRANZ FERDINAND

Count von Czernin ~ A-H Ambassador to Romania 1913-16
Imperial Foreign Minister of A-H 1916-18

Archduke Franz Ferdinand d'Este
with his daughter Princess Sophie

The Archduke and heir to the throne was a man of a very peculiar nature. The main feature of his character was a great lack of balance. He knew no middle course and was just as eager to hate ~ as to love. He was unbalanced in everything; did nothing like other people, and what he did was done in superhuman dimensions. His passion for buying and collecting antiquities was proverbial and fabulous. A first-rate shot, sport was for him a question of murdering en masse, and a few years before his death he shot his 5,000th stag. His ability as a good shot was phenomenal. When in India, staying with a Maharajah, an Indian marksman gave an exhibition of his skill. Coins were thrown into the air that the man hit with bullets. The Archduke tried and beat the Indian. He scorned all modern appliances for shooting, such as telescopic sights or automatic rifles.

In regard to foreign policy, he saw great danger to the friendly relations between Russia and ourselves in the rivalry between Vienna and Petersburg in the Balkans, and contrary to the reports that have been spread about him, he was rather a partisan than an opposer of Serbia. He was in favour of the Serbians because he felt assured that the petty agrarian policy of the Magyars, was responsible for the constant annoyance of Serbians. He favoured meeting Serbia halfway, because he considered that the Serbian question was a source of discord between Vienna and Petersburg. Another reason was that he was no friend of King Ferdinand of Bulgaria, who constantly pursued an anti-Serbian policy. I believe that if those who were responsible for the organization of the assassination of the Archduke had known what little justification there was for supposing him to be the man they thought him, they would have desisted. He was opposed to any closer combine with Germany.

I was in Constantinople shortly before the outbreak of war, and while there had a lengthy discussion of the political situation with the Marquis János de Pallavicini, our most efficient and far-seeing ambassador there. He looked upon the situation as being extremely grave ~ that if a rapid change in the entire course of events did not intervene, we were making straight for war. He explained to me that the only possibility of evading war with Russia lay in our definitely renouncing all claims to influence in the Balkans and leaving the field to Russia. Pallavicini was clear in his own mind that such a course would mean our resigning the status of a Great Power, but apparently to him even so bitter a proceeding as that was preferable to the war which he saw was impending. Shortly afterwards I repeated this conversation to the Archduke Franz Ferdinand, and saw that he was deeply impressed by the pessimistic views of Pallavicini, of whom, he had a very high opinion. The Archduke promised to discuss the question with the Emperor. I never saw him again. That was the last conversation I had with him.

One fine quality in the Archduke was his fearlessness. He was quite clear that the danger of an attempt to take his life would always be present, and he often spoke quite simply and openly of such a possibility. He also told me that when he went to the coronation in Spain he was to have made the journey with a Russian Grand Duke, but shortly before the train started the news came that the Grand Duke had been murdered on the way. When at St. Moritz, news was sent him that two Turkish anarchists had arrived in Switzerland intending to murder him. Every effort was being made to capture them. The Archduke showed me the telegram at the time. He laid it aside without the slightest sign of fear. The Duchess suffered in her fears for his life, and I think that in imagination the poor lady often went through the catastrophe of which she and her husband were the victims. Out of consideration for his wife's anxiety, he tolerated the presence of a detective to calm her fears.

The Archduke appears finally to have undertaken this particular trip more from a sense of duty than from the desire, as usually stated to have an opportunity to have his wife received with royal honours by his side. They decided to travel to Sarajevo by a different route, and in the last weeks had some doubt about going at all, because of his health. His private secretary Nikitsch-Bouilles noted several remarks that indicated that Franz Ferdinand was the opposite of enthusiastic about the trip. On 23rd June, the Imperial railway carriage regularly reserved for him had a fault, so that he and his wife had to travel in an ordinary first-class compartment after leaving their three children at Chulmetz. Franz Ferdinand said, *Well, the journey is beginning in a right promising fashion!* A little later, when told that the train by which he and his wife intended to leave Sarajevo on 29th June would have to start at 5 AM instead of 6 AM as originally planned, he exclaimed, *Tell Colonel Bardolff that if he continues daily to make the Bosnian trip still more disgusting with new difficulties and unpleasantnesses he can hold the* manœuvre *alone, and I will not go down there at all!* His private secretary Nikitsch-Bouilles wrote : *The idea that the Archduke himself wanted the trip to Bosnia to provide a triumphal journey is pure invention.*

However, apart from these few annoyances; and the fact that the electricity gave out in the train in which they rode from Vienna to Trieste when the electric lights fused, so that Franz Ferdinand and his éntourage would travel through the night by the funereal gleam of candles ~ the rest of the journey passed off excellently and the Archduke was in the very best of humour. He was greeted with enthusiasm at all of the railway stations along the way from the Adriatic to Sarajevo, and arrived on the afternoon of 25th June at the resort of Ilidze, near Sarajevo, where they were to stay for their visit. The A-H Army manœuvres passed off satisfactorily in spite of heavy rain, and the Archduke complimented General Oskar Potiorek on the spirit and training of their troops.

On the afternoon 26th of June, after returning from the first day's manœuvres, Franz Ferdinand and his wife motored to Sarajevo to do some preliminary shopping in the old bazaars. The Mayor of the town had already issued a proclamation expressing the loyalty of the population to Franz Josef and their pleasure that he had sent his Heir to visit Bosnia. He urged the people to decorate the stores and houses with flags and flowers, and this was done. Everywhere his picture was in the windows. On this afternoon, Franz Ferdinand was in uniform, and acclaimed with loyal shouts of *Zivio!*

On Sunday morning of 27th June 1914, the Archduke telegraphed to his children at Chulmetz. *Everything is going well with Papi and Mami and we are looking forward to seeing you again on Tuesday.* These were the last words Franz Ferdinand ever wrote. Sophie then toured the old Sarajevo bazaar, (which is still in use today) and together they officiated at several functions without any concern. When Sophie expressed her pleasure at the public reaction to an aide, her husband ominously replied: *I pray to God that when I have the honour tomorrow night of seeing you, you can repeat those words to me.* June 28th was the 14th anniversary of Franz Ferdinand's pledge that enabled him to marry Sophie. That the couple loved each other dearly is reflected in their comments. Sophie: *I wish I could relive every single day again.* Franz: *Soph, I have often found, that there are things in life we would do differently if we could do them again. But if I had to marry again, I would do what I have done, without change.*

The police of Sarajevo, who had prepared a special report on the activities of the *Young Bosnians* were rebuked by the Governor-General of Bosnia General Oskar Potiorek for having, *a fear of children.* On the morning of 28th June, the police again warned that the Archduke should not visit Sarajevo, however the chief of the reception committee rejected their final warning by saying ~ *Do not worry. These lesser breeds would not dare to do anything.* One policeman replied : *Security measures on 28th June will be in the hands of Providence.* On their own initiative, the police issued orders to their men to turn their faces toward the crowd during the passage of the Imperial party through the streets of Sarajevo. But 120 police would be of limited effectiveness, along a route five kilometres long.

The immediate cause of the Great War was the murder of the Austrian Archduke at Sarajevo. Had it not occurred, there would have been neither an Austro-Serbian War, nor would the Great War have flashed across the globe in the summer of 1914. But the murder of Franz Ferdinand and his wife ignited material which would not otherwise have detonated as it did ~ or perhaps at all. The details of the crime, the motives of the waiting assassins, and who their instigators and accomplices were have remained dark and difficult questions, neglected by historians who have devoted relatively little attention to them. Persistent mis-information and propaganda have survived for a longer time on this subject than any other aspect of those desolate days, which would set much of our world aflame.

The Archduke arrives in Sarajevo by train ~ Franz Ferdinand was so acutely aware of danger on the Balkan Peninsula, that before departing for Sarajevo he asked an aide to look after his children should he not return. Sophie however was so concerned about her husband's safety that she decided it better for them both if she joined him, as Sophie told a Countess prior to their leaving ; It is a dangerous undertaking, and I will not leave the Archduke to face it alone.

Archduke Franz Ferdinand greeted by city officials after arriving in Sarajevo

PART II

AFTER SARAJEVO

IMMEDIATE CAUSES OF THE GREAT WAR

THE BOILING POINT.

THE BLACK HAND ASSASSINATION PLOT

Apis in Belgrade

Trifko Grabez, Nediljko Cabrinovic, and Gavrilo Princip in Kalemegdan Park, Belgrade May 1914.

THE BLACK HAND ASSASSINATION PLOT

While the diplomats of Vienna were fully occupied throughout this period guarding the best interests of the Hapsburg Monarchy, the fractured provinces of the Ottoman Empire were the scene of many diplomatic tussles between A-H and other Empires, and yet these conditions seemingly eased enough to permit an arrangement between Vienna and St. Petersburg in March of 1913, when they agreed on the demobilization of their respective forces along the A-H-Russian border. While the Balkan Question continued to be the source of considerable anxiety and some diplomatic conferences, the political horizon of Europe appeared comparatively cloudless ~ to all but the keenest observers...

In 1894, Gavrilo Princip was born a serf's son in the village of Obljaj in the remote western hill country of Herzegovina, a place so poor that six of Princip's brothers and sisters died in childhood. Like many of his generation, he made his way to school in the old Balkan market town of Sarajevo, and then to Belgrade to improve his position through education. Such students often lived in poverty; sometimes on the edge of starvation, and ill health was the natural consequence of such dire poverty. In 1914, Gavrilo Princip was a 19-year student living in Belgrade when he heard of the coming visit of Archduke Franz Ferdinand and his wife, Duchess Sophie to Sarajevo. Princip and his classmate Trifko Grabez were joined by Nedeljko Cabrinovic, and these three Bosnian teenagers led by Princip, resolved to obtain arms and assassinate the Archduke during his ceremonial visit in June 1914.

In Sarajevo, Princip told his friend Danilo Ilic of his plan, who recruited a second troika of killers : two teenage students Cvjetko Popovic and Vaso Cubrilovic, plus 26-year old Mehmed Mehmedbasic. Among the radical Bosnian students in the spring of 1914, it was surprisingly easy for Ilic to recruit his accomplices. In Belgrade and Sarajevo, one student merely approached another to just ask;
Do you know that the Archduke Franz Ferdinand is coming to Sarajevo? We must lie in wait for him!
One evening, Cvjetko Popovic strolling in a Belgrade park agreed to Danilo Ilic's invitation.

Yes, he asked me and I instantly agreed. Look, you have to realize what it was like to live under foreigners. There we were, in our own country, and these fellows were telling us what to do. We students led the demonstrations against the Austrians. I had been in jail already for agitation in the Pjanic-Ljubibratic affair. Right down the street here in front of the cathedral we had burned the Hungarian flag. We were steeped in the literature of the Russian Social Revolutionaries who practiced assassination. When we met, when we walked in the evenings in the park, we would ask each other ~ Have you read this or that book? And if our friends hadn't, we would pass the book from hand to hand. We didn't have to spell things out. We knew instantly what each other had in mind. Look, there is no way of telling you what it is like to live under foreigners. You have to have experienced it yourself. We hated them with a burning hatred ~ Yes, I agreed instantly. Cvjetko Popovic

The members of the *Black Hand* were generally not known to each other personally, but were designated by secret numbers. Only the Central Committee led by *Apis* in Belgrade knew their names. The name of *Apis* appeared on the list as *No. 6* while his chief aide was Major Voija Tankosic at *No. 7.* The initiation of a new member took place in a darkened room, lit only by a candle on a small table covered with a black cloth, on which lay a crucifix, a dagger, and a loaded revolver. The candidates took the oath. *By the sun that warms me, by the Earth that nourishes me, before God, by the blood of my ancestors, on my honour, and on my life, that I will from this moment till my death be faithful to the laws of this organization, and I will always be ready to make any sacrifice for it.* The morbid seal of the *Black Hand* depicted an unfurled flag decorated with a skull and cross-bones, dagger, bomb, and bottle of poison, with an inscription referring to their vision of a Greater Serbia, *Ujedinjenje ili Smrt ~ Union or Death.*

When *Black Hand* assassin Bogdan Zerajic, a Herzegovinian Serb (*No. 111* on the covert list) unsuccessfully fired five pistol shots in an attempt to kill A-H Governor of Bosnia General Marijan Veresanin, the failed assassin committed suicide on the spot. As Zerajic lay sprawled and splashed with mud and blood on the Sarajevo Bridge, the unscathed Governor spurned the corpse with his boot and ensured that his subsequent burial was in the corner of the cemetery where only suicides and criminals were interred. Zerajic's grave was kept fresh with flowers, and became a site of pilgrimage for disaffected Bosnian youth and a morbid inspiration for others to imitate and avenge him in future. Here, Gavrilo Princip promised that before any attempt to assassinate one of the enemy, that he would place flowers of respect on Bogdan Zerajic's grave, and swear by it ~ *My hand will not waver!*

SERBIAN CABINET KNOWLEDGE OF BLACK HAND ASSASSINS

Professor Ljuba Jovanovic
Serbian Minister of Education

Nikola Pasic
Prime Minister of Serbia

Gavrilo Princip
Bosnian Assassin

I was Minister of Education in Nikola Pasic's Cabinet. I do not remember whether it was at the end of May or the beginning of June, when one day Pasic said to us that certain persons were making ready to go to Sarajevo to assassinate Franz Ferdinand, who was to go there to be solemnly received on St Vitus' Day. They told me that this plot had been hatched by a group of secretly organized persons of the Black Hand, and by some patriotic Bosno-Herzegovinian students in Belgrade. Pasic and Minister of the Interior, Stojan Protic both agreed to issue instructions to the frontier authorities on the River Drina to prevent the crossing over of the youths who had already set out from Belgrade for that purpose. But the frontier guards themselves belonged to the Black Hand, and did not carry out Stojan's instructions, but reported to him that the instructions had reached them too late, because the youths had already crossed over. Ljuba Jovanovic

From this it appears that members of the Serbian Cabinet knew about the proposed assassination plot a month or so before Franz Ferdinand had planned to visit Sarajevo ~ but took no effective measures to prevent it. The Serbian Cabinet was thus criminally negligent, to say the least. Not having nipped in the bud the plot prepared in their capital, by one of their own General Staff officers, and not having prevented the youths from crossing over into Bosnia, either because Pasic did not give his instructions in time or more probably because the frontier authorities themselves belonged to the secret organization of the Black Hand, the Serbian Government should at once have notified the Austrian authorities, giving the names of the criminals and all other details which might have led to their arrest before the execution of the plot. But Pasic and his Cabinet did nothing of a kind. Furthermore, after the crime had been committed, they made no searching inquiry into the incriminated secret organization in Serbia, and no attempt to arrest the accomplices who had helped hatch or carry out the plot. Instead, they sought to conceal every trace of it, and denied all knowledge of it, in the hope that Austria would be unable to discover their complicity. S.B. Fay, Professor of History, Harvard and Yale

The truth that Nikola Pasic knew of the plot is evidenced by his order to stop the assassins from crossing over from Belgrade into Bosnia, but the order was not carried out because the Captain of the frontier guards at Loznica was one of the *Black Hand.* This is further confirmed in Princip's transcript and in the papers of the frontier guard Captain Kosta Todorovitch, whom the Austrians captured.

We crossed over the Drina into Bosnia and went to Sabac to see Captain Popovic to ask him to put us across somewhere. When we got to Sabac, we found him at a café and went with him to his headquarters. He said to us; I can send you to the captain of the border guards. Then in the evening when it was dark we crossed the border. Grabez and I carried all the weapons. Then we turned toward some woods and hardly came out because it was so dark. We passed the night in a stable on the edge of the forest. We set out by day. Not exactly secretly, but not by an open road either. Then we came to Obren, and Jakov went to see who was at home. We waited. It was cold and we went into the house. There they offered us coffee. Because we had heavy things, each of us had two bombs and two revolvers, we asked for bags, and wrapped the weapons in rags. Then all three of us set out. We carried our things in two bags, and with great caution, we went by forest roads. Because we saw some footprints, perhaps of gendarmes, we were constantly checking out of caution. *Gavrilo Princip*

BELGRADE. Major Voijaslav Tankosic, sniper extraordinaire and commander of all *Black Hand* training immediately set out from Belgrade to test the weapons accuracy of his newest recruits in a secluded dark wood. Cabrinovic, Grabez, and Princip all now suffering from terminal tuberculosis, knew they would not live long and were willing to give their life for their dream of a Greater Serbia. With the mild exception of Princip, they were not impressive. Princip, who had originally planned to join a secret guerilla detachment of the *Black Hand,* had been rejected because of his weak physicality. His rejection was the motive that propelled him to do something spectacular to prove to the others that he was their equal. Despite their disheartening display of incompetence, the *Black Hand* equipped the seven assassins with bombs, revolvers, and prussic acid suicide capsules from the Serbian Army Kragujevac Arsenal. Princip then defaced a wall writing ~ *Our shadows will walk through Vienna!*

SARAJEVO for 700 years the capital of Bosnia, was crowded into a narrow valley at the foot of high hills. Thru its centre flows the putrid brown waters of the Miljachka River, half dry in summer. On 28th June 1914, the day began as a lovely Sunday nearing the close of the Viennese social season. The streets were flagged in the Archduke's honour and his portrait displayed in many shop windows. Crowds were gathering in the streets of Sarajevo to see them pass. Curiously with no precautions made to keep the crowd back, such as forming a line of soldiers, as had been done before in 1910, when Franz Josef had visited the city. Such precautions were relaxed for Franz Ferdinand's visit by the regional Governor General Oskar Potiorek. Either Potiorek was incompetent as evidenced in his future deployment of A-H forces in the field or as afterwards asserted by one of the Hapsburg family, that Potiorek was in cahoots with Conrad to remove Franz Ferdinand's ~ *I consider war to be lunacy!* opposition to any war against Serbia. Potiorek later defended himself from the charge of slackness by claiming that the same measures were taken as for the Emperor's visit in 1910 when a double cordon of troops guarded the Emperor's route, however on 28th June when there were thousands of A-H troops in the vicinity, the streets of Sarajevo were improperly only lined with a mere 120 policemen. Seven *Black Hand* assassins were now armed and waiting at several points along the well-publicized route of the royal motorcade of the Archduke, who in full uniform sat beside his wife Sophie.

At 10.10 AM, the assassin Nedjelko Chabrinovic knocked the cap off his bomb against a post, stepped forward, and hurled his bomb at the second vehicle flying the Hapsburg pennant carrying the Imperial couple. Franz Ferdinand seeing the hurtling black bomb flying through the air raised his arm to deflect it away from Sophie. The bomb glanced off Franz Ferdinand's arm and bounced off the convertible's fabric cover to detonate under the following car, slightly injuring Count von Waldeck, Colonel Morizzi, and six spectators. The inept assassin swallowed his old suicide pill, but its vintage poison only made him vomit. Trying to drown himself, Chabrinovich dived into the River Miljachka. But too shallow, the crowd pulled him out and gave him a severe thrashing. The Archduke ordered all cars to stop to see what damage was done. Having made sure the wounded had been dispatched to hospital, Franz Ferdinand said ~ *Come on. The fellow is insane. Gentlemen, let us proceed with our program. Hang him as quickly as possible or Vienna will give him a decoration! Do you think Sarajevo full of assassins?*

So, the party drove on to the Town Hall, at first rapidly, and then at the Archduke's order, more slowly so that the people could see him better. When the Archduke arrived at the Sarajevo Town Hall, Duchess Sophie met a deputation of Moslem women. Franz Ferdinand interrupted the Mayor furiously shouting, *Herr Mayor, I come to Sarajevo on a visit and I get bombs thrown at me. It is outrageous!* All gathered, and then waited for the Archduke's speech... for his papers still splashed with blood, to be retrieved from the second bombed car. The day's schedule was cleared for Franz and Sophie to visit the wounded at the hospital. Continuing on, the motorcade rounded a corner at 10.45 just outside of Moritz Schiller's food store, when the seventh assassin emerged eating a sandwich. Princip pulled a Browning revolver from his pocket. Cocking it, he took a step toward the car and fired two bullets. The first hit Archduke Franz Ferdinand, slicing through his jugular before lodging in his spine, so that blood spurted from his mouth. The second bullet struck the abdomen of Sophie, perhaps taking a bullet for Franz who died pleading, *Sophie dear! Stay alive for our children...* Both were dead in minutes. When apprehended by the police, Princip was beaten and taken to the police HQ, where he lamented; *I did not wish to kill her... I killed her accidentally.* The sinister climax to many years of covert conniving had come, with Serbian officers once more resorting to support assassins to spur their nationalist aims, when dark elements of the Serbian state; military officers, intelligence operatives and the criminal anarchist underworld all conspired to assassinate Archduke Franz Ferdinand. This crowning act in a series of provocations now confronted the leaders in Vienna with the choice of accepting the beginnings of disintegration or to investigate the crime of the assassination to its complete conclusion; even if drawing the sword in defence of their imperiled sovereignty became necessary against the rogue state of Serbia, to whom the continuance of peace was only a green light for more conspiracies.

Archduke Franz Ferdinand and his wife, Duchess Sophie departing the train station for the Sarajevo City Hall in a Gräf & Stift landau.

Crowds converge at the site of the bombing…

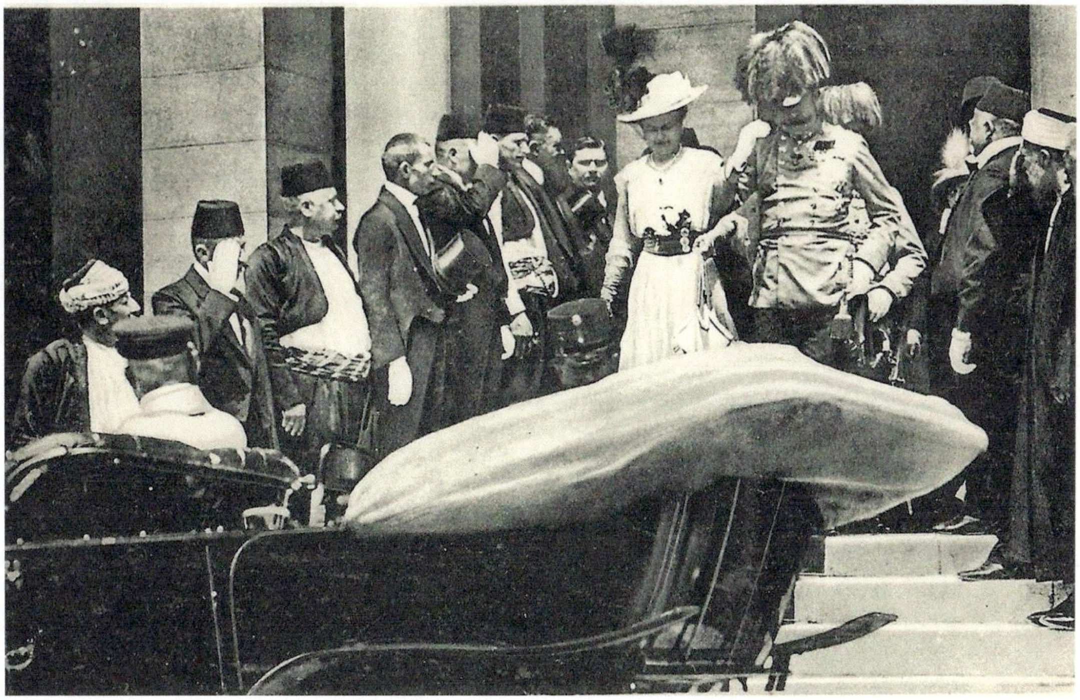

The Archduke Franz Ferdinand and Duchess Sophie depart the Sarajevo Town Hall…

~ the assassin strikes!

LA DOMENICA DEL CORRIERE

	NEL REGNO	ESTERO
Anno	L. 5 —	L. 10 —
Semestre	» 2,50	» 5 —

Si pubblica a Milano ogni Domenica
Supplemento illustrato del "Corriere della Sera"

Uffici del giornale: Via Solferino, N. 28 MILANO

Per tutti gli articoli e illustrazioni è riservata la proprietà letteraria e artistica, secondo le leggi e i trattati internazionali.

Anno XVI. — Num. 27. 5 - 12 Luglio 1914. Centesimi 10 il numero.

L'assassinio a Serajevo dell'arciduca Francesco Ferdinando erede del trono d'Austria, e di sua moglie.

(Disegno di A. Beltrame).

Sir Maurice de Bunsen to Sir Edward Grey — *Vienna*

Dictated 4.10 P.M. ~ Received 6.00 P.M. — *28th June 1914*

Vice-Consul at Sarajevo telegraphs Archduke Franz Ferdinand and Duchess Sophie Hohenberg assassinated this morning at Sarajevo by means of explosives. From another source I hear that bomb was first thrown at their carriage on their way to the town hall several persons being injured and later young Serbian student shot them both with a revolver as they were returning to the Garni Hotel Konak in Sarajevo

At Moritz Schiller's Delicatessen ~ the same place that Princip carried out his assassination on 28th June 1914, this particular photograph of Sarajevo six years previous shows citizens reading of the notification of Austria's annexation of Bosnia Herzegovina on the proclamation posters circulated from 7th October 1908. (Today, this building hosts the Museum of Sarajevo 1878-1918)

Count Franz von Harach, the owner of the car in which the Imperial party were travelling, and who was himself in the front of the car with the chauffeur, could do nothing. He recalled the scene in the car in this way ~ *As I was drawing out my handkerchief to wipe away the blood from the Archduke's lips, her Highness cried out: For God's sake! What has happened to you? Then she sank down from her seat with her face between the Archduke's knees. I had no idea that she had been hit and thought she had fainted from shock. His Royal Highness said: Soferl, don't die. Live for my children. I seized the Archduke by the coat collar to prevent his head from sinking forward and asked him: Is your Highness in great pain? To which he answered: It is nothing. His face was slightly distorted, and he repeated six or seven times, every time losing more consciousness and with a fading voice; It is nothing. Then came a brief pause followed by a convulsive rattle in his throat, caused by loss of blood. This ceased on arrival at the governor's residence. The two unconscious bodies were carried into the building where their death was soon established.*

On Sunday, 28th June 1914, we had taken our children in our car from Vienna to Székesfehérvár to visit my brother who was the officer commanding the 13th Hussars in that town. We were met by my brother and his wife in the courtyard ~ both looking extremely preoccupied. On our enquiring what the matter was, they replied that; A friend of ours, has just told us that… the heir to the throne, and his wife have been assassinated at Sarajevo. At first, we refused to believe the news. We could not credit that the necessary security measures had not been taken to guard against an assassination attempt ~ admittedly, I had wondered why Archduke Franz Ferdinand had chosen to visit to a town so near the border for Vidovdan on St. Vitus's Day ~ the Serbian national day, on which the Battle of Kosovo in 1389 was commemorated and on which national passions were liable to flare. We soon learned that the rumour was indeed true. We realized that this political murder was bound to have international repercussions. It seemed out of the question that it had been the unaided act of a single individual; it is now historical fact that Belgrade had a hand in it and that the plotters had the assurance of Russian support.

Korvettekapitan Miklos Horthy, ADC to Emperor Franz Josef

On 28th June 1914, I was at my country house at Senjak. At 5 PM an official telephoned to me from the Press Bureau telling me what had happened at Sarajevo. Although I knew what was being prepared there, yet as I held the receiver, it was as though someone had dealt me a heavy blow. When the news was later confirmed, a heavy anxiety oppressed me. I saw that the position with regard to other Governments would be very difficult, far worse than after the killing of King Alexander on 29th May 1903. Ljuba Jovanovic, Minister of Education

BUCHLAU CASTLE Monravia. Count Leopold von Berchtold and his wife Nandine having come to their Buchlau estate for the weekend, soon heard the news from Sarajevo. Almost immediately, Berchtold's staff ask him by telegram at 3.30 PM to return to Vienna. As they board their train at Ungarisch-Hradisch near Buchlau to arrive in Vienna at 9.32 PM, all previous certainties are suspect.

On the train journey back to Vienna, Berchtold surely reflected upon his abrupt change of fortune. Personal grief, worries about the Danubian Monarchy's future, concern for the dead couple's children, anger at the assassin's act, and worry about the monarchy's future were surely some of his thoughts. He now faces a fateful decision. Whether to engage in more militant diplomacy with Belgrade or to instigate a limited war with the neighboring kingdom? After all, twice during the crises of May and October 1913, Berchtold had been ready for action in the field to decide the issue of Serbian defiance. Thus, the A-H Foreign Minister entered the current crisis in a less tolerant, far less patient mood. Earlier when he had threatened war, first the Montenegrins, and then the Serbs had capitulated. But those threats and diplomatic wins had not prevented the assassinations : Is it now time for war? In reviewing Vienna's' few options ~ Count Leopold von Berchtold can no longer evade this question.

The 28th of June 1914, was a Sunday, a day when newspapers in Russia are usually scarce of news, and weary members of staff are able to seek a few hours' repose from their duties, which in summer are most tedious. I was hoping to enjoy a quiet evening at home, when I was startled by the telephone bell. My editor rang me up to tell of the tragedy which had taken place that same afternoon at Sarajevo, news of which had just reached him. To say that I was stunned is saying little. I had known Count and Countess Chotek, the parents of the unfortunate Duchess of Hohenberg. Fortune had transformed Sophie Chotek from a lady-in-waiting on the haughty Archduchess Isabella of Austria into the consort of the heir-presumptive to the Habsburg monarchy. I had seen her as a girl just out of her teens, in her mother's house in Brussels, where her father was Austrian Ambassador. I had entirely lost sight of her in after years. No one could have foreseen the strange freak of destiny which was to associate her name with one of the greatest dramas history will ever have to record, to which her assassination was to make a fitting prologue. In Petersburg, the news of his murder was received with horror. Strange as it may seem now, it is an undoubted fact that at the time, Kaiser Wilhelm was not half so much dreaded as the Archduke Franz Ferdinand, who was supposed to be a rabid Russophobe, whilst the Head of the Hohenzollern dynasty was credited with much wisdom, as well as with a sincere desire to uphold the peace which he had succeeded in preserving during the 25 years of his reign.

The whole world was convinced that the accession of Franz Ferdinand would mean war within a short time. This opinion had originated from Count Aerenthal, and had been so cunningly put into circulation that there are a good many people who to this day are convinced that the one bellicose element in Austria was the unfortunate Archduke who fell at Sarajevo. In reality, things were entirely different, and it is pretty certain that had Franz Ferdinand not been murdered in such a dastardly manner he would have done much to consolidate the peace of Europe. He had a will of his own, and would never have become a puppet in the hands of his Ministers and advisers, and he was far too good a husband and father to endanger the existence of his wife and children in an adventure which, in any case and whatever the ultimate results, could only have done them harm. His assassination was one of those misfortunes of which the magnitude can only be appreciated long after it has occurred. It left the field free to all the intrigues of the military party in Austria and of the partisans of war in Germany, and it afforded the pretext, which the latter had long been looking for, to put into execution the programme which has plunged the whole of Europe into chaos and calamity. Once the Archduke disappeared, there remained no strong man or character in the Imperial Family. Princess Catherine Radziwill

The news that an assassin's hand has struck down the successor to the Imperial throne of Austria, and his wife, burst upon Europe on the afternoon of Sunday, 28th June 1914, like a thunderclap in the midst of a calm summer's day. The ruthless militarists of Austrian society now pounce on the situation to pressure the antiquated form of autocratic government within the Dual Monarchy to carry out their policy of domination, and the next day Vienna sends Belgrade an immediate request for satisfaction if any Serbian subjects are among those anarchists who have carried out the savage plot. The German militarists similarly seething with ambition take full advantage of the deaths to further encourage A-H to punish and make extreme demands on Serbia, for its suspected role in the killings. Serbia leans toward Imperial Russia for advice and assistance. Russia is tied by treaties to support Serbia and France, which by extension are bound to the British Empire. Mired in the impenetrable diplomatic web of opposing alliances, the Empires of Europe now drift toward crisis.

Kaiser Wilhelm's Imperial Yacht ~ the Hohenzollern (Willy Stower 1900)

KIEL BAY. But what of Kaiser Wilhelm, Franz Ferdinand's hunting partner and personal friend? On 28th June, as the Kaiser is cruising through northern waters, all eyes turn towards Kiel Bay. Aboard his newest American-built yacht, *Meteor V* he is participating in the annual Kiel Week regatta. Seasoned with the invigorating salt air of the Baltic naval port, Wilhelm is striving to make Kiel Week a rival of the week at Cowes, the British yachting centerpiece. He is excited, directing boat maneuvers. The launch *Hilda* appears at 2 PM steaming straight for *Meteor* at full speed signaling that she urgently needs to contact the Kaiser, but Wilhelm waves her off with an abrupt imperious gesture. The launch, steams nearer and when a short distance separates it from the *Meteor,* Admiral Georg von Müller standing beside the pilot, waves a piece of paper high in the air that he needs to deliver. Placing the fatal telegram of the Sarajevo news in his cigarette case, he tosses it onto the foredeck of the *Meteor*.

A sailor picks it up and gives it to the Kaiser. Wilhelm opens the cigarette case to unravel the crumpled telegram, and he dramatically unfolds it. Stunned ~ the Kaiser turns pale and murmurs: *So my work of the past twenty-five years will have to be started all over again!* To the British Ambassador Sir Edward Goschen, who is at Kiel with a British squadron of Dreadnoughts ~ he unburdens himself. *Es ist ein Verbrechen gegen das Deutschtum! It is a crime against Germanity!* Wilhelm departs the regatta. His mind turns from regattas to regrets, as he boards the launch to sit alone in silence, and to simply stare straight ahead. Wilhelm boards the *Hohenzollern ~ The regatta will continue!*
All dancing to cease... Flags half-mast. Ensigns half-mast for the murder of the Austrian heir.

When Vice-Admiral Sir George Warrender, commanding the visiting British squadron of dreadnaughts came on deck, he told me of the consequences that the assassination might have. He bluntly expressed his fears that this crime would mean war between Serbia and Austria. That Russia would then be drawn in, and thus Germany and France could not remain onlookers. He said quite openly, that this murder could certainly result in a wider European war. I recorded this conversation in my official report. *Korvettekapitän* Georg von Hase

LONDON. When word comes from the British Embassy in Vienna, the news of the assassination causes a deep impression in London. King George V dispatches telegrams of sympathy and orders Court mourning for a week. The subsequent A-H heir to the Hapsburg throne, the Archduke Karl is a popular figure in Britain who speaks English and German fluently.

28th JUNE 1914 ~ GRAND PRIX DE PARIS

The day on which this abominable crime was committed, was in Paris that of the Grand Prix de Paris. A-H Ambassador Count Scezsen, was standing by my side in the presidential grandstand at Longchamp when the telegram announcing the death of the Archduke and the Duchess of Hohenberg was delivered to me. I of course, expressed to the ambassador the indignation that this horrible crime could not fail to excite in France, and my profound condolences for the grief that had fallen upon the Imperial family. As I left the grandstand before the end of the racing, I shook hands with the Romanian Ambassador Alexandru Lahovary, a shrewd observer of Balkan events, who remarked to me, with a preoccupied air ; This unhappy event may have very serious consequences. Raymond Poincaré ~ President of France

28th JUNE 1914 ~ GRAND PRIX DE PARIS

Top : Sardanapale, Baron Maurice de Rothchild's magnificent horse was one of the leading French three-year-olds of the spring of 1914. In early June, Sardanapale won the Prix d'Hedouville and Prix du Jockey Club over 2400 metres on the Chantilly Racecourse. On 28th June, Sardanapale faced the Grand Prix de Paris over 3000 metres at Longchamp Racecourse starting the 1.95/1 favourite in a twelve-runner field. La Farina led from the start and when George Stern moved Sardanapale up to challenge 800 metres out, the two colts drew away from the rest of the field. After a prolonged struggle, Sardanapale got the better of his rival 100 metres from the finish and won by a neck, with a gap of four lengths back to third place. After the race Sardanapale was described in the press as ~ possibly the best three-year-old in the world.

THE END OF ALBERTO POLLIO

The death of Italian Supreme Commander General Alberto Pollio who suffered a heart attack on the same day that the Archduke Franz Ferdinand was assassinated in Sarajevo is of some significance in the course of European events. Rome renewed its participation in the Triple Alliance with Germany and A-H (on 5th December 1912), but the British Foreign Office, bolstered by Edward VII's frequent meetings with Italian royalty in the first decade of the century, knew through secret Anglo-Italian agreements that the Italian Government intended to observe a neutral stance if a war ever broke out in the event that A-H / Germany were the aggressors.

However, such secretive agreements did not guarantee the actions of prominent individuals, and General Alberto Pollio apparently remained loyal to Rome's stated commitment to Berlin and Vienna. During the Italo-German military discussions of December 1912 after renewing the Triple Alliance, Pollio promised von Moltke that Italy would mobilize her forces if, and as soon as, war was declared. The Italian clearly intended to honour what he saw as Italy's international commitments and assured the Germans in February 1914 that he would send two cavalry divisions and ten infantry divisions for the defence of the Upper Rhine to help the Germans implement their secret Schlieffen Plan. In March, Pollio further agreed that the Italian IIIrd Army could even serve under German command to bring any future war to a swift and successful conclusion. German Army optimism for a second front along the Italian border with France was based on the assurances of Pollio and his strength of character. (Despite Alberto Pollio being particularly pro-German, pro-Austrian, and having an Austrian wife, he nevertheless also ordered alternate plans for a mobilization to be drawn in case of a war against A-H).

But on the same day as the Sarajevo assassinations, Pollio's heart condition was misdiagnosed as a gastric obstruction, and the unfortunate Italian was given a strong purgative. He died on 1st July, after suffering a massive heart attack, and Italy's part in the Schlieffen Plan also perished with him. Pollio was the one man who could have stood up to the Italian Government and made a strong case to fight for the honour and dignity of the nation ~ yet the circumstances of his death from a gastric ailment in the Turin Palace Hotel, led some to suggest evidence pointing to his assassination using a morphine overdose or various other poisons. With the demise of Alberto Pollio, the Italian Cabinet will experience minimal opposition to their decision to adopt a neutral stance.

VIENNA 29th June. In the first hours after returning to Vienna on Sunday, Berchtold appears to have been careful not to commit himself, though others were pressing for immediate war with Serbia. When Count Berchtold saw Conrad later that day, (the general just back from the Bosnian maneuvers) he heard the familiar refrain, *Krieg, Krieg, Krieg.* The Foreign Minister at this point told the military commander that he first had to meet with Kaiser Wilhelm when he arrived for Franz Ferdinand's funeral to see what the German reaction would be ~ then they would decide.

That the murder at Sarajevo would have to be avenged was manifest. Indignation at the plot instigated by the Serbians was intense and unanimous throughout Europe. This reaction could have been utilized to avoid a world war without loss of prestige. Had the representatives of the Courts of the Great Powers, as was customary, been invited to Vienna for the appropriate obsequies, the work of the diplomats would have been made easier and might have met with success. The Prince of Wales and the Russian Grand Duke Nikolai Nikolayevich expressed their willingness to undertake the journey. Kaiser Wilhelm II had announced his intention and that of the Princes of the German States to be present. But owing to the influence of the Lord Steward of the Household, Prince Montenuovo, Spanish etiquette and dynastic rules were adhered to, because of the morganatic nature of the Archduke's marriage which forbade royal mourning ceremonies. The formal excuse given to the world was the advanced age of the sovereign, and every offer to attend the obsequies was met with refusal. Miklos Horthy

BERLIN 30th June. German Foreign Under-Secretary Arthur Zimmermann gives the Serbian envoy to Germany some advice to avert complications. *Here at the Foreign Office, we hope that Serbia will now neglect nothing in order to call to account those persons guilty of the conspiracy. We call to your attention the consequences to which a Serbian refusal in this direction might lead, and furthermore will suggest to the Russian Ambassador that he get his Government to give the same advice to Belgrade. I offer this counsel on the ground that no one can tell what might happen should the Serbian Government fail to fulfill its obligations, considering the wrath the deed has aroused in Austria-Hungary.*

This is apt advice, considering it in the context that Kaiser Wilhelm has expressed: *Whatever action Austria-Hungary might take against Serbia ~ she should not delay. She should take it as quickly as possible, while the sentiment of Europe, shocked by the horrible crime at Sarajevo, is still in sympathy with the Hapsburgs and indignant at the regicidal Serbs.* So, when Wilhelm reads a report from his German Ambassador in Vienna, Count Heinrich von Tschirschky, detailing how just two days after the murders in Sarajevo, *I have used every opportunity to warn Austria calmly ~ but energetically and earnestly against overhasty steps.* Kaiser Wilhelm writes the marginal notes ~ *Now or never! Who authorized him to do this? It is very stupid! It's none of his business, for it is purely Austria's affair to consider what to do in this matter, for it will be said afterwards, if things go wrong, that Germany was not willing! Will Tschirschky please stop this nonsense. Matters must be cleared up with the Serbs, and that soon. That's all self-evident and the plain truth.*
Wilhelm advises A-H Ambassador to Berlin Count László Szogyény-Marich.
I would be sorry if you left unused the present moment, which is so favourable to us.

VIENNA. During the summer of 1914, the prevalent mood is much less of Belgrade's celebration of the dark deed than of a sharpening of Austro-Serbian relations. This atrocity is strong evidence of the worsening Serbian danger, and the strategists of Austria-Hungary conclude that the continued existence of an expansionist Serbia poses an unacceptable threat to the future of their Dual Monarchy. Several times, Austria has been ready to wage war on provocative Serbia, but has thus far been held back by concessions from Belgrade (in March 1909, October 1913 or restrained by Berlin in July 1913). However, this time, Vienna accepts the opportunity for a localized war on Serbia afforded by the assassination of the Heir to the Hapsburg throne. Such is the significance of the *Black Hand* arranged assassinations at Sarajevo that the action precipitates the Austro-Serbian conflict and the *July Crisis* involving the power and prestige of the Triple Alliance and Triple Éntente ~ but this time the Balkan problem is more complex, and is more incapable of a peaceful solution than ever before; as apparent from the A-H Emperor Franz Josef's words to Count Heinrich von Tschirschky. *I see a very dark future. What is particularly disquieting to me is the Russian practice mobilization which is planned for autumn, just at the time when we are shifting our recruit contingents. Hartwig is master at Belgrade, and Pasic does nothing without consulting him. Everyone is dying around me…* Tuesday 30th June ~ noon. Berchtold sees a shaken and worried Franz Josef at Schönbrunn where they meet for an hour to discuss the perils of passivity. In his memorandum notes, Berchtold wrote of advising Franz Josef, *I fear of the consequences of further inaction. If reports link Belgrade with Sarajevo, then an action program will be necessary.*

Count Conrad von Hotzendorf at the map table…

THE JULY CRISIS

Kaiser Wilhelm believes that assassins represent a profound threat to the monarchial principle, and this is confirmed by Grand Admiral Alfred von Tirpitz, who writes that Wilhelm has told him: *The Russian Tsar Nicholas II will not support the assassins of royalty.* In early July, Germany strongly supports Austria-Hungary as they seek justice in the Balkans. Such is the atmosphere when the Kaiser makes his decision. *Austria-Hungary must judge what is to be done to clear up her relation to Serbia. Whatever Austria's decision may turn out to be, Austria can count with certainty upon it, that Germany will stand behind her as an ally and friend. Any action that Vienna decides to take ~ should be done speedily.* Indeed, the Kaiser's remark as he departs to resume sailing ~ *I cannot imagine that the old gentleman in Schonbrunn will go to war, and certainly not over the Archduke Franz Ferdinand* ~ evidences that he does not know what Vienna's response will be through this gesture to his assassinated friend. Wilhelm thinks he has extended words that imply some choice. Tsar Ferdinand of Bulgaria: *I certainly do not like my cousin Kaiser Wilhelm, but I feel sorry for him all the same. Now he will be dragged into the whirlpool, become entangled, and he will have to fight whether he wants to or not.* Berlin has just empowered Vienna with a free hand in what becomes known as ~ *The Kaiser's Blank Cheque.* Professor S.B Fay observed : *Germany made the grave mistake of putting the situation outside of their control and into the hands of a man as reckless and unscrupulous as Berchtold ~ they had committed to a leap in the dark.*

Most governments of the world (even Serbia) begin to send their condolences to Vienna in the wake of Sarajevo ~ but St. Petersburg has not. Embassies throughout Europe now lower their flags to half-mast in respect to the grief of Vienna ~ again the flags of Russia appear a notable exception. Vienna suspects Russia of an active hand in the assassination itself, believing that Belgrade is still being guided behind the scenes by Hartwig. And Russia's approval of a shipment of 120,000 rifles for the Serbian Army two days after the assassination clearly signifies their position in support of Serbia. King Nikola of Montenegro having himself just taken delivery of 10,000 Russian rifles and 56 modern artillery pieces further exacerbates Vienna's concern for Balkan issues when he further purchases four heavy siege mortars from the stocks of the Italian War Ministry. Being allied to A-H and Germany, the Italians naturally claim ignorance of the sale. Also worrisome, is the mounting struggle with Italy for control and influence in the newly created Albanian state. Friends can be more trouble than enemies.

VIENNA 1st July. Conrad once more sees Berchtold; again, pressing for action. This time he finds the foreign minister ready to agree and that he believes that the old A-H Emperor shares this view. But Berchtold wants (as do others) as he tells Conrad ~ *to have the results of the investigation in hand before moving against Serbia. I am convinced that this information will justify a war with Serbia.*

VIENNA 2nd July, Berchtold visits Franz Josef to review the latest situation. The veteran Emperor is prepared for action, yet wants more proof of links to Belgrade and assurances of German support. Franz Josef and Berchtold knowing they must have clear evidence of German support, together draft personal letters to the Kaiser. It is planned to give both to the Kaiser when he arrives for the funeral of Franz Ferdinand. But their plans are interrupted when security concerns in Berlin about Wilhelm's safety in Vienna lead to a cancellation of his trip. Now Berchtold has to move quickly to catch Wilhelm before he departs for his annual North Sea cruise ~ more so when the German Ambassador Count Heinrich von Tschirschky expresses general support for the Habsburg position, and yet observes with much surprise, that Vienna has not presented Berlin with a definite plan of action.

VIENNA Saturday 4th July. Berchtold knowing he needs evidence of German support to gain the Emperor's approval, as with Berlin's backing, he can hope to convince Tisza as well. Thus he quickly decides to send his memorandum calling for action with a personal letter from Franz Josef to Berlin, along with an oral explanation of Vienna's intentions. With deliberate guile, Berchtold sends Tisza the proposed details going to Berlin behind schedule : timed to prevent him from making any changes in the text before Hoyos leaves on Saturday's overnight train to Berlin. As Berchtold astutely anticipates, Tisza wants to lighten the tone of the Emperor's letter, but it is too late.

In arranging the approach to Berlin, Berchtold prepares two poignant precautions in addition to outmaneuvering Tisza. He orders Ambassador Ladislaus Szögyény to give both items personally to Kaiser Wilhelm II ~ and then to see Chancellor Bethmann Hollweg in person. Having had some experience of the Kaiser's expansive agreements and their later softening by Chancellor Bethmann, Berchtold wants clear assurances of actual German support from the two key principals in Berlin.

BERLIN Sunday 5th July. Kaiser Wilhelm II pledges his support to A-H Ambassador to Germany Count Ladislaus Szogyeni-Marich that afternoon; while fully recognizing the danger posed by Russia. The first details of German support reach Vienna at 10 PM on Sunday night. The key aspect of the Ambassador's telegram reads ~ *If we have really recognized the necessity of warlike action against Serbia, Kaiser Wilhelm would regret it if we do not make use of the present moment, which is all in our favour.*

Late on Sunday, the Kaiser leaves for his annual cruise on *Hohenzollern* up the Norwegian coast. Seeing that he is anxious, Chancellor Bethmann advises ~ *Undertake this voyage to avoid the attention that will be aroused by giving up an outing that you have for years been accustomed to take at this time of year.* To give a clear sign that the storm clouds that have so nearly burst over Serbia might now be wafting beyond the Danube, the Kaiser's customary telegram for the King of Serbia's birthday is dispatched when he begins his cruise. While Wilhelm soothes his nerves in the northern sea breezes of the fjords, he reads in a borrowed Norwegian newspaper ~ not from Berlin ~ of an A-H note to be sent to Serbia. Alarmed at being too distant to exercise a direct influence on the crisis, he turns the *Hohenzollern* from Norwegian waters and steams home to Germany. *We are not at war yet, and if I can, I shall prevent it!* Admiral Alfred von Tirpitz: *If the Kaiser had remained in Berlin, and if the normal Government machinery had been at work, he would have found ways and means of evading the danger of a wider war.*

During the interval while Kaiser Wilhelm is away, his A-H *Chargé d'Affaires* Ritter von Storck, makes inquires of the rogue Serbian régime. *What police measures have you taken or do you propose to take, to follow up the clues to the crime, which notoriously are partly to be found in your Serbia?* Von Storck is told: *The matter has not yet engaged the attention of the Serbian police. Up to the present time nothing has been done, and the matter does not concern the Serbian Government.* Storck: *I am astonished that any Government which is always expressing its desire to live on good terms with its neighbours should exhibit such indifference.*

LONDON Monday 6th July. In light of recent events, professional diplomats at the British Foreign Office show slight concern. H.G. Nicolson to the British Ambassador Maurice de Bunsen in Vienna: *Albania apart, there are no urgent and pressing questions to preoccupy us in the rest of Europe.* Lloyd George addressing the Guildhall suggests; *An implacable revolutionary ferment that threatens the disruption of British industry poses a more direct threat to Britain than any foreseeable diplomatic contretemps in Europe. The assassination at Sarajevo is no more than a very small cloud on the horizon, and you never get a perfectly blue sky in foreign affairs.* That day from Belgrade, *Chargé d'Affaires* Sir Dayrell Crackenthorpe cables; *News of the assassination has produced in Belgrade a sensation rather of stupefaction than one of regret.*

VIENNA. With the crucial news of support from Berlin, Berchtold has received German support. When Berchtold returns early Monday morning to Vienna after the weekend at Buchlau, the A-H Foreign Minister meets with Franz Josef for 40 minutes, advising him of Berlin's backing and winning his agreement for strong action. Noon ~ Berchtold meets with Conrad and Count Johann von Forgách, (the second section chief in the Foreign Ministry). Berchtold requiring further proof to link Belgrade to the crime (and to sideline Tisza) expects that German support will convince the Hungarian leader, therefore wires him the Kaiser's recent words : *We should not leave the present favorable moment unused. Russia is not ready to fight and Germany stands in total alliance loyalty on our side.* In one of the few surviving fragments of Berchtold's July diary : *Spoke about eventual action against Serbia on the basis of the telegrams received from Szögyény about yesterday's conversation with Kaiser Wilhelm. 4 PM Saw Tschirschky.*

There is the German Ambassador at Vienna, Count Heinrich von Tschirschky ~ obviously a firebrand ~ enjoying, one imagines, the confidence of the war parties in both capitals : also apparently a busy intriguer; acting behind the back of the Berlin Foreign Office, and communicating direct with the Kaiser. We gather very clearly that he egged on the statesmen of Vienna, with great diligence and success, to press Serbia to extremes. Russia, he assured them, would never carry her opposition to the point of war. Even if she did so, he argued with much plausibility, she would be negligible. For she stood midway in a great military and naval reformation, from which no situation is more deplorable for the purposes of carrying on a campaign. Frederick Scott Oliver

BERLIN. Szögyény and Hoyos meet with Bethmann and Under-Secretary Zimmermann at the Wilhelmstrasse Foreign Ministry. Once again A-H diplomats receive further confirmation of support. The Hoyos telegram to Vienna reads ; *I ascertained that the Imperial Chancellor like his Imperial master considers immediate action on our part as the best solution of our difficulties in the Balkans.*
The *Hoyos Mission* has achieved its goal.

A-H Foreign Minister Count Leopold von Berchtold

Berchtold's social and political assets are significant. A noted sportsman, especially fond of both horse racing and hunting ~ as were Franz Josef and Franz Ferdinand. As an avid collector of art, he moves comfortably in the world of the secessionist movement in Vienna. Berchtold's Moravian title assures his place in the upper circles of Viennese society, while his Hungarian wife, the Countess Ferdinandine de Nagy-Károly ~ called Nandine by Berchtold and their friends, buttresses this social position, and helps his political position in Hungary as well. Their marriage reinforces Berchtold's own Hungarian connection, and leads him in 1911 to opt to sit in the Hungarian upper house. Berchtold brings more than social and political credentials to the Ballhausplatz. His eight years in St. Petersburg exposed him to Tsar Nicholas II, the Russian court, and leading Russian political figures. Moreover, since he has also served in London and Paris, Berchtold clearly sees the Triple Éntente and the problems it poses for Berlin and Vienna. Berchtold, contrasting from his repeated resistance to the military to his conversion to their view in 1914, becomes a variable in Franz Josef's decision for war.

Das interessante Blatt

Abonnementspreise mit wöchentlicher Postversendung: für Oesterreich-Ungarn: vierteljährig 3 Kronen 20 Heller, halbjährig 6 Kronen 40 Heller, ganzjährig 12 Kronen 80 Heller; für Deutschland: vierteljährig 4 M., halbjährig 8 M., ganzjährig 16 M.; für alle übrigen Länder des Weltpostvereines: vierteljährig 7 Francs, halbjährig 14 Francs, ganzjährig 28 Francs. — Einzelne Nummern 24 Heller.

Redaktion und Administration: Wien, III. Rüdengasse 11 (Telephone 4199, 9767 und 9540), Stadtbureau: Wien, I. Schulerstraße 18.

Nr. 29. Erscheint Jeden Donnerstag. Wien, 16. Juli 1914. Abonnements durch jede Buchhandlung und Postanstalt. XXXIII. Jahrg.

Der Schritt in Belgrad.

Vorkonferenz der gemeinsamen Minister mit dem Generalstabschef und dem Stellvertreter des Marinekommandanten.

Council of Ministers: Minister-President of Cisleithania Count von Stürgkh, Finance Minister Count Ritter von Bilinski, War Minister Alexa von Krobatin, Foreign Minister Count Leopold von Berchtold, Hungarian PM Count István Tizsa, Chief of Staff A-H Army Count Franz Conrad von Hötzendorf, Contre-Admiral von Raller.

The A-H Council do not know of Conrad's own pessimistic calculations mid-point in the crisis, with neither FM Berchtold nor his ministerial colleagues ever critiquing Conrad's bland assurances. Even the ever-turbulent Tisza does not force a more meaningful explanation of military plans once he had Conrad's assurances about the defense of Transylvania. The Council accepts the general's confident assessment ~ even if his assessments are open to serious challenge, and opt for war because they believe that their large A-H forces can quickly defeat Serbia, bring régime change in Belgrade, and quell the challenge to Habsburg rule in Bosnia and Herzegovina. In reaching this conclusion assurances of German support are crucial. But Germany did not push the Ministerial Council to war; they used Berlin's backing to get the war they believed would rescue A-H from the Serbian menace. War may achieve what their diplomacy has not ~ security against the troublesome, intriguing Serbia.

VIENNA 7th July. Despite Berlin's pledge of support, this is not the aspect to convince Tisza. Berchtold invites German Ambassador von Tschirschky to a private briefing of Tisza and the Austrian Premier Karl Stürgkh, regarding the German view. But this scene gets off to a terrible start when the Hungarian leader learns that Hoyos has discussed a possible partition of Serbia while in Berlin. Tisza is so furious, that Berchtold has to force Hoyos to say that this was only his private opinion.

Talks are only slightly better during a closed meeting of the A-H Common Ministerial Council...

With his fellow ministers on the A-H Common Ministerial Council, Count Leopold von Berchtold is quite clearly the first among equals ~ however the influence which he previously held, has been challenged since mid-1913 by István Tisza's appointment as Prime Minister of Hungary and with this, his seat on the Ministerial Council. From the start, Tisza ensures his input into the Monarchy's foreign policy. Council relations are often fractured from militarist dramas ; from resisting the pressure of Conrad for war with Serbia or from War Minister Alex Krobatin's insistent demands for more men. What especially now worries Berchtold, are two deeply disturbing reports : of Tsar Nicholas II having visited Romania in early June 1914, and of Russian Foreign Minister Sazonoff having crossed into the contested land of Transylvania held by Hungary. This is seen by Vienna as a deliberate provocation, as are Russian propaganda activities among the A-H Empire's restless Ruthenian population.

Although Count Berchtold feels personally confident about his place and his political future in the senior leadership, he is uncertain about two serious issues that have plagued Habsburg diplomacy for years : relations with Russia, and control of the western Balkans against the scheming of Serbia. In the six years since Aehrenthal's dramatic demarche of annexing Bosnia-Herzegovina, relations with Russia have become increasingly precarious. While Tisza and Franz Ferdinand often discussed the need for improved ties with Russia, no openings appeared.

Berchtold now realizes that the Archduke's death has dramatically altered his decision matrix. Since 1912, Franz Ferdinand had been a powerful voice for peace and restraint, a person with whom the Foreign Minister had to seriously reckon, but who was also an ally for peace. Now the Archduke's murder has become the potential occasion for a limited war with Serbia, which the Archduke had particularly warned against. If the assassinated victim at Sarajevo had been the Governor-General of Bosnia-Herzegovina General Oskar Potiorek, who was also in the car, likelihood of a war with Serbia would have been minor ~ but this is not the case, and Berchtold grasps the enormity of the murders. This time extremists have moved across state borders to murder members of a ruling family in another country. Evidence now links the assassins with officers of the Serbian Government, and it is clearly suggestible, to the senior leadership in Belgrade.

In earlier crises, Count Berchtold has had to consider three viewpoints : the views of Franz Josef and Franz Ferdinand ; those of his fellow ministers (including the generals and Tisza in October 1913), and those of the ruling Prussian élite in Berlin. That equation is now minus the Archduke, but includes a new variable; reports of unrest in Bosnia and Herzegovina generated by the assassinations. Berchtold opens the meeting:

Berlin has promised support in the eventuality of a warlike complication with Serbia. Conrad: *The moment has arrived, to render Serbia innocuous once and for all ~ by a display of force?* Tisza: *What would war achieve? I want no surprise attacks and only diplomatic pressure at first, there should be no attempt to use mobilization as a diplomatic weapon, as was the case during the Balkan wars. If war comes, no Serbian territory can be added to the monarchy. The provoking of such a war with Serbia would be a fatal mistake ~ it will pillory Austrians before the world as disturbers of the peace, besides beginning a war under the most unfavourable circumstances.*

After a briefing by Conrad, few doubt that war is far off, as the close of the Council minutes state: *Still an agreement has been arrived at, since the propositions of the Hungarian Premier will in all probability lead to a war with Serbia, the necessity of which he and all the other members of the Council have understood and admitted.* Even after a lengthy six-hour meeting of the Council, the others still cannot convince Tisza of the need for an effective response. Tisza still objects, and his objections are significant. Berchtold has to overcome Tisza, if he expects Emperor Franz Josef to agree to armed action.

VIENNA Wednesday 8th July ~ 6 PM. Berchtold meets with Conrad, Forgách, Burián, and Hoyos, to discuss the terms of a Note of Demands and its delivery in 14 days after *harvest leave* and an investigation are finalized. When Berchtold expresses the desire for both Conrad and War Minister Krobatin to *go on leave to give the appearance that nothing is happening,* Berchtold's aide Forgách taunts; *I wonder if you will really act.*

That afternoon Berchtold meets Tisza, who has visited the Emperor. While Tisza appreciates the dangers of inaction, he still presses for a more resolute diplomatic approach. Berchtold resists this as having scant prospects of success. This meeting disturbs Tisza, and the next day (when in Budapest) he drafts a persuasive memorandum to Franz Josef still opting for diplomacy, not war. He writes; *German considerations, should not dictate Habsburg policy. I have grave doubts about using the Sarajevo events as the occasion for a reckoning with Serbia, and suggest a need of more proof of Belgrade's role. After painfully scrupulous reflection, I cannot share responsibility for the proposed military aggression against Serbia. I must declare that in spite of my devotion to Your Majesty, or I should say more correctly precisely because of this devotion, I could not share the responsibility for the exclusively warlike solution and the intention of making the horrible crime of Sarajevo the occasion for the final reckoning with Serbia.*

Franz Josef is appalled by the assassination of his Heir and Sophie in its assault on the Monarchy and his Empire. The Emperor now dreads to end his long reign as it began, in war and revolution. Berchtold puts firm pressure on him to consent to a military response to confront the provocations, even if it means war. Franz Josef agrees that Serbia has gone too far but insists that Berchtold must gain unanimous consent from his cabinet before any military measures can receive his final approval, by which he means securing the approval of Hungarian PM Count István Tisza. Tisza favours having Bulgaria and Romania in an open alliance to surround Serbia with daunting neighbours. Tisza is the only voice not in favour of war against Belgrade, and a great deal of pressure and reason will be needed to convince the Hungarian to consent, as Count Tisza knows that his acceptance of Berchtold's proposal will make war against Serbia inevitable ~ whereas his proposal may parry or even prevent it.

Berchtold transmits to Emperor Franz Josef a telegram to outline Tisza's position and to report: *The rest of the ministers want to use the opportunity for warlike action against Serbia. Such an action will deal with the Serbian problem while protecting the monarchy's control in the two provinces. I hope to come to see you immediately at Bad Ischl but will wait so that I can bring a memorandum from Tisza.* Berchtold then takes Tisza's memorandum with him when he travels overnight to Bad Ischl. Given these differences, Berchtold realizes, that he and Tisza are in a struggle to shape the direction of Habsburg policy. Meanwhile, Berchtold is forcibly reminded of the need for clarity from Berlin, when Count Heinrich von Tschirschky speaks of *the need for immediate action,* whereas Szögyény on the contrary reports from Berlin that Under-Secretary Zimmermann urges that no humiliating demands be made of Serbia.

BAD ISCHL Thursday morning 10th July ~ 7 AM. Foreign Minister Berchtold arrives at the Emperor's favorite resort in a rainstorm. Later in the morning he meets with Franz Josef for an hour. Once again Berchtold's brief diary entry presents it well: *I read the Tisza memorandum to the Emperor. I sought conditions that would humiliate Serbia and create practical control over Serbia. The Emperor approved. He was concerned that a weak stance would discredit our standing with Germany. Still, the Monarch wants proof of Serbia's involvement in the murders and pressed me to bridge the differences with Tisza.* Of this, Conrad wrote in his diary; *Berchtold, just back from Bad Ischl, found the ruler to be for action against Serbia and only had concerns about unrest in Hungary.* Berchtold begins supervising the preparatory steps to war by sending Friedrich Wiesner to Sarajevo to gather evidence to link the conspiracy to Belgrade.

BERLIN 11th July. Kurt Riezler: *Our situation is frightful. Kiderlen always said we would have to fight. Berchtold is now considering the timing ~ before or after Poincaré's journey to St. Petersburg ~ before is better. There will be a greater chance that France, confronted suddenly by the prospect of war, will take fright and urge peace in St. Petersburg. Austria has just decided this. But first, the Hungarian harvest must be brought in.*

BELGRADE 11th July. Although the Russian Ambassador to Serbia, Nikolai Hartwig has enormous knowledge of the *Black Hand* and its methods, and his influence on the Serbian Government has remained significant, it appears that his personal outlook has changed since the assassination. This is understandable, following the unpredictable dangers of his establishing the Balkan League. However, the night Nikolai Hartwig unexpectedly visits Baron von Giesl to express his condolences for the outrage, he slumps to the floor. When a doctor examines him a few minutes later, he declares that the Russian has died from a heart attack. (Hartwig, now obese has long suffered angina pectoris). A dramatic scene ensues with the arrival of his daughter, Ludmilla Hartwig. She brusquely repulses the sympathy expressed by the von Giesl's to poke about the room sniffing at an eau de cologne, and rummage thru two large Japanese vases. As her father only smoked his own Russian cigarettes, Ludmilla wraps up two butts and puts them in her bag. With unveiled suspicion she sharply asks: *Did my father have anything to eat or drink?* In the tense atmosphere of the time, Hartwig's death causes rumours in Belgrade that makes a poisoner of the prudent Baron von Giesl, who is even accused of having brought back from Vienna a secret electric chair that can instantly kill anyone sitting in it. Because of the surprisingly close friendship that developed between von Giesl and Nikolai Hartwig, the death of the Russian Ambassador at this crucial moment represents a catastrophe for Europe. Perhaps these two seasoned diplomats could have even prevented the war, if they had talked more. Baron von Giesl: *If Nikolai Hartwig had lived beyond the critical 25th of July, the war would not have occurred.* British Chargé d'Affaires Dayrell Crackenthorpe: *Reports circulating that Russian Ambassador Hartwig had taken ~ a cup of tea ~ at the Austrian Legation before his sudden death. I merely mention this as affording an indication of the somewhat medieval morals prevailing this city.* On Hartwig's death, ADC to Franz Josef, Miklos Horthy comments: *The Russian Ambassador Hartwig, who was rumoured to know all the intrigues, had a fatal heart attack during his visit of condolence to Ambassador Baron Giesl ~ so great was his agitation.*

Hartwig terminated his career at Belgrade. While holding this post he did not hide his antipathy to Austria. His dislike was so openly manifested, that public opinion accused him (wrongly) of having celebrated the death of the Archduke Franz Ferdinand, by holding a demonstrative reunion at the A-H Imperial Legation. To explain matters, he called one evening on his colleague the Baron Giesl just as he returned from the Archduke's funeral. It was during that visit that he fell dead from a stroke of apoplexy. This sudden death gave rise to another improbable rumour ~ that he had been poisoned by Austrian cigarettes. Alexander Savinsky

Baron Nikolai Hartwig ~ New Belgrade Cemetery 14th July 1914 ~ the bearded Serbian PM Nikolai Pasic can be seen on the right

BELGRADE 13th July. Friedrich Wiesner reports by telegram as evidenced fact that three of the conspirators were armed in Serbia, smuggled across the border, and that two Serbian officials are responsible for supplying weapons. He will name them in the dossier as among those to be punished.

VIENNA. Governor-General Oskar Potiorek disagrees with Wiesner's decision not to describe Serbian Government participation in the plot. In a message to Krobatin ~ *There must have been someone!* Tisza reverses course and agrees to war against Serbia. Why he alters his stance remains less clear. Almost certainly Franz Josef's views have some sway, as do assurances that the A-H Army can protect Transylvania from a possible Romanian attack. In addition, Berlin's strong support has influence, as does the German argument that an unchastened Serbia could turn its attention to Hungary or even support Romania against Hungary. Tisza now agrees to investigative demands with a 48-hour limit for Belgrade's reply ~ on the condition that it will not involve Serbian territorial concessions.

Now, all that remains is to deliver the note to Belgrade and in this Vienna has now acquired some advantage. In the week that Tisza has taken to give his consent, French President Poincaré and his éntourage have voyaged to St. Petersburg to visit Tsar Nicholas II. Vienna therefore decides to wait until Poincaré has departed St Petersburg, and is inaccessible upon the sea returning to France; otherwise, Russia under influence of the champagne mood of Franco-Russian toasts and festivities at St Petersburg might offer Serbia far stronger military support and thwart A-H plans. Therefore, Berchtold resolves to wait until Poincaré has departed Russia and is far out on the North Atlantic, before delivering their note to Belgrade.

The German Ambassador in Vienna Count von Tschirschky wires to Berlin that Count Berchtold is now in agreeance, and Kaiser Wilhelm underlines Berchtold's words : *The speediest action is desirable.* Wilhelm with his natural impetuosity again urges determined action from the Viennese Ballplatz, whatever it might be to be taken *as quickly as possible*. Yet for more than two weeks, Europe waits for Vienna to decide and act. Berchtold parries repeated German questions of *when will something happen?* citing the need to gather more evidence of the assassin's criminal plot. Berchtold does not disclose that the A-H Empire's pre-occupation with *harvest leave* has made a direct mobilization impossible.

The Austro-Hungarian Embassy in Belgrade

The situation becomes much clearer when the *Wiesner-dokument* arrives from Sarajevo in Vienna. Dr. Friedrich Wiesner, a legal counselor of the A-H Foreign Office, has carried out his preliminary investigation seeking any evidence to link Serbia to the assassination. In his report, Dr. Wiesner states; *There is hardly a doubt that the crime was resolved upon in Belgrade, and prepared with the cooperation of Serbian officials. There is direct evidence of complicity by members of the Serbian military and border guards. Milan Ciganovic, a railroad official in Serbia, assisted in bringing the armed assassins into Bosnia, and then Major Voijaslav Tankosic provided weapons from the Kraguljevac Armory and trained the killers. Assistance was provided to the conspirators by border guards at Sabac, and Serb peasants volunteered their houses to shelter the conspirators and hide their weapons. Therefore, responsibility can be focused on Belgrade.*

By mid-July, A-H Foreign Minister Berchtold has reason for caution vis-à-vis Berlin. On 16th July he learns that the German Foreign Office has informed its Ambassador to Italy Hans von Flotow, of Vienna's intentions. Flotow, who is at an Italian spa, then tells Italian Foreign Minister, the Marquis Antonino San Giuliano of Vienna's plans. The Italian wastes no time in telegraphing this intelligence to his envoys in Belgrade, St. Petersburg, and even Vienna where the codes have already been broken. Berchtold concludes (correctly) that St. Petersburg knows of his intentions and perhaps even Belgrade. Henceforth a virtual wall of silence descends upon Viennese-Berlin communications, even to the point of refusing to give Berlin a copy of the Note until almost the eve of its delivery. Count von Berchtold now acts to block ~ if at all possible, any further indiscretions from Berlin, and for the remainder of the July Crisis, Vienna exercises extreme communication caution ~ and this impacts heavily on events.

VIENNA 16th July. Conflict in Vienna is confirmed by the former A-H Ambassador to Italy, Count Heinrich von Lützow over lunch at his country house with his new friend, British Ambassador to A-H Sir Maurice de Bunsen who lives near the estate of Count von Lützow. The two have become firm friends during the spring. Worried about a conversation with Berchtold, Lützow asks de Bunsen: *Do you realize how grave the Balkan situation now is? Force will be used to compel Servia to behave as expected and Berchtold is certain of German support.* Maurice de Bunsen records and then telegraphs to London: *Austria is not going to stand Servia's insolence any longer and demands Servia take effective measures to prevent the manufacture and export of bombs, and put to an end to the insidious and murderous propaganda against the Hapsburg Monarchy. Austria will not tolerate futile discussions ~ if Servia does not at once cave in, force will be used to compel her. Austria is determined to have its way, and refuses to be deterred by anybody.*

Sir Maurice de Bunsen to Sir Edward Grey — *Vienna*
D. 01.50 PM   R. 03.15 PM   Confidential — *16th July 1914*

From language held by the A-H Minister for Foreign Affairs to a friend of mine, who has repeated it to me, I gather that situation is regarded at the A-H Ministry for Foreign Affairs in a serious light and that a kind of indictment is being prepared against the Servian Government for alleged complicity in the conspiracy which led to assassination of the Archduke. Accusation will be founded on the proceedings in the Sarajevo Court. My informant states that the Servian Government will be required to adopt certain definite measures in restraint of nationalist and anarchist propaganda, and will insist on immediate unconditional compliance, failing which force will be used. Germany is in agreement with this procedure, and it is thought that Europe will sympathize with A-H in demanding that Servia adopt in future more submissive attitude. Austria-Hungary will go ahead regardless of results. She will lose her position as a Great Power if she stands any further nonsense from Servia.

Grey presumes that Vienna will not act until they have first disclosed to the public their case against Serbia, founded upon what they may discover at the trial of the assassins, expecting this will make it easier for the other powers, such as Russia to counsel moderation in Belgrade. However Grey, with scant perception of the actual depth of Russian intrigue in Belgrade, supposes that the more reasonable the demands of Vienna, the easier it will be to secure peace. *I hate the idea of a war between any of the Great Powers, and that any of them should be dragged in to a war by Servia. It would be detestable.*

VIENNA Sunday 19th July. Three weeks after Sarajevo, the Ministerial Council meet incognito at Count Berchtold's private residence to review their Note of Investigative Demands and supporting documentation that accuse Serbian officials of complicity in the murder. Berchtold advises the group that their Note of Demands will be handed over at 5 PM on 23rd July. General Conrad outlines some of his war plans, focusing attention on mobilization details. Not on the question of what Russia may do. Later that afternoon, Berchtold visits to the residence of British Ambassador Maurice de Bunsen where his wife Berta, notes in her diary : *Berchtold came at 6 ~ and stayed to dine. He was very pleasant, but with M, was as secretive as with all Ambassadors, and let no word fall as to his intentions towards Servia.*

FRENCH PRESIDENT RAYMOND POINCARÉ VISITS RUSSIA

Princess Militza of Montenegro
Grand Duchess Militza of Russia

Grand Duke Nicholas of Russia
Brother of Tsar Nicholas, husband of Anastasia
Commander of the Imperial Guard

Princess Anastasia of Montenegro
Grand Duchess Anastasia of Russia

15th July. The French President Raymond Poincaré and his Foreign Minister René Viviani embark from Dunkerque on the cruiser *La France,* to steam into the vast Gulf of Finland on 20th July. The weather of the day is cloudy… After a fresh breeze from the open sea delivers a heavy shower, the sun bursts forth through pearl-grey clouds, and *La France* enters the Russian estuary of the Neva.

At 11.30 AM in the harbour of Peterhof, Tsar Nicholas II's new yacht *Alexandria* is under steam. The Tsar, in the uniform of an admiral, has at least two hours until the arrival of *La France.* As he likes to linger over his meals, there are long intervals between the courses, in which he chats and smokes. The French Ambassador to Russia Maurice Paléologue is seated on his right, Sazonoff on his left and Tsar Nicholas II tells Paléologue of his delight to be receiving the President of the French Republic. *We shall have weighty matters to discuss, I'm sure we shall agree on all points… But there's one question which is very much in my mind ~ our understanding with England. We must get her to come into our alliance. It would be such a guarantee of peace!* Paléologue: *Yes, Sire, the Triple Éntente cannot be too strong if it is to keep the peace.* Nicholas: *I've been told that you yourself are uneasy about Germany's intentions.* Paléologue: *Uneasy? Yes, Sire, I am uneasy although at the moment I have no particular reason to anticipate a war in the immediate future. But the Kaiser Wilhelm and his Government have let Germany get into a state of mind such that if some new dispute arose, in Morocco, the East ~ anywhere ~ they could neither give way nor compromise. A success is essential at any price and to obtain it they'll risk some adventure.* The Tsar reflects for a time… *I can't believe the Kaiser wants war… If you knew him as I do! If you knew how much theatricality there is in his posing! Perhaps I am doing my cousin Kaiser Wilhelm too much honour in thinking him capable of willing or simply accepting the consequences of his acts. But if war threatened, would he, and could he prevent it?* Paléologue: *No, Sire, I don't think so, honestly, I don't.* Tsar Nicholas sits silent and puffs his cigarette, then suggests: *It's all the more important for us to be able to count on England in such an emergency. Unless they have gone out of their mind altogether, Germany will never attack Russia, France and England combined.*

Coffee arrives, just as the arrival of the French squadron is signaled. The Tsar and his group ascend to the bridge with him to see in a quivering light, *La France* surging forward over the waves, leaving a long white wake behind her. Then she stops. The formidable warship is met in the harbour, by the welcoming guns of warships and shore batteries firing. The crews all cheer. The *Marseillaise* answers the Russian national anthem and the cheers of thousands of spectators, who have come from St. Petersburg on decorated pleasure boats. Shortly after, President Poincaré steps onto the *Alexandria.* The Tsar receives him at the gangway, and the Imperial yacht sets course for Peterhof. Sitting in the stern, Tsar Nicholas and President Poincaré enter into a discussion. The Tsar often nods in approval, and they both radiate a fresh confidence. Before long, the diplomatic entourage arrives at the Peterhof. Through its grand trees and sparkling fountains, the favourite palace of Tsarina Catherine the Great appears above a long terrace from which a foaming cascade pours its waters. Carriages ascend the drive leading to the palace entrance, and in due course all retire for the night.

21st July ~ morning. President Raymond Poincaré is in conference with the Tsar reviewing the Austro-Serbian dispute, a problem that becomes more worrying every day owing to the mysterious, and as yet unrevealed attitude of Vienna. Poincaré insists that the only way of saving the peace is an open discussion between all European Empires, taking care that one group is not opposed to another. *It's the method that served us so well in 1913. Let's try it again!* Tsar Nicholas agrees.

At 4 PM, an Imperial procession takes Poincaré to the Winter Palace for a diplomatic festivity. He receives an enthusiastic welcome all along the route. The Russian police have arranged it all, so at every street corner, a group of poor wretches cheer noisily ~ under the watchful eye of a policeman. At the Winter Palace, it is a full-dress occasion. Etiquette requires that the Ambassadors will be introduced one by one to President Raymond Poincaré, who has Foreign Minister Viviani on his left. It is the function of Ambassador Maurice Paléologue to present his foreign colleagues…

The first to enter is German Ambassador Count Friedrich von Pourtalès. President Poincaré receives him with the greatest affability. He asks him about the French origin of his family, his wife's relationship to the Castellanes, and a motor tour which the Count and Countess are proposing to make through Provence. Not a word of politics. Paléologue next presents the Japanese Ambassador Baron Ichiro Motono. Their conversation is short but not without importance. Then comes the turn of British Ambassador Sir George Buchanan to be introduced. Poincaré assures him that the Tsar is determined to show himself conciliatory, and adds their hope that the British Government may soon realize the necessity of transforming the Triple Éntente into a Triple Alliance.

Last appears the Austro-Hungarian Ambassador Count Szapary, a typical Hungarian nobleman. For two months, he has been away from St. Petersburg. He came back unexpectedly the night before. Many infer from his sudden return, that the Austro-Serbian difference is growing more acute, that there is going to be a rupture, and the ambassador must be at his post to play his part in the dispute. Paléologue tells Poincaré what he thinks, and the President replies ~ *I'll try and clear up this business.*

When President Poincaré meets with A-H Ambassador Szapary, after a few words of sympathy on the assassination of Archduke Francis Ferdinand, he tries to draw him out as to Berchtold's intent, and to caution him against Vienna holding Belgrade too responsible for recent events in Sarajevo. Poincaré asks Szapary: *Have you any news from Serbia?* Szapary: *The judicial enquiry is proceeding.* Poincaré continues: *Of course, I'm anxious about the results of this enquiry, Monsieur l'Ambassadeur. I can remember two previous enquiries, which did not improve your relations with Serbia.* Szapary replies sternly: *Monsieur le Président, we cannot suffer a foreign government to allow murderous plots against our sovereignty to be hatched on its territory!* President Poincaré tries a more conciliatory tone, to be doubly prudent. *With a little goodwill, this Serbian affair is easy to settle. But it is easy also for it to become most envenomed. Serbia has very warm friends in the Russian people. And Russia has an Ally, France. What complications are to be feared here!* Poincaré thanks the Ambassador for his call. Szapary bows and departs without a word. When Poincaré, Viviani, and Paléologue are alone, Poincaré : *I'm not satisfied with this conversation. The A-H ambassador has obviously been instructed to say nothing. Austria has a coup in store for us. Sazonoff must be firm and we must back him up.* At midnight, Poincaré returns to Peterhof.

Wednesday 22nd July ~ midday. Tsar Nicholas gives a luncheon in the Peterhof Palace to the President of the Republic and the officers of the French squadron. At 3.30 PM, they travel by train for the military camp at Krasnoe Selo. A blazing sun lights up the vast undulating plain bounded on the horizon by wooded hills. The Tsar and Tsarina, President Poincaré, the grand dukes and grand duchesses, along with the entire Imperial Staff then inspect the cantonments of the assembled troops. Meanwhile Ambassador Paléologue waits for them, with the ministers and civil functionaries on a slight hill on which tents have been pitched, where the élite of St Petersburg society are now crowded. As the sun descends towards the horizon in a sky of purple and gold, on a hand signal from the Tsar an artillery salvo signals it time for evening prayer. The bands play a hymn, and everyone uncovers. An officer recites the *Pater* in a loud voice. Tens of thousands pray for the Tsar and all Holy Russia. The silence and composure of the multitude on the great plain, the magical atmosphere of the hour, and the aura of the alliance, gives the grand ceremony a touching majesty.

From the military camps, the Imperial family and the French éntourage return to the village of Krasnoe Selo, where the Commander of the Imperial Guard, Grand Duke Nicholas Nicholaevich gives a banquet in honour of President Poincaré and the sovereigns. Here the war spirit and champagne mood stirred by the presence of the French guests much strengthens the militarist clique in Russia, headed by Grand Duke Nicholas, who wants Sazonoff to exert pressure on his peace-loving brother, the Tsar. Three long tables have been set in half-open tents around a garden, which is in full flower. The beds have just been watered and from them the fresh scent of flowers wafts into the warm air.

The French Ambassador to St. Petersburg, Maurice Paléologue arrives early to see the mysterious Montenegrin Princesses, Anastasia and Militza, wives of Grand Duke Nicholas and Grand Duke Peter respectively, decorating the tables with Lorraine thistles. Anastasia begins to talk with him excitedly…

Do you realize that we're passing through historic days, fateful days! At the review tomorrow the bands will play nothing but the Marche Lorraine and Sambre et Meuse. I've had a telegram in the proper pre-arranged code from my father today. He tells me that we shall have war before the month is out. What a hero, my Father! He is worthy of the Iliad. Just look at this little box I always take about with me. It's got some Lorraine soil in it, real Lorraine soil I picked up over the frontier when I was in France with my husband two years ago. And now look at that table of honour! It is decorated entirely with thistles. I would not have any other flowers put on it. Now then. They are thistles from Lorraine! I picked a few stalks on the territory annexed by Germany, brought them here and had the seeds sown in my garden. Militza, talk to the Ambassador some more; tell him all this day means to us, while I go and receive the Tsar.

During the dinner, Maurice Paléologue is seated on the left of the Grand Duchess Anastasia and her rhapsody continues, interspersed with profound theatrical prophecies. *There is going to be war. Nothing will be left of Austria. You will take back Alsace-Lorraine. Our armies will both meet in Berlin. Germany will be destroyed. I must control myself, the Tsar is looking at me ~ he has his eye on me.*
Under the stern gaze of Tsar Nicholas, the Montenegrin sibyl lapses into silence.

KRASNOE SELO Thursday 23rd July ~ morning. A review of 60,000 Russian troops takes part in a magnificent pageant of might and majesty, where the Tsar's Imperial Guard indeed march past to the strains of the *Marche de Sambre et Meuse* and the *Marche Lorraine.* The Tsar is mounted at the foot of the mound on which is the Imperial tent. Poincaré is seated on the Tsarina's right in front of the tent. That evening, they are all enjoying a farewell dinner on *La France.* The moment it is over, the French squadron will prepare to leave, and set course for Stockholm. Tsarina Alexandra has come with the Tsar, all of the grand dukes, and the grand duchesses are now present.

About 7 PM a momentary squall does some slight damage to the floral decorations of the deck, but the table looks splendid all the same. It even has a kind of terrifying grandeur when the French warship *La France's* four massive 30-cm guns raise their huge muzzles above the heads of the guests. The sky clears, a light breeze wafts over the waves, and the moon now rises above the horizon. Conversation between the Tsar and the President never ceases, while Grand Duchess Anastasia raises her champagne glass towards the French leaders more than once, indicating with a sweep of her arm the warlike tackle all about them. As the entrée is about to be served, a servant brings Paléologue a note from René Viviani, he has scribbled on a menu. *Be quick and prepare a communiqué for the press.* Admiral Ivan Grigorovitch whispers in his ear ~ *It seems to me you're not left in peace for a minute!*

Paléologue takes his own and the Admiral's menu and composes a note for the Havas Agency. *The visit which the President of the Republic has just paid to H.M. the Tsar of Russia has given the two friendly and allied governments an opportunity of discovering that they are in entire agreement in their views on the various problems, which concern for peace and the balance of power in Europe has laid before the Powers, particularly in the East.*

Excellent! Said René Viviani. Viviani and Paléologue immediately venture to discuss the note with Tsar Nicholas, Serge Sazonoff, Alexander Isvolsky, and President Poincaré. All four unreservedly approve the draft and Paléologue sends it at once to the Havas Agency. As the time for departure approaches, the Tsar tells Poincaré he would like to continue the discussion for a few minutes longer. *Suppose we go onto the bridge, Monsieur le President? It will be quieter.*

Minutes later, Ambassador Paléologue finds himself alone on deck with the Tsarina, who asks him to take a chair on her left. She seems worn out, and with a forced smile she says in a tired tone, *I'm glad I came tonight. I was afraid there would be a storm. The decorations on all the ships are magnificent. The President will have lovely weather for his voyage.* At 11 PM preparations for the departure are in progress. The guards shoulder arms. Commands ring out. The *Alexandria's* launch greets *La France.* Farewells are said to the strains of the Russian national anthem and the *Marseillaise.* The Tsar speaks warmly to President Raymond Poincaré. Ambassador Paléologue bids goodbye to President Poincaré who asks him to call on him in Paris ~ in a fortnight's time.

As Paléologue bows to Tsar Nicholas at the top of the gangway, he asks the French Ambassador. *Will you come with me, Monsieur l'Ambassadeur? We can talk undisturbed on my yacht. You'll be taken straight back to St. Petersburg.* From *La France* they transfer to the *Alexandria.* It is a magnificent night. With the stars above stretching into space, and not a breath of wind, *La France* and her escorting squadron steam towards the west, leaving behind them a wake of foam glistening in the moonlight. Nicholas: *It's a wonderful night. Suppose we go for a sail!* And *Alexandria* steers for the coast of Finland. The Tsar asks Ambassador Maurice Paléologue to take a seat with him in the stern of the *Alexandria* and tells him of the conversation, he has just had with President Poincaré.

I'm delighted with my talk with the President. We see absolutely eye to eye. I am no less peace loving than he, and he is no less determined than I to do all necessary to prevent the cause of peace from being compromised. He fears some Austro-German manœuvre against Serbia and thinks we should reply with the united front of a common diplomatic policy. I think the same. We must show ourselves firm and united in our efforts to find possible solutions and the necessary adjustments. The more difficult the situation becomes the more important will unity and firmness become. This policy seems to me the essence of wisdom.
I'm afraid we shall have to resort to it before long. You are still uneasy?

Maurice Paléologue replies: *Yes, sire.* Nicholas: *Have you any fresh reason for your apprehension?* Paléologue: *I have at least one ~ the unexpected return of Ambassador Szapary, and the air of cold and hostile reserve he adopted towards the President the day before yesterday. Germany and Austria are preparing a shock for us.* Nicholas: *What can they want? A diplomatic success at the expense of Serbia? To score a point off the Triple Éntente? No, no, notwithstanding appearances, Kaiser Wilhelm is too cautious to launch his country on some wild adventure, and Franz Josef's only wish is to die in peace.*

For a minute the Tsar sits in silence, lost in thought as if following up some vague line of thought. Then he rises and paces the deck. Around them the Grand Dukes are standing, and waiting for him. He calls them up in turn, and seems to show them unrestrained frankness; an affectionate familiarity, as if he wants them to forget that he usually keeps them at a distance, never to talk politics with them. The Grand Dukes Nicholas and Paul, along with the Grand Duchess Marie Pavlovna congratulate Ambassador Paléologue. *The visit has been a supreme success, and that Nicholas II is well pleased.*

Then, the mysterious Grand Duchesses Anastasia and Militza confide Paléologue into a corner. *What a glorious speech the President made. It was just what wanted saying, just what we've been waiting for so long! Peace in strength, honour and self-respect. Remember those words, Monsieur l'Ambassadeur; they will mark a date in the history of the world.* At around midnight, the *Alexandria* drops anchor in Peterhof Bay. After leaving Tsar Nicholas and Tsarina Alexandra, Ambassador Paléologue transfers to the smaller escort yacht, *Strela* (Russian : *Arrow*) to return to St. Petersburg where they expect to arrive at 2.30 AM As he steams up the Neva, he thinks of the profound prophecies of the Montenegrin sibyls.

The *Okhrana* are now searching for an Austrian agent, who was asking several days ago on the docks of St. Petersburg to confirm the precise date of the French warship's departure from Russia. The festivities continue in St. Petersburg, unaware that Vienna has deliberately delayed the presentation of their Note of Demands to Belgrade until late on 23rd July, by which time President Poincaré and his entourage will have departed, to avert any co-operation between France and Russia. The Montenegrin Princesses, also known as the *Black Peril* from their notorious interest in the occult; as collectors of seers, mediums, the cult of St Seraphin of Sarov and other para-ecclesiastical aspects will indirectly contribute much to the decline of Imperial Russia, when Grand Duchess Militza presents the decadent influence of the idiot passing as a saint, Grigori Rasputin to Tsarina Alexandra.

VIENNA. On the A-H diplomatic silence, Ambassador Sir Maurice de Bunsen wires to London. *It might have been supposed that Duc Guiseppe Avarna as Ambassador of the Italian Kingdom which bound to be so closely affected by complications in the Balkans would have been taken fully into the confidence of Count Berchtold during this critical time. In point of fact, His Excellency has also been left completely in the dark.*

The A-H Government in an effort to dupe the Triple Éntente, sends every important diplomat and military official on leave, publicizes their departure, and will not recall those already on holiday. In this way, Austria and Germany (whose Kaiser is still sailing off Norway) keenly seek to exploit the diplomatic calendar and aristocratic notions of leisure to reduce the chance that the Éntente will intervene if A-H decides to invade Serbia. Well into the *July Crisis,* Count Leopold von Berchtold takes a more active part in events. He is no longer vacillating between the two opposing voices of opinion represented respectively by Conrad and Tisza. After Sarajevo, Berchtold decides to use the crime as the final justification to clear up all old issues with Serbia. Now totally agreeing with Conrad, Berchtold drafts a letter for Emperor Franz Josef to sign and dispatch to Kaiser Wilhelm.

I am sending you a memorandum, drawn up by my Minister for Foreign Affairs ~
The attack on my poor nephew is a direct result of the agitation of the Russian and Serbian Pan-Slavs, whose single aim is the weakening of the Triple Alliance and the disruption of my Empire. According to all indications, the crime of Sarajevo is not the deed of a single individual, but the result of a well-arranged plot whose threads reach to Belgrade; and though presumably it will be impossible to prove the complicity of the Servian régime, there can be no doubt that its policy of uniting all the South Slavs under the Servian flag promotes such crimes, and that a continuation of the situation spells lasting danger for my dynasty and for my territories.

The aim of my Government must henceforth be to isolate and diminish Servia. This will only be possible when Serbia is eliminated as a political factor in the Balkans. After the last frightful events in Bosnia, you too will be convinced that a friendly settlement of the antagonism that divides Austria-Hungary from Servia is no longer to be thought of, and that the peace policy of all European monarchs is threatened so long as the source of criminal agitation in Belgrade lives on unpunished. Emperor Franz Josef

The A-H Ambassador to Serbia Baron von Giesl, travels to Vienna to personally report on the situation in Serbia, and when he speaks with Berchtold and Hungarian PM Count Tisza, Giesl emphasizes the importance of not presenting demands to Belgrade that may be negotiated down. Count Berchtold: *You can be assured that will not happen.* On recent developments Franz Josef remarks : *Serbia at present constitutes the pivot of Panslavist policy, and is disposed as a power-factor in the Balkans. Be convinced after the recent fearful event in Bosnia that a reconciliation is not to be thought of, and that maintenance of the peace will be threatened if this hotbed of criminal agitation in Belgrade goes unpunished.* Returning to Belgrade, Giesl becomes convinced that the strife with Serbia can only be settled with force, adopts a stronger attitude, and expresses his opinions in a letter to Count Berchtold on 21st July.

Before the assassination, as in fact since the annexation crisis, relations between the monarchy and Serbia have been poisoned on the part of the latter through national chauvinism, enmity and an effective propaganda in our Serbian inhabited lands ~ since the last two Balkan wars Serbian success has raised this chauvinism to paroxysms whose outbreaks at times carry the stamp of madness. The policy of Serbia is built on the detachment of the South Slav provinces and, as a further result, on the destruction of the Dual Monarchy as a Great Power.

Serbian newspapers without fear of reprimand, discuss the decrepitude of the Dual Monarchy, and insult her officials, and the exalted person of our ruler. The press is the educator of the Servian people; it promoted the great Serbian propaganda, from which sprang the crime of Sarajevo. Political parties and governmental policy are wholly subservient to it. Its accusations that the death of Hartwig was due to poison, verge on insanity. The London Times called them ravings. The people, in gratitude for the past, and in anxiety for the future, outbid one another in servility to Russia. The serious words of our statesmen are regarded as bluff. This picture leads up to the conclusion that a reckoning with Serbia, a war for the position of the Dual Monarchy as a great power, even for its existence as such, cannot be permanently avoided. Should we put forward far-reaching requirements joined to effective control ~ for this alone could clear the Augean stable of great Serbian intrigues ~ then all possible consequences must be considered, and from the beginning there must be a strong and firm determination to carry through the matter to the end. Half measures in the presentation of demands followed by long discussions and ending only in an unsound compromise, would be the hardest blow that could be directed against Austria-Hungary's reputation in Serbia and her position in Europe. Baron Wladimir von Gieslingen

Following the withdrawal of expected Russian support during the dangerous 1908 Bosnian Crisis, the Belgrade Cabinet on 31st March 1909 provided Austria-Hungary with the pivotal promise that now appropriately forms the premise of A-H's expectant Note of Investigative Demands to Belgrade. Serbia has failed to abide by their 1909 commitment to end Belgrade-based efforts to revolutionize Bosnian Serbs. King Petar and his Government being unwilling or unable to suppress the *Black Hand*.

On the 31st of March 1909, the Serbian envoy in Vienna, on the instructions of the Serbian Government, made the following declaration to the Imperial and Royal Government: Serbia recognizes that the fait accompli regarding Bosnia has not affected her rights. In deference to the advice of the Great Powers, Serbia undertakes to renounce from now onwards the attitude of protest and opposition which she has adopted with regard to the annexation since last autumn. She undertakes moreover, to modify the direction of her policy with regard to Austria-Hungary and to live in future on good neighborly terms with the latter.

The history of recent years, and in particular the painful events of the 28th June, have shown the existence of a subversive movement with the object of detaching a part of the territories of Austria-Hungary from the Monarchy. The movement which had its birth under the eye of the Serbian Government has gone so far as to make itself manifest on both sides of the Serbian frontier in acts of terrorism, outrages and murders.

Far from carrying out the formal undertakings contained in the declaration of the 31st March 1909, the Royal Serbian Government has done nothing to repress these movements. It has permitted the criminal machinations of various societies and associations directed against the Monarchy, and has tolerated unrestrained language on the part of the press, the glorification of the perpetrators of outrages, and the participation of officers and functionaries in subversive agitation. It has permitted an unwholesome propaganda in public instruction; in short, it has permitted all manifestations of a nature to incite the Serbian population to hatred of the Monarchy and contempt of its institutions. This culpable tolerance of the Royal Serbian Government had not ceased at the moment when the events of the 28th June proved its fatal consequences to the whole world.

It results from the depositions and confessions of the criminal perpetrators of the outrage of the 28th June that the Sarajevo assassinations were planned in Belgrade; that the arms and explosives with which the murderers were provided had been given to them by Serbian officers and functionaries belonging to the Narodna Odbrana; and finally that the passage into Bosnia of the criminals and their arms was organized and effected by commanding officers of the Serbian frontier service. The above-mentioned results of the magisterial investigation do not permit the Austro-Hungarian Government to pursue any longer the attitude of expectant forbearance which they have maintained for years in the face of the machinations hatched in Belgrade, and thence propagated in the territories of the Monarchy. The results, on the contrary, impose on them the duty of putting an end to the intrigues which form a perpetual menace to the tranquility of the Monarchy,

To achieve this end, the Imperial and Royal Government see themselves compelled to demand from the Royal Serbian Government a formal assurance that they condemn this dangerous propaganda against the Monarchy; in other words, the whole series of tendencies, the ultimate aim of which is to detach from the Monarchy territories belonging to it, and that they undertake to suppress by every means this criminal and terrorist propaganda.

In order to give a formal character to this undertaking the Royal Serbian Government shall publish on the front page of their Official Journal of July the following declaration:

The Royal Government of Serbia condemn the propaganda directed against Austria-Hungary: the general tendency of which the final aim is to detach from the Austro-Hungarian Monarchy territories belonging to it, and they sincerely deplore the fatal consequences of these criminal proceedings.

The Royal Government regret that Serbian officers and functionaries participated in the above-mentioned propaganda and thus compromised the good neighborly relations to which the Royal Government were solemnly pledged by their declaration of 31st March 1909.

The Royal Government, who disapprove and repudiate all idea of interfering or attempting to interfere with the destinies of the inhabitants of any part whatsoever of Austria-Hungary, consider it their duty formally to warn officers and functionaries, and the whole population of the kingdom, that henceforward they will proceed with the utmost rigour against persons who may be guilty of such machinations, which they will use all their efforts to anticipate and suppress. This declaration shall simultaneously be communicated to the Royal Serbian Army as an order of the day by His Majesty the King and shall be published in the Official Bulletin of the Army. The Royal Serbian Government further undertake:

1. To suppress any publication which incites to hatred and contempt of the Austro-Hungarian Monarchy and the general tendency of which is directed against its territorial integrity;

2. To dissolve immediately the society styled: Narodna Odbrana. To confiscate all its means of propaganda, and to proceed in the same manner against other societies and their branches in Serbia which engage in propaganda against the Austro-Hungarian Monarchy. The Royal Government shall take the necessary measures to prevent the societies dissolved from continuing their activities under another name and form;

3. To eliminate without delay from public instruction in Serbia, both as regards teaching and also the methods of instruction, everything that serves or might serve, to ferment propaganda against Austria-Hungary;

4. To remove from the military service, and from the administration all officers and functionaries guilty of propaganda against the Austro-Hungarian Monarchy whose names and deeds the Austro-Hungarian Government reserve to themselves the right of communicating to the Royal Government;

5. To accept the collaboration in Serbia of representatives of the Austro-Hungarian Government for the suppression of the subversive movement directed against the territorial integrity of the Monarchy;

6. To take judicial proceedings against accessories to the plot of the 28th June who are on Serbian territory, delegates of the Austro-Hungarian Government will take part in the investigation relating thereto;

7. To proceed without delay to arrest Major Voija Tankosic and Milan Ciganovic, a Serbian State employee, who have been compromised by the results of the magisterial inquiry at Sarajevo;

8. To prevent the cooperation of the Serbian authorities in the illicit traffic in arms and explosives across the frontier, to dismiss and punish severely the officials of the frontier service at Schabatz and Loznica guilty of having assisted the perpetrators of the Sarajevo crime by facilitating their passage across the frontier;

9. To furnish the Imperial and Royal Government with explanations regarding the unjustifiable utterances of high Serbian officials, both in Serbia and abroad who have not hesitated since the crime of 28th June to express themselves in interviews in terms of hostility to the Austro-Hungarian Government; and finally,

10. To notify the Imperial and Royal Government without delay of the execution of the measures comprised under the preceding heads.

The Austro-Hungarian Government expect a reply by 6 o'clock on Saturday evening the 25th July.

ANNEX : The criminal enquiry opened by the Court of Sarajevo against Princip and his accessories in, and before the assassination committed by them on 28th June, has to the present, led to the following conclusions:
1. The plot, having as its object the assassination of the Archduke Franz Ferdinand at the time of his visit to Sarajevo, was formed at Belgrade by Gavrilo Princip, with the assistance of Major Voija Tankosic.
2. The six bombs and four Browning pistols and ammunition with which the guilty parties committed the act were delivered to Princip and others by the man Milan Ciganovic and Major Voija Tankosic at Belgrade.
3. The bombs came from the arms depot of the Serbian Army at Kragujevac; where they were manufactured.
4. In order to ensure the success of the act, Ciganovic and Tankosic taught Princip, Cabrinovic, and Grabez how to use the bombs, and gave lessons in firing Browning pistols to Princip and Grabez in a forest near Topschider.
5. To enable Princip, Cabrinovic, and Grabez to cross the frontier of Bosnia-Herzegovina, and to smuggle in their arms secretly, a secret system of transport was organized by Milan Ciganovic. The entry of the criminals with their arms into Bosnia and Herzegovina was effected by the frontier captains of Shabatz (Rade Popovic) and of Loznica, as well as by the customhouse official Rudivoy Grbic of Loznica with the aid of several other persons.

* * *

(The Weisner Dossier was not presented to the British Government until 29th July, the day after war had been declared by A-H. Grey to Bunsen, 29th July 1914 in telegram No.282. Sazonov told the A-H Ambassador in St Petersburg, when the later promised to communicate the document ~ We need not had given ourselves the trouble, since we had sent an ultimatum. Szapary to Berchtold 24th July 1914 Diplomatiische Aktenstucke 1919)

The Crown Prince was at first favourably disposed toward the society ~ Union or Death or Black Hand. He contributed 26,000 dinars to support the paper Piyemont and paid the expenses of Apis's illness in 1912, but the refusal to admit him to membership that would have involved his becoming its head, wounded his pride and turned him against the Black Hand, so that in 1914 he appears to have taken the side of the Government.

M. Bogicevic
K.S.F. IV 678 1926

Baron Wladimir Giesl von Gieslingen presents the Note to Lazar Paču

BELGRADE 23rd July ~ 6 PM. Baron Giesl nears the Yellow House of Serbia's Foreign Ministry, the hour having been fixed to ensure Poincaré and Viviani will be safely aboard the cruiser *La France* and steaming for the open sea, to deliver Berchtold's ~ *Befriste Démarche* ~ Note with a Time Limit. Pasic is not in his office, so Serbian Finance Minister Lazar Pacu is the one who receives Giesl. Baron Wladimir von Giesl is astonished. *With the Russian Ambassador dead, the French Ambassador crazy, and the English envoy ill ~ the Triple Éntente is barely represented in Belgrade!* When Pacu sees the paper in Giesl's hand, he knows what it is and refuses to accept it. The Finance Minister repeats over and over, *I do not have the authority to accept this note.* Giesl simply places the note on his desk, and walks out.
On 23rd July at 6 o'clock, I gave the note to Minister Pacu. The nest day, the swarm of bees was very active.

LONDON. Foreign Secretary Sir Edward Grey arranges for an interview with A-H Ambassador to Britain, Count Albert von Mensdorff, to request an extension of the 48-hour time limit. Sir Edward views the limit as likely to inflame Russian opinion. *If war should occur between the four Great Powers, it will result in the complete collapse of European credit and industry. Irrespective of who might be the victors, many things might be completely swept away.* In accurate response, Mensdorff is quick to evidence the bad faith of Belgrade in failing to fulfill their 1909 promise to live on neighbourly terms with A-H. *On the contrary, Serbia has agitated to disintegrate Austria-Hungary. Austria-Hungary has to protect herself.* Sir Edward makes no offer of reply to this just comment.

Count Mensdorff did not demur to this statement of the possible consequences of the present situation, but he said that everything depends on convincing Russia of the justice of Austria's demands and on whether the demands can be accepted by Serbia. I made the remark that it was just as true to say that it required two to keep the peace as ordinarily it took two to make a quarrel. I hoped very much that, if there were difficulties, Austria and Russia would be able in the first instance to discuss them directly with each other. Count Mensdorff said ~ I hope this will be possible, but the attitude in St Petersburg has not been favourable recently. Sir Edward Grey

PARIS Friday 24th July. German Ambassador to France Baron Wilhelm von Schoen meets with French Acting Foreign Secretary Jean-Baptiste Bienvenu-Martin and French Deputy-Head of Foreign Ministry (and Balkan specialist) Philippe Berthelot to read an official communiqué from Berlin, backing Vienna. *If the conflict does not remain localized, the most serious consequences are to be feared.* Phillipe Berthelot: *Russia might be unable to resist the pressure to defend Serbia and I therefore hope Vienna will discuss any items to which Belgrade cannot agree.* Bienvenu-Martin: *Should we give to the Austrian note, the character of a simple diplomatic de mise en demeure that permits discussion or an ultimatum?* Schoen: *Personally, I have no views.* Schoen reports to Berlin: *The minister said that it was the duty of Serbia to proceed energetically against any accomplices of the murderers of Sarajevo ~ a duty which she cannot escape. The French minister is visibly relieved at our idea that the Austro-Serbian conflict is one to be settled exclusively by the two participants. French Government sincerely shares the wish that the conflict remain localized.*

J.B Bienvenu-Martin, Philippe Berthelot & Baron von Schoen

David Lloyd George & Winston Churchill

Paris Chamber of Deputies

French Ministry of Foreign Affairs on the Quai D'Orsay

British Ambassador to Russia ~ Sir George Buchannan

Captain William (Blinker) Hall
British Director of Naval Intelligence 1914-19

Monsieur Julien Davignon
Belgian Minister of Foreign Affairs 1907-16

Albert I
King of the Belgians 1909-34

Lieutenant Winston Churchill
4th Queen's Own Hussars 1895

Winston Churchill
Elected Member of Parliament 1904

Raymond Poincaré
President of France 1913-1920

Prince Karl Max von Lichnowsky
German Ambassador to Great Britain 1912-14

Helmuth von Moltke
Chief of German General Staff 1906-14

Count Albert von Mensdorff
A-H Ambassador to Great Britain 1904-14

David Lloyd George
British Chancellor of the Exchequer 1908-16
Prime Minister of Great Britain1916-22

King George V
King of the U.K. & British Dominions

Venetia Stanley & Herbert Henry Asquith
Prime Minister of Great Britain 1908-16

Pierre-Paul Cambon
French Ambassador to Britain 1898-1920

Sir Edward Grey
British Foreign Secretary 1905-16

Sir Maurice de Bunsen
British Ambassador to Vienna 1913-14

Count Istvan Tisza
PM of Hungary 1903-05 & 1913-17

Sir Edward Goschen
British Ambassador to Germany 1908-14

Sir George Buchanan
British Ambassador to Russia 1910-18

LONDON. The European conflict that has always been dreaded, with the perfect hope that it will never break out is fast becoming an alarming reality. By Friday 24th July all of Europe is buzzing with rumours of a pending Austrian strike against Serbia. As war approaches, it is let free upon the world by the blunders of leaders of nations, some of whom stumble and stagger into a wider war; almost literally. The often-inebriated British Prime Minister, Herbert *Squiffy* Asquith who sways on his feet when speaking or answering questions in the House of Commons, becomes the muse of a popular London songwriter. *Mr. Asquith says in a manner sweet and calm, another little drink won't do us any harm.* As such dire airs waft through the music halls of Britain, PM Asquith assures; *We are within measurable distance of a real Armageddon. Happily, there seems to be no reason why we should be more than spectators.* Lord Lovat tartly observes ; *The PM, what with drink, and holding girls' hands, is now incapable of doing anything except drift.* And as the unruffled British Cabinet prepare to depart London for the weekend, *The Times* declares ~ *As the London season nears its end, the call of the country begins to grow imperious.*

Sir Edward Grey leaves London for his fishing lodge in Hampshire. Others follow his example and leave London too. The German Ambassador to Great Britain, Prince Max von Lichnowsky hears that the German Army has cancelled all officers' leave. He also notes *the growing difficulty of doing anything until tomorrow, since there is no one to speak to in the Foreign Office.* The timing of the Vienna Cabinet and Sir Edward Grey's love of leisure prove disadvantageous during the international crisis. Often, Sir Edward appears to prefer the natural world to the sphere of diplomacy, as he once spoke of an interview with a particular ambassador as, *of less importance than an appointment with a wild duck.*

He was the most insular of our statesmen, and knew less of foreigners through contact with them than any other Minister in the Government. He rarely, if ever, crossed the seas. Northumberland was good enough for him, and if he could not get there and needed a change, there was his fishing lodge in Hampshire. This was a weakness ~ and it was a definite weakness in a Foreign Secretary. He had no real understanding of foreigners. Moreover, when a Conference in some foreign capital might have saved the situation, his dislike of leaving England stood in the way. David Lloyd George

VIENNA. Berchtold requests the A-H Ambassador to Britain, von Mensdorff to clarify with Grey: *The note presented to Serbia was not a formal ultimatum, but merely a note with a time limit that if not acceded to will be followed only by Austria's breaking off diplomatic relations and the beginning of military preparations. If Serbia is to give way only under the pressure of our military preparations, we shall indeed have to demand that she should make good the expenses which we will incur to mobilize, because of Serbia.*

LONDON. The Austrian stance is not completely condemned. The British are convinced of Vienna's righteousness in this matter. Even so, British Foreign Secretary, Sir Edward Grey takes a wider view of the big picture and understands that the machinations of Berchtold in the Ballplatz can have consequences. One option that Vienna may well accept, is a proposal made by Sir Edward Grey. *The four nations not immediately concerned ~ England, Germany, France, and Italy ~ should undertake to mediate between Russia and Austria.* Grey's initiative for a Conference of Ambassadors to pressure Belgrade to give Vienna full satisfaction is welcomed in all capitals, but not the two most crucial; Vienna and St. Petersburg. Berchtold refuses to state the future of Austro-Serbian relations, and Russia declines to acquiesce to the destruction of Belgrade. Sir Edward therefore contrives the *Stop at Belgrade* option, whereby Austrian troops may occupy the Serbian capital with the approval of Europe until the Serbian Government complies with the demands of the A-H Note. However, *Stop at Belgrade* stops short of granting Vienna her full satisfaction for military action to subdue the rogue state of Serbia.

BERLIN. French Ambassador Jules Cambon calls on Foreign Minister von Jagow, who asserts: *Germany was not aware of the terms of the A-H Note before they were published, though we do support them. The procedure and demands of Austria are equitable and moderate, and both Vienna and Berlin desire the localization of the conflict ~ any interference by another power may be followed by incalculable consequences. Serbia's friends should give her wise advice.* Jules Cambon: *Berlin should give similar wise advice in Vienna.* Jagow: *The problem has to be localized between Austria-Hungary and Serbia.* Later at the French Embassy... Cambon: *I believe the Austro-Hungarians are using the assassination to salvage their position in the Balkans. Germany is committed to supporting Austria. Berlin will support them because Germany does not want Austria-Hungary weakened any further. Vienna and Berlin are playing a dangerous game of bluff, and they think they can carry matters through with a high hand.* Visiting the French Embassy at the time, British *Chargé d'Affaires* Sir Horace Rumbold hears this, and reports Jules Cambon's words to London.

VIENNA. Morning ~ Count von Berchtold seeking Russian and Bulgarian neutrality in the hope to localize the impending war with Serbia, asks Russian Embassy official Prince Nicholas Koudacheff to call on him to advise : *The purpose of the Note is to stop Belgrade from supporting the Greater Serbia movement, and enable Austria-Hungary to check that it is doing so.* Berchtold instructs him to tell Sazonoff: *Vienna does not covet Serbian possessions or intend to infringe the sovereignty of Serbia. Nothing could be further from our minds than to wish to humiliate Belgrade. We have not the slightest interest in this. Make it clear that Vienna will go to extreme lengths to obtain fulfillment of its demands, and it will not recoil from the possibility of European complications.* Prince Nicholas Koudacheff additionally reports to St Petersburg. *Vienna considers her demands can be met, and is prepared to risk armed conflict in the event of rejection.*

St PETERSBURG 10 AM. FM Serge Sazonoff receives the text of Vienna's Note to Belgrade.
and after reading it in its entirety to his staff, he declares ~ *C'est La guerre Europénne!*
Immediately reporting, and for the first time ever on the telephone to Tsar Nicholas, Sazonoff advises; *Given the ultimatum's demands and its wording, Austria must know that it cannot be complied with by Serbia. I suggest Vienna must have received prior German approval ~ and that Vienna intends to attack.* Calmer and less pessimistic than his FM, the Tsar orders his Council of Ministers to assemble and discuss Russia's response, then directs Sazonoff to call an urgent meeting with the British and French Ambassadors.

Sir George Buchanan: *I was sitting in my study in the morning musing on all that I was going to do during my approaching holiday, when I was roused by the ringing of the Embassy telephone ~ Who's there?*
It is I, Sazonoff. Austria has presented an ultimatum at Belgrade couched in terms, which means war.
Please meet me at the French Embassy in an hour's time, as I must discuss matters with you and Paléologue.
After their lunch and a toast to the Tsar, the French Ambassador advocates a policy of firmness.
Serge Sazonoff: *But suppose that policy is to lead to war?*
Maurice Paléologue: *It will only lead to war if the Germanic powers have already made up their minds to resort to force to secure the hegemony of the East. Firmness does not exclude conciliation. But it is essential for the other side to be prepared to negotiate and compromise. You know my own views as to Germany's designs. The Austrian note seems to me to provoke the dangerous crisis I have anticipated for a long time. Henceforth we must recognize that war may break out at any moment. That prospect must govern all our diplomatic action.*
Buchanan: *Direct British interests are nil in Serbia. Assuming that the British Government desires to remain neutral, I am apprehensive that France and Russia will be crushed by the Triple Alliance.*
Sazonoff: *At the present juncture England's neutrality would be tantamount to her suicide!*
Buchanan: *I'm certain of that. But I'm afraid public opinion with us is still far from realizing, what our national interests so imperiously require.*
Sazonoff: *Great Britain cannot afford to efface herself from the problems now at issue. Russia wants Britain to join in warning Austria that her intervention in Serbia will not be tolerated. Suppose Austria wars in Serbia?*
A council of ministers is being held this afternoon on mobilization.
At a meeting tomorrow, where the Tsar will preside, a decision will be come to.
Paléologue: *England might play a decisive part in quenching Berlin's warlike stance. Tsar Nicholas himself expressed to me four days ago : Unless they have gone out of their minds altogether, Germany will never attack Russia, France and England combined. It is urgently necessary for the British Government to announce its adhesion to our cause, which is the cause of peace.*
Buchanan: *I promise to make a strong representation in favour of resistance to Germanic arrogance.*

2.00 PM. Sazonoff departs for Ielaguin Island where the Council of Ministers has been summoned.
Sir George Buchanan cables to Sir Edward Grey: *Sazonoff has said that the A-H demands clearly mean war. Austria's demands are provocative, some being impossible of acceptance. She would never have taken such action unless Germany had first been consulted. I hope Great Britain will proclaim her solidarity with Russia and France. France will fulfill her treaty obligations with Russia and support Russia in all diplomatic negotiations.*

3 PM. The meeting of the Imperial Russian Council of Ministers begins to decide the Russian response to Vienna's Note of Demands. It will last for over two hours and considering the importance of this meeting, it is worth recalling some of the statements of Russia's Ministers that afternoon. Chairman of the Council of Ministers Goremykin asks the Ministers to speak.

FM Serge Sazonoff speaks first.

German preparations are calculated to increase her power in Central Europe in order to enable her to carry out her wishes, not only as regards matters in the Near East, but in all international questions, without taking into consideration the opinion and influence of the powers not included in the Triple Éntente. In the course of the last decade Russia has shown great moderation and made many concessions wherever her interests and Berlin's came into conflict, but Germany has looked upon our concessions as so many proofs of our weakness and far from having prevented our neighbours from using aggressive methods, we have encouraged them.

Now comes the moment to make a stand. The Austrian ultimatum to Belgrade is beyond question drawn up with German connivance, and will if accepted, turn Serbia into a de facto protectorate of the Central Powers. In the past Russia has made immense sacrifices to secure the independence of the Slav peoples and if she now abandons under threat her historic mission, she will be considered a decadent state and will henceforth have to take second place among the powers, losing all authority and allowing Russian prestige in the Balkans to collapse utterly. Nor will concessions now ensure peace in the near future since there is nothing to stop Germany from mounting still further challenges to Russian interests. A firm stand will, however, mean a real risk of war with the Central Powers, whose consequences are all the more dangerous since it is not known what attitude Great Britain will take in the matter.

Minister of Agriculture Alexander Krivoshein is next to speak.

Only the Army's loyalty saved our régime from collapse in 1905. Since then a constitutional system has been established, Russia's financial position vastly improves and major reforms in the armed forces have been undertaken. However, our rearmament programme has not been completed and it seems doubtful whether our Army and our Fleet will ever be able to compete with those of Germany or Austria-Hungary as regards modern technical efficiency, since in cultural and industrial terms Russia is far behind the Central Powers.

On the other hand general conditions have improved a great deal in Russia in the past few years, and public and parliamentary opinion will fail to understand why, if at a critical moment involving Russia's vital interests, the Imperial Government is reluctant to act boldly. Our exaggeratedly prudent attitudes have unfortunately not succeeded in placating the Central European Powers. The disastrous consequences of the Russo-Japanese War has shown the grave danger which Russia will run in case of hostilities.

Consequently, our policy should aim at reducing the possibility of a European War, but if we remain passive we will not attain our object. War can break out in spite of our efforts at conciliation. Stronger language than that we have used hitherto is desirable. All factors tend to prove that the most judicious policy Russia can follow in present circumstances is a return to a firmer and more energetic attitude towards the unreasonable claims of the Central European Powers. Although the policy I advocate entails serious risks, the only hope of influencing the Germans is to show them, by making a firm stand that we have come to an end of the concessions we are prepared to make. In any case, we should take all steps which will enable us to face an attack.

Finance Minister Peter Bark writes : *Krivoshein's declaration made a profound impression on the Cabinet and was most instrumental in influencing our decisions.* Chairman Goremykin turns to prompt Minister for War Soukhomlinoff for his view.

Although great improvements in the armed forces have occurred since 1905, the programme of rearmament is not complete and Russian military superiority over the Central Powers cannot now be assumed. Nevertheless, hesitation is no longer appropriate as far as the Imperial Government is concerned. We see no objection to a display of great firmness in our diplomatic negotiations.

Finance Minister Peter Bark ; *Although Kokovtsov's careful management has built up a surplus in the Treasury, an inevitable excessive wartime expenditure could see the financial and economic stability of the Empire endangered. Since further concessions in no way guarantee peace, and since the honour, dignity and authority of Russia are at stake, as Finance Minister, I will adhere to the opinion of the majority of the Cabinet.*

Chairman Goremykin summarizes:

It is the Imperial Government's duty to decide definitely in favour of Serbia ; that firmness rather than conciliation is likely to secure peace ; but if it fails to do so, Russia should be ready to make the sacrifices required of her. Vienna should be asked to extend her time limit ; that Belgrade be urged to show a desire for conciliation and to fulfill the Austrian Government's requirements in so far as they do not jeopardize the independence of the Serbian state, and that the defence ministers shall request Imperial permission for the mobilization, if events should require it, of the Moscow ; plus the Odessa, Kiev, and Kazan Military Districts facing Austria-Hungary.

Tsar Nicholas accepts the Council policy that if Austria-Hungary takes action against Belgrade, to escalate to a partial mobilization of all military districts facing Austria-Hungary. The Tsar is to be formally asked to approve the measure the next day. Serbia is to be advised if they are not able to resist the expected attack, they should offer no resistance and appeal to the Great Powers for help. Sazonoff is to contact the other major powers and seek an extension of the deadline so each Empire can take a view on Vienna. Sazonoff calls in General Yanushkevich and tells him : *The Russian Army should prepare for a partial mobilization against Austria-Hungary ~ but not as yet against Germany.*

Evening ~ 7 PM. FM Serge Sazonoff sees Serbian Ambassador Miroslav Spalajkovic to enlighten; *No sovereign state can accept some parts of it. Serbia can rely on Russia's help. I disagree with the A-H and German view that the dispute should remain localized as it only concerns Vienna and Belgrade. Our view is that there should be international arbitration. If Austria-Hungary devours Serbia, we will go to war with her.* Sazonoff's views are passed on to Belgrade, where news of the continued support of Imperial Russia comes as music to the ears of Pasic and Prince-Regent Alexander, who while horrified at the prospect of war, are now emboldened to convince the Belgrade Cabinet to reject Point Six of Vienna's Note and to return a more equivocal response to some other demands than they would otherwise have done.

As a result of the decisions of the Imperial Council of Ministers being confirmed by the Tsar, the Russian armed forces very quietly begin their preparations for war. Magazines and supply depots are readied, railway repairs completed, and all units and departments check their instant readiness to carry out their tasks during a mobilization. Railway personnel are brought up to full complement for mobilization, while covering troops discreetly take up positions on threatened fronts, and in border areas, some reservists are recalled not to their units but to training camps. At a conference of officers of the General Staff, General Yanushkevich suggests; *In the energetic fulfillment of your tasks, if necessary it is permissible to slightly overstep the strict limitations of the law in preparing the army for war.*

Russia's key decision has been taken ~ to support Serbian independence ~ even at the risk of war. Knowing Vienna's need to subdue Serbian terrorism, and Berlin's willingness to back A-H, even if war with Russia and France should eventuate, Russia's stand makes a European conflict probable. Russia's head-of-mission in Sofia, A.A. Savinsky wires the following message to FM Serge Sazonoff;

In the event of a capitulation our prestige in the Slav world and in the Balkans will perish and never return. From Constantinople, head-of-mission M.N. Griers cables to FM Serge Sazonoff; *If Russia recognizes the irresistible nature of German power the Turks and Balkan states will unfailingly swing into the camp of the Central Powers, which will result in the total destruction of our prestige and our position in the Near East.*

With Russia deciding to back Belgrade, the only course for the Triple Éntente to evade war is by prompt action in London, and patience in St. Petersburg. But in this respect, Russia is severely at fault because ~ *there was no need to begin any mobilization moves against Vienna* ~ since a mere threat to do so would have been a wiser warning measure to direct at Vienna to deter A-H from invading Serbia.

Similarly, it was unwise for any precipitate Russian mobilization to be seen by A-H or Germany, as the more committed Hapsburg forces become involved in Serbia, the better it will be for subsequent Russian operations. Pre-war Russian generals had been determined that Vienna should divide their strength between the Serbian and Russian fronts ~ as evidenced at a 1912 conference of Chiefs of Staff of Russia's military districts. *Only if Russia stands aside in the initial phase of an Austro-Serb conflict will Austria be able to transfer overwhelming forces to the south in order to settle accounts with the Serbs by rapid and decisive blows. In view of this, the declaration of war to the calculation that the actions of the Russian armed forces should be fully developed at a time when Austria has still not finished its struggle with Serbia.*

St PETERSBURG 8 PM. Maurice Paléologue visits the Russian Foreign Office where Sazonoff is closeted with the German Ambassador. A few minutes later, Count Friedrich von Pourtalès emerges, his face purple and his eyes flashing. He furtively shakes Paléologue's hand as he enters the Foreign Minister's office. Sazonoff is still visibly anxious over the dispute in which he had just been engaged. He has quick, nervous movements. His voice is dry and jerky.

Paléologue: *What's happened?*
Sazonoff: *As I anticipated, Germany supports the Austrian cause. Not the slightest suggestion of conciliation. So I told Pourtalès quite bluntly that we should not leave Serbia to settle her differences with Austria alone. Our talk ended in a very acrimonious tone.*
Paléologue: *Really?*
Sazonoff: *Yes... Can you imagine what he had the audacity to tell me? He reproached me, and all Russians, with disliking Austria and having no scruples about troubling the last years of her aged Emperor. Of course I retorted: No, of course we don't like Austria. Why should we like her? She has never done us anything but harm. As for her aged Emperor Franz Josef, he owes it to us that he still has his crown on his head. Just remember how he showed his gratitude in 1908. What! Reproach us with not liking Austria! That's a bit too much!*

Paléologue: *It's a bad business Minister. If conversations between Petersburg and Berlin are to continue in this strain they won't last long. Very soon we shall see Kaiser Wilhelm rise in his shining armour. Please be calm. Exhaust every possibility of compromise! Don't forget that my government is based on public opinion and can only support you effectively if it has public opinion behind it. And think of English opinion also.*
Sazonoff: *I shall do everything possible to avoid war. But like you I am uneasy about the turn events are taking.*
Paléologue: *Can I give my government an assurance that you have not yet ordered any military preparations?*
Sazonoff: *None whatever. All we have decided is privately to withdraw the 80 million rubles we have on deposit in the German banks. I will endeavour to obtain from Count Berchtold an extension of the time fixed for the Serbian reply, so that we might have an opportunity of finding some peaceful solution.*
Our Russian ministers are to meet again tomorrow with the Tsar presiding.
Paléologue: *I recommend the greatest caution as to the advice you might give.*
Sazonoff: *You needn't fear! Besides you know the Tsar's caution. Berchtold has put himself in the wrong.*
It's our business to make him solely responsible for everything that comes. I even consider that if the Vienna Cabinet resorts to action the Serbians ought to let their territory be invaded and confine themselves to denouncing Austria's infamy to the civilized world.

PARIS. The French Minister of Justice J.B. Bienvenu-Martin, who receives information as to the contents of Vienna's Note to Belgrade in the morning papers, calls to see A-H Ambassador to France Count Nikolaus Szécsen de Temerin who duly explains. *For Austria-Hungary, it is an act of self-defence. The question is one which must be brought to an issue directly between Serbia and us, but it is in the general interests of Europe that the trouble which for years past has been kept up by the Serbian intrigues against us should make way for a clear situation. As friends of peace, I place France in the first rank of these to therefore give serious advice to Serbia to completely change her attitude, and to satisfy our just demands.* Count Niklaus telegraphs Bienvenu-Martin's response to Vienna. *Energetic action by Austria-Hungary can be understood and it is Serbia's duty to take energetic steps against possible accomplices of the murderer of Sarajevo and I express the hope that the dispute will be peacefully settled in a manner agreeable to our wishes.*

Madame Waddington, a great friend of the late King Edward VII, told me that feeling was so strong against the Austro-Hungarian Ambassador, Count Niklaus Szécsen de Temerin, that one evening, as he was about to sit down to dinner with his fellow diplomatist, Romanian Ambassador Alexandre Lahovary at Cercle de l'Union, one of the most select clubs of Paris, the secretary of the club requested that he should no longer take his meals at the club nor appear on the premises, because his presence under prevailing political conditions rendered him an undesirable personage. Count de Temerin, who had just ordered a fine bottle of Mouton Rothschild claret for his dinner, at once left the club. Charles Inman Barnard ~ *New York Tribune*

BELGRADE. British *Chargé d'Affaires* Dayrell Crackenthorpe wires Sir Edward Grey in London. *Austrian demands are considered absolutely unacceptable by the Serbian Government, who earnestly trusts that his Majesty's Government may see their way to somehow induce the Austrian Government to moderate them. This request was conveyed to me by Serbian Prime Minister Pasic, who returned early this morning to Belgrade. His Excellency is dejected, and is clearly very anxious as to developments that may arise.*
He should be anxious, since his orders are coming from Russia and France.

LONDON. A-H Ambassador Count Albert von Mensdorff tells Grey; *On expiry of the time limit, Austria will immediately begin military preparations, but not military operations.* Sir Edward then forewarns Prince Max Lichnowsky. *We shall soon be faced by a mobilization of Austria and Russia. Our only chance of peace is for Germany, France, Italy, and Great Britain to keep together and join in asking Austria and Russia not to cross the frontier till we have time to try and arrange matters.* Lichnowsky forcefully states to Berlin. *Unless Germany drops its pretence of innocence and agrees to help mediate, Great Britain will soon turn hostile.* He makes several strong appeals for immediate British-German cooperation. Grey then wires British *Chargé d'Affaires* Sir Horace Rumbold in Berlin, that Lichnowsky is personally favourable to possible mediation between A-H and Russia.

I impressed upon Ambassador Lichnowsky that, in the event of Russian and Austrian mobilization, the participation of Germany would be essential to any diplomatic action for peace. Alone we can do nothing. The French Government are travelling at the moment, and I have had no time to consult them, and cannot therefore be sure of their views, but I am prepared, if the German Government agree with my suggestion, to tell the French Government that I now think it the right thing to act upon.

Sir Edward Grey telegraphs the British Ambassador in Vienna Sir Maurice de Bunsen to ask the A-H Government for an extension of the 48-hour time limit for the Serbian reply. *A refusal would be against international ethics.* Sir Edward also asks Sir Maurice de Bunsen to support the Russian position. *I trust that if the Austro-Hungarian Government considers it too late to prolong the time limit, they will at any rate give time in the sense and for the reasons desired by Russia before taking any irretrievable steps.*

VIENNA. Russian *Chargé d'Affaires* Prince Koudacheff sends dual requests for an extension of time for the Serbian reply to Count Berchtold as two telegrams. One addressed to him on his journey, and the other to his destination in Ischl. A-H Foreign Office Under-Secretary Baron Karl von Macchio: *I received with icy coldness the prince's expostulation that the submission by A-H of grievances against Serbia without permitting time for their examination was not consonant with international courtesy. One's interests sometimes exempt one from being courteous. The A-H Government will accept no intervention from any power until the blow has been delivered and received full in the face by Serbia.* At noon, Count von Berchtold telegraphs from Lembach to direct Baron Karl von Macchio that he should punctually confirm that: *An extension of the time limit in the note to Serbia will not be granted, and also to inform Prince Koudacheff that even after breaking-off of diplomatic relations, Serbia can still secure peace by complying unconditionally with our demands, in which case Serbia must then pay the cost of all A-H military measures.* Count Berchtold then sends this pointed telegram to A-H Ambassador, Count Frigyes von Szápáry in St. Petersburg.

Prince Koudacheff has based his request on being taken by surprise in the demands on Serbia, and therefore Russia should have more time to consider the evidence in Austria-Hungary's dossier. These grounds however, rest on Prince Koudacheff's mistaken hypothesis. Our note to the powers was in no way intended to invite them to make known their own views on the subject, but merely bore the character of a statement for information, the communication of which we regarded as a duty laid on us by international courtesy. We regarded our action as concerning us, and Serbia alone. Any further toleration of Serbian intrigues would undermine our existence as a state and our position as a great power, thus also threatening the balance of power in Europe. We are convinced, that the existing European balance of power ~ of such importance for the peace of the world, must be maintained. Our action against Serbia ~ its object is the necessary preservation of our position in Europe.

BELGRADE. A-H Ambassador Baron Wladimir Giesl von Gieslingen wires to Count Berchtold. *The Belgrade Cabinet is preparing their reply to the note and will deliver it before the time limit expires. Preparations are now being made by the Serbian Government to depart for the interior and to reform at Nish. The Russian and some other foreign legations are now packing up. Our own legation will be ready to leave Belgrade on the 6.30 evening train to Vienna.*

VIENNA. British Ambassador, Sir Maurice de Bunsen telegraphs to Sir Edward Grey in London. *Servian concessions are all a sham. Servia has proven that she well knew that they are likely to be insufficient to satisfy the legitimate demands of Austria-Hungary by the fact that before making her offer she had ordered the mobilization and retirement of Government from Belgrade.*

FRANCE. Foreign Secretary Bienvenu-Martin: *Not until today was the Austrian note communicated to Italy, and that she was neither consulted nor informed on it. Vienna still nurses the illusion that Russia will not hold fast. Italy is only bound by the engagements of the Triple Alliance ~ if she has been consulted beforehand.*

BELGRADE. With its 48-hour deadline, the A-H Note demands a raft of Serbian concessions; most notably, a commission to investigate the assassination which could reveal the Serbian régime's complicity in the assassinations. Vienna's suspicion is fully focused upon Belgrade for complicity in, (if not the actual responsibility for) the assassinations. And the leaders of Austria-Hungary are correct. When one considers the number of victims murdered in the region by the roving associates of the *Black Hand*, the sharp 48-hour time limit given by Vienna during the holiday season to the Belgrade Cabinet is as George Bernard Shaw declares; *A decent time in which too ask a man to pay his hotel bill.* Serbian Education Minister Professor Ljuba Jovanović: *We have no other choice than to fight it out!*

The Royal Serbian Government have received the communication of the Imperial and Royal Government of Austria-Hungary and are convinced that their reply will remove any misunderstanding which may threaten to impair the good neighbourly relations between the Austro-Hungarian Monarchy and the Kingdom of Serbia.

Conscious of the fact that the protests, which were made both from the tribune of the National Skupchtina and in the declarations and actions of the responsible representatives of the State ~ protests which were cut short by the declarations made by the Serbian Government on the 31st March, 1909 ~ have not been renewed on any occasion as regards the great neighboring Monarchy, and that no attempt has been made since that time, either by the successive Royal Governments or by their organs, to change the political and legal state of affairs created in Bosnia and Herzegovina, the Royal Government call attention to the fact that in this connection the Imperial and Royal Government have made no representation except one concerning a school book, and that on that occasion the Imperial and Royal Government received an entirely satisfactory explanation. The Royal Government cannot be held responsible for manifestations of a private character, such as articles in the press and the peaceable work of societies ~ manifestations which take place in nearly all countries in the ordinary course of events, and which, as a general rule, escape official control. The Royal Government are all the less responsible, in view of the fact that at the time of the solution of a series of questions which arose between Serbia and Austria-Hungary they gave proof of a great readiness to oblige, and thus succeeded in settling the majority of these questions to the advantage of the two neighboring countries.

For these reasons the Royal Government have been pained and surprised at the statements, according to which members of the Kingdom of Serbia are supposed to have participated in the preparations for the crime committed at Sarajevo; the Royal Government expected to be invited to collaborate in an investigation of all that concerns this crime, and they were ready, in order to prove the entire correctness of their attitude, to take measures against any persons concerning whom representations were made to them. Falling in, therefore, with the desire of the Imperial and Royal Government, they are prepared to hand over for trial any Serbian subject, without regard to his situation or rank, of whose complicity in the crime of Sarajevo proofs are forthcoming, and undertake to publish on the first page of the Journal official, on the date of 26th July, the following declaration:

The Royal Government of Serbia condemn all propaganda which may be directed against Austria-Hungary, that is to say, all tendencies aimed at ultimately detaching from the Austro-Hungarian Monarchy territories which form part thereof, and they sincerely deplore the baneful consequences of all these criminal movements. The Royal Government regret that, according to the communication from the Imperial and Royal Government, certain Serbian officers and officials should have taken part in the above-mentioned propaganda, and thus compromised the good neighbourly relations to which the Royal Serbian Government was solemnly engaged by the declaration of the 31st March 1909, which declaration disapproves and repudiates all idea or attempt at interference with the destiny of the inhabitants of any part whatsoever of Austria-Hungary, and they consider it their duty to warn the officers, officials, and the entire population of the kingdom that henceforth they will take the most rigorous steps against all such persons who are guilty of such acts.

This declaration will be brought to the knowledge of the Royal Army in an order of the day, in the name of His Majesty the King, by His Royal Highness the Crown Prince Alexander, and will be published in the next official army bulletin. The Royal Government further undertake :

1. To introduce at the first regular convocation of the Skuptchina a provision into the press law providing for the most severe punishment of incitement to hatred or contempt of the Austro-Hungarian Monarchy, and for taking action against any publication the general tendency of which is directed against the territorial integrity of Austria-Hungary. The Government engage at the approaching revision of the Constitution to cause an amendment to be introduced into article 22 of the Constitution of such a nature that such publication may be confiscated, a proceeding at present impossible under the categorical terms of article 22 of the Constitution.

2. The Government possess no proof, nor does the Imperial and Royal Government furnish them with any, that the Narodna Odbrana (Defence of the People) and other similar societies have committed up to the present any criminal act of this nature through the proceedings of any of their members. Nevertheless the Royal Government will accept the demand of the Imperial and Royal Government, and will dissolve the Narodna Odbrana Society and every other society which may be directing its efforts against Austria-Hungary.

3. The Royal Serbian Government undertakes to remove without delay from public educational establishments in Serbia all that serves or could serve to foment propaganda against Austria-Hungary, whenever the Imperial and Royal Government furnish them with facts and proofs of this propaganda.

4. The Royal Government also agree to remove from military service all such persons as the judicial inquiry may have proved to be guilty of acts directed against the integrity of the territory of the Austro-Hungarian Monarchy, and they expect the Imperial and Royal Government to communicate to them the names and the acts of these officers and officials for the purposes of the proceedings which are to be taken against them.

5. The Royal Government must confess that they do not clearly grasp the meaning or the scope of the demand made by the Imperial and Royal Government that Serbia shall undertake to accept the collaboration of the organs of the Imperial and Royal Government upon their territory, but they declare that they will admit such collaboration as agrees with the principle of international law, with criminal procedure, and with good neighbourly relations.

6. It goes without saying that the Royal Government consider it their duty to open an inquiry against all such persons as are or eventually may be, implicated in the plot of the 28th June, and who happen to be within the territory of the Kingdom. As regards the participation in this inquiry of Austro-Hungarian agents appointed for this purpose by the Imperial and Royal Government, the Royal Government cannot accept such an arrangement, as it would be a violation of the Constitution and of the law of criminal procedure; nevertheless, in concrete cases communications as to the result of the investigation in question might be given to the Austro-Hungarian agents.

7. The Royal Government proceeded, on the very evening of the delivery of the note, to arrest Commandant Voija Tankosic. As regards Milan Ciganovic, who is a subject of the Austro-Hungarian Monarchy and who up to the 28th June was employed (on probation) by the directorate of railways, it has not yet been possible to arrest him. The Austro-Hungarian Government are requested to be so good as to supply as soon as possible, in the customary form, the presumptive evidence of guilt, as well as the eventual proofs of guilt which have been collected up to the present, at the inquiry at Sarajevo for the purposes of the later inquiry.

8. The Serbian Government will reinforce and extend the measures, which have been taken for the illicit traffic of arms and explosives across the frontier. It goes without saying that they will immediately order an inquiry and will severely punish the frontier officials on the Schabatz-Loznitza line, who have failed in their duty and allowed the authors of the crime of Sarajevo to pass.

9. The Royal Government will gladly give explanations of the remarks made by their officials, whether in Serbia or abroad, in interviews after the crime, which, according to the statement of the Imperial and Royal Government, were hostile toward the Monarchy, as soon as the Imperial and Royal Government have communicated to them the passages in question in these remarks, and as soon as they have shown that the remarks were actually made by the said officials, although the Royal Government will itself take steps to collect evidence and proofs.

10. The Royal Government will inform the Imperial and Royal Government of the execution of the measures comprised under the above heads, in so far as this has not already been done by the present note, as soon as each, measure has been ordered and carried out.

**If the Imperial and Royal Government are not satisfied with this reply, the Serbian Government, considering that it is not to the common interest to precipitate the solution of this question, are ready, as always, to accept a pacific understanding, either by referring this question to the decision of the International Tribunal of The Hague, or to the great powers which took part in the drawing up of the declaration made by the Serbian Government on the 31st March 1909.*

Belgrade, 25th July 1914

BELGRADE 5.55 PM. Five minutes before the deadline expires, Pasic delivers the Serbian reply to Baron Giesl. The Cabinet of Belgrade has accepted all Austrian demands ~ with the exception of one. Knowing that any inquiry will eventually lead to *Apis,* the Serbian régime dare not agree to Vienna's reasonable proposal of investigators coming into Belgrade to investigate the Sarajevo assassinations. Pasic: *We have accepted most of your demands. For the rest, we place our hopes on your loyalty and chivalry as an Austrian general. We have always been very satisfied with you.* The Austrian rightly suspects Pasic's régime of some burden of Serbian responsibility in the Sarajevo murders ~ even if he cannot prove it. Giesl feels the assassination has been the work of connivers , those who plan the ruin the A-H Empire. The work of extremists in Serbian and Russian pay : the result of a conspiracy against Austrian unity. Serbia having already ordered a full military mobilization and by now transferring their Government from Belgrade to Nish, prove they know it to be insufficient to satisfy the legitimate demands of A-H. Seeing this disappointing situation, Baron von Giesl declares ~ *Do you understand that your response will be rejected as unsatisfactory, and that Austria will break off all diplomatic relations? This now is a fait accompli!*

From the Austrian viewpoint, the Note is not an ultimatum for war, but of investigative demands, and Serbia has rejected their demands to properly investigate the assassinations. Baron Giesl having expected Pasic's noncompliance, now hastens back to the embassy where his staff have packed all their essentials, burned the embassy cipher book and sensitive diplomatic papers. As a group, they depart for the railway station to board the last train for Semlin, severing all diplomatic relations between A-H and Serbia. On the viaduct above the River Savé, the 6.30 Belgrade-Vienna train crosses the railway bridge into Hungary. Reaching Semlin, Giesl makes his way to the local garrison HQ where the waiting Hungarian PM Count Tisza meets him and asks: *Did it have to be?* Baron Giesl: *Yes.* Giesl then phones the Ballplatz to inform Berchtold. *I have broken off all diplomatic relations with Serbia.*

BERLIN. Warning to Russia is delivered in the way often favoured by the German Foreign Office; semi-official communications. News agencies announce: *Germany will take no part in the Austro-Serbian quarrel unless some other power interferes to prevent the Danubian Monarchy from obtaining satisfaction for the murder of Franz Ferdinand and his wife. If any other power interferes, Germany will do her duty by her A-H ally.* With no hesitation, Russia takes up the challenge. From St Petersburg comes confirmation that Russia has decided to intervene in the Austro-Serbian quarrel.

VIENNA. Russian Chargé d'Affaires Prince Koudacheff on hearing this telegraphs St. Petersburg. *At the last moment we are assured that the Austrian Ambassador has just left Belgrade hurriedly. He must have thought the Serbian Government's acceptance of the conditions imposed by his Government inadequate.*

St PETERSBURG. In the afternoon, a Ministerial Council is held at which Tsar Nicholas presides. German Military Attaché Major Eggeling writes ; *I intuitively feel the impression that the Ministerial Council are considering to mobilize the Russian Army. Perhaps they have even decided to order it in the four Southern Military Districts facing A-H.* More revelatory are the unguarded words of General Alderberg, the Governor of St. Petersburg who by a slip of the tongue when talking to Kaiser Wilhelm's personal representative to the Tsar, General Oskar von Chelius. Alderberg spoke of ~ *measures for mobilization.*

Later that day, a Russian artillery officer looks at his watch and comments to General Chelius. *The cannon on the Danube will have begun to fire by now, for one does not send such an ultimatum except when the cannon are loaded.* At an evening banquet, the Tsar's Grand Equerry Baron Grunwald who is sitting next to Chelius states: *The situation is very serious. What was decided at noon, I am not permitted to tell you. You yourself will soon learn it. But take it from me ~ it looks very serious.* He touches glasses with Chelius, and drinks to his health ~ *Let us hope, we shall see each other again in better times.*

The prescient notion that mobilization and war are imminent is furthered by the immediate promotion of all St. Petersburg Military Academy cadets to officer status in the Russian Army, instead of later in the year as standard. Following the Tsar's celebratory address to all these new appointees, German Military Attaché Major von Eggeling wrote: *These young officers openly expressed their joy to me, that now at least they are starting something special against Austria. Others aired their rage against Austrian presumption. Even Prince Peter of Montenegro, who was present at the time, told me that in his country there was an enthusiasm for war and mobilization was in full progress.* Spalajkovic: *In all circles without exception, the greatest resolve and jubilation reigns, on account of the stance adopted by the Tsar and the Government.*

St PETERSBURG 3 PM. Foreign Minister Serge Sazonoff receives Sir George Buchanan and Maurice Paléologue, to inform his friends. *An extraordinary Ministerial Council has been held this morning at Krasnoe Selo, with Tsar Nicholas presiding, and His Majesty has decided in principle to mobilize the 13 army corps provisionally listed for operations against Austria-Hungary.* Serge Sazonoff turns to Sir George. *England should hesitate no longer to range herself on the side of Russia and France in a crisis in which the stake is not merely the European balance of power but the very liberties of Europe itself.*

French Ambassador Maurice Paléologue supports the effort of Serge Sazonoff and points to the portrait of the old Chancellor, Prince Gortchakoff that adorns the room in which they are now talking. *In July 1870, on this very spot, my dear Sir George, Prince Gortchakoff said to your father, Sir Andrew Buchanan then Ambassador to St. Petersburg, who warned him of the danger of Imperial German ambition; There's nothing to worry Russia in the increase of German power.*
Don't let England make the same mistake today, which cost Russia so dear then!
Sir George Buchanan with a weary smile: *You know, you're preaching to the converted.*
Sazonoff: *Russia has no aggressive intentions, and she will take no action until it is forced on her.*
Buchanan: *My earnest hope is that Russia will not precipitate a war by mobilizing. All I can do is to impress prudence and warn that if Russia mobilizes Germany will not be content with mere mobilization or give Russia time to carry out hers, but will probably declare war at once!*
Sazonoff: I *assure you that we do not wish to precipitate a conflict, but unless Germany can restrain Austria you should regard the situation as desperate. Public feeling is rising every hour. This following note has been communicated to the Press: The Imperial Government is closely following the development of the Austro-Serbian conflict which cannot leave Russia indifferent.*

Later in the afternoon, German Ambassador, Count Friedrich von Pourtalès advises Sazonoff.
As Austria's ally, Germany naturally supports the most legitimate claims of the Vienna Cabinet against Serbia.

At seven o'clock I was at the railway station to say goodbye to Alexander Izvolsky, who was returning to his Paris post in hot haste. On the platforms, there was a great bustle. The trains were crowded with officers and soldiers. This already looks like mobilization. We exchange our impression, and come to the same conclusion... Cette fois, c'est la guerre ~ This time, it's war.

When I returned to the embassy I was informed that the Tsar had just ordered measures preliminary to mobilization in the military areas of Kiev, Odessa, Kazan and Moscow. Further, the cities and Governments of St. Petersburg and Moscow have been declared in a state of siege. Lastly, from the camp at Krasnoe Selo, this evening the troops are being sent back to their usual garrisons. Maurice Paléologue

BERLIN. The French Ambassador Jules Cambon discovers through Belgian Ambassador Beyens.
The Italian Ambassador, who had just cut short his leave to return to his post, said that Italy was surprised, to say the least, at having been kept in the dark regarding the whole affair by her two allies. In no way were we consulted, we were told nothing. Italian opinion is now hostile to Austria in this affair.

ROME. In a brief conversation with French Ambassador Barrére, Signor Salandra clearly explains, *We shall make the greatest efforts to prevent the rupture of peace: our situation is somewhat similar to that of England. It is possible that we might be able to take some action with England toward maintaining the peace.*
Premier Giovanni Giolitti declares his approval of the Italian Government's policy of neutrality thus :

I feel it my duty to recall a precedent showing how correct was the interpretation of the alliance by the Government when the conflict began. During the Balkan War, on 9th August 1913, being absent from Rome, I received the following telegram from the Marquis di San Giuliano: Austria has communicated to us and Germany that it has been the intention to act against Servia, defining such action as defensive and hoping for an application of a casus foederis by the Triple Alliance, which I consider inapplicable. I am trying to agree with Germany concerning efforts to prevent Austrian action, but it may be necessary to say clearly that we do not consider such eventual action as defensive, and therefore, do not think that there exists a casus foederis. I then answered Marquis di San Giuliano thus: If Austria goes against Servia, a casus foederis evidently does not exist. It is an action she accomplishes on her own account. It is not defensive, because nobody thinks of attacking her. It is necessary to declare this to Austria in the most formal manner, hoping that Germany will act to dissuade Austria from a very dangerous adventure. This was done, and our interpretation of the treaty was accepted by our allies, our friendly relations not being in the least disturbed. Thus, the declaration of neutrality, made at the beginning of this conflict, is according to the spirit and letter of the treaties.

VIENNA. On his return, Baron von Giesl contributes to the Ballplatz talks over the Serbian reply. *If we accept the Serbian answer, relations will only get worse. Of the Serbian assurances and reservations in relation to our demands, practically nothing is achieved with this Serbian pronouncement and all will remain as before. To be satisfied with a basically worthless sham, success would be self-deception. I want a declaration of war as soon as possible.* Berchtold calls in Conrad and a dispute ensues on the question of when they should begin operations in Serbia. Berchtold: *We should like to deliver the declaration of war on Serbia as soon as possible so as to put an end to diverse influences.* Conrad: *When do you want the declaration of war?* (Having just read a report estimating that the artillery of the A-H Army have just 500 shells per gun). *Only when we have progressed far enough for operations to begin immediately ~ on approximately 12th August* Berchtold: *The diplomatic situation will not hold as long as that.* Conrad additionally wants to delay the proposed invasion of Serbia to enable his reservists time to finish gathering in the strategic harvest. *I suggest that we wait and see how the diplomatic situation develops, to gain a better understanding of Russia's attitude before deploying the army.* No decision is made but preparations begin for a declaration of war. Berchtold suggests to Giesl, *I think Serbia will accept their demands unconditionally after a limited military demonstration, perhaps also involving the occupation of Belgrade.* At 4.30 PM telegrams are sent to A-H ambassadors in Berlin, Rome, London, and Paris to warn: *War is imminent because Austria-Hungary is faced with the necessity of enforcing on Serbia by the sharpest means a fundamental change in its attitude.*

BALTIC SEA. President Poincaré and René Viviani are at sea on *La France* nearing Stockholm. Communications are difficult, but through broken messages they learn of the Austro-Hungarian Note. On reaching Stockholm, Poincaré and Viviani realize if Germany is insisting the dispute between Austria-Hungary and Serbia remain localized, then Germany must be supporting Austria-Hungary. Paris advises Poincaré: *Abandon state visits to Denmark and Norway and return home.* He wires to assert: *As regards the precautionary and defence measures which Russia believes it necessary to adopt she should not take any immediately step which might offer to Germany a pretext for a total or partial mobilization of her forces.*

LONDON. A-H Ambassador Count Albert von Mensdorff explains to Sir Edward Grey: *If Servia, in the interval that had elapsed since the murder of the Archduke Franz Ferdinand, had voluntarily instituted an enquiry on her own territory; all this might have been avoided. In 1909 Servia said in a note that she intended to live on terms of good neighbourhood with Austria-Hungary, but she has never kept her promise, she has stirred up agitation to disintegrate Austria, and it is now absolutely necessary for Austria to protect herself.*

PM Asquith writes a letter to his friend Venetia Stanley about the developing European crisis. *Russia is trying to drag us in. The news this morning is that Serbia has capitulated on the main points, but it is very doubtful if any reservations will be accepted by Austria, who is resolved upon a complete, final humiliation. The curious thing is that on many, if not most of the points, Austria has a very good and Serbia a very bad case. But the Austrians are quite the stupidest people in Europe (as the Italians are the most perfidious), and there is a brutality about their mode of procedure, which will make most people think that is a case of a big Power wantonly bullying a little one. Anyhow, it is the most dangerous situation of the last 40 years.*

Sir Edward Grey: *I propose a Conference of Ambassadors of the Four Powers in London immediately, if Austria and Russia will hold back. While the conference is in meeting, all are to suspend military operations.* Telegrams are sent to the British ambassadors in Paris, St Petersburg, Vienna, Berlin, Rome, and also to Great Britain's Chargé d'Affaires in Nish, instructing them to ask their respective foreign ministers if they can agree to a timely Conference of Ambassadors in London to prevent further complications. When approached, Paris and Rome promptly agree to Sir Edward's proposal.

German Ambassador Prince Max von Lichnowsky receives a message from Bethmann Hollweg. *Russia might be calling up reserves without declaring a mobilization. We therefore request Sir Edward Grey to use his influence at St Petersburg.* Lichnowsky prepares a note to Sir Edward. *If Russia calls up reserves Germany will mobilize. My Government instructs me to request you to use all your influence in St. Petersburg.* In the evening, Lichnowsky takes this note to the British Foreign office. In Grey's absence, he sees Nicolson and Sir William Tyrrell, who tell Lichnowsky more about the proposed conference. Lichnowsky is delighted by this conference concept, and wires von Jagow to urge his support for it. The British diplomats then warn Prince Lichnowsky that if A-H attacks Serbia, a wider European war is inevitable. The localization of the Balkan conflict as hoped for by Berlin now appears impossible. Prince Max concludes ~ *My government accepts your suggested mediation à quatre.*

PARIS. German Ambassador to France, Baron von Schoen asks Bienvenu-Martin and Berthelot. *Is France willing to advise Russia to keep out of the conflict, as Vienna has said it will not annex Serbian territory?* Paris asks Berlin to exercise restraint in Vienna, as Serbia has accepted most of the Note. Schoen telegrams Berlin. *Bienvenu-Martin is willing to exercise a quietening influence in St Petersburg now Austria declares that no annexation is intended. The conditions for doing so have been created and he considers Sazonoff's idea that all the powers acting together can pass judgment on Serbia is juridicially hardly tenable.*

After conveying his report of this to Berlin, von Schoen returns to the Quai d'Orsay, to suggest that Berthelot and Bienvenu-Martin release a press statement to say: *Germany and France are acting in an identical spirit of peaceful co-operation to find ways of preserving peace.* But, Phillipe Berthelot declines. *I think that Vienna would not be acting the way it is without German approval. To my simple mind Germany's attitude is inexplicable. If it is not aiming at war, Germany is not trying to change Vienna's stance.* Schoen: *Austria-Hungary has only rejected formal mediation and a conference where it might be arraigned before what could be interpreted as a European tribunal, however Germany will not refuse to give advice to Vienna in all circumstances.* Berthelot comments to his staff, *I think Austria-Hungary and Germany are aiming at a brilliant diplomatic victory, but not at war at any price, although in the extreme case they will not recoil from it.*

St PETERSBURG. The French military attaché in St Petersburg confirms to the French Ministry of War that the Russian Army is mobilizing in the military districts of Kiev, Odessa, Kazan and Moscow and secretly making further preparations in the military districts of Warsaw, Vilna, and St Petersburg. The Russian Minister of War says he is determined to leave to Germany the eventual initiative of an attack on Russia. German Ambassador Pourtalés and British Ambassador Sir George Buchanan are feeling increasingly apprehensive of the danger of even a partial Russian mobilization against Austria as a means of exerting diplomatic pressure. If Russia is attempting a bluff of this kind, militarists everywhere will gain increased influence and soon take a policy of threats beyond the control of the diplomats. French Ambassador Paléologue asks his friend Serge Sazonoff: *Honestly between ourselves, do you think we can still save peace?* Serge Sazonoff: *If we had only Austria to deal with I should be hopeful... But there is Germany. She has promised her ally a great personal triumph. She is convinced that we dare not resist her to the bitter end, and that the Triple Éntente will give way as it has always given way. But this time we cannot give way, on pain of ceasing to exist. We shall not avert war. It's terrible to think of what's to come.* At 9.30 PM, Count Friedrich von Pourtalés receives this telegram from Chancellor Bethmann Hollweg.

After Count Berchtold has declared to Russia that Austria does not aim at any territorial acquisitions in Serbia, but only wishes to secure repose, the maintenance of the peace in Europe now depends on Russia alone. We trust in Russia's love of peace and in our traditional friendly relations with her, that she will take no step which would seriously endanger the peace of Europe.

Having heard too many rumours of Russian preparatory mobilization measures, Pourtalés deems it wise to give Serge Sazonoff a friendly, but firm warning. *Concerning news current among the foreign military attachés to which it is supposed that mobilization orders have been issued to several Russian Army Corps on the Western Frontier ~ I call your attention to the great danger of such measures, which might easily call forth counter-measures.* Sazonoff: *I can guarantee that no mobilization order of the sort has been issued. On the contrary, the Ministerial Council decided to delay any such order until Austria-Hungary adopts a hostile attitude toward Russia. There have been military measures taken in order not to be taken by surprise.* Sazonoff then telephones to Russian Minister of War General Vladimir Soukhomlinoff. *Please explain, to the German Military Attaché, as one military man speaking to another, that nothing is contemplated except measures preparatory to a contingent partial mobilization against A-H. Militärattaché* Bernhardt von Eggeling is therefore invited to an interview with General Soukhomlinoff, after which von Eggeling drafts and wires his report to Berlin at 1 AM. (Berlin receives this report on 31st July)

The Minister of War gave me his word of honour that no sort of mobilization order had yet been issued. For the present, merely preparatory measures are being taken. No horse has been recruited, nor a reservist called in. If Austria crosses the Serbian frontier, Military Districts directed against Austria: Kiev, Odessa, Moscow, and Kazan will be mobilized. Under no circumstance will those on the German front, Warsaw, Vilna, St. Petersburg. Peace with Germany is earnestly desired. I get the impression of great nervousness, and anxiety. I consider military statements in so far correct, that complete mobilization has probably not been ordered, but preparatory measures are very far reaching. They are evidently striving to gain time for new negotiations and for continuing their armaments. The general feeling is of hope from Germany and for the mediation of the Kaiser.

Oberleutnant zur See Karl Lody

Grossadmiral Prince Henry of Prussia

Foreign Secretary Gottlieb von Jagow

LONDON. As German leaders continue to anticipate British neutrality, the Kaiser's brother Prince Henry of Prussia, meets with King George V at Buckingham Palace to ask the British Monarch; *What will England do in the event of a European war?* King George answers. *I don't know what we shall do, we have no quarrel with anyone, and I hope we shall remain neutral. But if Germany declares war on Russia, and France joins Russia, then I am afraid we shall be dragged into it. But you can be most sure, that I, and my Government will do all we can to prevent a European war.* Prince Henry continues to be convinced that: *King George's statement was made in all seriousness, and that England shall remain neutral at the start.* Germany therefore hopes to preserve this neutral attitude of England, which they much misinterpret and believe has been promised to Prince Henry.

King George's supposed assurance is telegraphed to the Berlin Naval Office by the German Naval Attaché in London: *King of Great Britain said to Prince Henry of Prussia that England will maintain neutrality in case war should break out between Continental Powers.* Kaiser Wilhelm II is much encouraged to receive a message from Prince Henry, suggesting George V intends to keep Britain out of any war.

H.R.H. Prince Henry was received by his Majesty King George V in London, who empowered him to transmit to me verbally, that England would remain neutral if war broke out on the Continent involving Germany and France, Austria and Russia. This message was telegraphed to me by my brother from London after his conversation with H.M. the King. Kaiser Wilhelm II

When British PM *Squiffy* Asquith is told by King George V that he has remarked to Prince Henry, *I hope we shall remain neutral* ~ he points out the danger ~ *Russia says to us; if you won't say that you are ready to side with us now, your friendship is valueless, and we shall act on that assumption in the future.* Meanwhile, the revelation the next day that the British Fleet will not be dispersed makes it doubtful to Berlin whether King George V's misinterpreted words can be relied upon. Prince Henry suggests, *I doubt whether England will remain neutral permanently, on account of her relations with France.*

Grand-Admiral Prince Henry of Prussia, who enjoys many warm friendships in England and visits the country at frequent intervals, spent a week in London. He saw the King, called on Prince Louis of Battenberg, and paid his respects to Mr. Winston Churchill, First Lord of the Admiralty. Englishmen can only conjecture how he put in the rest of his time. Perhaps an episode in the trial of Karl Lody, the German naval spy who was executed at the Tower of London has its place in the unrecorded history of Prince Henry of Prussia's visit. Lody confessed to military judges at Middlesex Guildhall that he received his orders to report on British naval preparations from, a distinguished personage. Give us his name, commanded presiding officer Lord Cheylesmore. I would rather not tell it in open court, pleaded the prisoner whom Scotland Yard, the day before, had asked me to look at, with a view to possible identification with certain Berlin affiliations. Lody added ~ I will write his name on a piece of paper for the court's confidential information, His request was granted. James W. Gerard

VIENNA. The week of fate opens on Monday 27th July 1914, amid expectations that the worst is now becoming inevitable. Nervous depositors besiege banks, and gold begins to fade from circulation.

Sir Maurice de Bunsen: *Country has gone wild with joy at prospect of war with Serbia.*

Count Mensdorff told me earlier today that; The Serbian Government have not accepted the demands which the A-H Government are obliged to address to them in order to secure permanently vital Austrian interests. Serbia has showed that she does not intend to abandon her subversive aims towards continuous disorder in the Austrian frontier territories and their final disruption from the Austrian Monarchy. Very reluctantly, and against their wish, the A-H Government are compelled to enforce a fundamental change of the attitude of enmity pursued up to now by Serbia. As the British Government knows, the A-H Government has for many years endeavoured to find a way to get on with this turbulent neighbour, though this has been made very difficult by the continuous provocations of Serbia. The Sarajevo murder has made clear to everyone what appalling consequences the Serbian propaganda has already produced, and what a permanent threat to Austria it involves.

LONDON. Sir Edward Grey replies:

We would understand that the Austrian Government must consider that the moment has arrived to obtain, by means of the strongest pressure, guarantees for the definite suppression of the Serbian aspirations and for the security of peace and order on the southeastern frontier of Austria. As the peaceable means are now exhausted, the Austrian Government must at last appeal to force. Their action, which has no sort of aggressive tendency, cannot be represented otherwise than as an act of self-defence. Also, they think that they will serve a European interest if they prevent Serbia from being henceforth an element of general unrest such as she has been for the last ten years. Finally Count Mensdorff added that the repression of Serbian subversive aims is a matter of self-defence and self-preservation on Austria's part. He reiterated that Austria has no intention of taking Serbian territory or has aggressive designs against Serbian territory. The Serbians have refused the precise one thing ~ the cooperation of Austrian officials and police ~ which would be a real guarantee that in practice the Serbians will not carry on their subversive campaign against Austria.

Sir Edward Grey forwards another peace proposal through Prince Max Lichnowsky asking Berlin to use its influence on Vienna for peace. Sir Edward warns Prince Max. *If Austria continues with its aggression against Serbia, and Germany with its policy of supporting Austria, then Great Britain will see no other choice, but to side with France and Russia.*

PARIS. The French Foreign Minister René Viviani, informs German Ambassador von Schoen. *France is anxious to find a peaceful solution, and I am prepared to do my utmost with our current influence in St. Petersburg ~ if Germany can counsel moderation in Vienna ~ since Serbia has fulfilled nearly every point.*

LONDON. Sir Edward Grey addresses the British Cabinet. *We now have to discuss and decide soon whether to choose neutrality if war comes, or to enter the conflict.* While the Cabinet is still undecided about what course to choose, Winston Churchill places the Royal Navy on alert. His order reads :

SECRET ~ European political situation makes war between Triple Alliance and Triple Éntente by no means impossible. This is not the Warning Telegram, but be prepared to shadow possible hostile ships of war. Measure is purely precautionary.

Grey urges Berlin to influence Vienna to accept some reply from Belgrade; implying that only in this case can London ask for moderation in St Petersburg. Grey aims the onus on Berlin to prevent wider war, suggesting that ~ *Germany can stall Austria-Hungary's foolhardy policy if she wants to.*

BERLIN. Foreign Minister Gottlieb von Jagow: *Germany will await the outcome of the exchange of views between Austria and Russia. Austria is partially mobilizing. If Russia mobilizes only in the south, Germany will not mobilize, but Russian mobilization is so complicated that it might be difficult to exactly locate her mobilization. Germany will have to be very careful not to be taken by surprise.* Berlin militarists send messages to Vienna urging Berchtold to present a declaration of war on Serbia before Franz Joseph for his signature to thwart Grey and Sazonoff, who are thought by to be playing for time. As for Sir Edward Grey's latest suggestion for mediation between Vienna and Belgrade, Bethmann Hollweg edits the British message to forestall any possibility that A-H will refrain from attacking Serbia, by deliberately omitting the last line of the British Foreign Office note that originally read when sent :
I hear from my colleagues that the key to the situation lies in Berlin and that if Berlin seriously wants peace it will prevent Vienna from following a foolhardy policy.

BERLIN. Kaiser Wilhelm and his éntourage return from their northern cruise of Norway to meet with his optimistic Berlin GHQ. One comment he heard, *I have the impression that it will all blow over.* But the Kaiser is not told that his Chancellor has rejected Sir Edward Grey's proposed ambassador's mediation conference. When Wilhelm arrives at the Potsdam Palace train station late that evening, he is met by a pale, agitated, and somewhat fearful Chancellor. Bethmann Hollweg's apprehension stems from his dread of the Kaiser's wrath when the extent of his deception is revealed. From Vienna, a telegram arrives. *They have decided here to send out the declaration of war tomorrow, or the day after tomorrow at the latest, to frustrate any attempt at intervention.* Kaiser Wilhelm's words to Bethmann are suitably brusque. *How did it all happen?!* Rather than try to explain, Imperial Chancellor Bethmann offers his resignation by way of apology. Kaiser Wilhelm is furious, refuses to accept it, and snaps. *You've made this stew. Now you're going to eat it!* Berlin belatedly directs Lichnowsky to inform London: *The German Government now accepts in principle the idea of mediation by the Four Powers, reserving of course, its right to support Austria-Hungary if attacked.* Chief of the Naval Cabinet, Admiral Georg von Müller: *The mood is decidedly milder. It appears that the neutral tendency of England is having a very cooling effect on Russia and France. On the other hand, Austria will not have completed her mobilization till 12th August. Tendency of our policy : to keep quiet, letting Russia put herself in the wrong ~ but not shying away from war.*

LONDON. Sir Edward Grey replies: *The Serbian reply should be treated as a basis of discussion, and the German Government should continue to urge this in Vienna.* Edward Grey to Ambassador de Bunsen, *I cannot understand how the Austrian Government can treat Serbia's reply as a refusal. It forms at least a basis of negotiations. Austria seems to imagine that she can make war on Serbia without bringing Russia in. If she can, well and good ~ if not, the consequences will be incalculable. The Serbian reply was meant to diminish tension, but Russia finds that on the contrary there is increased tension. The situation will become increasingly serious with great anxiety in Europe. For example, our fleet was to have dispersed, but we have kept it mobilized. This is not a threat, but an illustration of our anxiety.*

St PETERSBURG. Serge Sazonoff telegraphs to Sir Edward Grey in London. *From interviews with German Ambassador appears that Germany is, if anything, in favour of Austria's extreme demands. Germany has exercised no influence of moderating kind in Vienna. This attitude of German Government is most alarming. Can England possibly influence German Government? The key to the situation is at Berlin.*
British Ambassador, Sir George Buchanan telegraphs to Sir Edward Grey.

The Minister for Foreign Affairs Sazonoff said that the Austrian Government has now definitely declined direct conversation between Vienna and St. Petersburg. The Minister for Foreign Affairs said he had proposed such exchange of views on advice of German Ambassador. He proposed, when informing German Ambassador of this refusal of Austria's, to urge that a return should be made to your proposal for a conference of our Ambassadors, or at all events for an exchange of views between the three Ambassadors less directly interested, yourself, and also the Austrian Ambassador, if you thought it advisable. Any arrangement, approved by France and England, would be acceptable to him, and he did not care what form such conversations took.

No time is to be lost, and the only way to avert a wider war is for you to succeed in arriving, by means of conversations with Ambassadors either collectively or individually, at some formula which Austria can be induced to accept. Throughout Russian Government has been perfectly frank and conciliatory, and has done all in their power to maintain peace. If their efforts to maintain peace fail, the Russian Minister for Foreign Affairs trusts that it will be realized by the British public that it was not the fault of the Russian Government.

BERLIN. British Ambassador, Sir Edward Goschen suggests. *Since German Government now accepts principle of mediation, might we ask Germany to suggest the lines on which she would consent to work with us?* Sir Edward Grey transmits his approval from London.

BERLIN ~ night. Chancellor Bethmann Hollweg once again refuses to co-operate in any way in a conference, stating that Austria-Hungary's quarrel with Serbia is a purely Austria-Hungary's concern, and that Russia has nothing to do with it; and yet he states, *I strongly desire peace, and am doing my best at Vienna and St. Petersburg to get the two Governments to discuss the situation directly with each other in a friendly way.* British Ambassador in Vienna Maurice De Bunsen emphatically denies this: *Germany is, as yet, using no such influence. Such influence, if any, seems to be extraordinarily timid and is neutralized by German Ambassador at Vienna, acting with the war party at home.*

BERLIN. As Kaiser Wilhelm wakes on Tuesday morning, he receives a copy of the Serbian reply. While reading, he writes in the margin ~ *A brilliant performance for a time limit of 48 hours. This is more than one could have expected! A great moral success for Vienna, but with it every reason for war drops away, and Giesl ought to have remained in Belgrade! After such a thing, I would never order a mobilization!* ~ *W.*

10 AM. Kaiser Wilhelm II abandons his long-standing pressure on Austria-Hungary to invade Serbia in favour of a major policy-shift that now aims for Vienna to accept mediation and negotiate a settlement to avoid war. Wilhelm writes to von Jagow and forwards a copy to General von Moltke.

I am convinced that on the whole the wishes of the Danubian Monarchy have now been acceded to. The few reservations that Serbia makes can be settled by negotiation. It contains the announcement orbi et urbi of a capitulation of the most humiliating kind, and as a result, every cause for war falls to the ground. Nevertheless, the piece of paper, like its contents, is of little value so long as it is not translated into deeds. The Serbians are Orientals, therefore lying, deceitful, and masters of evasion. In order that these beautiful promises may be converted into reality and deeds, and in order to give the army, now mobilized to no purpose for the third time, the external satisfaction d'honneur of an ostensible success, Austria should be given temporary military occupation of Belgrade as a pledge. I propose that we say to Austria : Serbia has been forced to retreat in a very humiliating manner, and we offer our congratulations. Naturally, as a result ~ no more cause for war exists; but a guarantee that the promises will be carried out, is probably necessary. This can probably be secured by a temporary military occupation of a portion of Serbia; similar to the way we left troops in France in 1871 until the billions were paid. On this basis, I am ready to mediate for peace with Austria. Submit a proposal to me, along the lines sketched out to be communicated to Vienna. Wilhelm II

Thus, unaware that Austria-Hungary is about to declare war, Kaiser Wilhelm proposes to act as a mediator and advises Vienna to abandon the prospect of war with Serbia. He urges a *Halt in Belgrade* order for A-H to seize Belgrade as a guarantee till the Serbians carry out their promises. Waiting to see whether Vienna and St Petersburg accept the Kaiser's *Halt in Belgrade* proposal to avert a wider war in Europe, Chancellor Bethmann Hollweg's position remains the same. *If Austria is to vanish as a power, Germany will be threatened by rampant Pan-Slavism supported by growing Russian power in the east and by French revanchism in the west. By prompting the Austrians to attack Serbia immediately, the conflict can be localized and the Serbian menace nullified. If Austria can act swiftly to present Europe with a rapid fait accompli, war can be confined to the Balkans, and the intervention of third parties avoided as much as possible.*

The Kaiser begins to seriously realize that Russia may not remain quiet. Since the assassination, Berlin's main position has constantly been for the *localization of the conflict.* The Kaiser telegraphs his views to Vienna, which arrives late in the afternoon. *We are aware that the eventual acts of hostility by Austria against Serbia might bring Russia on the scene and drag us into a war in conjunction with our Ally. But we cannot, knowing that the vital interests of Austria are at stake, either advise our Ally to condescension incompatible with her dignity, or refuse her our support at this difficult juncture.* In the same way, British Ambassador to Berlin Sir Edward Goschen telegraphs to Sir Edward Grey his clear impressions of Chancellor Bethmann Hollweg on the perils of preliminary Russian mobilization.

If the news is true which I have just read in the papers, that Russia has mobilized fourteen army corps in the South, I think the situation is very serious, and I will be in a very difficult position, as in these circumstances it will be out of my power to continue to preach moderation at Vienna. Austria, who as yet is only partially mobilizing, will have to take similar measures, and if war is to result, Russia will be entirely responsible. Goschen reports an interview with Jagow. *He begged me to impress on you the difficulty of Germany's position in view of Russian mobilization and military measures which he hears are being taken in France.*

VIENNA. Austria refuses to enter into any general European discussions. Count von Berchtold is so confident that Germany will stand by Austria because of their military treaty, that Berlin is being so left in the dark that they really do not know what the A-H Foreign Minister is planning or even what the official stand by Vienna is. From the start, Berchtold has taken a one-track line that A-H was wronged by Serbia and has the right to respond militarily. Europe may have once been united in agreement to compel a correct Serbian attitude toward Austria, and perhaps even to punish Belgrade for allowing anti-Austrian propaganda and actions to be encouraged on her soil. But A-H has been so provoked by Serbia, Vienna appears prepared to ignore greater issues involved, and insists on action.

To my peoples! It was my fervent desire to dedicate the years, which by the Grace of God are still left to me to works of peace and to preserve my peoples from the heavy sacrifice of war. Divine Providence has decreed it otherwise. The intrigues of a hate-filled enemy force me after long years of peace to reach for the sword to preserve the honour of my Monarchy, to defend its prestige and position, to secure its property. A halt must be called to these intolerable proceedings, and an end put to the incessant provocations of Servia. The honour and dignity of my monarchy must be preserved, and its development guarded from continual shocks. Franz Josef

VIENNA. By the time of the Sarajevo assassination, the 84-year old Emperor Franz Josef is an exhausted man who has suffered many personal tragedies. His brother, Maximilian was shot by revolutionaries in Mexico. His only son, Rudolf committed suicide and his beloved wife Elizabeth was assassinated by an anarchist. Franz Josef saw the assassination of his heir Archduke Franz Ferdinand as a personal attack on the royal family of the Hapsburg Monarchy, requiring a profound response. His view intensifies, given the direct knowledge of the Serbian Army and of the Belgrade Cabinet. Franz Josef's moment of decision comes at 11 AM, when the A-H declaration of war is laid in front of him for his signature. Franz Josef signs the declaration of war on the strength of von Berchtold's report to him that Serbian troops on a river steamer have fired on Imperial forces near Temes-Kubin.

The next day, Count von Berchtold admits that the report of Serbian firing was unconfirmed, and that he has therefore, struck reference to it from the declaration of war. It appears that Berchtold has been misled. As the A-H Army has not yet been mobilized, and as no military operations are contemplated for two weeks, Berchtold still believes that some last-minute miracle will stall war from really breaking out. The diplomats of the Ballplatz and Wilhelmstrasse realize, as soon as war is formally declared, they will relinquish effective control to the military, whose hand will wreak havoc.

For this reason, and not as vengeance for the assassination, Austria-Hungary must draw the sword against Serbia. It is not a question of a knightly duel with ~ poor little Serbia ~ as she likes to call herself, nor of punishment for the assassination. It is more the highly practical importance of the prestige of a Great Power. The Monarchy has been seized by the throat, and has to choose between allowing itself to be strangled and making a last effort to prevent its destruction. Conrad von Hötzendorf

In order to bring to an end the subversive intrigues originating from Belgrade and aimed at the territorial integrity of the Austro-Hungarian Monarchy, the Imperial & Royal Government delivered to the Royal Serbian Government a note in which a series of demands were formulated, for the acceptance of which a delay of 48 hours was allowed. The Government of Serbia not having replied in a satisfactory manner to the note remitted to it by the A-H Ambassador in Belgrade on 23rd July 1914, the Imperial & Royal Government finds itself compelled to proceed to safeguard its rights and interests, and to have recourse to force of arms. Austria-Hungary considers itself, therefore, from this moment in a state of war with Serbia. Count Leopold von Berchtold

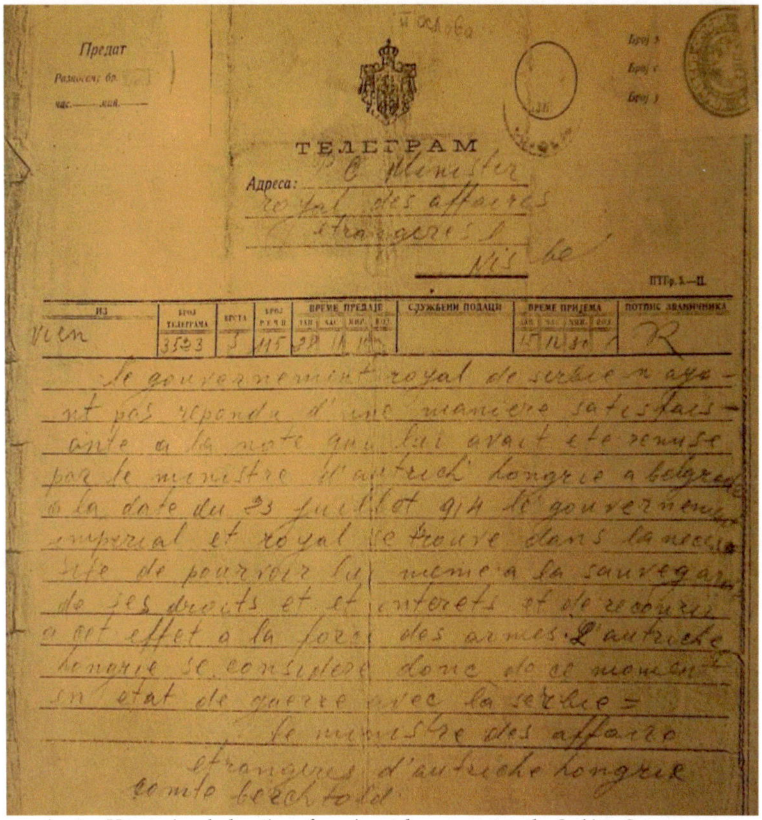
Предат
Разносач бр.
час. мин.

ТЕЛЕГРАМ

Адреса: le Ministre royal des affaires etrangeres Nish

ПТТр. X—II.

ИЗ	БРОЈ ТЕЛЕГРАМА	ВРСТА	БРОЈ РЕЧИ	ВРЕМЕ ПРЕДАЈЕ	СЛУЖБЕНИ ПОДАЦИ	ВРЕМЕ ПРИЈЕМА	ПОТПИС ЗВАНИЧНИКА
Vien	3523	3	115	28 11 [illegible]		15 12 30	R

le gouvernement royal de serbie n aya-
nt pas repondu d une maniere satisfais-
ante a la note qui lui avait ete remise
par le ministre d autriche hongrie a belgrade
a la date du 23 juillet 914 le gouvernement
imperial et royal se trouve dans la neces-
site de pourvoir lui meme a la sauvegarde
de ses droits et interets et de recourir
a cet effet a la force des armes. L'autriche
hongrie se considere donc de ce moment
en etat de guerre avec la serbie =
le ministre des affaires
etrangeres d'autriche hongrie
comte berchtold

Austro-Hungarian declaration of war in a telegram sent to the Serbian Government on 28th July 1914, signed by Imperial Foreign Minister Count Leopold von Berchtold.

A PURELY DEFENSIVE WAR ~ *The Evening News* Newark, N.J., U.S.A.

Austria-Hungary looks upon this war as a purely defensive one, which has been forced on her by the agitation directed by Russia against her very existence. Austria-Hungary has given many proofs in late years of her peaceful intention. She refrained from any interference with arms in the Balkan war, though her interests were at stake. Subsequent events have proved what a serious danger the increase in territory and prestige which it brought Servia were for Austria-Hungary. Servia's ambitions have since grown and have been solely directed against the Dual Monarchy. Russia has tacitly approved of Servia's action because Russian statesmen wish to form an iron ring of enemies around Austria-Hungary and Germany in order that Russia's grasp on Constantinople, the Straits, and on Asia should never again be meddled with. Austro-Hungarian soldiers are fighting for their homes and for the maintenance of their country, the Russians are fighting to help the Russian Tsar to gain the rule of the world, and to destroy all his neighbors who may be dangerous to Russian ambitions. England is helping Russia to oust her German rival. She feared for some time that German culture and German scientific methods would prove the stronger in a peaceful competition, and she now hopes to crush Germany with the help of Russia and France. And France is fighting to win back Alsace-Lorraine, to take her revenge on Germany, which the French nation has been aiming at for the last forty-four years. That is how we look upon the war. We never wished for territorial increase, we wished for peace and that our people should develop in safety.

Germany equally had nothing to gain by a war, but Germany knows that Austria-Hungary's enemies are her enemies and that the dismemberment of the Hapsburg Monarchy would mean the isolation of Germany. And so, after all efforts to keep Russia and England from breaking the peace of Europe had failed, she drew her sword to defend her and her allies' interests. Truth and honor are on the side of the two empires in this war, the inventions and prevarications published by the French, Russian, and English press must prove to the American people who can provide the truth in this war. The Austro-Hungarian and German people have a clear conscience and fear no misrepresentation of their action. Count Leopold von Berchtold

BERLIN. Confirmation of the Austro-Hungarian declaration of war on Serbia finally arrives. Hearing that the Emperor Franz Josef has formally declared war against Serbia, Kaiser Wilhelm has to accept the consequences his pledge to Vienna. As the determined Austrian policy begins to induce a partial military mobilization across Russia, Berlin continues to restrain her ally and attempt to modify the foreign policy of Count von Berchtold.

LONDON 6.45 PM. A telegram sent by Ambassador Lichnowsky from London arrives in Berlin. *Although Count Berchtold has now on several occasions disclaimed any intention to annex Serbian territory, and has declared on several occasions Austria's territorial disinterestedness, nevertheless the Austrian Ambassador in London confided the following to me. Serbia is to be beaten to death, and it is the intention to present portions of Serbia to Bulgaria and Albania. This duplicity of Austria is intolerable. At St Petersburg they are lambs with not a wicked thought in their hearts. In London their Embassy now talks of giving away portions of Serbian territory to Bulgaria and Albania.*

BERLIN. Bethmann facing masked duplicity from Vienna, now prudently pursues the Kaiser's proposed *Halt-in-Belgrade Plan.* At 10.15 PM, Bethmann wires the German Ambassador in Vienna Count Heinrich von Tschirschky complaining about the lack of information from Vienna regarding their true intentions in Serbia. Bethmann now severely errs in distorting Kaiser Wilhelm's contention by omitting his crucial sentence : *That war is no longer necessary* : and directs Count von Tschirschky to ;

Advise Berchtold ;

Moderate Vienna's policy, lest it incur the odium of having been responsible for a world war.

To accept the Kaiser's Halt-in-Belgrade Plan and announce that Austria has no interest in acquiring Serbian territory and that its occupation of Belgrade is temporary and contingent on Serbian compliance with the terms of the Austrian note of demands. Aside from a declaration to Russia that it intends no territorial acquisition in Serbia, the A-H Government, in spite of repeated questions as to its purposes, has left us in the dark. The reply now at hand of the Serbian Government to the Austrian Note makes it evident that Serbia has met the Austrian demands in such a wide-ranging a manner that if the A-H Government adopts a wholly uncompromising attitude, a revulsion of public opinion against it in all of Europe shall have to be reckoned with.

Russia will presumably be satisfied if the Vienna Cabinet repeats in St. Petersburg the definite declaration that territorial acquisitions in Serbia lie far from its purpose, and that its military measures aim solely at a temporary occupation of Belgrade and other definite points of Serbian territory in order to compel the Serbian Government to a complete fulfillment of the demands, and to serve as guarantees for future good behaviour, to which Austria-Hungary unquestionable has a claim after her experiences with Serbia. The occupation could be regarded like the German occupation in France after the Peace of Frankfurt, as security for the demand of the war indemnity. As soon as Austrian demands are fulfilled, a withdrawal should follow.

You are immediately to express yourself emphatically in this sense to Count Berchtold, and have him take proper steps in St. Petersburg. You are carefully to avoid giving the impression that we wish to hold Austria back. It is solely a question of finding a method which will make possible the accomplishment of Austria's purpose of cutting the nerve of Greater Serbian propaganda, without at the same time unchaining a world war, and in the end, if this is unavoidable, of improving as far as possible the conditions under which it is to be waged. Wire reply.

Bethmann Hollweg

Count Heinrich von Tschirschky does not convey the Kaiser's present position that his proposed *Halt in Belgrade Plan* is the right policy (or knows Wilhelm's viewpoint : *That war is no longer necessary*).

* * *

Not a single one of the men who had real power was wise enough and strong enough to arrest the military demon that was about to bring upon us all the most awful catastrophe in human history. And after this war had commenced, though very many of them from motives either of fear or of humanity desired to see it ended, they had so committed themselves to one another or were so distrustful of each other's private intentions that they could not close the conflict for the origin of which they had been themselves responsible. Meanwhile the guiltless peoples were destroyed.

Earl Robert Loreburn

BERLIN. Bethmann's telegram recognizes that A-H is at war with Serbia, and how he now aims to induce the Austrians to *Halt in Belgrade*. But, his prose is too affable to convey the Kaiser's decisive directive that Vienna now appreciate his view that *every reason for war drops away*. Bethmann appears too cautious of offending Austria and more concerned with preventing the responsibility for war from falling on Berlin or Vienna, rather than the higher responsibility of preventing any war altogether. While Bethmann advises British Ambassador Sir Edward Goschen, *I am doing my very best both at Vienna and St Petersburg to get the two governments to discuss the situation directly with each other in an open and friendly way. I have great hopes that such discussions will take place and lead to a satisfactory result. A war between the Great Powers must be avoided.* However before Bethmann tells either Russia or Britain the terms of the Kaiser's *Halt in Belgrade Plan,* he first wishes to learn ~ *Will it be acceptable in Vienna?*

Chancellor Bethmann telegraphs von Tschirschky in Vienna. *Answer by wire immediately whether telegram of yesterday concerning the Pledge Plan has arrived. I await immediate carrying out of telegram.* Tschirschky promptly carries out his instructions, but confirms Austrian evasion. *Count von Berchtold is ready to repeat his previous declarations of territorial disinterestedness, but as to the further declaration concerning military measures, Berchtold says that he is not in a position to give me a reply at once. In spite of my representations as to the urgency of the matter, I have up to this evening received no further communication.*

Bethmann, now badly shaken and desperately seeking to save the situation sends another telegram at 2.55 AM urging Vienna to enter into negotiations with Russia to secure British neutrality. Five minutes later his suggestion is rejected ~ because, even if the A-H Army *Evidenzbureau* cannot distinguish between the specific involvement of the Belgrade Cabinet and unofficial nationalist groups, investigators have confirmed that the assassin's weapons came from Serbian Army sources. Bethmann dispatches yet another desperate telegram to Tschirschky in Vienna.

The expressions of Austrian diplomats must be regarded as indications of more recent wishes and aspirations. I regard the attitude of the Austrian Government and its unparalleled procedure toward various Governments with increasing astonishment. In St Petersburg they declare their territorial disinterestedness. They leave us wholly in the dark as to its programme. Rome they put off with empty phrases. In London Count Mensdorff hands out parts of Serbia to Bulgaria and Albania and places himself in contradiction with Vienna's solemn declaration at St. Petersburg. From these contradictions, I must conclude that Austria is harbouring plans which it sees fit to conceal from us in order to assure itself in all events of German support and to avoid the refusal which might result from a frank statement. A refusal to hold an exchange of opinion with St Petersburg will be a serious error and a direct provocation of Russia. Talk to Count Berchtold at once, with all impressiveness and great seriousness.

Chief of General Staff Helmuth von Moltke now advises Foreign Secretary Gottlieb von Jagow. *There is no alternative but to fight a preventative war so as to defeat the enemy while we can still emerge fairly well from the struggle* ~ unaware of the advice of the Chancellor Otto von Bismarck who once warned: *Preventative war is like committing suicide out of fear of death.* From the position of Belgrade's *Black Hand,* as soon as Austria-Hungary formally declares war on Serbia, they have played into the hand of *Apis,* who thereafter can count on the support of Russia in their quest for a Greater Serbia.

ROME. In view of Italy's nationalist hostility to Austria, and her own ambitions in the Balkans which now conflict with those of Austria, and her secret agreement with Russia (Racconigi in 1909), Italy now inclines more to side with the Éntente, than to remain allied with Austria and Germany. This attitude of Italy emanates from the wars of 1859 and 1866, when she had less territory to the north and east than desired. The districts of Trent and Trieste, including the harbour of Trieste are predominantly Italian in population and sentiment, but Italy has been forced to leave them in the hands of Austria-Hungary. *Italia Irredenta,* as these territories are called in Italy, forms a pertinent issue in the politics of the Italian Peninsula. Patriots look forward to a time when these places will be reunited with Italy ~ and yet it is accepted, that Vienna will never yield these lands except by force. The city of Trieste is especially valuable to Austria-Hungary as her only sizeable seaport. In 1902, when Italy renewed the Triple Alliance, the Italians secretly gave to France through Signor Prinetti a written assurance that they will observe neutrality in any war in which France is not the aggressor. The Italians admirably keep their word, and the British Ambassador in Rome telegraphs to London. *As Austria did not consult Italy before delivering the Note, and as her mode of her attack on Serbia would be constructively provoking Russia, the casus foederis contemplated by the Triple Alliance will not arise.*

VIENNA. Russian Ambassador states: *Since Austria has declined Russia's proposal for conversations, the only hope remaining is a conference of the four less-interested Powers in London. Russia will stand aside.* While the British seek to persuade Vienna to accept mediation or arbitration, some German militarists continue to secretly advise the Austrians not to do so. Despite a series of frantic appeals from Bethmann to heed a second *Stop in Belgrade* proposal from London that has already been endorsed by Russia for suspending the war against Serbia once the A-H Army have occupied Belgrade, Conrad rushes into the Ballplatz brandishing a telegram received during the night from von Moltke in Berlin, urging them to reject the British proposal and to mobilize at once against Russia. Conrad meets with Berchtold and reads out the message from von Moltke now urging A-H to mobilize against Russia. But these words contrast starkly with Bethmann's appeals, and the Kaiser's latest message.
Berchtold exclaims: *How odd ~ Who runs the German Government... Moltke or Bethmann?!*
Conrad wires Moltke: *The A-H mobilization order has been issued for activation on 31st July.*

Tension rises between Vienna and Berlin. Bethmann's unwavering line has always been that the conflict must be localized. And if this is impossible, then Russia must be branded as the aggressor to maintain British neutrality. However, the hope of British neutrality is an illusion, which apparently no German realizes except their Ambassador in London Prince Max von Lichnowsky, who with personal feelings for England has achieved a prestigious position for himself in British official and social circles. Prince Max's concern is that despite his repeated serious warnings, Berlin confidently continues to count on British neutrality. Nevertheless, in the evening ~ Bethmann writes to the Kaiser proposing that he send a personal message to the Tsar : *Germany seeks direct talks between A-H and Russia.*
A telegram of this kind would if war were to come about, place the guilt of Russia in the strongest light.

Wilhelm II to Nicholas II — Berlin, 28th July 1914
D. 10.45 PM

Dear Nicky ~ I have heard with the greatest anxiety of the impression which is caused by the action of Austria-Hungary against Serbia. The unscrupulous agitation, which has been going on for years in Serbia, has led to the revolting crime of which Archduke Franz Ferdinand has become a victim. The spirit which made the Serbians murder their own king and his consort still dominates that country. Doubtless you will agree with me that both of us, you as well as I, and all other sovereigns, have a common interest to insist that all those who are responsible for this horrible murder shall suffer their deserved punishment. On the other hand, I by no means overlook the difficulty encountered by you to stem the tide of public opinion. In view of the cordial friendship which has joined us both for a long time with firm ties, I shall use my influence to induce Austria-Hungary to obtain a satisfactory understanding with Russia. I hope confidently that you will support me in my efforts to overcome all difficulties which may yet arise ~ Your most sincere and devoted friend and cousin. ~ Willy

Nicholas II to Wilhelm II — Peterhof Palace 29 July 1914
D. 1.00 AM R. 1.10 AM

Dear Willy ~ I am glad that you are back in Germany. In this serious moment I ask you earnestly to help me. An ignominious war has been declared against a weak country and in Russia the indignation which I share is tremendous. I fear that very soon I shall be unable to resist the pressure exercised upon me and that I shall be forced to take measures which will lead to war. To prevent a calamity as a European war would be, I urge you in the name of our old friendship to do all in your power to restrain your ally from going too far. ~ Nicky

Wilhelm II to Nicholas II — Berlin, 29th July 1914
D. Night

Dear Nicky ~ I have received your telegram and I share your desire for the conservation of peace. However, I cannot ~ as I told you in my first telegram ~ consider the action of Austria-Hungary as an ignominious war. Vienna knows from experience that the promises of Serbia as long as they are merely on paper are entirely unreliable. According to my opinion the action of Austria-Hungary is to be considered as an attempt to receive full guarantee that the promises of Serbia are effectively translated into deeds. In this opinion, I am strengthened by the explanation of the Austrian Cabinet that Austria-Hungary intend no territorial gain at the expense of Serbia. I am therefore of opinion that it is perfectly possible for Russia to remain a spectator in the Austro-Serbian war without drawing Europe into the most terrible war it has ever seen. I believe that a direct understanding is possible and desirable between your Government and Vienna, an understanding which I endeavour to aid with all possible effort. Naturally military measures by Russia, which might be construed as a menace by Austria-Hungary, would accelerate a calamity which both of us desire to avoid and would undermine my position as mediator ~ which upon your appeal to my friendship and aid ~ I willingly accepted. ~ Willy

An artist's view of Semlin and Belgrade at the confluence of the Danube and Savé

BELGRADE. With the time limit of the Note expired, Belgrade's leaders expect the A-H Army to launch an offensive and invade at any moment. The Cabinet of Belgrade ask Crown Prince Alexander to send a personal telegram to his uncle, King Vittorio Emanuele III of Italy in the hope of trying to persuade Vienna to extend the expired 48-hour time limit. Even so, the populace of Belgrade depart as the Government evacuates in stages from Belgrade, to the second largest city within Serbia ~ Nish. Pasic announces: *No decisions will be taken until the Russians make their views known. If war is unavoidable, then we shall fight, though Belgrade may not be defended.* He cables Serbian Ambassador in St Petersburg, Miroslav Spalajkovic asking him to ascertain the views of the Russian Government. The reply from Ambassador Spalajkovic arrives in two parts ~ the first part arriving at 4 AM gives a promise of Russian support for Serbia. The second arriving at 10 AM ~ reasons that Serbia should not offer any resistance to an A-H attack, but to give up Belgrade, and then to appeal for international support. With a firm promise of Russian support the Cabinet of Belgrade order the Serbian Army to initiate all measures for a mobilization. Crown Prince Alexander signs the order for an *immediate* mobilization : becoming the first mobilization of the Great War to come into instantaneous and active effect.

St PETERSBURG. *Even if Austria-Hungary goes no further than the occupation of Belgrade by troops thrown across the Sava at Semlin, Russia will declare for a general mobilization. Already rapid mobilization is proceeding in the west and southwest virtually from the German frontier to the Black Sea.* Press Dispatch

In the beginning the attitude of the English press toward the Austrians was quiet and friendly, as the murder was condemned. Little by little however, voices increased in number insisting that, however necessary the punishment of a crime might be, no elaboration of it for a political purpose could be justified. Austria was urgently called upon to act with moderation. The whole world outside Berlin and Vienna understood that it meant war, and world war. Sir Edward justly pointed out that even without annexations of territory a country can be humiliated and subjected, and that Russia would regard this as a humiliation, which she would not stand. The urgent appeals and definite declarations of Serge Sazonoff, and the positively humble telegrams of the Tsar, the repeated proposals of Sir Edward, the warnings of Italian Foreign Minister San Giuliano and Ambassador Bollati in Berlin, and my urgent advice were all of no use, for Berlin went on insisting Serbia be massacred. The more I pressed, the less willing they were to alter their course. So, Grey resolved upon his warning. I replied that I had always reported that we should have to reckon upon English hostility if it came to war with France. Sir Edward said to me repeatedly ~ If war breaks out, it will be the greatest catastrophe the world has ever seen!

Prince Max von Lichnowsky

First act of war ~ Serbian Army destroys the A-H Semlin-Belgrade Bridge

The A-H deadline has expired. Belgrade has rejected the conditions of Vienna's Note of Demands. Early on the morning of 29th July, the mobilized Serbian Army commences hostilities by blowing up the Semlin-Belgrade Bridge near the confluence of the Savé and Danube in a symbolic *no return* action. The news of this action is not well received in the capitals of Europe. Several hours later, three swift monitors of Austria's Danube Flotilla steam toward Belgrade, firing their 125mm guns at the barracks, two banks, and the Customs House. They keep this up for an hour, and then withdraw to Semlin. Their bombardment, in response to the Serbian Army demolition of the A-H Semlin-Belgrade Bridge is the second military action so far ~ but these two actions send shockwaves across Europe.

Serbian troops board a Belgrade train…

BERLIN. Considering what Belgian Ambassador Beyens believes of Kaiser Wilhelm II's personal inclinations, he no longer has any doubt ~ no hope can now be entertained of a peaceful arrangement. Baron Beyens makes an evening appointment to visit French Ambassador Jules Cambon to discuss his dismal forecast. It also transpires that Cambon too has few illusions. That night, Beyens writes to his Foreign Minister to acquaint him with his fears and to urge him to be on his guard. The report is entrusted to Beyens' First Secretary who departs for Brussels on the overnight Berlin-Brussels express. In the morning, his report is hand delivered to Foreign Minister Julien Davignon, who then opens it…

The ultimatum to Serbia, is a blow; contrived by Vienna and Berlin, or rather contrived here and carried out at Vienna. Requital for the assassination of the Austrian heir-apparent and the Pan-Serb propaganda serves as a stalking-horse. The real aim, apart from the crushing of Serbia and the stifling of Slav aspirations, is to deal a deadly thrust at Russia and France, with the hope that England will stand aside from the struggle. In order to vindicate this theory, I beg to remind you of the view prevailing in the German General Staff, namely, that a war with France and Russia is unavoidable and close at hand ~ a view which the Kaiser has been induced to share. This war, eagerly desired by the military and Pan-German party, might be undertaken today under conditions extremely favourable for Germany, conditions that are not likely to arise again for some time to come.
We, too, have to ask ourselves these harassing questions, and keep ourselves ready for the worst.

Other heads of legations, while sharing Beyens' anxieties, do not agree with pre-meditation of the Kaiser. After sending his report to Brussels, Beyens visits Italian Ambassador Count Ricardo Bollati who thoroughly versed in German statecraft, holds the opinion that those in Berlin, while agreeing in principle with the Vienna Cabinet as to the strategic necessity for chastising Serbia, had not known beforehand the terms of the A-H Note. Ricardo Bollati believes that Vienna and Berlin are wrongly convinced that Russia ~ is in no position to engage in a seriously challenging war, and that Russia will not dare to embark upon such a perilous adventure.

Berlin has false ideas as to the course that the Tsar will adopt. Imperial Russia will find itself forced into drawing the sword, in order to maintain its prestige in the Slav world. Its inaction, in face of Austria's entry into the field would be equivalent to suicide. A widespread conflict would not be popular in Italy. The Italian people have no concern with Russia who is Austria's enemy. Italy wishes to devote all its attention to other problems, more absorbing from its own point of view.

At present, feeling runs so high in Vienna that all calm reflection goes by the board. Moreover, in seeking to annihilate Serbia's military power, the Vienna Cabinet is pursuing a policy of personal revenge. It cannot realize the mistakes that it made during the Balkan War or remain satisfied with the partial successes then gained with our aid, were certainly diplomatic victories. All that Count Berchtold sees today is Serbia's insolence and the criticism he has had to endure in Austria. By this bold stroke, very unexpected from a man of his stamp, he hopes to turn the criticism into applause. Count Ricardo Bollati

Chancellor Bethmann tries to pin British Ambassador Sir Edward Goschen down on a British neutrality agreement by promising not to seek French territory after a victorious war. While making no promises about French colonies, Bethmann proposes ; *If Great Britain remains neutral, Germany will make no territorial aggressions at the expense of France. Ever since I have been Chancellor, the object of my policy has been to bring about an understanding with London for a general neutrality agreement between Germany and Great Britain.* But now, the Chancellor has imprudently revealed the German option to violate French territory. Sir Edward Grey calls Bethmann's proposal, *an infamous offer*. Now the British Government know for certain, that Germany has prepared military plans to implement a wider war. This is seen in London as the end of the façade of German foreign policy, previously presented to the world as their hope for a limited war, and it now seems apparent that the hawks of Berlin are pursuing a parallel path for possible action in the west ~ if their hope for limited war diminishes.

ATHENS. The Serbian Legation receives the following telegram from Nish.
During the night Belgrade was bombarded. Shells fell in various quarters of the town, causing great damage. Several shells fell on the Franco-Serbian and Andrevitch banks. Andrevitch of the banking firm was wounded. An artillery duel is proceeding at Vichnitza, about three and a half miles down the river from Belgrade.

St PETERSBURG 28th July. It becomes clear that Russia will support Serbia regardless of events. Tsar Nicolas: *Any state should not present any sort of demands to another, unless, of course, it is bent on war.* Serge Sazonoff informs Tsar Nicholas of Austria-Hungary's declaration of war against Serbia and proposes the preparation of two army orders; one for a partial, and another for a general mobilization, so that a decision on which to use can be left to the last moment in light of the latest developments. The Tsar agrees to this and to the declaration of a partial mobilization the following day.

The Tsar and Sazonoff now intend, *to see the effect of the announcement of a partial mobilization.* Sazonoff telegraphs to Berlin, and repeats to Vienna, Paris, London, and Rome. *In consequence of the Austrian declaration of war on Serbia, we shall tomorrow (the 29th July) proclaim mobilization in the districts of Odessa, Kiev, Moscow, and Kazan. Inform the German Government of this and lay stress on the absence of any intention on the part of Russia to attack Germany.* Sazonoff telegraphs to Count Benckendorff in London.

It will be necessary for England with all speed to take action in view of mediation and for Austria at once to suspend military measures against Serbia. Otherwise mediation will only furnish a pretext for delay in bringing the matter to a decision and make it meanwhile possible for Austria to annihilate Serbia completely. Confirming Russian mobilization affects the Austrian frontier and not the German frontier, a correspondent wrote:

In Russian eyes the die is cast, and only a miracle can avert war. Austria's proposal to discuss terms, when Belgrade is occupied is regarded as inadmissible. The general opinion in Russia is that the only hope of preventing a conflagration lies in the combined efforts of the Powers to check Austria. It is positively stated that Russia will take to the field the moment Austria crosses the Servian border. Russia is unswerved in her determination to defend Servia. Partial mobilization has already been ordered and there is every indication that the whole vast military machinery will soon be in motion. The Russian public is prepared to accept war.

BERLIN. As the partial Russian mobilization does not necessarily imply war between Russia and Germany; Foreign Secretary Gottlieb von Jagow informs both French Ambassador Jules Cambon and British Ambassador Sir Edward Goschen : *Germany will not mobilize against Imperial Russia if the Russian mobilization is only directed against Austria–Hungary ~ but will counter-mobilize if Russia subsequently orders a direct mobilization against Germany.* Vienna and not Berlin is again the focus of decision-making; just as long as Tsar Nicholas II declines to order a full mobilization.

VIENNA. After disregarding all last-minute pleas from Berlin to accept mediation, reports arrive in Vienna of the Russian partial mobilization ~ consequently A-H counter mobilizes.

St PETERSBURG. Vienna's refusal of direct talks and the A-H retaliatory shelling of Belgrade (after the Serbian Army committed the first act of the war demolishing the Semlin-Belgrade Bridge) induce Sazonoff to meet with Minister of War Vladimir Soukhomlinoff and Chief of General Staff, General Nikolai Yanuskevich. They decide, *in view of the small probability of avoiding war with Germany* to recommend an immediate general mobilization to the Tsar, who over the phone consents to the full deployment of Russian forces against both A-H and Germany. After receiving an alarming telegram from the Kaiser, and fully realizing the terrible consequences, he halts it shortly after authorizing it. *Everything possible must be done to save peace, I will not become responsible for a monstrous slaughter.*

The Russian High Command having other aspirations, refuse to yield and win Foreign Minister Serge Sazonoff over to their side to further their own strategic interests. While Russia is not ready, Soukhomlinoff, Sazonoff, and Isvolsky are sure of France and Britain's involvement, and as others have designs on large swaths of Austrian territory and now wait for an opportune moment to annex Galicia and other parts of the fragile A-H Empire, many Russians see gain in both safeguarding Serbia and simultaneously expanding the lands of Russia. Therefore, many in the Russian military hope to sway Sazonoff to win over the Tsar. Sazonoff, accompanied by a staff officer, boards a carriage for an appointment at Peterhof, to convince the Tsar that a general mobilization can no longer be delayed. Chief of Staff General Yanuskevich asks Sazonoff to telephone him at once from Peterhof to let him know whether or not he has succeeded. If Sazonoff's news is positive, the Chief of Staff has agreed to: *Convert the partial mobilization to a general one and thereafter retire from sight, smash my telephone and take all measures so that I cannot be found to give contrary orders for a new postponement of general mobilization.*

The psychologically interesting point is that Tsar Nicholas, who helped prepare the Great War and already ordered a partial mobilization, wished to recede at the last moment. My earnest, warning telegram, it seems, made him realize clearly for the first time the colossal responsibility, which he was bringing upon himself by his warlike preparations. Therefore, he then wished to stop the war machine, the murderer of entire peoples, which he had just set in motion. This would have been possible, if Sazonoff had not frustrated his wish. Wilhelm II

St PETERSBURG. Tsar Nicholas is sitting behind his fine bronze-trimmed mahogany desk, now covered with maps in his office on the ground floor of Peterhof. As he looks out his window at the Gulf of Finland, he hardly seems to hear what Sazonoff is saying. The Tsar's bearded face, now pale with fatigue, remains expressionless, and his dreamy eyes remain fixed on the remote blue horizon. *Think of the responsibility you are asking me to take if I follow your advice! Think what it means to send thousands and thousands of men to their death.* Sazonoff's aide General Tatisheff then chooses to agree. *Yes, it is a terrible decision to take.* Tsar Nicholas snaps ~ *I am the one who decides!* Attentive to Sazonoff, the Tsar is impressed by his view (an erroneous one as we now know) that Germany is bent on war. *Russia cannot allow Austria-Hungary to crush small Serbia and become the predominant power in the Balkans.* After a terrible inner struggle, the Tsar gives in. *All right Serge, telephone the Chief of the General Staff that I give the order for general mobilization.* Sazonoff in one hour has convinced him, and so the Tsar cannot rescind the order if he has another change of heart, Sazonoff telephones the General as agreed:
Now you can smash your telephone. Give your orders General, and disappear for the rest of the day.

At 6 PM the St Petersburg Central Telegraph Office transmits the following message:
HIS IMPERIAL MAJESTY ORDERS : ARMY AND NAVY TO BE PLACED ON WAR FOOTING.
TO THIS END RESERVISTS AND HORSES ARE TO BE CALLED UP
ACCORDING TO THE MOBILIZATION PLAN OF THE YEAR 1910

This profound action of calling up troops for border security does much to escalate the crisis. The provocation begun by covert forces in Sarajevo is now continued by St Petersburg's unwavering support for Belgrade and its unwillingness to negotiate with Berlin and Vienna : because Russia who had been so humiliated by the forcible annexation of Bosnia in 1908, will not yield a second time. Therefore, Tsar Nicholas having ordered the full mobilization of Russian forces assumes added responsibility for the rupture. While many Germans continue to assume that Russia is deterred by Berlin's strong support of Vienna, Kaiser Wilhelm receives a series of reports that Russian troops have set fire to their frontier posts, torn up railway tracks and posted red mobilization notices on the walls of public buildings across Russia. Kaiser Wilhelm telegraphs his cousin, Tsar Nicholas at 1.20 AM.

Dear Nicky, Count Pourtalès has instructions to direct the attention of your Government to the dangers and grave consequences of a mobilization. Austria-Hungary has mobilized only against Serbia, and only a part of her army. If, as it is now the case that Russia mobilizes against Austria-Hungary, the part of the mediator with which you have entrusted me in such friendly manner and which I have accepted at your express prayer, will be endangered, if not made impossible. The whole weight of decision now rests upon your shoulders…
You have to bear the responsibility for war or peace. *~ Willy*

The Tsar is unemotional about his rupture with cousin Willy. Late that night, after drinking a glass of tea and chatting with the Tsarina, who is already in bed, Tsar Nicholas decides to take a bath. As he lowers himself into the tub, a footman knocks on the door to inform him that there is an urgent personal telegram from the German Kaiser. *I read the telegram, and then I repeated it out loud to myself, but still I do not understand. Wilhelm portends that it is still in my power to avoid war He implores me not to let my troops cross the border. Have I gone mad? I return to the Tsarina and read her Wilhelm's telegram.*
She wants to read it herself, and then said ~ You will not reply to it, will you?
No. I felt that all was finished between Wilhelm and myself. I slept soundly ~ N

BRUSSELS. Minister for Foreign Affairs Julien Davignon telegraphs to Belgian Ambassadors in Berlin, Vienna, Paris, London, St. Petersburg, Rome, and to Luxemburg: *The Belgian Government has decided to place the army upon a strengthened peace footing. This step should in no way be confused with mobilization. Owing to the small extent of her territory, all Belgium consists, in some degree, of a frontier zone. Her army on the ordinary peace footing consists of only one class of armed militia. On the strengthened peace footing, her army divisions and her cavalry division comprise effective units of the same strength as those of the corps permanently maintained in the frontier zones of the neighboring powers.*

BERLIN 9 AM. British Ambassador Sir Edward Goschen dispatches his report of an interview in which Chancellor Bethmann Hollweg *again* makes his duplicitous proposal for British neutrality. Sir Eyre Crowe comments : *These astounding proposals reflect discredit on the statesman who makes them.* Goschen concludes that Germany is determined to go to war, and that the sole restraining influence is Berlin's apprehension that Great Britain will join the defence of France and Belgium.

He proceeded to make the following bid for British neutrality. He said that it was clear that Great Britain would never stand by and allow France to be crushed in any conflict there might be. That however, was not the object at which Germany aimed. Provided that neutrality of Great Britain were certain, every assurance would be given to the British Government that Imperial Germany aimed at no territorial acquisitions at the expense of France should they prove victorious in any war that might ensue. I questioned his Excellency about the French colonies, and he said that he was unable to give a similar undertaking in that respect. As regards Holland, however, his Excellency said that, so long as Germany's adversaries respected the integrity and neutrality of the Netherlands, Germany was ready to do likewise. It depended upon the action of France what operations Germany might be forced to enter upon in Belgium, but when the war was over, Belgian integrity would be respected if she had not sided against Germany. He had in mind a general neutrality agreement between England and Germany. In reply to his Excellency's inquiry how I thought his request would appeal to you, I said that I did not think it probable that at this stage of events you would care to bind yourself to any course of action and that I was of opinion that you would desire to retain full liberty. Sir Edward Goschen

LONDON. Sir Edward Grey is aghast at Bethmann's proposal and knows that Britain can never stand by if France is attacked. He therefore states in unmistakable terms, that Britain will aid France. Grey's ire is apparent in his wire to Goschen. *His Majesty's Government cannot entertain the Chancellor's proposal. What he asks us in effect is to stand by while French colonies are taken and France is beaten, so long as Germany does not take French territory as distinct from the colonies. It would be a disgrace for us to make this bargain with Germany at the expense of France; a disgrace from which the good name of this country would never recover. The Chancellor also in effect asks us to bargain away whatever obligation or interest we have as regards the neutrality of Belgium. We cannot entertain that either. The way for Germany to preserve good relations with England is that they should continue to work together to preserve the peace of Europe.*

BERLIN. In the early morning, Kaiser Wilhelm receives a telegram from Tsar Nicholas explaining: *Russia's measures started five days before are only aimed at Austria-Hungary.* Kaiser Wilhelm then receives a report from Military Attaché Bernhardt von Eggeling in St Petersburg of his interview (sent 26th July) with the Chief of Staff of the Russian Army who assured him that no mobilization has begun on the German border, but Attaché von Eggeling advises that in view of the information he has about the current calling out of Russian reserves, he considers the conversation to be an attempt to mislead him. Russian military preparations are far more advanced than officially stated. Now the Kaiser realizes, that Germany is being tricked. The Russians were mobilizing even when Nicholas asked him to speak to Vienna. Bethmann is sent a copy of the Tsar's telegram from Potsdam, with the Kaiser's assertion: *Now Russia is much ahead of Germany, I must mobilize too! I regard my mediation action as brought to an end!* Bethmann feeling most apprehensive, advises the Kaiser not to end his mediation while there's still no answer from Vienna. He suggests the Kaiser sends another telegram to the Tsar, explaining how his mobilization is weakening his mediation. *This telegram will become an important document for history.*

Prince Henry of Prussia telegraphs to King George V:

I arrived here yesterday and have communicated what you were so good as to say to me at Buckingham Palace last Sunday to Wilhelm, who was very thankful to receive your message. Wilhelm, who is very anxious, is doing his utmost to comply with the request of Nicholas to work for the maintenance of peace. He is in continual telegraphic communication with Nicholas, who has today confirmed the news that he has ordered military measures which amount to mobilization, and that these measures were taken five days ago. We have also received information that France is making military preparations. While we have not taken measures of any kind, we may be obliged to do so at any moment if our neighbours continue their preparations. This would then mean a European war. If you seriously and earnestly desire to prevent this terrible misfortune, may I propose to you to use your influence on France, and also on Russia that they should remain neutral. In my view this would be of the greatest use. Germany and England should now more than ever give each other mutual support in order to prevent a terrible disaster, which otherwise appears inevitable. William is inspired by the greatest sincerity in his efforts for the maintenance of peace. But the military preparations of his neighbours may end in compelling him to follow their example for the safety of his own country, which otherwise would remain defenseless. ~ *H*

LONDON 8.45 PM. King George V: *I am very glad to hear of William's efforts to act with Nicholas for the maintenance of peace. I earnestly desire that such a misfortune as a European war ~ the evil of which could not be remedied ~ may be prevented. My Government is doing the utmost possible in order to induce Russia and France to postpone further military preparations, provided that Austria declares herself satisfied with the occupation of Belgrade and the neighbouring Serbian territory as a pledge for a satisfactory settlement of her demands, while at the same time the other countries suspend their preparations for war. I rely on William applying his influence in order to induce Austria to accept this proposal. In this way he will prove that Germany and England are working together to prevent what would be an international catastrophe. Please assure William that I am doing all I can, and will continue to do all that lies in my power, to maintain the peace of Europe.*

BERLIN. King George's telegram from London now shocks the entire German Army stratagem. The hope of British neutrality, if Russia can be shown to be the aggressor, is now seen as a delusion. Kaiser Wilhelm: *If the Russian mobilization can no longer be retracted ~ which is not true ~ why, then did the Tsar appeal for my mediation three days afterward without mention of the issuance of his mobilization order? His frivolity and weakness are to plunge the world into the most frightful war, which aims for the destruction of Austria and Germany.* Bethmann is alarmed and renews pressure on Vienna to negotiate with Russia.

VIENNA. Count Heinrich von Tschirschky receives a crucial message from Bethmann warning; *Great Britain will not remain neutral. We should be two against four Great Powers. Germany, as the result of England's hostility, will have to bear the brunt of the fighting. We must urgently and emphatically recommend to the consideration of the Vienna Cabinet the acceptance of mediation on the honorable terms indicated.* Tschirschky shows this to Berchtold, who is shocked. *I will have to talk with the Emperor*. Berchtold and Conrad meet the Emperor to brief him on the new messages from Berlin and the military position. They reply to Berlin: *Serbia must meet the demands of the Note in full, and pay the cost of the mobilization, and all military operations.* Despite German pressure, A-H resolve to continue the war against Serbia, and to give a courteous reply to the English proposal ~ without accepting it.

Tschirschky continues talking with Berchtold and his colleagues in the Ballplatz, who are so far unwilling to limit their military action against Serbia. At 5 PM, Tschirschky telegraphs to Berlin: *Instructions executed. Count Berchtold will reply by return ~ after receiving Emperor Franz Josef's commands.* In the evening, after Baron von Stumm telephones Tschirschky to confirm Vienna's determination to reject all compromise and mediation, Tschirschky telegraphs to Berlin. *I begged them to bear in mind the incalculable consequences of a rejection of mediation. Conrad von Hötzendorf this evening was to submit the order for general mobilization to the Emperor as the answer to measures already taken by Russia. It was not quite certain whether in the present situation mobilization is still the right course.*

7.30 PM. In response to Helmuth von Moltke's question: *What Austria-Hungary will do in response to Russian partial mobilization*; Conrad: *On the basis of His Majesty's decision the resolve is to go forward with the war against Serbia. To mobilize remainder of army and assemble in Galicia. First day of A-H mobilization is 4th August. Request intimation of your first mobilization day.* Moltke tells the A-H liaison officer that the Russian partial mobilization is no reason for Germany to mobilize. This will only occur if Russia declares war against A-H. Moltke wires Conrad: *Stand firm against Russian mobilization. Austria-Hungary must be preserved, mobilize at once against Russia. Germany will mobilize soon.* This is the exact opposite to what Chancellor Bethmann is trying to achieve ~ and reveals how many German GHQ Staff officers who report directly to the Kaiser, are now acting independent of their civilian leaders.

LONDON. Sir Edward Grey again advises Prince Max. *So long as the conflict remains confined to Austria and Russia we can stand aside, but if Germany and France should be involved, then the British Government will be forced to make up its mind quickly.* And if the Tsar, had stuck to his refusal not to let the Russian mobilization go beyond the limited call-up in the south, peace might still have been saved. The Russian military have already surreptitiously exceeded the terms of the *ukase*.

BERLIN. Bethmann drafts new instructions for Heinrich von Tschirschky in Vienna advising: *Germany must decline to be drawn wantonly into a world conflagration without having any regard paid to our counsels.* The Chancellor makes a pathetic confession of failure to the Prussian Council of Ministers. *All governments, including Russia's, and the great majority of their peoples are peacefully inclined, but direction has been lost, and the stone has started rolling.* Prince Bernhard von Bülow in desperation asks: *How did it all happen?* The Chancellor replies, raising his arms in despair ~ *Ah, if only one knew...*

Franz Josef receiving Kabinettsdirektor Baron von Schiessl-Pernstorff in his study at Bad Ischl. The Kabinettsdirektor was head of the Kabinettskanzlei liaison body between the A-H Emperor and the Government, who also coordinated all of his contact with his (mostly loyal) subjects.

VIENNA. As four weeks have passed since the assassinations ~ ample time for the application of statecraft to address events ~ and as the Kaiser has always urged immediate action, he wants to know Franz Josef's decision regarding the recent *Halt in Belgrade* proposal. Berchtold convenes the Ministerial Council, and they eventually decide that mediation will inevitably work against them. From this time on, German pressure has near no influence upon Austrian diplomacy, because the Austrians believe that Germany will stick by them. To avoid further problems arising with Serbia in future, the Austrians are intent on destroying the Serbian Army, as is clear in their reply to Berlin: *War operations against Serbia must be continued. (2) We cannot negotiate on the English proposal unless Russian mobilization is suspended. (3) Our terms must be accepted by Serbia. We cannot consent to negotiation.* Conrad: *The war with Serbia will continue; the English proposals will receive an obliging answer without taking any position; general mobilization will be ordered on 1st August with the 4th of August the first day.*

Emperor Franz Josef signs the order for a general mobilization at 12.20 PM. He dispatches the order to Vienna and telegraphs to Berlin. *I have ordered the general mobilization of A-H forces following news of Russia's partial mobilization. The latest British mediation proposal has come too late. A-H Army operations against Serbia can suffer no interruption and any fresh rescue of Serbia by Russian intervention would have serious consequences for Austria-Hungary, and therefore Vienna cannot possibly permit such intervention. Threatening attitude of Russia will not stop A-H acting against Serbia.*

BERLIN. Kaiser Wilhelm telegrams to Franz Josef. *Germany is preparing to mobilize and will fulfill its alliance obligations. It is of the greatest importance that Austria directs her chief force against Russia and does not split it up by a simultaneous offensive against Serbia.* Conrad telephones from Vienna 12 minutes later: *Austro-Hungarian mobilization against Russia is only for the purpose of taking precautions against attack from Russia, without any intention of declaring or beginning war.* Moltke sends a telephone message to Conrad: *Germany will proclaim mobilization of entire military forces probably on 2nd August and open hostilities against Russia and France. Will Austria leave her in the lurch?* Conrad: *The Austrian leadership is not sure yet whether Russia is merely threatening ~ therefore we cannot allow ourselves to be diverted from the action. Austria-Hungary will go to war with Russia as well as Serbia.* Sir E. Goschen to Grey: *German Secretary of State again assured me that both he and the Kaiser, at the request of the Tsar and the German Foreign Office, have even up till last night been urging Austria to show willingness to continue discussion and communications from Vienna had been of a promising nature ~ but ~ Russia's mobilization has spoiled everything.*
So ended the last chance of peace.

BERLIN. *Kaiser Wilhelm II to Nicholas II* *Neuen Palais D. Evening*

Dear Nicky ~ On your appeal to my friendship and your call for assistance to mediate between your and the Austro-Hungarian Government. While this action was proceeding, your troops were mobilized against Austria-Hungary, my ally. Thereby, as I have already pointed out to you, my mediation has been made almost illusory. I have nevertheless continued my action. I now receive authentic news of serious preparations for war on my Eastern frontier. Responsibility for the safety of my Empire forces preventative measures of defence upon me. In my endeavours to maintain the peace of the world I have gone to the utmost limit possible. The responsibility for the disaster that is now threatening the whole civilized world will not be laid at my door. In this moment it still lies in your power to avert it. Nobody is threatening the honour or power of Russia who can well afford to await the result of my mediation. The peace of Europe may still be maintained by you, if Russia will agree to stop the military measures, which must threaten Germany and Austria-Hungary. ~ Willy

St PETERSBURG. *Nicholas II to Wilhelm II* *Peterhof Palace D. 1.20 AM R. 1.45 AM*

Dear Willy ~ I thank you heartily for your mediation which begins to give one hope that all may yet end peacefully. It is technically impossible to stop our war preparations, obligatory owing to Austria's mobilization. We are far from wishing war. As long as the negotiations with Austria on Servia's account are taking place my troops shall not make any provocative action. I give you my solemn word for this. I put all my trust in God's mercy and hope in your successful mediation in Vienna for the welfare, and for the peace of Europe. ~ Nicky

BERLIN. German intelligence receives a report that Russia's general mobilization is underway. Moltke asks for evidence, and by noon he is handed a red Russian mobilization notice that had been posted in a frontier village. With confirmation of the Russian mobilization, von Moltke meets with Bethmann and Falkenhayn to expedite an immediate German mobilization.

It seems clear that, if Germany resolved to do so, she could quite safely have abstained from encouraging Austria in a mad adventure. The reason why the war came appears to have been that the German Government finally laid the reins on the necks of men whom up to then it had held in restraint. The decision appears to have been allowed at this point to pass from civilians to soldiers. I do not believe that the German Government as a whole intended deliberately to invoke the frightful consequences of war. But in departing from the appearance of self-restraint, which they had displayed well up to then, the Kaiser and his Ministers misjudged the situation. They did not foresee the crisis to which their policy was conducting, and when that crisis arrived they lost their heads and blundered in trying to deal with it. British Lord Chancellor, Viscount Richard Haldane

BRUSSELS. Foreign Minister Julien Davignon telegrams to legations abroad. *The Minister of War informs me that the Belgian mobilization has been ordered and that Saturday 1st August will be the first day. Belgium will do her utmost to maintain her neutrality, and that she desires and expects that the other powers will respect and maintain it.* Julien Davignon then contacts the German Ambassador von Below Saleske. *The situation is a consequence of our desire to fulfill our international obligations, and in no way implies an attitude of distrust toward our neighbours. I will remind you, German Ambassador of the instructions your Chancellor Bethmann Hollweg once gave to your predecessor, Herr von Flotow. Bethmann Hollweg declared that Germany has no intention of violating Belgian neutrality. Since 1913, Herr von Jagow has made reassuring declarations respecting the maintenance of Belgian neutrality.* Ambassador von Below Saleske: *I know of the conversation with Herr von Flotow. I am certain that the sentiments expressed at that time have not changed.*

BERLIN. Moltke instructs the diplomats to activate protocols to enable German forces now poised in the West, to strike without incurring the odium of unprovoked aggression. Germany sends an ultimatum to Russia. *If Russia does not halt its mobilization within 12 hours and completely demobilize, Germany will mobilize.* Kaiser Wilhelm's issue of the decree of *Kriegsgefahrzustand Imminence-of-War* is the first phase of a German general mobilization ~ the rule of martial law to substitute military for civil authority as regards all means of communication, telegraphs, telephones, and post. This fateful decision is revealed via an express edition of the *Berliner Lokalanzeiger* distributed at street corners.

- RUSSIA WANTS WAR! -

From official sources, we have received (at 2 PM) the following report, pregnant with consequences: The German Ambassador at St. Petersburg sends word today that a general mobilization of the Russian Army and Navy had previously been ordered. That is why His Majesty Kaiser Wilhelm II has decreed an Imminence-of-War. His Majesty will take up his residence in Berlin today. Imminence-of-War is the immediate prelude to a general mobilization, in answer to the menace that hangs over Germany today, owing to the step taken by the Tsar.

In the evening, the Kaiser speaks from the balcony of his *Neuen Palais*. The people welcome him and respond with wild cheering. *A grievous situation has come upon Germany. Envious nations on all sides are forcing us to justified defense. They are forcing the sword into my hand. If my attempts are not successful in bringing our opponents to their senses and in keeping peace at the eleventh hour, I hope that with God's help we may so use the sword, that we may be able to sheathe it again with honor. Enormous sacrifices in life and property would be demanded from the German people by a war; but we would show the enemy what it means to attack Germany. And now I bid you go to church, bow down before God, and ask His help for our brave army.*

LONDON. Lichnowsky calls on Grey with news that Berchtold has authorized the resumption of talks between Vienna and St Petersburg. He assumes Berchtold is sincere, so telegraphs to von Jagow. *In order to remove the Russian mistrust of Austria, and the Austrian mistrust of Serbia, the Four Powers should undertake (1) that Serbia shall give Austria full satisfaction, provided she respects Serbian sovereignty and integrity; and (2) Austria shall respect the said sovereignty. Further, if Germany can get any reasonable proposal put forward, which makes it clear that Germany and Austria are striving to preserve European peace and that Russia and France will be unreasonable if they reject it, I will support it at St. Petersburg and Paris, and go to the length of saying, that if Russia and France will not accept it I will have nothing to do with the consequences. Otherwise, if France is drawn in, we shall be drawn in.* Jagow: *I cannot answer this new appeal, because Germany has just sent an ultimatum to Russia, requiring her to demobilize within 12 hours.*

BERLIN 8.30 PM. Bethmann telegrams Prince Lichnowsky in London. *Russia's general mobilization has cut short Austria's pending reply to our mediation proposal. We have told Russia we should have to mobilize, which will mean war, unless within twelve hours, the military preparations against Austria-Hungary and ourselves are suspended. Please use every means to insure this course of events is duly recognized in the English press.* Goschen asks Jagow: *In the event of war, will Germany respect the neutrality of Belgium?* Jagow answers in a *Freudian Slip ~ I cannot answer such a question, as it would reveal Germany's intentions.*

LONDON. Late in the evening, Lichnowsky receives a telegram from Bethmann informing him of the German ultimatum to Russia to stop mobilizing ~ and if Germany has to mobilize ~ it means war. Grey is unavailable, so Lichnowsky gives a copy to Tyrrell, who late at night delivers it to Asquith. They prepare a message for King George V to send to the Tsar, appealing to him to stop the Russian mobilization. In the interval, a telegram arrives from Kaiser Wilhelm to be delivered to the King. Thereupon, PM Asquith and Tyrell drive to Buckingham Palace to rouse King George V from his bed.

Many thanks for your friendly communication. Your proposals coincide with my ideas and with the communication which I have this evening received from Vienna, and which I have passed on to London. I have just heard from the chancellor that intelligence has just reached him that Nicholas this evening has ordered the mobilization of his entire army and fleet. He has not awaited the result of the mediation in which I am engaged, and he has left me completely without information. I am traveling to Berlin to assure the safety of my eastern frontier, where strong Russian forces have already taken up their position. *Wilhelm II*

Dear Wilhelm ~ Many thanks for your telegram of last night. I have sent an urgent telegram to Nicholas in which I have assured him of my readiness to do everything in my power to further the resumption of the negotiations between the powers concerned. *~ George V*

BERLIN. Imperial Chancellor Bethmann telegraphs Ambassador Pourtalés in St. Petersburg.
In spite of negotiations still pending and although we have up to this hour made no preparations for mobilization, Russia has mobilized her entire army and navy, hence also against us. On account of these Russian measures, we have been forced, for the safety of the country, to proclaim the threatening state of war, which does not yet imply mobilization. Mobilization however, is bound to follow if Russia does not suspend every war measure against Austria-Hungary and ourselves within twelve hours, and notifies us definitely to that effect. Please communicate this at once to Sazonoff and wire hour of notification.

St PETERSBURG ~ Midnight. Count Pourtalés delivers Berlin's ultimatum, and Sazonoff replies. *This is a technical impossibility, and Germany is overestimating the significance of our Russian mobilization. Is the proposed German mobilization equivalent to war?* Pourtalés: *We find ourselves on the brink of war.* Sazonoff: *I give you the Tsar's assurance on his word of honour that the Russian Army will not move…*
Though, it will continue to mobilize. Count von Pourtalés ~ *Such an assurance is worthless!*
The only thing which might yet prevent war, is a withdrawal of Russia's mobilization order.

France is prepared to enter a wider war because her leaders do not deem that they can take the risk of losing their Franco-Russian Alliance, by either too feebly supporting or too harshly restraining Russia in a venture to which the Russians appear utterly committed. This illuminates the comparative passivity of French diplomacy during the July Crisis, as all of Europe drifts ever nearer to a wider war. It is not that the French wish for a military confrontation with a formidable opponent who is stronger than they, but France does not consider she is strategically free to choose. Thereafter, the heart of French diplomacy becomes completely concerned with consolidating the cohesion between her allies.

France should encourage Russia not to assume the absurd and obsolete attitude of being protectress to all Slav States ~ this will lead to war. French President is convinced a declaration by Britain in support of France will almost certainly prevent Germany from embarking on war. But if we gave assurance of armed resistance to France and Russia now, Russia would become more exacting and France would follow in her wake. Sir F. Bertie

At 11.45 PM on 29th July, news that Russia has fully mobilized is telegraphed from Berlin to Paris. Then, in the small hours of 30th July, the leaders of France are woken from their beds with an urgent request from Ambassador Izvolsky to know their attitude to the recent Russian partial mobilization. Working from inadequate information ; as the Russians have been slow to brief the Quais d'Orsay, and when they have, they transmit their cables via a circuitous Scandinavian route, fearing that if they wire them via Germany, they may be deciphered. When it is revealed that the Russians on 26th July had decided to prepare for a partial mobilization ~ and to then present France with a *fait accompli;* Minister for War Adolphe Messimy asks with integrity ~ *Is Russia's action compatible with our alliance?* It is amid this duplicitous act, that the French discussion for their own mobilization now commences.

The Russian decision to mobilize now compels the French leaders to face their commitments. When the Paris Cabinet meet at 9.30 PM. their main concern with Russia is not to stop the Russian military preparations, but to ensure for the sake of public opinion in France, and the support of Britain that Germany is seen as the first country to take military action. Premier René Viviani, *haunted by the fear that war might erupt by accident, through a black look, a brutal word, or a shot,* takes the astonishing step of ordering a ten-kilometer withdrawal of all French forces along the entire French-German border. Viviani: *France takes a chance never before taken in history. In doing so we have no other reason than to prove to British public opinion and the British Government that France, like Russia, will not fire the first shot. Therefore, we agree for covering troops to take up positions but to approach no closer than ten kilometres to the frontier, to avoid all contact with any German patrols.*

LONDON. On the instructions of Premier Viviani in Paris, French Ambassador Paul Cambon notifies Sir Edward Grey of the comparative facts of recent French and German military preparations: *To show that if France is resolved ~ it is not she who is taking aggressive steps.*

Although Germany has made her covering dispositions a few hundred meters from the frontier, along the whole front from Luxemburg to the Vosges, and has transported her covering troops to their war positions, we have kept our troops ten kilometers from the frontier and forbidden them to approach nearer. By leaving a strip of territory undefended against sudden aggression of the enemy, the Government of the Republic hopes to prove that France does not bear, any more than Russia, the responsibility for any attack.

In order to be convinced of this, it is sufficient to compare the steps taken on the two sides of our frontier. In France soldiers who were on leave were not recalled until we were certain that Germany had done so five days before. In Germany, not only have the garrison troops of Metz been pushed up to the frontier, but they have been reinforced by units transported by train from garrisons of the interior such as Trèves or Cologne; nothing like this has been done in France. The arming of frontier defenses; clearing of trees, placement of field batteries, and the strengthening of wire entanglements was begun in Germany on the 25th ~ with us, it is going to begin now. France can no longer refrain from taking similar measures.

Finally, in Germany the reservists have been recalled and those living abroad have also been recalled, the officers of the reserve have been summoned; in the interior roads are closed, motorcars circulate with permits. It is the last stage before mobilization. None of these measures has been taken in France. The German Army has its outposts on our frontier; on two occasions yesterday German patrols penetrated our territory. The whole Sixteenth Army Corps from Metz, reinforced by part of the Eighth from Trèves and Cologne, occupies the frontier from Metz to Luxemburg; the Fifteenth Army Corps from Strasbourg is massed on the frontier. Under penalty of being shot, the inhabitants of the annexed Alsace-Lorraine are forbidden to cross the frontier.

General Joseph Joffre

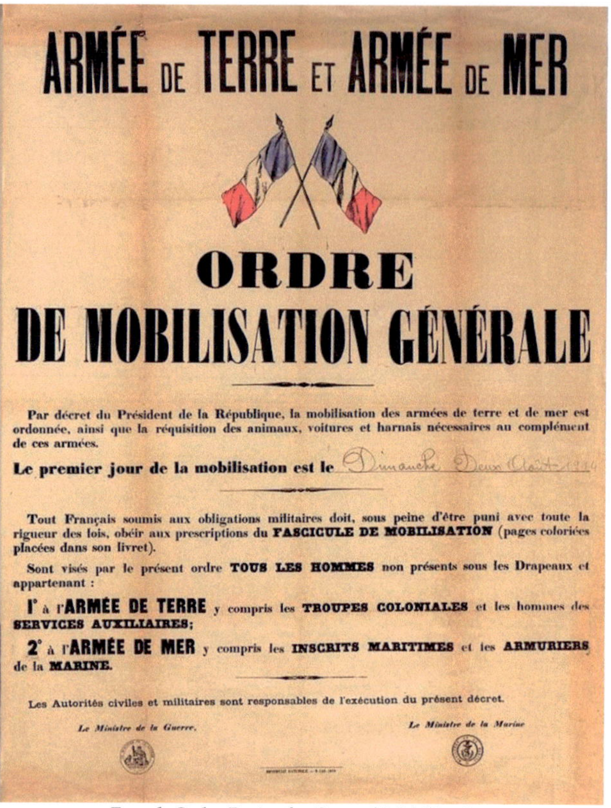

French Order Poster for General Mobilization

While the French Commander in Chief General Joseph Joffre agrees in principal, he still however prefers to seize the initiative and implores Premier Viviani to mobilize. *Every delay of 24 hours before a calling up reservists means a retardation of the concentration of forces~ that is the initial abandonment of ten to twenty kilometers of French territory for every day of delay.* Joseph Joffre refuses to carry this responsibility, and claims Germany is secretly mobilizing. In the afternoon, the Paris Cabinet approve the positioning of covering forces by train, but the reservists are not to be called up. In the evening, the Paris Cabinet again meet to discuss Joffre's demand for an immediate mobilization and agree to wait till 4 PM the next day. This is the latest time at which the announcement can be made with Monday 2nd August being the earliest practicable date for the first day of a French mobilization. As the British urgently need to know if the French will respect Belgium neutrality. Sir Edward Grey telegraphs Ambassador Francis Bertie in Paris to enquire of France respecting Belgian Neutrality. Viviani is called out of the cabinet meeting to see Ambassador Bertie. As the Paris Cabinet is keen to be seen as responding to German action rather than initiating military measures, Sir Francis Bertie cables to Sir Edward Grey :

French Government are resolved to respect the neutrality of Belgium, and it would be only in the event of some other power violating that neutrality that France might find herself under the necessity, in order to assure defense of her own security, to act otherwise. This assurance has been given several times. President of the Republic spoke of it to the King of the Belgians, and the French Minister at Brussels has spontaneously renewed the assurance to the Belgian Minister for Foreign Affairs today.

Viscount Sir Francis Bertie
British Ambassador to France

General Joffre makes an appointment to see War Minister Messimy to bluff him into action. *Germany will be entirely mobilized by 4th August even without an order for mobilization having been issued.* Joffre now threatens to resign, if the French mobilization is not ordered before 4 PM on 1st August. The Paris Cabinet meets with Joffre present where he repeats his argument that the Germans are well advanced with their plans and are secretly mobilizing. Premier Viviani authorizes the full mobilization of French forces to begin the following day. The yellow *DE MOBILISATION GÉNÉRALE* posters with crossed tricolour flags are then prepared, printed, and pasted upon the walls of France. Because of General Joseph Joffre's initiative, France will begin to mobilize one hour before Germany.

At 4 PM this afternoon I was standing on the Place de la Bourse when the mobilization notices were posted. Paris seemed electrified. All cabs were immediately taken. I walked to the Place de l'Opéra and Rue de la Paix to note the effect of the mobilization call upon the people. Crowds of young men with French flags promenaded the streets, shouting Vive La France! Bevies of young sewing girls, collected at the open windows and on the balconies of the Rue de la Paix, cheering and waving their handkerchiefs at the young patriots below, and throwing down on them handfuls of flowers and garlands that had decked the fronts of the shops. The crowd was not particularly noisy or boisterous. No cries of On to Berlin! Or Down with the Germans! were heard.
The shouts that predominated were simply, Vive La France! Vive l'Armée! and Vive l'Angleterre!
British flags were also borne along beside the French tricolour. Charles Inman Barnard ~ *New York Tribune*

Early in the morning of Friday 31st July, the British Ambassador in Paris Viscount Francis Bertie wires to London: *French President Raymond Poincaré has expressed: I believe the preservation of peace is now in the hands of Britain. If Britain announces it will come to the aid of France in any conflict between France and Germany, Germany will modify her attitude, and perhaps there will be no war.* If PM Asquith can only do as Poincaré and Sazonoff urge and declare: *Britain will go to the aid of France in the event of a conflict between France and Germany as a result of the present differences between Austria and Serbia.* (Bertie to Grey) *The German politicians have some hope of holding their militarists in check.* But PM Asquith has only said ~ *I have not made up this mind what to do.* His humiliating confession, is all the more inept for Asquith knows that Britain will have to stand by France and Russia, whether he announces it or not.

Secret talks with France since 1906 (kept secret from most members of the British Cabinet) have often discussed plans for sending an expeditionary force to France if war broke out with Germany. Grey having committed Britain to this course with the consent of the Inner Cabinet renders it meaningless to tell Poincaré, *we should have to decide what to do.* The course of action has been decided long before. Yet Asquith is too timid, too constitutionally incapable of action. Instead he seems to hope a solution can come of its own accord. All thru the last week of July, Grey repeatedly tries in vain to gain Cabinet consent to threaten Germany with British intervention, and yet he knows that if London gives some clear and timely warning that they will stand with France and Russia if war breaks out, whilst it may have had some deterrent effect on Berlin, the Foreign Office are also well aware that it will encourage the Russians to pursue a more intransigent policy, than the one they are following.

German strategic designs on her western neighbour of France are not as yet divulged ~ only the self-denying undertaking that the Wilhelmstrasse are prepared to give in order to secure British neutrality, and this does not extend to France's fine colonies. Yet another sinister proposal lurks within Berlin's secret instructions to Ambassador von Schoen in Paris. On 31st July, von Schoen is instructed to inform the French Premier that Russia's mobilization has compelled Berlin to proclaim an *Imminence-of-War* which must be followed by mobilization ~ unless Tsar Nicholas suspends all Russian military measures against Germany within 12 hours ~ Russia's mobilization will incur war. Ambassador Sir Francis Bertie: *If we gave an assurance of armed resistance to France and Russia now, Russia would become more exacting and France would follow in her wake. It seems incredible that the Russian Government should plunge Europe into war in order to make themselves the protector of the Serbians.*

A communiqué from German Chancellor Bethmann Hollweg on which Ambassador von Schoen in Paris will act is contained in a specially complicated and new secret cipher ~ the German message: *If the French Government declares that it will remain neutral, Your Excellency will be good enough to inform it that, as guarantee of this neutrality, we must insist on the handing over to us of the fortresses of Toul and Verdun, which we shall occupy, and which we shall restore after the completion of the war against Russia.* The German Ambassador is to ask Premier Viviani what the attitude of France will be in the event of war between Russia and Germany. Whether France will remain neutral in a Russo-German war. France's answer is requested in 18 hours. In the conversation that von Schoen has with Premier Viviani in the evening of 31st July, the German begins ~ *After the Russian mobilization and our ultimatum, which will probably not be accepted, our decision now depends mainly on France's statement, as to which I therefore asked for France's attitude. If France decides on a neutral attitude, we should be prepared to do the same; further details can then be discussed.* Viviani: *I have no information of a Russian general mobilization against Germany, and only know of a partial mobilization against Austria-Hungary and of precautionary measures.* Baron von Schoen: *I must point out, that not only the whole Russian Army, but their fleet have also been mobilized, a clear proof that the measures are also directed against Germany.* Viviani makes no reply…

French Caudron G3

Viviani's ignorance of the Russian general mobilization seems remarkable, in view of the fact that it had been publicly proclaimed in the early morning of that day in St. Petersburg, had been ordered the evening before, and undoubtedly decided on earlier. In view off the close relationship of alliance, and of the circumstance that the Premier had just returned from St. Petersburg, and therefore ought not to have been in ignorance of what was intended; it must either be assumed that he knew all about it, or that the Russian Government had misled him, in the same way as it has tried to mislead our representatives in St. Petersburg. Baron Wilhelm von Schoen

That night, Premier René Viviani telegraphs St. Petersburg advising that he proposes to respond to Berlin's enquiry regarding her neutrality by answering ~ *France will be guided by her own interests, and that France does not owe any account of her actions to anyone, but her ally.*

When Baron von Schoen calls the next day for an answer (before the expiration of the 18 hours), Viviani will not be induced to give any further explanation. *France will do whatever her interests demand. I have reason to believe the general situation has changed and is considerably easier. Russia has accepted in principle a fresh proposal made by Sir Edward Grey as to the suspension of military preparations all round, and Austria-Hungary has again stated that she will not encroach on the Serbian territorial status and Serbian sovereignty. This revives the hope of a peaceable settlement, which has been abandoned.* Baron von Schoen: *France's attitude is not doubtful.* After their meeting, the French President Poincaré addresses his staff... *That, was the reward that was to be offered to us in the event of our repudiating our alliance with Russia. It is significant of the psychology of the directors of German policy that they could imagine that a proud country, even if tempted to desert an ally, could entertain a demand to hand over her fortresses as a pledge of her good faith, and to place herself at the mercy of the Power which was engaged in the meantime in crushing that ally.*

British Ambassador to France, Viscount Francis Bertie telegraphs London: *The French Government, whose wishes are markedly pacific, sincerely desire the preservation of peace and do not quite despair, even now, of its being possible to avoid war. French orders for general mobilization were given at 3.40 PM in answer to the German Kriegsgefahrzustand (Imminence-of-War) which by calling troops up to war strength was tantamount to mobilization. Minister of War Messigny explained that this act of mobilization purely for defensive purposes.*

VIENNA ~ Early morning. Russian Ambassador Nikolai Shebeko arrives at the Ballplatz to see Count Leopold von Berchtold, and to announce: *Russian military measures bare no hostile character. Austria-Hungary must not solve the conflict with Serbia without consulting Russia.* Conrad, more focused on the greater danger of Russia than Serbia assures von Moltke: *A-H will now employ the main weight of our strength in the north towards Russia, despite our difficulties in moving troops from the Serbian frontier.*

BELGRADE. British *Chargé d'Affaires* Dayrell Crackenthorpe: *Serbian plan of campaign is to draw into interior as large a portion as possible of Austrian Army to weaken Austria elsewhere. Considerable panic in Belgrade, in consequence of which refugees are crowding into British Legation to seek protection of our flag.*

ROME. King Vittorio Emanuele III of Italy is in a difficult position. While he is alliance bound to support Vienna and Berlin, he is married to Elena, the fourth daughter of King Nikola of Montenegro, whose interests are quite the opposite. Equally, the King of Italy is not enthused to succumb to a quarrel with the British Empire, which has always been a special friend to Italy. Thus in due course, Ambassador Barrère reports the welcome news of the Italian decision, to take no part in the war.

I went to see the Marquis di San Giuliano this morning, in order to get precise information from him as to the attitude of Italy in view of the provocative acts of Germany, and the results which they may have. The Minister of Foreign Affairs answered that he had seen the German Ambassador yesterday evening. Herr von Flotow had said to him that Germany had requested the Russian Government to suspend mobilization, and the French Government to inform them of their intentions; Germany had given France a time limit of 18 hours and Russia a time limit of 12 hours. Herr von Flotow as a result of these communiqués asked what were the intentions of the Italian Government. The Marquis di San Giuliano answered that as the war undertaken by Austria was aggressive and did not fall within the purely defensive character of the Triple Alliance, particularly in view of the consequences which might result from it according to the declaration of the German Ambassador, Italy could not take part in the war. French Ambassador to Italy, Camille Barrère

BERLIN 2.05 PM. The Kaiser receives a telegram from the Tsar.

Dear Willy, I received your telegram. I understand why you are obliged to mobilize, but I wish to have the same guarantee from you as that I gave you; that these measures do not mean war, and that we shall continue negotiating for the benefit of our countries and universal peace dear to all our hearts. Our long proved friendship must succeed, with God's help, in avoiding bloodshed. Anxiously, full of confidence await your answer. ~ Nicky

Dear Nicky,

Thank you for your telegram. I yesterday pointed out to your government the way by which war may be avoided. Although I requested an answer for noon today, no telegram from my ambassador conveying an answer from your government has reached me yet. I therefore have been obliged to mobilize my army. Immediate affirmative clear and unmistakable answer from your government is the only way to avoid endless misery. As Germany has not yet received a reply to the noon deadline demand that Russia stops mobilizing ~ I cannot discuss this telegram. I must request you immediately order your troops on no account to commit the slightest act of trespassing over our frontiers. ~ Willy

LONDON. Once again, von Jagow refuses to verify the German standpoint on Belgian neutrality. *If I did so, it could not fail in the event of war, to have the undesirable effect of disclosing to a certain extent, part of the German plan of campaign.* Belgian Foreign Minister Julien Davignon cables Sir Edward Grey. *Belgium will to the utmost of her power maintain neutrality. That the relations between Belgium and the neighboring Powers are excellent ~ no reason to suspect their intentions, but that Belgian Government believed that in case of violation they are in a position to defend the neutrality of their country.* Sir Edward Grey then advises Prince Max Lichnowsky. *The German position on Belgian neutrality is a matter of very great regret. The neutrality of Belgium so affects public opinion in Britain that if Germany cannot pledge not to violate the neutrality of Belgium, Britain cannot promise British neutrality.*

The French outlook darkens considerably after Sir Edward informs Ambassador Paul Cambon not to expect help from Britain. *The cabinet has agreed we could not propose to Parliament at this moment to send an expeditionary force to the continent. This does not mean under no circumstances will Britain assist France, but it does mean that France must take her own decision without reckoning on assistance from Britain.* Ambassador Paul Cambon reminds Sir Edward: *The French fleet is concentrated in the Mediterranean as a result of understandings with Britain, and the French Channel and Atlantic coasts are now undefended.* Cambon shocked and distressed, staggers from the interview muttering ~ *They are going to abandon us!*

Ambassador Paul Cambon sends the following report to Paris on the developing situation:

Sir Edward Grey will ask the Cabinet to authorize him to state on Monday in the House of Commons that the British Government will not permit a violation of Belgian neutrality. In the second place, the British fleet is mobilized, and Sir Edward Grey will propose to his colleagues that he should state that it will oppose the passage of the Straits of Dover by the German fleet, or if the German fleet should pass through, will oppose any demonstration on the French coasts. These two questions will be dealt with at the meeting on Monday. I drew the attention of the Secretary of State to the point that, if during this intervening period any incident took place, it was necessary not to allow a surprise, and that it would be desirable to think of intervening in time.

BERLIN. By 4 PM, Russia has yet to reply to the German ultimatum that expired at noon. Therefore, the German declaration of war is prepared, and Minister of War Erich von Falkenhayn asks Bethmann to go with him to see the Kaiser, and to ask him to promulgate the order for mobilization. Bethmann: *If the iron dice must fall, may God help us.* At 5.00 PM, the Kaiser is seated at a desk made from the wood of Nelson's flagship HMS *Victory* in the Star Room of his *Neuen Palais* to sign the order that will start his armies rolling into Luxembourg and Belgium. After Kaiser Wilhelm signs the paper, Erich Falkenhayn declares ~ *God bless Your Majesty and your arms, God protect the beloved Fatherland.* Then the declaration of war against Russia is signed. *His Majesty the Kaiser, my August Sovereign, accepts the challenge in the name of the Empire, and considers himself as being in a state of war with Russia.* This declaration of war is then dispatched by telegraph to the German Embassy in St Petersburg to be provisionally presented at 7 PM to Serge Sazonoff. As the Kaiser rises, he looks into the faces of his naval and military chiefs all standing respectfully around his desk ~ then in a harsh low voice...
Gentlemen, You will all live to regret this.

To further reveal the role of Helmuth von Moltke, U.S. Ambassador James Gerard recorded :

I do not believe that Wilhelm II deep down in his heart, craved for war. I can vouch for the literal accuracy of a hitherto unrecorded piece of ante-bellum history which bears out the Kaiser's immediate responsibility for the war, while it does not acquit him of supine acquiescence in, and to that extent abetting the War Party's plot. On the afternoon of Saturday, 1st August 1914, the wife of General Helmuth von Moltke, paid a visit to a certain home in Berlin, which shall remain nameless. The Frau was in a state of obvious mental excitement ~ Ach, what a day I've been through, Kinder! My husband came home just before I left. Dog-tired, he threw himself on to the couch, a total wreck, explaining to me that he had finally accomplished the three days' hardest work he had ever done in his whole life ~ he has helped to induce the Kaiser to sign the mobilization order!

There is the evidence, disclosed in the homeliest most direct, fashion, of the German War Party's inescapable culpability for the supreme crime against humanity. The sword had indeed, been forced into the Kaiser's hand. This is no brief for the Kaiser's innocence. No man did more than Wilhelm II himself, during twenty-six years of explosive reign, to stimulate the military clique in the belief that when the dread hour came the Supreme War Lord would be with his Army. Yet German officers, in those occasional moments when conviviality bred loquacity, were fond of averring, as more than one of them has averred to me : The Kaiser lacks the moral courage to sign a mobilization order. Die Post, a War Party organ, said as much during the Morocco imbroglio in 1911. Perhaps that is why General von Moltke had to force the pen, which for the nonce was mightier than the sword, into the reluctant hand of Wilhelm II.

The Kaiser was constitutionally addicted to swaggering war talk, but, in my judgment, he preferred the bark to the bite. Raised in the belief that the Hohenzollerns were divinely appointed to their Royal estates, Wilhelm II does not want to lose his throne. The Kaiser's personal interests, all lay along the path of peace. However, his concessions to the War Party were generous and frequent. He rattled the saber on its demand. He donned his shining armour at Austria's side when the Germanic Powers coerced Russia into recognition of the Bosnian annexation. He sent the Panther to Agadir harbor, because the War Party howled for deeds in Morocco. It hoped that history in N.W. Africa would repeat itself ~ that the Triple Éntente would yield to German bluff as it had yielded in S.E. Europe two years previous. It did not, and it was then that the German War Party swore a solemn vow of ~ Never Again! The days of the Kaiser who merely threatened war were numbered. The next time the sword was forced into his hand, he said : Before God and history my conscience is clear ; I did not will war. Herewith I place Frau von Moltke on the stand as chief witness in the Kaiser's defense. James W. Gerard

BELGRADE Sir Dayrell Crackenthorpe: *Shots are being fired over Belgrade... there is general panic, inhabitants hiding in cellars during bombardment. Austrian gunners are now bombarding Belgrade by sections, no serious attempts being made to occupy town. Ciphers and highly secret archives have been destroyed by fire.*

It was obvious that Russia, who had refused two days before to cease from her military preparations, would not accept the German ultimatum. As, no answer had come from St. Petersburg by the afternoon of 1st August, Herren von Jagow and Zimmermann rushed to see the Chancellor and the Kaiser, to request that the decree for a general mobilization might at least be held over until the following day. They urged that the telegraphic communication with St. Petersburg had presumably been cut, and that this would explain the silence of the Tsar. Their efforts made no headway against the opposition of the War Minister and the army chiefs, who represented to the Kaiser the dangers of 24 hours delay. The order for mobilization was then signed at 0700, on 1st August. At 0710 in the evening, the German declaration of war was forwarded to Russia. Baron Beyens

St PETERSBURG 7 PM. Ambassador Count Friedrich von Pourtalès solicits Serge Sazonoff. *Is the Russian Government disposed to give a favourable answer to the ultimatum presented by Germany yesterday ~ and timed to expire at noon today?* Sazonoff explains with some emotion his efforts for peace. *Russia wants to continue talks. No suggestion held out to me has been refused.* Count Pourtalès now in tears asks Serge Sazonoff twice more, but to no avail : *Russia cannot demobilize, but we are, as before, prepared to continue negotiations for a peaceful settlement.* Count Pourtalés fumbles in his pocket and breathing hard, draws out the German Government's declaration of war, and delivers his fateful words.

In that case, I am instructed to give you this note. Russia not having considered it necessary to answer this demand, and having shown by this attitude that her action was directed against Germany, I have the honour, on the instructions of my Government, to inform your Excellency as follows : His Majesty the Kaiser in the name of the German Empire, takes up the defiance, and considers himself in a state of war with Russia. Sazonoff: *This is a criminal act of yours. The curses of the nations will be upon you.* Then, losing all control of himself, Pourtalés walks to the window that looks across to the Winter Palace reddened by the evening sun, and turning his back on Sazonoff, bursts into tears. Sazonoff pats his shoulder, and the distressed Pourtalés cries ~ *Never did I think that I would have to leave St Petersburg under such conditions.* When he has finished reading, Sazonoff repeats. *This is a criminal act! You could have prevented the war by one word. You didn't want to. In all of my efforts to save the peace I haven't had the slightest help from you. But there's a divine justice!* Pourtalès with a look of desperation mutters a few incomprehensible words and staggers towards the grand window nearest the Winter Palace. There he leans, bursts into tears and stammers. *So, this is the result of my mission!* Finally, he rushes to the door, which he can hardly open with his trembling fingers, and walks out murmuring ~ *Goodbye! Goodbye!*

BERLIN 7 PM. An inaccurate report is received from an exhausted Prince Max Lichnowsky in London suggesting that Sir Edward Grey is now proposing that if Germany does not attack France, Great Britain will remain neutral in a Russian-German war and guarantee the neutrality of France. *The British Government is asking whether we would pledge not to enter French territory if England guaranteed France's neutrality in our conflict with Russia.* Kaiser Wilhelm, who still hopes for peace is euphoric and wires his telegram of acceptance at 7.20 PM. *We agree that if France and England remain neutral Germany will not countermand the mobilization on the French border, but will not enter French territory.*

Wilhelm grasps this hope of a limited one-front war and sends for von Moltke, who has just put the mobilization order into effect. As German troop trains steam toward Belgium, a car is sent to find and bring the perplexed chief of staff to the Kaiser's Palace. The Kaiser explains the situation to him… *Now we can go to war against Russia only. We simply march the whole of our army to the East.*
The exhausted von Moltke is stunned ~ *This is impossible!* ~ and suffers a coronary.

Kaiser Wilhelm orders his A.D.C. Prince Hans of Pless to telephone German Army HQ at Trier, near the Luxembourg border, where German troops are scheduled to cross the frontier. The Kaiser's halt order does not arrive in time, and one infantry company has crossed the border into Luxembourg and entered the town of Trois Vierges. Having just seized the railway station and telegraph office, more troops arrive to counter-order the invasion as a mistake ~ *A British peace proposal has arrived in Berlin. The invasion is therefore now halted.* When the Kaiser and von Moltke finally agree that the British proposal should be swiftly accepted, the Kaiser sends a personal message to his uncle, King George V, supporting the new British proposal. But when Wilhelm's message arrives at Buckingham Palace, King George V is perplexed. *There must be a misunderstanding. I have just received a telegram from the Kaiser saying that Germany agrees with the British proposal that Great Britain and France remain neutral in a war between Germany and Russia.* Grey is asked to present himself at Buckingham Palace, to explain the Kaiser's telegram and draft a reply. Grey advises there is no such proposal and drafts a reply to Berlin.

LONDON 10 PM. Sir Edward Grey, in a message to French Ambassador Cambon. *You should know that we have done a very important thing. We shall not allow the German Fleet to come into the Channel.* Cambon realizes the significance of this pledge. *The British Government promises to protect our coasts. That suffices to involve England against Germany. The English Fleets are mobilized.* Great Britain, whilst the leading sea power has neither conscription nor a large standing army ~ only an expeditionary force of a mere 160,000. This constitutes however, *a rapier among the scythes* as of all the European armies, Britain's alone has seen repeated action in myriad colonial campaigns. Sir Maurice de Bunsen reports from Vienna ~ *A general mobilization of all Austro-Hungarian forces has just been ordered.*

BERLIN 11 PM. King George's formal denial of an offer of British neutrality arrives in Berlin. *There must have been some misunderstanding as to a suggestion that passed in friendly conversation between Prince Lichnowsky and Sir Edward Grey this afternoon when they were discussing how actual fighting between German and French armies might be avoided while there is still chance of some agreement between Austria and Russia.* The deflated Kaiser, now wearing a military cloak over his nightshirt recalls von Moltke to growl at him ~ *Now you can do what you like!* Wilhelm's wife, the Kaiserin Augusta wrathfully adds; *Now there's nothing for it but war, and my husband and our six sons will be in the thick of it.*

When the Kaiser, after 25 years of his reign had passed, abruptly left the straight and tranquil path that he had marked out for the happiness of his subjects, his consort, we may be quite sure, made no effort to hold him back. In spite of her femininity, German patriotism, with its dreams of domination continually haunts her brain. The horrors of war, that bane to mothers bella matrihus detestata ~ do not dismay the wife of Wilhelm II. During the crisis of Agadir, when the Court of Berlin was chafing with impatience to measure its strength with France on another field than that of diplomacy, the Kaiserin shared the impulse she felt throbbing in the air around her. In a tone of reproach, she said to Herr von Kiderlen-Wächter, whom she disliked ~
Are we always going to retreat before the French and put up with their insolence? Baron Beyens

With all hope of peace now gone, von Moltke promptly contacts the 16th Division to reorder their swift attack in the west to secure the Duchy of Luxembourg ~ the railways of Luxembourg being essential to the German *Plan 1914* timetable, since they track through Belgium and into France. French Ambassador Jules Cambon, tells British Ambassador Sir Edward Goschen that there are perhaps just three people now in Berlin who regret that the war has started ~ *You, me, and Kaiser Wilhelm.*

Our entire diplomatic machine failed. The menace of war was not seen because the Foreign Office was so hypnotized with its idea of ~ surtout pas d'histoires ~ above all no histories ~ that its belief in peace at any cost, had indeed completely eliminated war as a possible instrument of Éntente statesmanship from its calculations, and therefore, did not rightly estimate the importance of the signs of war. The standpoint of the Foreign Office brought it to a certain extent into conflict with the General Staff and the Admiralty Staff, who uttered warnings, as was their duty, and wished to make preparations for defense. This conflict in views showed its effect for a long time. The German Army could not forget that, by the fault of the Foreign Office, it had been taken by surprise, and the diplomats were piqued because, in spite of their stratagems, war had ensued after all. Wilhelm II

St PETERSBURG ~ Peterhof Palace 10 PM. British Ambassador Sir George Buchanan delivers King George V's urgent message to Tsar Nicholas, which even if it could have had some influence, has come too late. Nicholas II then telegrams his reply to George V: *In this solemn hour, I wish to assure you once more that I have done all in my power to avert war. Now that it has been forced upon me, I trust your country will not fail to support France and Russia. God Bless and protect you.* The Tsar subsequently sent : *I was compelled to mobilize in consequence of A-H's complete mobilization.* This is deceptive and false for A-H has as yet, only mobilized against Serbia ~ and then only after Russia itself fully mobilized. Maurice Paléologue: *I am obliged to report that at the present moment, the Russian Empire is run by lunatics.*

Whilst Russia remains responsible for their pre-emptive order for a general military mobilization, the Tsar has sent a telegram to Kaiser Wilhelm explaining that the mobilization was aimed at A-H and not at Germany, and ultimately it is the German leadership that decides to declare war on Russia on 1st August 1914. News of the German declaration of war on Russia spreads quickly following the meeting between Count Pourtalès and Serge Sazonoff. Vast crowds soon began to gather on the Nevsky Prospekt. Women threw their jewels into collection containers to support the families of the reservists who have been called up. By 11 PM that night around 50,000 people surrounded the British embassy cheering, *God save the King, Rule Britannia, and Bozhe Tsara Khranie ~ God save the Tsar.*

1tes Extra-Blatt

der

„Kriegs-Verlag"

Aktien-Gesellschaft

Berlin W9. Potsdamer Platz 8.

Allgemeine Mobilmachung.

Berlin, den 1. August 1914, 6 Uhr abends.

Der Kaiser hat die Mobilmachung angeordnet.
Der Wortlaut des Kaiserlichen Erlasses im Reichsanzeiger heißt:
„Ich bestimme hiermit: Das Deutsche Heer und die Kaiserliche Marine sind nach Maßgabe des Mobilmachungsplanes für das deutsche Heer und die Kaiserliche Marine kriegsbereit aufzustellen. Der 2. August 1914 wird als erster Mobilmachungstag festgesetzt".

Berlin, den 1. August 1914.

Wilhelm I. R.
v. Bethmann Hollweg.

München, 1. August, 7 Uhr 20 Minuten abends.

Der König hat die Mobilmachung der bayrischen Armee befohlen.

Druck von Ernst Litfaß' Erben, Berlin, Adlerstr. 6.

Mobilization Proclamation of Imperial Germany

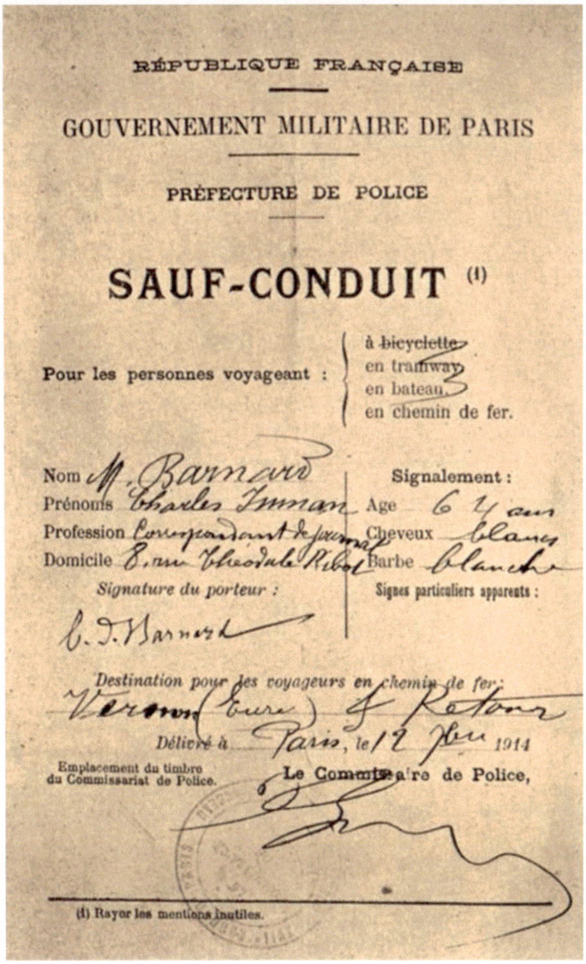

RÉPUBLIQUE FRANÇAISE

GOUVERNEMENT MILITAIRE DE PARIS

PRÉFECTURE DE POLICE

SAUF-CONDUIT (1)

Pour les personnes voyageant : ~~à bicyclette~~ / ~~en tramway~~ / ~~en bateau~~ / en chemin de fer.

Nom M. Barnard
Prénoms Charles Inman
Profession Correspondant de journal
Domicile 8 rue Théodule Ribot
Signature du porteur :
C. I. Barnard

Signalement :
Age 64 ans
Cheveux blancs
Barbe blanche
Signes particuliers apparents :

Destination pour les voyageurs en chemin de fer : Vernon (Eure) & Retour

Délivré à Paris, le 12 Xbre 1914

Emplacement du timbre du Commissariat de Police.

Le Commissaire de Police,

(1) Rayer les mentions inutiles.

SAUF-CONDUIT issued by France to all departing foreigners

Austro-Hungarian armoured train…

When all the signs that heralded the storm are remembered ~ in Russian military circles, especially among the immediate surroundings of the Grand Duke Nicholas who had long been the leader of the extreme War Party that clamoured for war with Germany. The War Party caused the Russian Government, during the Balkan wars, to support by its influence and its arms the Bulgarians and Serbs against Turkey. They were at the bottom of the strong manifestations that took place from 1912-1913 in Petersburg against Austria, the hereditary-enemy of the Slavs in Turkey and throughout the Balkans. At that time however, the head of the Russian Cabinet was still Count Vladimir Kokovtsov. With all his faults and inexperience of diplomacy, Kokovtsov possessed sufficient common sense, and knowledge of the resources of his country, to apply his energy to warding off such a calamity as war. The consequences he realized, if others did not, would be far more than could be foreseen or expected. He was seconded by Foreign Minister Sazonoff, who shared his opinion on that point, and to these two is owed the signing of the 1913 Bucharest Treaty, after which the world thought it could breathe freely again.

But, in 1914 things were different. President of the Council Kokovtsov, had been replaced by Goremykin; full of the best intentions, but an old man of 76 was not strong enough to show independence of character in presence of people like the Grand Duke Nicholas. The desire of the Tsar to preserve peace was so strong that he insisted on carrying patience to its utmost limits. A friend of mine, the correspondent of a French newspaper, who that day had talked with Goremykin, related to me that during a Cabinet consultation that same morning, the Tsar had said ~ For seven years I have endured the arrogance of Austria, but my patience is now exhausted!

It was only among the working class that loud murmurings were noticed. Indeed, some workmen in the composing room where the paper on the staff of which I was working, the St. Petersburg Courier was printed, declared to me that if they were asked to join their regiments they would never shoot at the enemy, but would discharge their rifles in the air. I hasten to add that this feeling existed only among a small minority ; but still it was there, and in responding to the order of mobilization, it is a symptom which deserves to be noted.

The Russian Government, far from exciting the crowds begged the Press to preserve its calmness and not to rouse the feelings of the mob. The Tsar, determined to do nothing which might compromise the dignity of Russia, made every possible personal effort to persuade Kaiser Wilhelm II to listen to reason, and was often heard to say; I cannot conceive how the Kaiser can believe that Russia really nurses aggressive designs against her neighbour. The whole controversy concerns only Austro-Russian relations, and has absolutely nothing to do with Prussia. The prevailing thought was that Germany had let herself unexpectedly and foolishly to be drawn into the vortex.

The text of the German ultimatum only became known to the general public on 1st August, the Government having purposely refrained from communicating it to the Press until just before the declaration of war. The news that war had been declared was at once communicated to the Press by the Russian Government, and at 8 o'clock big placards to this effect appeared in the windows of the offices of the Novoie Vremia on the Nevski Prospekt. Immense crowds assembled also in front of the Kazan Cathedral, where they requested the clergy to hold a service in the open air, in order to invoke the help of God for the triumph of Russian arms. The attitude of the public was extremely dignified and solemn. The soul of the nation had been moved. This was not the time to shout or hurrah, rather it was the hour to pray and to act. The streets were thronged by people exchanging impressions on the great event; but when the first troops for the front appeared on the Nevski no one cheered. Those who saw them pass uncovered themselves and made the sign of the Cross in silence. I spent the evening at the office of the Petersburg Courier, and returned home through the crowded streets at about midnight, feeling quite crushed under the weight of the terrible catastrophe which had fallen on civilization, and the consequences of which were to strike so deeply into my own heart by robbing me of the one whom I cherished most on earth.

Not many people slept that night in Petersburg. Scarcely one family had a son, brother, husband, father, relative or friend not starting for the front. This war had come so unexpectedly that no one was prepared for it, which made the blow harder to bear. In the first hours following upon the declaration of war, most were inclined to look at things through black spectacles. Consternation was general, but there was considerable dignity in the whole attitude of Russia on that eventful second day of August 1914. The mobilization went on quietly with the men going to their barracks with their wives or mothers. The streets of the capital presented a strange spectacle, with regiments continually parading from the Winter Palace to the Moscow railway station, singing hymns and patriotic anthems. No panic was observed anywhere. Even the workmen who, a few days before, had been on strike declared that they would stand by the government until the enemy was beaten ~ a resolution to which unfortunately, they did not adhere, as events unfolded, but which at the moment was undoubtedly sincere.

At eleven o'clock the Emperor came to Petersburg from Peterhof, accompanied by the Empress and their four daughters, and they proceeded at once to the Chapel of the Winter Palace, where the Metropolitan Vladimir celebrated a religious service to invoke blessings upon the Russian arms, after which a deacon read out in a loud voice the Imperial Manifesto announcing to the nation that Germany had attacked us. The church was full to overflowing with officers of high rank and representatives of all the Guards' Regiments stationed in the capital. In the hope of seeing the Tsar appear on the historical balcony from which all matters of importance that take place in the Imperial family are proclaimed to the population, an enormous crowd had assembled on the square and began singing prayers intermingled with national hymns. Flags were unfurled and icons were carried by women and priests. When the guns of the fortress announced that the Manifesto had been read, all crossed reverently, as Russians generally do on all grave occasions. Blessings, too, were invoked on the head of the Emperor, and wishes uttered that he might come out victorious from this new war.

Inside the palace, the enthusiasm was greater. Nicholas II spoke a few words that rang with an emotion he tried vainly to subdue. His speech ended with the declaration that, since he had been obliged to draw the sword, he would not sheathe it until the last enemy had been driven out of his dominions. At two o'clock, the great doors leading to the balcony facing the square in front of the Winter Palace were thrown open. The Tsar appeared, and behind him stood the Empress, with the young Grand Duchesses dressed in white. Nicholas II saluted his people and acknowledged their frenzied hurrahs. He spoke, but no one could quite catch his words ~ he then raised his hand in a gesture of benediction, whereon the multitude before him fell on its knees. The whole scene had a grandeur and gravity about it which made it very solemn and impressive, and the Monarch and his people were entirely in unison, and that, after so many years of estrangement, Russia and the House of Romanoff had at last become reconciled, and determined to work together in order to resist the enemy.

Everybody thought or believed that the campaign would be finished in a few weeks. In a wave of patriotism passing over the country, the different parties seemed united in a common effort against the common foe. Russia showed a most dignified attitude, and it is very much to be regretted that the government and the Press did not emulate it, but seemed to excite passions better left dormant. The first discordant note came from Novoie Vremia, the most Conservative and pro-government paper in Russia. It imagined that it was contributing its share to the public good by imploring its readers to hunt down the Germans. It is perfectly true that the various tales which reached us concerning the attitude of the populace of Prussia toward our compatriots travelling abroad when the war broke out were appalling. Everywhere Russians had been treated in the most shameful way, and insulted. Though this conduct called for reprisals, it was wrong to make it a pretext for encouraging acts of brigandage. All solemnity and dignity of the war, was compromised by hysterical Press outbursts, and it is regretted that the government, instead of stopping these rabid attacks, thought it right and proper to encourage them.

The first example of this anti-German feeling was the sacking of the German Embassy. One of the reporters of the Novoie Vremia appeared at the window and read aloud the official bulletin ~ The Dowager Empress has been stopped at Berlin on her return to Russia, and obliged to change her route. Her train has been diverted to the Danish frontier by orders of the German Kaiser. This news excited indescribable fury. One member of the staff of the Novoie Vremia shouted ~ Let us all go to the Embassy, and show these people that we are not going to allow the mother of our Sovereign to be insulted. The crowd collected on the Nevski Prospekt and soon one long cry was heard all along the street; The Embassy! The Embassy! The mob surged in the direction of Moika Street, and began to plunder the building. Being informed that an attack on the German Embassy by a riotous crowd was in progress, I went with a couple of friends, to see what was going on. The sight that met our eyes was not one to be proud of. A crowd had collected in front of the Embassy; a howling mob of such sinister figures as usually appear in large towns whenever rioting is in the air. An officer of police and a couple of policemen were calmly looking on while pieces of furniture, crockery and glassware, were flying in the air and coming down on the pavement with a crash, which was greeted with howls of delight by the crowd. Within half an hour, every window was broken, and everything of value in the Embassy had been thrown out and burnt in the street. The infuriated crowd searched every comer, and finally murdered an official who had not had time to make his escape. Having satisfied its lust of destruction, the crowd moved on in the direction of the Austrian Embassy. But there, Austria strangely not having yet declared war, and the Ambassador and his Staff still being present, necessary precautions for their protection had been taken and the crowd found all approaches to the Embassy building barred by troops. In deliberately allowing the dregs of the populace to have their way, the Government were already beginning to let the reins of power slip through their fingers. This disgraceful episode, although of no importance at the time, proved indeed, a mild foretaste of what was to come two years and a half later.

Princess Catherine Radziwill
St Petersburg Courier

SUNDAY 2nd AUGUST 1914 ~ BELGIUM MOBILIZES

Albert, King of the Belgians

BRUSSELS. *Last night when I went home, at about midnight, I found the police going about with orders for mobilization, ringing the doorbells, and summoning the men to the colours. There was no time to tarry, but each man tumbled out of bed into his clothes and hurried away to his regiment. Two of my neighbours were routed out a little after midnight, and got away within the hour.* Hugh Gibson
Secrétaire de la Légation U.S.A.

I was awakened suddenly out of a sound sleep this morning by loud, insistent knocks at my door. I got up, it was six o'clock, I opened the door and there stood Omer, in uniform ~ the rough blue tunic, linen pantaloons, and bonnet de Police. He stood at attention, his hand at the salute: C'est la guerre, Excellence. The words of course, were superfluous. Omer standing there, ready to depart, was somehow the symbol of the thing we had feared for a week. He was in a hurry; he had to get to town, report, and go to Liége at once. I fumbled through my clothes and gave him all the money I had, while he told me the latest news; the Germans had entered Luxembourg and were throwing down bridges. I dressed, breakfasted, and gave the order to move back to town. We drove in, the motor piled with bags, with a little silk flag fluttering from the roof of the car. We passed some mounted troops in the Avenue Tervueren ~ mobilization well under way. Germany has declared war on Russia, and Luxembourg has been invaded ~ these are the features of today's news. The whole world is now mobilizing. We are very tired tonight. U.S. Ambassador Brand Whitlock

LONDON. *Daisy,* Princess of Pless: *A telegram from Pless Castle came to Saville Row at 8.30 to say that our motorcar with Schulz the chauffer, was to return to Germany. I was surprised that no message came for me. I was in England for only a month, and the day I left they all thought I should never get over the Dutch border. This horrible situation has arisen so terribly suddenly. I went straight to the shipping office to get my passage to Germany on a Hamburg-Amerika liner. I also went to the German Embassy. Poor Prince Lichnowsky saw me. He was terribly distressed and pale and I had to hide a smile when he said to me with his two hands stretched out ~ Just look how things are after all I have done ~ Meaning of course, that he thought England and Germany had become friends. Poor man, he really thinks he can stop a European war because they like him in England, him, and his dear wife. Lichnowsky then gave me of course a laisser-passer and I left on Sunday morning 2nd August. We did not know whether the train would be allowed to pass the Dutch frontier, but the sleeping-car portion did so and we arrived in Berlin, six hours late. When I arrived in Berlin I told my maids to stay on the platform, while I went into town to find Hans…*

To think that we should bring such sorrows to an innocent, happy people! I have always hated and loathed our Kaiser ~ have I not said so a thousand times? His friends are all brutes. Princess Mechtilde Lichnowsky

General von Moltke before some of his countrymen has said ~ We must throw overboard, all the stock commonplaces about the responsibility of the aggressor. As soon as there is a ten-to-one chance in favour of war, we must forestall our opponent, commence hostilities without more ado, and mercilessly crush all resistance. It was not merely a rapid assault that the General recommended, but a surprise attack before the declaration of war, as if in a duel, one were to strike one's opponent before he had had time to assume a posture of defence. The sudden violation of Belgian neutrality, after our Government had been allowed one night to think matters over, was one of these murderous blows approved by the German Chief of the General Staff. Baron Beyens

BERLIN 2.30 AM. German troops enter Luxembourg as a preliminary move to invading Belgium. Although France and Belgium have mobilized, neither they nor Britain are yet involved in the conflict. Bethmann is meeting with Moltke, Tirpitz, and Jagow ~ when War Minister Falkenhayn advocates : *The war is here and the actual question of a declaration of war on France is a matter of indifference.* The militarists ardently oppose a declaration of war on France, and want to invade. Bethmann: *It is a legal requirement, and Berlin's ultimatum to Belgium makes no sense ~ unless Germany is at war with France.* Moltke: *I do not consider it necessary yet to deliver the declaration of war to France ~ rather I am counting on the likelihood that, if it is held back for the present, France on her part, will be forced by public opinion to organize warlike measures against Germany.* Naval Minister Alfred von Tirpitz is astonished to learn that the old Schlieffen inspired *Plan 1914* is their army's only option. The Russian Army then launches a surprise dawn attack on Germany ~ *Russian troops have crossed the frontier at Schwidden S.E. of Bialla.*

Sir Edward Goschen to Sir Edward Grey *D. 10.30 AM R. 11.00 AM*
Secretary of State for Foreign Affairs Herr von Jagow has just informed me that, owing to Russian troops having crossed frontier, Germany and Russia are now as in a state of war.

Sir Francis Villiers to Sir Edward Grey *D. 10.50 AM R. 11.45 AM*
Belgian Government have official confirmation German forces have entered Grand Duchy of Luxemburg.

Julien Davignon: *When I met Herr von Below, he thanked me for this attention, and added that up to the present he had not been instructed to make us an official communication, but that we knew his personal opinion as to the feelings of security, which we had the right to entertain toward our eastern neighbours. I at once replied that all that we knew of their intentions, as indicated in numerous previous conversations, did not allow us to doubt their perfect correctness toward Belgium. I added, however, that we should attach the greatest importance to the possession of a formal declaration, which the Belgian nation would hear of with joy and gratitude.*

BERLIN 2nd *August, I called in the morning to say goodbye to the Russian Ambassador. His Embassy was filled with unfortunate Russians who had gone there to seek protection and help. Right and left, men and women were weeping and the whole atmosphere seemed that of despair. I sent the Russian Ambassador my automobile to take him to the station. The chauffeur and footman reported to me that the police protection was inadequate, that the automobile was nearly overturned by the crowd, and that men jumped on the running board and struck the Ambassador and the ladies with him in the face with sticks. The Russian Ambassador reciprocated by confiding to me a Russian Princess who had intended to go out with him but intimidated, perhaps by the scenes on the way to the station, had lost her nerve at the railway station and refused to depart with the others. She remained a while in Berlin, and after a week, recovered sufficient courage to make the trip to Denmark.* James W. Gerard

PARIS. The French Minister for Foreign Affairs dutifully notifies all foreign diplomatic staff.

This morning, French territory was violated by German troops at Ciry and near Longwy. They are marching on the fort, which bears the name. Elsewhere the Custom House at Delle has twice been fired upon. German troops have also violated this morning the neutral territory of Luxemburg. You will at once use this information to lay stress on the fact that the German Government is committing itself to acts of war against France without provocation on our part, or any previous declaration of war, whilst we have scrupulously respected the zone of ten kilometers which we have continually maintained between our troops and the frontier.

WESTERN FRONT. In a skirmish between a German mounted patrol and a few French sentries, German *Leutnant* Albert Mayer becomes the first combat death in France, some 30 hours before the two countries are officially at war. French PM Viviani protests to German Embassy officials in Paris. *On Sunday, 2nd August, early in the morning the German Army entered Luxembourg by the bridges of Wasserbourg and Remleh, proceeding toward the south of the country. They have also brought armoured trains, with troops and munitions of war. Further, the special French Commissioner at Petitcroix has announced to the Sûreté Générale that the Germans have just opened fire on the frontier station of Delle-Petitcroix.*
A German cavalry officer has just been killed ten kilometers on our side of the frontier.

LONDON. Sir Edward Grey telegraphs to Sir Francis Bertie in Paris. *After the Cabinet meeting this morning I gave French Ambassador in London Paul Cambon, the following memorandum ~ Cambon asked me about the violation of Luxemburg. I told him of the doctrine on that point laid down by Lord Derby in 1867. He asked me what we should say about the violation of the neutrality of Belgium. I said that was a much more important matter; we were considering what statement we should make in Parliament tomorrow ~ in effect, whether we should declare violation of Belgian neutrality to be a casus belli.*

VIENNA. German Ambassador Count Heinrich von Tschirschky relays Berlin's latest report of Russian troops crossing the frontier at Schwidden to Count Berchtold with the Wilhelmstrasse wish: *We expect of Austria fulfillment of her allied obligations and immediate vigorous intervention against Russia.* Moltke telegraphs Conrad: *Austro-Hungarian military efforts should concentrate against Imperial Russia, the mortal enemy. Serbia can be kept in check with limited forces.*

Ambassador Count Heinrich von Tschirschky calls on Count Berchtold to persuade Austria-Hungary to declare war on Russia, but Berchtold does not want to do this until A-H Army troops have returned from the Serbian front in mid-August. Following further pressure from Berlin, Vienna agrees to declare war on Russia before 5th August.

LONDON 6.30 PM. The British Cabinet meet and agree to a formal statement regarding Belgium. *It should be made evident that a substantial violation of the neutrality of that country would place us in the situation when interference with Belgian independence would compel us to take action.* It now seems certain that Germany will invade, the Belgians will resist; and that if Belgium resists, the majority of the British Cabinet believes Britain must enter the war. Two cabinet members announce their resignation.

Prince Lichnowsky calls on Prime Minister H.H Asquith, who informs Prince Max that a war between Great Britain and Germany is unthinkable, but the outcome rests largely with those in Berlin. To make British armed intervention impossible, Germany needs not to invade Belgium, nor send her fleet into the Channel to attack the unprotected north coast of France. Prince Max cables Berlin:
It is most likely Britain will oppose Germany if our forces violate Belgian territory.

BRUSSELS. Davignon informs Sir Edward Grey. *We do not intend to appeal to affirm our neutrality, and assume that we will need no foreign aid to repel the invaders.* German Ambassador von Below has confirmed instructions to open the secret sealed envelope he received on 29th July in Berlin and deliver the ultimatum inside (drafted by von Moltke) to the Belgian Government at 7 PM that evening. Berlin's note falsely claims the French are to attack Germany via Belgium, and that the German Army require access to march through Belgium to repel this attack. Von Below is to give the impression that he received the note that afternoon. The Belgian reply is to be in Berlin by 1 PM Monday 3rd August.

The story is around town this afternoon is that the Germans have already crossed the frontier without the formality of a declaration of war ~ but that remains to be seen. Brussels was put under martial law last night, and is now patrolled by grenadiers and lancers. On the way back from the Legation this evening, I saw the German Ambassador driving home. He passed close to me, and I saw that perspiration was on his forehead. He held his hat in his hand and puffed at a cigarette like a mechanical toy, blowing out jerky clouds of smoke. He failed to give me his usual ceremonious bow. He continues to figure in the newspapers stating that all is well, that Germany has no intention of setting foot on Belgian soil, and that all Belgium has to do is to keep calm. In an interview given to Le Soir he sums up by saying; Your neighbour's house may burn, but yours will be safe.

Hugh Gibson ~ Secrétaire de la Légation U.S.A.

BRUSSELS 7 PM. German Ambassador Herr von Below returns to see Belgian Foreign Minister Julien Davignon in his office. The German looks pale, leans on a table, putting a hand to his chest. Davignon asks ~ *What's wrong? Are you ill!?* Von Below replies ~ *I think I just climbed the stairs too fast, It's nothing. I have brought a confidential note.* The German opens the envelope and presents its contents. Davignon reads it, he turns ashen ~ *No, it's not possible!!* …and the German's note falls to the floor. Both men are upset. Julien Davignon with vivid indignation grows angrier, as he disbelieves the deceitful German claim that France is about to attack. Davignon advises ~ *The note will be dealt with.* As the German Ambassador departs, still wobbling… Davignon's son and private secretary Hénri, opens the door ~ *My Father, had the pallor of a dead man.*

King Albert II with his son Crown Prince Leopold III ~ before heading to the front.

King Albert refuses to give his approval for the German Army right of passage thru his Kingdom, and expressly calls for a late evening Crown Council with his ministers who meet through the night. There is heated argument. Not on whether or not to accept the German ultimatum, as they all agree it must be rejected, therefore King Albert with the full support of his people rejects Berlin's demand. Their only argument, is how best to resist the imminent invasion of Germans. Should they fight near the frontier or should they withdraw into the interior of Belgium? Knowing the best railways and roads from Berlin to Paris run through his kingdom, Albert orders the Belgian Army to begin their planned response of demolishing the border railway tunnels to bar enemy trains from their homeland. They decide to ask for diplomatic support ~ but not for military support until the expected invasion begins. Therefore, when France offers Belgium five entire Army Corps for the defence of her territory, the Belgium Government decline the offer ~ for the present. Julien Davignon telegraphs Belgian envoys in Paris, London, Berlin, Vienna, and St. Petersburg that he has emphatically warned Germany that the Paris papers will publish Berlin's note regarding their expectations of Belgian neutrality.

7 AM ~ an emissary of the Belgian Foreign Ministry delivers their reply to Berlin.

Were Belgium to accept the proposals that lay before it, the Belgian Government would sacrifice the nation's honour while being false to its duties towards Europe. Belgium rejects the German demand and will oppose by force any violation of her neutrality. Belgium, which since the establishment of her kingdom, has taken every care to assure the protection of her dignity and of her interests, and has devoted all her efforts to the peaceful development of progress, strongly protests against any violation of her territory from whatever quarter it may come; and supposing the violation takes place, she will know how to defend with energy her neutrality, which has been guaranteed by the powers, and notably by the King of Prussia.

After an interview with Ambassador von Below, General Baron van der Elst wrote the following:

The German Ambassador has informed the Belgian Government that French dirigibles have thrown bombs, and that a French cavalry patrol has crossed the frontier in violation of international law, seeing that war has not been declared. The secretary general asked von Below where these incidents had happened, and was told that it was in Germany. I then observed, that in that case I could not understand the object of this communication. Herr von Below stated that these acts were calculated to lead to the supposition that other acts, contrary to international law, would be committed by France.

BRUSSELS. FM Julien Davignon replies to the German note.
This note has made a deep and painful impression upon the Belgian Government. If Belgian neutrality should be violated, Belgium intends to fulfill her international obligations and the Belgian Army will offer the most vigorous resistance to the invader. The attack upon her independence with which the German Government threatens her, constitutes a flagrant violation of international law. No strategic interest justifies such a violation. Conscious of the part which Belgium has played for more than 80 years in the civilization of the world, we refuse to believe that the independence of Belgium can only be preserved at the price of the violation of her neutrality. The Belgian Government firmly resolve to repel, by all the means in their power, every attack upon their rights. Baron von Below telegraphs Berlin. *Belgium has rejected our demands and will oppose by force any violation of her neutrality. Feeling towards Germany bad.* Berlin decides not to formally declare war on Belgium.

LONDON 10 AM. Prince Lichnowsky gives Sir Edward Grey the latest assurance from Berlin. *We can definitely state a threat to the French north coast on our part will not take place as long as England remains neutral. Germany will maintain the integrity of Belgium after the war.* Lichnowsky takes his leave.

11 AM. The British Cabinet reconvene to earnestly approve Winston Churchill's unauthorized mobilization of the Royal Navy. Churchill also reasons that Britain has obligations to France, and that Great Britain cannot stand aside to see Belgium occupied and France defeated. Four ministers are prepared to resign over the issue of British intervention. Their discussion lasts for three hours, at the end of which they agree on the line to be taken by Sir Edward Grey when he addresses the House of Commons at 3 PM. Grey: *The Cabinet was very moving. Most of us could hardly speak at all for emotion.* A note is handed to Sir Edward from the Belgian Legation that summarizes the German ultimatum, declaring that Belgium has rejected the ultimatum, and is resolved to repel aggression by all means. Asquith describes to Venetia Stanley the German ultimatum ~ *a piece of almost Austrian crassness.*

At 3 PM Sir Edward Grey, who has been Foreign Secretary since 1905, makes a statement to a packed House of Commons regarding the relationship with France, Russia, and Belgium, explaining that while Britain has no binding military or naval agreements with France; any French involvement in a war with Germany arises out of the current French alliance with Russia. *No country has less desire to be involved in war over a dispute with Austria and Serbia than the Government and the country of France. France is involved in it because of her obligation of honour under a definite alliance with Russia. It is only fair to the House to say that that obligation cannot apply in the same way to us. We are not parties to the Franco-Russian alliance. We do not even know its terms.* The House of Commons votes to approve Royal Navy protection of the French coast and for Grey to warn Berlin ~ *Britain will take action if Belgium is invaded.* Count Lalaing wires Davignon: *Grey confirms. If your neutrality is violated it means war with Germany.*

COLOGNE 3 PM. The Belgian vice-consul in the Rhine city of Cologne reports to Brussels that since 6.00 A.M. he has seen German troop trains leaving Cologne's central station every five minutes, headed not west towards France, but for Aix-la-Chapelle and the nearby Luxembourg-Belgian border.

LONDON 5.03 PM. Aldershot HQ receives a one-word telegram ~ *Mobilize.*
6 PM Sir Arthur Nicholson congratulates Sir Edward Grey on the success of his speech…
He did not answer… the Secretary of State leaned gloomily by the window. He moved to the centre of the room and raised his hands with clenched fists above his head. He brought his fists down with a crash upon the table. Sir Edward groaned ~ I hate war, I hate war…

BRUSSELS ~ U.S. Legation. *According to the news which was given me when I got out of bed this morning ~ during the night the garrison left and the Garde Civique came on duty to police the capital. As nearly as we can make out the Germans have sent patrols into Belgian territory. There has been a battle at Visé and they have driven the Germans back with heavy losses. All the regular troops have been withdrawn from this part of the country and dispatched to the front, leaving the protection of the capital to the Garde Civique, who are now patrolling the streets of Brussels.* Hugh Gibson ~ Secrétaire de la Légation U.S.A.

NEW GERMAN RAIL APPROACHES TO BELGIUM ~ SINCE 1909

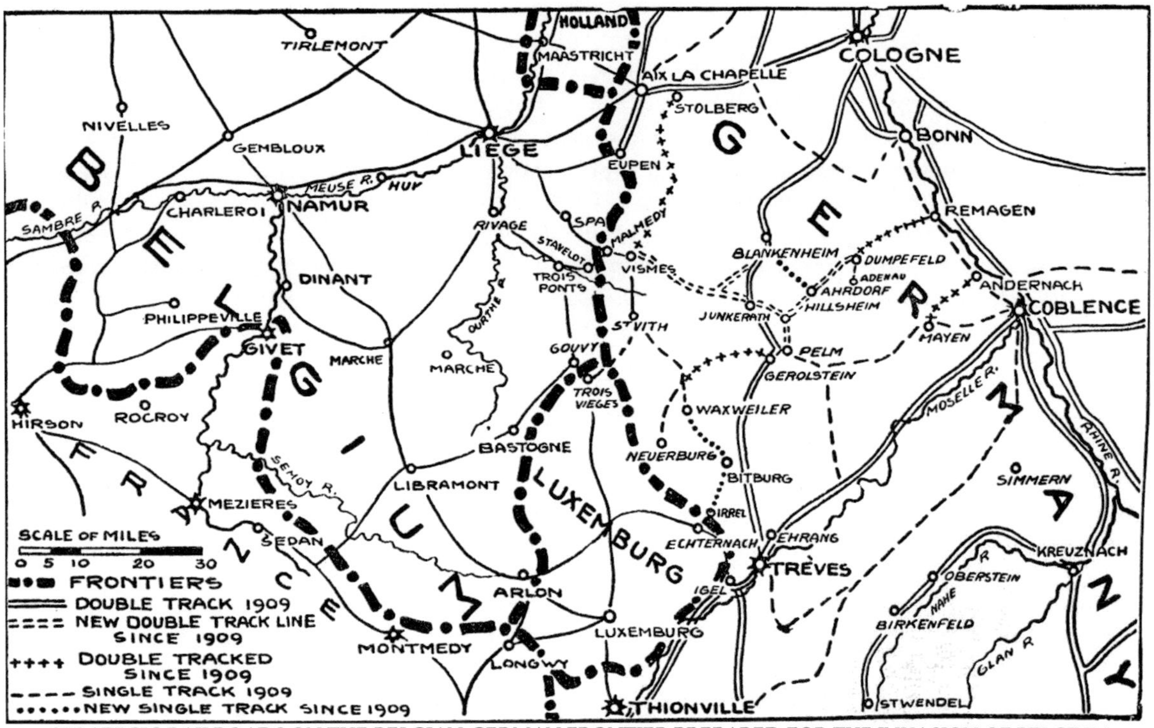

STRATEGIC RAILROADS ON THE BELGIAN-GERMAN FRONTIER PREPARED FOR THE INVASION OF BELGIUM

Nearly four million German troops have to be transported from every part to the borders, so the entire railway service is at their disposal. Hundreds of military trains are set to arrive at stations without delay, and from the first flush of mobilization, the German Army are directed through the vast interacting series of preparations that form *Plan 1914* for an attack to the west. Hundreds of trains steam over different tracks, to be disembarked by road to their appointed place on the front. For a precise awareness into the German planning and preparedness apparent in this otherwise inexplicable network of railways that approaches the borders of Belgium and Luxemburg, Walter Littlefield wrote:

In the southwest corner of Prussia, is a rectangular piece of territory, the western and eastern sides of which are formed respectively by the Belgian and Luxemburg frontiers and the River Rhine. Five years ago, this little corner of Prussia had about 15 miles of railway to every hundred square miles of territory. At the opening of 1914 this had increased to 28. In five years, without any apparent industrial and commercial demand for it, this traction has been increased to nearly twice its length. Villages of less than 1,300 inhabitants have been linked up with double-track lines. For example, Pelm is three miles from Gerolstein ~ a town principally of comic-opera fame, and yet over this short distance, between the two villages, there are laid down six parallel lines of rail, besides numerous additional sidings. Few of these lines cross the frontier. Three of them lead to blind terminals within a day's march of it ~ the double lines from Cologne to Weiwertz, to St. Vith, and via Pelm to Pronsfeld.*

Another point that is noticeable, is that provision exists everywhere at these new junctions and extensions for avoiding an upline crossing a down line on the level. The upline is carried over the down line by a bridge, involving long embankments on both sides (so new that as yet nothing has had time to grow on them) at great expense, but enormously simplifying traffic problems, when it comes to a question of full troop trains pushing through at the rate of one every quarter of an hour, and the empty cars returning eastward at the same rate. The detraining stations are of sufficient length to accommodate the longest troop train (ten cars) easily, and they generally have at least four sidings apart from the thru up and down lines. Moreover, at almost every station there are two lines of sidings long enough for troop trains, so that they can be used to some extent as detraining stations, and so that a couple of troop trains can be held up at any time while traffic continues uninterrupted.

Such facts of railway preparedness explain, the means whereby Germany was able to launch upon the Belgian, Luxemburg, and French frontiers such a vast array of fully equipped troops almost at the moment of the outbreak of the war. The absolute facts declare that German mobilization was directed, not at the French frontier, but at the frontier of Luxemburg and Belgium, especially at the great Belgian plain, commanded and dominated by the great fortress of Liége. As bearing upon the subject of mobilization, however, it is to be remembered that at this point, Belgium, and not at France, was directed the main first mobilization of the German Army.

**La Grande Duchesse de Gerolstein* ~ Offenbach

PARIS. Premier Viviani warns that German Ambassador von Schoen said at the Quai d'Orsay: *Yesterday 80 French officers in Prussian uniform attempted to cross the German frontier in cars near Walbeck. Be good enough to urgently contradict this news, which is pure invention, and to draw the attention of the British Foreign Office to the German campaign of false news which is beginning.* Viviani challenges the false German allegations and reports to all French diplomats overseas. *German troops have now crossed the Belgian frontier at Gemmerich near the Dutch border. I request that you protest in writing against the violation of the neutrality of Luxemburg by German troops against the ultimatum addressed to the Belgian Government by the German Ambassador in Brussels to force upon them the violation of Belgian neutrality and to require of that country that she should facilitate military operations against France on Belgian territory; finally against the false allegation of invasion of these two countries by French armies, by which he has attempted to justify the state of war which he declares henceforth exists between Germany and France.*

6 PM. Ambassador Wilhelm von Schoen delivers the subsequent German declaration of war.
French military aviators have committed flagrantly hostile acts on German territory, throwing bombs on rail lines near Karlsruhe, Nuremberg, and Wesel; and have also violated the neutrality of Belgium by flying over Belgian territory. I am instructed, and I have the honor to inform your Excellency, that in the presence of these acts of aggression the German Empire considers itself in consequence in a state of war with France.

Ambassador Wilhelm von Schoen: *As I was starting for the Ministry, a man jumped on to the footboard of my motor in front of the Embassy and forced his way into it, gesticulating wildly, and using threatening language. He was immediately followed by a second. As I was alone, I had to invoke the help of the police stationed at the street corner to get rid of the intruders. Two other men then got into the vehicle, and a third seated himself beside the chauffeur. They explained that they were detectives by whispering the word ~ service. They then remained with loaded revolvers till the moment of my getting into the train, and behaved admirably.*

The departure from Paris took place very quietly from a station near the Embassy, late in the evening, and without any contretemps. We started for Cologne, via Belgium. At the Belgian frontier however I was told that the train could not go on to Cologne, as the tunnels and bridges were destroyed. That we escaped this is perhaps only due to the fact that we travelled through Belgium by night. With some difficulty, the train was finally taken via Brussels and Antwerp through Holland to the German frontier, which we reached at Goch.

Having reached the German frontier on the way home, difficulties arose in the way of getting any farther. Our march westwards was in full swing, and the transport of troops made such claims on the line that no opportunity could be given me of continuing my journey to Berlin. There was nothing for it but to take the French special train through to Berlin, assuming that it would be brought back from there at once. But in order to do this it was repeatedly necessary to overcome the opposition of the officers in command of the stations, who insisted on commandeering the train and arresting the two French guards. Finally, I was obliged to resort to the drastic measure of putting on my military uniform to ensure my being treated with respect. In Berlin, I even had to ask the Imperial Chancellor to intervene, to ensure the train being given up and sent back.

On my arrival in Berlin, I was immediately given an opportunity of reporting my recent experiences and my impressions to the Kaiser and the Chancellor. In doing so I expressed an opinion that the war would be a long and unspeakably serious struggle for existence. In view of the immense superiority of our opponents in resources of all kinds, it appeared doubtful whether we should be able to do more than maintain our position and our honour, in spite of our extraordinary efficiency. I referred at some length to the fact, which had struck me, that there was a tendency on our side to underrate the French military and national power of attack and resistance, a view which had never been encouraged by the Embassy reports. The Kaiser's mood was grave, but absolutely calm and confident. He was particularly pleased to hear of the uniform enthusiasm and determination I had noted on my way back. That was my last interview with him.

LONDON. Sir Edward Grey was standing at a window of his room in the Foreign Office, looking from his office window out across St James Park, watching the lamps being lit below along the Mall. *It was getting dusk, and the lamps were being lit in the space below on which we were looking. My friend recalls that I remarked on this. The lamps are going out all over Europe; we shall not see them lit again in our lifetime.* John Alfred Spender was the friend to whom Sir Edward had spoken. *I had two short talks with Grey: on the stairs of the Foreign Office on 1st August and in the evening at his room at the Foreign Office on 3rd August, and he said to me ~ The lamps are going out all over Europe, and we shall not see them lit again in our lifetime. We were standing together at the window looking out into the sunset across St James Park, and the appearance of the first lights along the Mall suggested the thought.* J.A. Spender

BRUSSELS 8 AM. Belgian Lieutenant Anton Picard peering dutifully through his field glasses, sees a troop of German Uhlan cavalry crossing over the border, entering into Belgium from Germany. As they canter over the fields of Belgium, their follow-up infantry prepare to assault on a 25km front.

Secrétaire de la Légation U.S.A., Hugh Gibson describes the day.

We got about early and made for the Chamber of Deputies to hear King Albert's speech. The Ambassador and I walked over together and met a few straggling colleagues headed in the same direction. Most of them had got there ahead of us, and the galleries were all jammed. The Rue Royale, from the Palace around the park to the Parliament building was packed with people, held in check by the Garde Civique. There was a buzz as of a thousand bees and every face was ablaze. The Garde Civique had two bands in front of the Senate, and they tried to play the Brabançonne in unison. Neither of them could play the air in tune, and they were about a bar apart all the time. They played it through and then began to play it over again without a pause between. They blew and pounded steadily for half an hour, and the more they played, the more enthusiastic the crowds became.

When I saw how crowded the galleries were, I went out onto a balcony with Webber of the British Legation to see the arrival of the King and Queen. When the King and Queen departed the Palace, a roar of cheering came surging across the Park. The procession came along at a smart trot, and although it was hidden from us by the trees we could follow its progress by the steadily advancing roaring of the mob. When they turned from the Rue Royale into the Rue de la Loi, the crowd in front of the Parliament took up cheering to make the windows rattle.

First came the staff of the King and members of his household. Then the Queen, accompanied by the royal children, in an open daumont. The cheering for the Queen was full throated and with no sign of doubt because of her Bavarian birth and upbringing ~ she is looked on as a Belgian Queen and nothing else. After the Queen came a carriage or two, with members of the royal family and the Court. Finally, the King on horseback. He was in the field uniform of a Lieutenant General, with no decorations and none of the ceremonial trappings usual on such occasions as a speech from the Throne. He was followed by a few members of his staff who also looked as though they were meant more for business than for dress parade. As the King drew rein and dismounted, the cheering burst forth with twice its former volume, and in a frantic demonstration of loyalty, hats and sticks were thrown into the air. Two bands played on manfully, but we could hear only an occasional discord.

Just as the King started into the building an usher came out, touched me on the arm and said something, beckoning me to come inside. One of the galleries had been locked by mistake but had now been opened, and Webber and I were rewarded for our modesty by being given the whole thing to ourselves. Our places were not ten feet from the Throne, and we could not have been better placed.

The Queen came in quietly from one side and after acknowledging a roaring welcome from the members of the two Houses, she took a throne to the left of the tribune, When the cheering had subsided, the King walked in alone from the right, bowed gravely to the assembly and walked quickly to the dais above and behind the tribune. With a business-like gesture, he tossed his cap on to the ledge before him and threw his white cotton gloves into it. Then he drew out his speech and read it. At first his voice was not very steady but he soon controlled it and read the speech to the end in a voice that was vibrating with emotion but without any oratory or heroics. He went straight to the vital need for union between all factions and all parties, between the French, Flemish, and Walloon races, between Catholics, Liberals, and Socialists in a determined resistance to the attack upon Belgian independence. The House could contain itself for only a few minutes at a time, and as every point was driven home they burst into cheering. When the King, addressing himself directly to the members of Parliament, said ~ Are you determined at any cost to maintain the sacred heritage of our ancestors? The whole Chamber burst into a roar, and from the Socialist side came cries of ~ At any cost, by death if need be! It was simple and to the point, a manly speech. And as he delivered it he was a kingly figure, facing for the sake of honour what he knew to be the gravest danger that could ever come to his country and his people. When he had finished, he bowed to the Queen, then to the Parliament, and then walked quickly out of the room, while the assembly roared again. The Senators and Deputies swarmed about the King on his way out, cheering and trying to shake him by the hand, and none were more at pains to voice their devotion than the Socialists.

After he had gone the Queen rose, bowed shyly to the assembly, and then withdrew with the royal children. She was given a rousing ovation as everybody realized the difficulty of her position and was anxious to show her all their confidence and affection. The whole occasion was moving, but when the little Queen acknowledged the ovation so shyly and so sadly and withdrew, the tears were pretty near the surface ~ my surface at any rate.

For several minutes after the Queen withdrew, the cheering continued. As we came out, some colleagues were gathered about debating whether they should go over to the Palace and ask to take leave of the King. They were saved that labour, however, for the King had stepped into a motor at the door and was already speeding to the GHQ which has been set up nobody knows where. That looks like business.

When I got back to the Legation I found von Stumm, Counselor of the German Legation, with the news that his chief had received his passports and must leave at once. He had come to ask that the American Ambassador take over the care of the German Legation and the protection of the German subjects who had not yet left the country. We agreed to accept provisionally on grounds of humanity, until we receive specific instructions as to who would be definitely entrusted with the protection of German interests. In case of need, we shall be asked to take over certain other Legations and shall have our hands more than full.

At five o'clock we went over to the German Legation, which we found surrounded by a heavy detachment of Garde Civique as a measure of protection against violence. We drew up, signed and sealed a protocol accepting what is known as la garde des clefs et des sceaux, until such time as definite arrangements might be made. The Ambassador and von Stumm were nearly unstrung. They had been under a great strain for some days and were making no effort to get their belongings together to take them away. They sat on the edge of their chairs, mopped their brows, and smoked cigarettes as fast as they could light one from another. I was given a lot of final instructions about things to be done at once, as the German Army would doubtless be in Brussels in three days. While we were talking, the chancellor of their Legation, Hofrat Grabowsky, a typical white-haired German functionary, was pottering about with sealing wax and strips of paper, sealing the archives and answering questions in a deliberate and perfectly calm way. It was for all the world like a scene in a play. The shaded room, the two nervous diplomats registering anxiety and strain, the old functionary who was to stay behind to guard the archives and refused to be moved from his calm by the approaching cataclysm. It seemed altogether unreal, and I had to keep bringing myself back to a realization of the fact that it was only too true and too serious.

They were very ominous about what an invasion means to this country, and kept referring to the army as a steamroller that will leave nothing standing in its path. Baron von Stumm kept repeating… Oh, the poor fools! Why don't they get out of the way of the steamroller? We don't want to hurt them, but if they stand in our way they will he ground into the dirt. Oh, the poor fools! Before leaving, von Below gave out word that we would look after German interests, and consequently we have been deluged with frightened people ever since. The Red Cross is getting ready to handle the wounded, and everybody is doing something. Nearly everybody with a big house has fitted it in part as a hospital. Others are rolling bandages and preparing all sorts of supplies. Hugh Gibson

Secrétaire de la Légation U.S.A.

Menace d'Allemands arrêté s par les autorités Belges ~ Germans of menace arrested by the Belgian authorities.

PARIS. President Poincaré is so worried that London has made no commitment as yet to send the British Expeditionary Force to France, that he writes to King George V requesting him to send British troops to cover the French left flank, as the German Army is now attacking strongly through Belgium. French Ambassador Paul Cambon needing to know how Britain might support France, asks Grey: *How will you fight the war? When will you send your Expeditionary Force?* To which Sir Edward replies: *We shall blockade the German ports. We have not yet considered sending a military force to the Continent.* Cambon: *Public opinion will force Britain to intervene on the continent, and to be effective it must be soon. Here is a map of our French Army deployments, to illustrate the need for British forces on the left of the French ~ all as agreed by the French and British GHQs if France is attacked through Belgium.*
Please advise Prime Minister Asquith, and the British Cabinet of these considerations.

LONDON. King George V makes an appeal to Ambassador Sir E. Goschen in Berlin, instructing him to ask for an assurance that Germany will not proceed with their demands made upon Belgium. Therefore at 9.30 AM, Grey instructs Goschen to protest against the German violation of Belgian neutrality, and to request an assurance that demands made to Belgium will not be proceeded with. Goschen calls on von Jagow with the British request sent that morning that Germany will no further violate Belgium neutrality. *Britain is bound to protest against this violation of a treaty to which Germany is a party in common with themselves, and requests an assurance that the demand made upon Belgium will not be proceeded with, and that her neutrality will be respected by Germany. This fait accompli of the violation of the Belgian frontier renders the situation exceedingly grave ~ is there still time to draw back and avoid the possible consequences, which both you and I would deplore?*

The German answer must be No, in consequence of the German troops having already crossed the Belgian border this morning. Belgian neutrality has been already violated, and the Belgians are resisting. The Imperial Government has been obliged to advance into France by the quickest and easiest way, to be able to get well ahead with our operations and endeavour to strike some decisive blow as early as possible. It is a matter of life and death for us, as if we had gone by the more southern route we could not have hoped, in view of the paucity of roads and the strength of the fortresses, to have got through without formidable opposition entailing loss of time. This loss of time would have meant time gained by the Russians for bringing their troops to the German frontier. Rapidity of action is the great German asset, while that of Russia is an inexhaustible supply of troops.
For the reasons I have given you, it is now impossible to draw them back. Gottlieb von Jagow
Imperial German Foreign Secretary

BRUSSELS. Foreign Minister Davignon receives German Ambassador von Below, who cautions. *As the Belgian Government have now declined the well-intentioned proposals submitted to them by the Imperial Government, the latter will, much to their regret, be compelled to carry out, if necessary by force of arms, measures considered indispensable in view of the French menace.* Davignon urges Baron Gaiffier d'Hestroy: *Let us go to mass and offer prayers for our poor country. Never before has it stood in such great need of them!* Sir Francis Villiers telegraphs to London: *German troops have now entered Belgian territory. Liége has been summoned to surrender by small party of Germans who were repulsed.*

LONDON. Sir Edward Grey replies; *Great Britain expects the Belgian Government to resist by any means in their power the German violation of their neutrality and that Great Britain is prepared to join Russia and France in common action to resist the German invasion and guarantee Belgian independence and integrity. The invasion of Belgium is the direst crime that ever stained the pages of history.*

At noon, Prince Lichnowsky presents the British Foreign Office with a message from von Jagow repeating Berlin's assurance. *In the armed conflict with Belgium, Germany will not annex Belgian territory.* Once again, von Jagow gives his false and misleading assertion. *German forces are compelled to invade Belgium, to forestall a French attack on Germany through Belgium.* Lichnowsky receives a wire from Berlin.

Please dispel any mistrust that may subsist on the part of the British Government with regard to our intentions, by repeating positively our formal assurance that, even in the case of armed conflict with Belgium, Germany will, under no pretence whatever, annex Belgian territory. Sincerity of this declaration is borne out by the fact that we solemnly pledged our word to Holland strictly to respect her neutrality. Please do impress upon Sir E. Grey that the German Army cannot be exposed to a French attack across Belgium, which was planned according to absolutely unimpeachable information. Germany has consequently to disregard Belgian neutrality, it being for her a question of life or death to prevent a French advance. Gottlieb von Jagow

LONDON. A-H Ambassador Count Albert von Mensdorff intriguingly advises Sir Edward Grey: *There is no cause for Britain and Austria to quarrel as long as Austria does not go to war with France.*

Sir E. Grey said to me that at present there was no reason why he should make any communication to the Imperial and Royal A-H Government, and there was no cause why a conflict should arise between us, so long as we were not in a condition of war with France. In any case, he hoped that we would not begin hostilities without the formality of a previous declaration of war. He does not intend to recall Sir Maurice de Bunsen. Should we be at war with France, it would indeed be difficult for Great Britain, as the ally of France, to cooperate with her in the Atlantic, and not in the Mediterranean. Count Albert von Mensdorff

LONDON 2 PM. German troops having entered Belgian territory; Sir Edward Grey's response is to telegraph an ultimatum via British Ambassador Sir E. Goschen in Berlin, and to request a reply.

We hear that Germany has addressed a note to Belgian Minister for Foreign Affairs stating that German Government will be compelled to carry out, if necessary by force of arms, the measures considered indispensable. We are also informed that Belgian territory has been violated at Gemmenich.

In these circumstances, and in view of the fact that Germany declined to give the same assurances respecting Belgium that France gave last week in reply to our request made simultaneously at Berlin and Paris, we must repeat that request, and ask that a satisfactory reply to it and to my telegram of this morning be received here by 12 o'clock tonight. If not, you are then instructed to ask for your passports, and to say that ; His Majesty's Government feel bound to take all steps in their power to uphold the neutrality of Belgium and the observance of a treaty to which Germany is as much a party as ourselves.

BERLIN. As Sir E. Goschen proceeds to call upon von Jagow and arrives in the Wilhelmstrasse waiting room, he sees U.S. Ambassador James Gerard already there and hoping for some news. Goschen tells Ambassador James Gerard ~ *Perhaps I am here for the purpose of asking for my passports!* (As he speaks in English, he is overheard by a man sitting in the room ~ a lurking a newspaperman). Goschen is then admitted to read Britain's ultimatum to von Jagow.

Unless the Imperial Government can give assurance by 12 o'clock this night that they will proceed no further with their violation of Belgian frontier and stop their advance, I have been instructed to demand my passports and inform the Imperial Government that His Majesty's Government will take all steps in their power to uphold neutrality of Belgium.

To this, Gottlieb von Jagow responds: *To my great regret, I can give no other answer than that which I have had given you earlier in the day, namely, that the safety of Germany renders it absolutely necessary that Imperial troops should advance through Belgium.* Goschen: *Whether in view of the terrible consequences which will necessarily ensue, is it not possible even at this last moment that your answer should be reconsidered.* Jagow: *If the time given were even 24 hours or more, the answer must be the same. We regret the crumbling of our entire policy which has been to make friends with Great Britain, and through Britain, to get closer to France.*

Sir Edward Goschen concludes: *In that case, I should have to demand my passports. This sudden end to my work in Berlin is to me also a matter of deep regret and disappointment, but you must understand that under the circumstances his Majesty's Government could not possibly have acted otherwise than they have done.*
I should like to see the Chancellor. It might be the last time I should have an opportunity of seeing him.

Sir Edward Goschen meets with the Imperial Chancellor Theobald von Bethmann Hollweg...
I found the Chancellor very agitated. He at once began a harangue, which lasted for about twenty minutes. Bethmann: *The step taken by his Majesty's Government is terrible. Just for a word ~ neutrality, a word which in wartime has often been disregarded. Can this neutrality which we violate only out of necessity, fighting for our very existence really provide the reason for a world war?! Compared to the disaster of such a holocaust does not the significance of this neutrality dwindle into a scrap of paper? Just for a scrap of paper Great Britain is going to make war on a kindred nation who desires nothing better than to be friends with her. All my efforts in this direction have been rendered useless by this last terrible step, and the policy to which I have devoted myself since taking office have tumbled down like a house of cards. What you have done is unthinkable. It is like striking a man from behind while he is fighting for his life against two assailants. I hold Britain responsible for all the terrible events that might happen.*

Goschen: *In the same way as you and Herr von Jagow wish me to understand that for strategical reasons it was a matter of life and death to Germany to advance through Belgium and violate her neutrality, so I wish you to understand that it is, so to speak, a matter of life and death for the honor of Great Britain that she should keep her solemn engagement to do her utmost to defend Belgium's neutrality if attacked. This solemn compact has to be kept, or what confidence can anyone have in engagements given by Great Britain in the future?*
Bethmann: *But at what price will that compact have been kept? Has the British Government thought of that?*
Goschen: *Fear of consequences can hardly be regarded as an excuse for breaking solemn engagements.*
Goschen leaving: *Unfortunately, notwithstanding our efforts to maintain peace between Russia and Austria, the war has spread and has brought us face to face.* After this meeting, Goschen writes a report for Grey. Goschen's telegram is handed in at Berlin's Central Telegraph Office, but never dispatched.

LONDON 3 PM. In a meeting between Sir Edward Grey and U.S. Ambassador W.H. Page.
Page enters… Sir Edward, a tall and now worn and pallid figure is standing against the mantelpiece. He greets Page with a solemn handshake. Sir Edward then refers to the German invasion of Belgium.

The neutrality of Belgium, is assured by treaty. Germany is a signatory Power to that treaty. It is upon such solemn compacts as this that civilization rests. If we give them up or permit them to be violated, what becomes of civilization? Ordered society differs from mere force only by such solemn agreements or compacts. But Germany has violated the neutrality of Belgium. That means bad faith. It means also the end of Belgium's independence. And I fear it will not end with Belgium. This very morning the Swedish Ambassador informed me that Germany has made overtures to Sweden to come in on Germany's side. The whole plan is thus quite clear. This one great military Power means to annex Belgium, and to subjugate France.

England would be forever contemptible, if it should sit by and see this treaty violated. Its position would be gone if Germany were permitted to dominate Europe. I have therefore asked you to come, to tell you that this morning we sent an ultimatum to Germany. We have told Germany that if this assault on Belgium's neutrality is not reversed England will declare war.
Page: *Do you expect Germany to accept it?*
Grey: *No. Of course, everybody knows that there will be war. Yet we must remember there are two Germanys. There is the Germany of men like ourselves, of men like Count Lichnowsky and Herr von Jagow.*
Then there is the Germany of men of the war party. The war party has the upper hand.
Grey's eyes fill with tears. *Thus the efforts of a lifetime go for nothing. I feel like a man who has wasted his life. Please explain the situation to President Woodrow Wilson, expressing the hope that the United States will take an attitude of neutrality, and that Great Britain might look for the Courtesies of Neutrality from his country. Such a war will cause President Wilson and the American people sincere pain.*
Page: *I came away, with a sort of stunned sense of the impending ruin of half the world.*

As a result of anticipated, suspected, and actual mobilizations, Germany and Russia have been brought to the brink of war. With the hope of preserving peace almost gone, the contagion of *mobilitis* overspreads much of Europe and now threatens within hours, to traverse the English Channel as well. The first and fundamental consideration that German GHQ always keep in view, is their rapid speed of mobilization ~ and they will not allow this advantage to escape them. Even if Berlin sincerely seeks peace, their principal hope of military success (as soon as war appears likely or inevitable) is in taking advantage of their perfected plans for mobilization, and striking before any of their rivals are ready. The German Army look upon this speed in mobilizing as their greatest military asset, and never feel that they can allow it to be taken from them. Hence, the very moment that Berlin perceives either Russia or France have begun mobilization, Berlin is expected to insist that it come to a sudden end, and if their demand is not immediately settled, they are almost certain to declare war in response. Otherwise, delay will see their forces at compounding disadvantage. To allow either Russia or France to commence, continue and complete mobilization will deny German forces much of their strength.

I swore to the colours when I came to the throne, that the bayonet and cannon would have to rest, but that bayonet and cannon, however, would have to be kept sharp and effective in order that envy from without should not disturb us in the development of our garden and our beautiful house. I have made a vow, as a result of what I have learned from history, never to strive for world-dominion. For what has become of the worlds empires? Alexander the Great, Napoleon I, all the great warriors have swum in blood and have left subjugated peoples behind them who at the first opportunity have risen up again and brought the empire to ruin. Wilhelm II

KAISER WILHELM II OPENS THE REICHSTAG 4th AUGUST 1914

Honoured Gentlemen ~ I have assembled the elected representatives of the German people about me. For nearly half a century we have been allowed to follow the ways of peace. The attempts to attribute to Germany warlike intentions and to hedge in her position in the world have often sorely tried the patience of my people. Undeterred, my government has pursued the development of our moral, spiritual, and economic strength as its highest aim, even under provocative circumstances! The world has been witness that during the last years, under all confusion, we have stood in the first rank in saving the nations of Europe from a war between the great powers. The most serious dangers to which the events in the Balkans had given rise seemed to have been overcome then suddenly an abyss was opened through the murder of my friend the Archduke Franz Ferdinand.

My lofty ally, the Emperor Franz Josef, was forced to take up arms to defend the security of his empire against dangerous machinations from a neighboring state. The Russian Empire has stepped in the way of the Dual Monarchy following her just interests. Not only our duty as ally calls us to the side of Austria-Hungary, but our great task to protect our own position and community of culture between the two empires against the attack of hostile forces. The Imperial Russian Government, yielding to the pressure of an insatiable nationalism, has taken sides for a state which through its sanctioning of criminal attacks has brought about the evils of this war. That France, too, should have taken sides with our enemy cannot surprise us. Too often our attempts to come to friendlier relationships with the French Republic have failed because of her old hopes.

Honoured Gentlemen, the hostility which has been making itself felt in the east and in the west for a long time past has now broken out in bright flame. The present situation is the result of an ill will which has been active for years against the power and the prosperity of the German Empire. No lust of conquest drives us on; we are inspired by the unalterable will to protect the place in which God has set us for ourselves and all coming.

From the documents which have been submitted to you, you will see how my government and especially my Chancellor have endeavoured to stave off the inevitable. In a defensive war that has been forced upon us, with a clear conscience and a clean hand we take up the sword. I issue my call to the peoples and stocks of the German Empire, that with their united strength they may stand like brothers with our allies in order to defend what we have created through the works of peace. Following the example of our fathers, staunch and true, earnest and knightly, humble before God, but with the joy of battle in the face of the enemy, we trust in the Almighty to strengthen our defense and guide us to good issue. Honored Gentlemen, the German people gathered about their leaders are today looking to you. Come to your decisions quickly and unanimously. Such is my earnest wish.

BERLIN. As A-H has surprisingly not yet moved to fight Russia, Bethmann telegraphs to Vienna :

We have been compelled to go to war on account of Austria's procedure, and have a right to expect that Austria should not seek to hide this fact, but will openly announce that the threat of interference in the Serbian conflict is forcing Austria to go to war. From Vienna's now extremely suspicious silence, it comes to pass that few have misjudged it all as hopelessly as Bethmann who at 3 PM addresses the Reichstag :

We wished to continue our work of peace, and like a silent vow, the feeling that animated everyone from the Kaiser down to the youngest soldier was this: Only in defence of a just cause shall our sword fly from its scabbard. The day has now come when we must draw it against our wish, and in spite of our sincere endeavours. Russia has set fire to the building. We are at war with Russia and France ~ a war that has been forced upon us. Germany has endeavored to localize the conflict between Austria-Hungary and Serbia. All other European Governments (particularly Great Britain) save one took the same attitude. Russia alone asserted that she had to be heard in a settlement of the matter. Thus, the danger of a European crisis has raised its threatening head. Russia began to mobilize. On this, Germany declared that Russian military measures against Austria-Hungary would find her on the side of her ally, and that she would take countermeasures, coming near to actual war. Russia assured us of her desire for peace, and declared that she was making no military preparations against us. In the interval, Great Britain, warmly supported by us, has tried to mediate between Vienna and St. Petersburg.

Germany influenced Austria-Hungary to resume the broken conversations with Russia. But before the final decision was taken at Vienna, the news arrived that Russia had mobilized her entire forces and that her mobilization was therefore directed against us also. The Russian Government, who knew from our repeated statements what mobilization on our frontiers meant, did not notify us of this mobilization, nor did they offer any explanation. It was not until the afternoon of July 31 that the Kaiser received a telegram from the Tsar in which he guaranteed that his army would not assume a provocative attitude toward us. But mobilization on our frontiers had been in full swing since the night of July 30, and France, though indeed not actually mobilizing, was admittedly making military preparations. What was our position? For the sake of the peace of Europe we had, up till then, deliberately refrained from calling up a single reservist. Therefore, on July 31, we called upon Russia to demobilize as the only measure which could still preserve the peace of Europe, and informed her that in case our demand met with a refusal, we should have to consider that a state of war existed. No answer was given, and we mobilized our forces on August 1, at 5 PM.

Gentlemen, we are now in a state of necessity, and necessity knows no law. Our troops have occupied Luxemburg and perhaps have already entered Belgian territory. Gentlemen, that is a breach of international law. It is true that the French Government declared at Brussels that France would respect Belgian neutrality as long as her adversary respected it. We knew, however, that France stood ready for an invasion. France could wait, we could not. A French attack on our flank on the lower Rhine might have been disastrous. Thus we were forced to ignore the rightful protests of the Governments of Luxemburg and Belgium. The wrong ~ I speak openly, the wrong we thereby commit we will try to make good as soon as our military aims have been attained. He who is menaced as we are and is fighting for his highest possession can only consider how he is to hack his way through. Gentlemen, we stand shoulder to shoulder with Austria-Hungary. Gentlemen, again the words of the Kaiser;

With a clear conscience, we are fighting for the fruits of our works of peace, for the inheritance of a great past and for our future. Our army is in the field, our navy is ready for battle ~ behind them stands the entire German nation united to the last man. Gentlemen, you know your duty and all that it means. The proposed laws need no further explanation. I ask you to pass them quickly.

BERLIN. *Round about that part of the Wilhelmstrasse in which the British Embassy is situated, a large crowd had forgathered. Respectfully dressed citizens of both sexes were bellowing out with frantic enthusiasm, their best-loved hymn, Deutschland uber Alles. A volley of catcalls succeeded the national anthem, after which came a shower of missiles bricks and lumps of coal, for no stones are found in the asphalt roadways of Berlin. The ground floor windows of the British Embassy were slivered to atoms, the two policemen posted on either side of the door making no attempt to interfere. I had seen and heard enough.* Baron N.E. Beyens

The British Embassy is attacked by a wild mob stirred into excitement by the *Berliner Tageblatt ; Great Britain has declared war on Germany.* Ambassador James Gerard and his British counterpart have no doubt the insight came from the fellow who overheard them in the Wilhelmstrasse waiting room. The German Government offers police protection and apologizes for what Gottlieb von Jagow calls ; *The indelible stain on the reputation of Berlin.* Ambassador Gerard describes the continuing incident.

I went for a walk about Berlin, soon becoming involved in the great crowd in front of the British Embassy on the Wilhelmstrasse. The crowd threw many stones, and managed to break all the windows of the Embassy. The Germans charged afterwards that people in the Embassy had infuriated the crowd by throwing pennies to them. I did not see any occurrences of this kind. As the Unter den Linden and the Wilhelm Platz are paved with asphalt, the crowds must have brought with them the missiles which they used, to smash the Embassy windows. A few mounted police made an appearance, but were insufficient to hold the crowd in check.

Afterwards I went around to the Unter den Linden where there was a crowd in front of the Hotel Adlon. A man standing on the outskirts of the crowd begged me not to go into the Hotel, as he said the people were looking for British newspaper correspondents. So threatening was the crowd towards the British and believing that possibly the British Embassy might be in such a condition that Sir Edward Goschen might not care to spend the night there, I ordered an automobile and went up through the crowd which still choked the Wilhelmstrasse.

Sir Edward and his staff were perfectly calm and politely declined the refuge offered them in our Embassy. I chatted with them for a while, and as I was starting to leave, a servant told me that the crowds in the street had increased and some were watching my automobile. I sent out word by the servant to open the landau, and to tell the chauffeur to drive very slowly. We drove slowly through the crowd, assailed only by the peculiar hissing word that the Germans use when they are especially angry ~ which is supposed to convey the utmost contempt. This word is Pfui, and has a peculiar effect when hissed out from thousands of Teutonic throats.

As we left the outskirts of the crowd, a man of respectable appearance jumped on the running board of the automobile, spat at me saying Pfui and struck Harvey in the face with his hat. I stopped the automobile, jumped out, chased him down the street, and caught him. My German footman came running up and explained that I was the American Ambassador, not an Englishman. The man who struck Harvey thereupon apologized and gave his card. He was a Berlin Lawyer, who came to the Embassy next morning and apologized again for his mistake.

* * *

LONDON 10 PM. Still no response from Berlin. Although Sir E. Goschen's messages were not transmitted to London, Grey learns from an intercepted message from Berlin to the German Embassy in London that Goschen has asked for his passports. They wait till 11 PM or midnight Berlin time.

If in a crisis like this we run away from our obligations of honour as regards the Belgian Treaty, I doubt whether, whatever material gain we might have at the end of it, it would be of very much value in the face of the respect that we should have lost. Sir Edward Grey

If I am asked what we are fighting for, I can reply in two sentences. In the first place, it is to fulfill a solemn international obligation, in the second place, we are fighting to vindicate the principle, which in these days, when sometimes material force seems to be the dominant influence and factor in the development of mankind, that small nationalities are not to be crushed, in defiance of international truth and good faith. H.H. Asquith

LONDON. J.A. Spender, who the night before had been at the Foreign Office with Sir Edward when he spoke his famous words, is now walking across the Horse Guards Parade toward the Admiralty with First Sea Lord, Winston Churchill. *At midnight, we shall be at war. Think of it if you can, the Fleet absolutely ready with instructions for every ship, and the word going out from that tower at midnight.*

That evening, some of us sat with the Prime Minister in the Cabinet Room in 10 Downing Street. I was there in touch with the Foreign Office to certify that no satisfactory reply had come from Berlin, though this was after all that had happened, a foregone conclusion and a matter of form. Churchill also was among those present, ready at the appointed hour to send out the war order that the Royal Navy were expecting. Midnight came. We were at war. Sir Edward Grey

At midnight, when the signals are transmitted, all foreign policy effectively passes into the hands of sailors, soldiers, and aviators for the next four years. When news spreads through the crowd that war has been declared on Germany, excitement mounts. When King George V and Queen Mary appear on the balcony of Buckingham Palace, the cheering is terrific. By the end of the day, five of the great Empires of Europe are at war, plus Serbia and Belgium. The tragedy of the Great War has begun.

HOW THE WAR AFFECTED THE UNITED STATES

LINERS ARRIVING AT NEW YORK WITH AMERICAN REFUGEES

THE OUTBREAK OF WAR COINCIDING WITH THE HEIGHT OF THE TOURIST SEASON, MANY THOUSANDS OF AMERICANS WERE HELD UP IN EUROPE AND SUFFERED MUCH DELAY AND INCONVENIENCE BEFORE THEY COULD GET HOME AGAIN. OUR ARTIST, WHO HAD CROSSED THE ATLANTIC IN THE ERIN ON A MISSION TO ILLUSTRATE THE AMERICAS CUP RACE, WHICH HAS NOW BEEN POSTPONED UNTIL NEXT YEAR, WRITES: "THE LACONIA, TEEMING WITH PEOPLE, GOT IN SHORTLY AFTER WE ARRIVED. SHE'S A CUNARDER, BUT SHE DISGUISED HERSELF BY PAINTING HER FUNNELS BLACK WITH A RED BAND, AND ALL WHITE PAINTWORK BEING TURNED TO GREY. THE SHAMROCK IS LAID UP IN BROOKLYN, WHILE THE VATERLAND AND MANY OTHER GERMAN LINERS ARE TIED UP AT HOBOKEN". ~ DRAWN BY CHARLES DIXON, R.I.

THE BEER FROM THE SPRING HARVEST HOPS HAS ARRIVED
THE ENEMY IS SURRENDERING TO US

HUNGARIAN POSTER ~ 1914

EXPANSION OF THE WAR

~ OTHER MOBILIZATIONS AND DECLARATIONS OF WAR ~

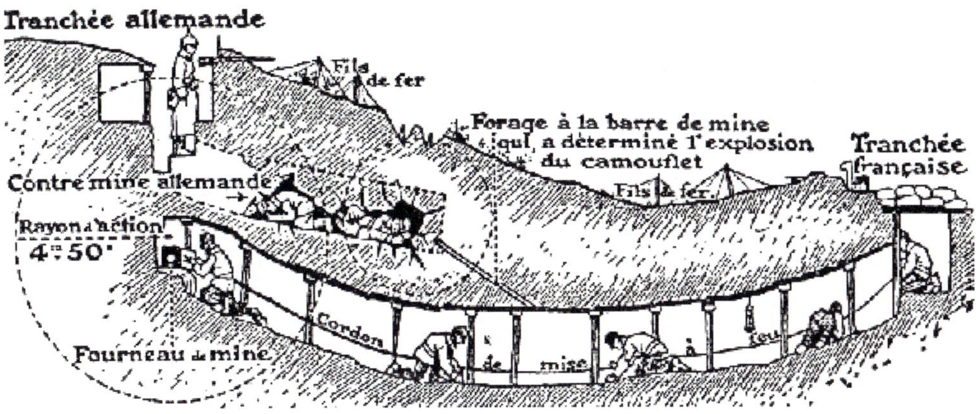

Belgian cavalry defend the route to the border

BRUSSELS. *German troops are said to have invaded Belgian soil on Tuesday morning at eleven o'clock. They entered near Dolhain. In the afternoon about four o'clock, they arrived in the region of Fléron, and it is said that the guns of the forts at Liége can be heard booming away. Patriotic fever is at its height. The Belgian Army has fought heroically today and the forts at Liége all hold. The French and English are expected to soon support. At the British Legation we had the word at dinnertime ~ just as we lifted our glasses to the Kingdom of Belgium, Gibson was told that many Germans were mowed down today by the Belgian guns of the Liége forts.*

Ah, the pity of it! All day the birds have been singing at our little Bois Fleuri and the two magpies there have good news. And over beyond the trees, Belgian and French and German boys are waiting to kill each other. They know not why. All they know is that some force has put them in clothes of a certain colour with trimmings, and therefore, when they kill each other, the deed is something other than killing! The whole fabric of civilization, so amazing, so delicately adjusted, all smashed... U.S. Ambassador Brand Whitlock

Great Britain is the only country that had no need to fight. She took up the sword for the sake of a principle and out of respect for her own signature. Her chivalry in maintaining the rights of Belgium has assigned to her the foremost place in Europe... Princess Catherine Radziwill

British officers near the front

BERLIN 11 AM. While the British Ambassador, Sir Edward Goschen is receiving his passports; U.S. Ambassador James W. Gerard spends the morning taking over all of British interests in Germany. Gerard and First Secretary Joseph Grew, first drive to the British Embassy where the official seals are archived and the American diplomats receive instructions and information (as which can be given) with reference to known British subjects in Germany and their interests. All British correspondents are gathered within the Embassy, and permission is obtained for them to leave on the diplomat's train. During the day, British subjects, without distinction of sex or age, are seized wherever found and transported to the fortress of Spandau. Gerard tells von Jagow ~ *Such a measure is from the Middle Ages.*

Kaiser Wilhelm sends one of his aides to the U.S. Embassy with the following message.
The Kaiser has charged me to express to your Excellency his regret for the occurrences of last night, but to tell you at the same time that you will gather from those occurrences an idea of the feelings of his people respecting the action of Great Britain in joining with other nations against her previous old Allies of Waterloo. His Majesty also begs that you will tell the King that he has been proud of the titles of British field marshal and British admiral, but that in consequence of what has occurred he must now divest himself of those titles.

PARIS. A message from President Poincaré is read before an extraordinary session of Parliament.
The violent and premeditated attack on France by Germany in insolent defiance of the law of nations was delivered before any declaration of war, as was the asking for his passports by the German Ambassador at Paris. I recount the pacific course of Frenchmen in burying at the bottom of their heart the desire for legitimate reparation, of the wrong done their country by Germany in 1871, and in using their rejuvenated strength in the interest of progress and for the good of humanity. France has made for peace since Austria's ultimatum to Serbia. France has made up to the last moment supreme efforts to avert the war now about to break out, the crushing responsibility for which the German Empire will have to bear before history. (Unanimous applause).

On the very morrow of the day when we and our allies were publicly expressing our hope of seeing negotiations begun under the auspices of the London Cabinet carried to a peaceful conclusion Germany suddenly declared war upon Russia; she has invaded the territory of Luxemburg; she has outrageously insulted the noble Belgian nation our neighbour and our friend, and attempted treacherously to fall upon us while we were in the midst of diplomatic conversation. But France is watching. As alert as she is peaceful, she is prepared; and our enemies will meet on their path our valiant covering troops, who are at their post and will provide the screen behind which the mobilization of our forces will be methodically completed. In the war which is beginning France will have right on her side, the eternal power of which cannot with impunity be disregarded by nations any more than by individuals. (Loud applause.)

She will be heroically defended by all her sons; nothing will break their sacred union before the enemy. Today they are joined together as brothers in a common indignation against the aggressor, and in a common patriotic faith. (Loud applause and cries of ~ Vive la France). She is faithfully helped by Russia. She is supported by the loyal friendship of Great Britain. (Loud applause). And already from every part of the civilized world sympathy and good wishes are coming to her. For today once again she stands before the universe for liberty, justice, and reason. Haut les cœurs et vive la France! (Prolonged applause).

Premier René Viviani speaks next before the Chamber of Deputies to recount the recent actions of Germany during the crisis on which the light of subsequent events has cast a sinister interpretation. The fabricated complaints against France for violating German territory offered by Baron von Schoen are offset by the true charges made by France ~ of the actual German violation of French territory.

At no time has any French aviator penetrated into Belgium, nor has any French aviator committed either in Bavaria or any other part of Germany any hostile act. The opinion of Europe has already done justice to these wretched inventions. The mobilization of the Russian Army also continues with remarkable vigour and unrestrained enthusiasm. The Belgian Army, mobilized with 250,000 men, prepares with a splendid passion and magnificent ardor to defend the neutrality and independence of their country. The entire British fleet is mobilized and orders have been given to mobilize the land forces (Loud cheers, all deputies rise to their feet).

VIENNA. After a significant lapse of time, Austria eventually decides to declare war on Russia. The A-H Army is defeated at Semendria by a force of veteran Serbian troops led by officers displaying their military experience gained in the recent Balkan Wars ~ something the A-H Army lacks.

BELGIUM. Baron Albéric du Fallon reports to Minister for Foreign Affairs Julien Davignon that Holland intends to institute war buoying on the Dutch River Scheldt leading to Antwerp in Belgium. The river will henceforth be closed at night only, and navigation by day will continue under the direction of Dutch pilots. Belgian lightships will also be withdrawn from Dutch territory to facilitate the maintenance of its neutrality. Davignon returns the passports of German Ambassador von Below who entrusts the custody of the German Legation, to U.S. Ambassador Brand Whitlock.

BERLIN. Belgian Ambassador Baron N.E. Beyens asks for his passports, but before leaving he telegraphs a report of the Chancellor's speech to the Reichstag on the violation of Belgian neutrality. *It is noteworthy that Herr von Bethmann Hollweg recognizes without the slightest disguise, that Germany is violating international law by her invasion of Belgian territory, and that he is committing a wrong against us.*

LONDON. Belgian Ambassador in Great Britain, Count Charles de Lalaing cables Brussels that Sir Edward Grey has advised British diplomatic staff in neutral Norway, Holland, and Belgium that :

Great Britain expects that these three kingdoms will resist German pressure and observe neutrality. Should they resist they will have the support of Great Britain, who is ready in that event, should the three mentioned Governments desire it, to join France and Russia in offering an alliance for resisting the use of force by Germany against them, and a guarantee to maintain the future independence and integrity of the three kingdoms.

BRUSSELS. Davignon telegraphs to Belgian Ambassadors in Paris, St. Petersburg, and London all important diplomatic events regarding Belgium from 31st July to current appeals to guarantee Belgian neutrality under present deliberation. Davignon appeals to Great Britain, France, and Russia to cooperate as guarantors of her territory and independence. *To employ concerted action to resist by force German violation, and at the same time to guarantee the future independence and integrity of Belgium.*
Belgium is happy to declare that she will undertake the defense of her fortified places.

King Albert addresses the Belgian Parliament, and closes his speech with :
The Belgian Army is equal to its task. The Government and myself have full confidence. The Government understands its responsibilities and will maintain them till the end to safeguard the supreme good of the country. If the stranger violates our territory he will find all Belgians gathered round their sovereign, who will never betray his constitutional oath. I have faith in our destinies. A country which defends itself, imposes respect on all and does not perish. God will be with us.

LONDON. German Ambassador to Great Britain, Prince Lichnowsky describes his departure.
Before my departure, Sir Edward Grey received me on 5th August at his house. I had gone there at his desire. Our departure was thoroughly dignified and calm. Before we left, the King had sent his equerry, Sir E. Ponsonby, to me, to express his regret at my departure and that he could not see me personally. Princess Louise wrote to me that the whole family lamented our going. Mrs. Asquith and other friends came to the embassy to say goodbye. A special train took us to Harwich, where a guard of honour was drawn up for me. I was treated like a departing sovereign. Thus, ended my London mission. It was wrecked, not by the perfidy of the British, but by the perfidy of our policy. At the railway station in London, Austrian Ambassador Count Albert Mensdorff appeared with his staff. He was cheerful, and gave me to understand that perhaps he would remain in London. To the English, he said ~ It was not Austria ~ but we, who had wanted the war. Prince Max Lichnowsky

The estimate for the attitude of Austria must be qualified. The estimate that Austria simply and foolishly hoped for a triumph over Serbia without the danger of war with Imperial Russia; that when she found out this to be a miscalculation, she tried to get out of it by diplomatic settlement with Russia direct ~ but found it too late. Austria's Ally, Germany, had gone to war with Russia and dragged Austria in. If this view is put to a German, I am told the answer is that Austria's delay in going to war with Russia was a pose, intended to clear Austria, even at the expense of Germany, of responsibility for war. The truth probably is, that there were sinister and reckless influences in Austria; and they were capable of any crime and any blunder. In Austria, as in Russia, there was no head with direction and grip of affairs. Sir Edward Grey

While these events were taking place, poor Princess Lichnowsky was walking in St James Park weeping, and the workmen were busily engaged taking down from the doors of the Embassy and Chancery the brass plates bearing the Imperial German arms. *Daisy* Princess of Pless

Prince Lichnowsky in his earlier years had held one or two unimportant diplomatic posts, but had for a long time led the retired life of a landowner and sportsman. He was a man whom it was impossible not to like, of most agreeable manners, a lover of hospitality, and capable of ready and sympathetic adaptation to the strange ways and customs of the English people. What was more important, he was a sincere friend of peace, anxious and eager to come to a settlement of all questions between Germany and Great Britain and, so far as one can judge, not unconscious to the real danger of his own countries enmeshment in the tangle of Austrian interests, Austrian ambitions, and Austrian intrigues in the Near East. The failure, tragic as it was, of Lichnowsky's honest efforts and good intentions was undoubtedly due to the fact that he never possessed the confidence, and was never made a party to the real designs of German policy in Berlin. It would seem that Lichnowsky never realized his own isolation. He was optimistic up to the end, and when at last he discovered that the ship was heading straight for the rapids and the cataract, his despair was pathetic to witness. H.H. Asquith
Prime Minister of Great Britain

BERLIN 6 AM. Sir Edward Goschen, all his staff, and the British correspondents depart Berlin in the early morning with no inconvenient incidents at the time of their departure. This however, cannot be said of the departure from Germany, of the French and Russian diplomatic staff.

During these first days of hostilities, there was a great spy excitement in Germany. People were seized by the crowds in the streets and, in some instances, on the theory that they were French or Russian spies, were shot. Foreigners were in a very dangerous situation throughout Germany, and many Americans were subjected to arrest and indignities. A curious rumour spread all over Germany to the effect that automobiles loaded with French gold were being rushed across the country to Russia. Peasants and gamekeepers and others turned out on the roads with guns, and travelling by automobile became exceedingly dangerous. A German Countess was shot, an officer wounded, and the Duchess of Ratibor was shot in the arm. It was sometime before this excitement was allayed, and many notices were published in the papers before this mania was driven from the popular brain. There were rumours also that Russians had poisoned the Muggelsee, the lake from where Berlin draws its water. There were constant rumours of the arrest of Russian spies disguised as women throughout Germany. Many Americans were detained under a sort of arrest in their hotels. James W. Gerard

Departing British Ambassador, Sir Edward Goschen describes his send-off from Germany.
Count von Wedell told me that he had been instructed to confer with me as to the route I should follow for my return to England. He said that he had understood that I preferred the route via the Hook of Holland to that of Copenhagen. I agreed to this and he said that I might be quite assured that there would be no repetition of the disgraceful scenes of the preceding night, as full precautions would be taken. He added that they were doing all in their power to have a restaurant car attached to the train, but it was rather a difficult matter. He also brought me a charming letter from Herr von Jagow couched in the most friendly terms.

In the morning, a strong force of police was posted along the usual route to the Lehrter Station, while the embassy staff were smuggled away in taxicabs to the station by side streets. We suffered no molestation whatsoever, and avoided the treatment meted out by the crowd to my Russian and French colleagues. Count von Wedell met us at the station to say good-bye on behalf of Herr von Jagow and to see that all arrangements ordered for our comfort had been properly carried out. A retired Colonel of the Guards accompanied the train to the Dutch frontier and was exceedingly kind in his efforts to prevent the great crowds which thronged the platforms at every station where we stopped from insulting us, but beyond the yelling of patriotic songs and a few jeers and insulting gestures we had nothing to complain of during our tedious journey to the Dutch frontier.

Also on the diplomatic train is Belgian Ambassador Baron Eugène Napoléon Beyens.
No one had laboured with more energy and skill to quench the flames lit by Austria and her ally than the representative of the French Republic at Berlin. Throughout these terrible days, nothing has been able to affect his coolness, his presence of mind, and his insight. Don't you think M. Cambon's attitude has been admirable? Remarked the British Ambassador to me, in the train that was whirling us far away from Berlin on 6th August. I cannot express my own admiration better than by repeating this verdict of so capable a diplomat as Sir Edward Goschen, who himself took a most active part in the vain attempt of the Éntente to save Europe from calamity.

THE BELGIAN FRONT. As three fully-equipped French cavalry divisions are entering Belgium, General Eric Ludendorff leads a heroic German pre-dawn attack to penetrate the ring of forts at Liége. British War Council order not the expected six ~ but four divisions to France to arrive on 9th August. In London, Field Marshal Lord Kitchener, as new Secretary of State for War stuns the British Cabinet by predicting that the war will last at least three years, and a force of millions will be required to win.

BERLIN. From the outbreak of war, the U.S. embassy is cut off for days from communication with the United States, but soon establishes two new lines of communication, through Italy and Denmark. Herr von Jagow asks Gerard ~ *To transmit through the American Legation a proposition offering Belgium peace and indemnity if no further opposition is made to the passage of German troops through Belgium.* Gerard forwards the belated proposition for peace to the Belgian Government, after which the U.S. State Department cables a message to Gerard from U.S. President Woodrow Wilson: *The United States stands ready at any time to mediate between the warring powers. Present such a proposition to the Kaiser.*

I therefore asked for an audience with the Kaiser, and received word that he will receive me at the Palais in Berlin on the morning of 10th August. I drove in a motor into the courtyard of the palace and was there escorted to the door which opened on a flight of steps leading to a little garden about fifty yards square, on the embankment of the River Spree that flows past the Royal Palace. As I went down the steps, the Kaiserin Augusta and her daughter, the Duchess of Brunswick, Viktoria Luise came up. Both stopped and shook hands with me, speaking a few words. I found the Kaiser seated at a green iron table under a large canvas garden umbrella. Telegraph forms were scattered on the table in front of him and basking in the gravel were two small dachshunds. I explained to the Kaiser the object of my visit and we had a general conversation about the war and the state of affairs. He said: Wait a moment; I shall write something for the President. *Then taking some telegraph blanks on the table, he wrote his reply to the President's offer. He wrote rapidly and fluently ~ always clear concise, consecutive and forceful. This reply, I cabled immediately to the State Department.* James Gerard

For the President of the United States Personally *10 VIII 14*

1. H.R.H. Prince Henry was received by his Majesty King George V in London, who empowered him to transmit to me verbally, that England would remain neutral if war broke out on the Continent involving Germany and France, Austria and Russia. This message was telegraphed to me by my brother from London after his conversation with H.M. the King, and repeated verbally on the 29th July.
2. My Ambassador in London transmitted a message from Sir E. Grey to Berlin saying that only in case France was likely to be crushed, England would interfere.
3. On the 30th my Ambassador in London reported that Sir Edward Grey in course of a private conversation told him that if the conflict remained localized between Russia ~ not Serbia ~ and Austria, England would not move, but if we mixed in the fray she would take quick decisions and grave measures; that is, if I left my ally Austria in the lurch to fight alone, England would not touch me.
4. This communication being directly counter to the King's message to me, I telegraphed to H.M. on the 30th thanking him for kind messages through my brother and begging him to use all his power to keep France and Russia ~ his Allies ~ from making any warlike preparations calculated to disturb my work of mediation, stating that I was in constant communication with the Tsar. In the evening the King kindly answered that he had ordered his Government to use every possible influence with his Allies to refrain from taking any provocative military measures. At the same time, H.M. asked me if I would transmit to Vienna the British proposal that Austria was to take Belgrade and a few Serbian towns and a strip of country to make sure that Serbian promises on paper should be fulfilled in reality. This proposal was in the same moment telegraphed to me from Vienna for London, in conjunction with the British proposal; besides, I had telegraphed to the Tsar the same as an idea of mine, before I received the two communications from Vienna and London, as both were of the same opinion.
5. I immediately transmitted the telegrams vice versa to Vienna and London. I felt that I was able to tide the question over and was happy at the peaceful outlook.
6. While I was preparing a note to the Tsar the next morning, to inform him that Vienna, London and Berlin were agreed about the treatment of affairs, I received the telephone from the Chancellor that in the night before the Tsar had given the order to mobilize the whole of the Russian Army, which of course, also meant against Germany; whereas up till then the southern armies had been mobilized against Austria.
7. In a telegram from London my Ambassador informed me he understood the British Government would guarantee neutrality of France and wished to know whether Germany would refrain from attack. I telegraphed to H.M. the King personally that mobilization being already carried out could not be stopped, but if H.M. could guarantee with his armed forces the neutrality of France I would refrain from attacking her, leave her alone and employ my troops elsewhere. H.M. answered that he thought my offer was based on a misunderstanding, and as far as I can make out, Sir E. Grey never took my offer into serious consideration. He never answered it. Instead, he declared England had to defend Belgian neutrality, which had to be violated by Germany on strategical grounds, news having been received that France (offered five Army Corps and) was preparing to enter Belgium, and the King of Belgians having refused my petition for a free passage under guarantee of his country's freedom.
I am most grateful for the President's message. *William IR*

For the President of the United States Personally

I

10/VIII 14

Telegraphie des Deutschen Reichs.

Telegramm

Sr. Majestät des Kaisers und Königs.

1) H.R.H. Prince Henry was received by his Majesty King George V. in London who empowered him to transmit to me verbally that England would remain neutral if war broke out on the Continent involving Germany & France, Austria & Russia. This message was telegraphed to me by my brother from London after his conversation with H.M. the King, & repeated verbally on the 29th of July.

2) My Ambassador in London transmitted a message from Sir E. Grey to Berlin saying that only in case France was likely to be crushed England would interfere.

3) On the 30th my ambassador in London reported that Sir E. Grey in course of a "private" conversation told him, that if the conflict remained localized between Russia – not Servia – & Austria

II

Telegraphie des Deutschen Reichs.

Telegramm

Sr. Majestät des Kaisers und Königs.

England would not move, but if we "mixed" in the fray she would take quick decisions & grave measures: i.e. if I left my Ally Austria in the lurch to fight alone England would not touch me.

4) This communication being directly counter to the Kings message to me I telegraphed to H.M. the thanking him for his kind message through my brother, & begging him to use all his power to keep France & Russia – his Allies – from making any warlike preparations calculated to disturb my work of mediation, stating that I was in constant communication with H.M. the Czar. In the evening the King kindly answered that he had ordered

III

Telegraphie des Deutschen Reichs.

Telegramm

Sr. Majestät des Kaisers und Königs.

his Government to use every possible influence with his allies to refrain from taking any provocative military measures. At the same time H.M. ~~proposal~~ asked me if I should transmit to Vienna the British proposal that Austria was to take Belgrade & a few other Servian towns & a strip of country as a "main-mise" to make sure that the Servian promises on paper should be fullfulled in reality. This proposal was in the same moment telegraphed to me from Vienna for London, quite in conjunction with the British proposal besides I had telegraphed to H.M. the Zar the same as an idea of mine, before I recieved the two communications from Vienna & London. As both were of the same opinion, I

IV

Telegraphie des Deutschen Reichs.

Telegramm

Sr. Majestät des Kaisers und Königs.

4) I immediately transmitted the telegrams vice versa to Vienna & London. I felt that I was able to tide the question over & was happy at the peaceful outlook.

5) While I was preparing a note to H.M. the Czar the next morning, to inform him that Vienna, London, & Berlin were agreed about the treatment of affairs I recieved the telephones from H.E. the Chancellor that in the night before the Czar had ~~mobilized~~ given the order to mobilize the whole of the Russian Army, which was of course also meant against Germany, whereas up till then the southern Armies had been mobilized against Austria.

6) In a telegram from London my Ambassador, ~~by another~~

V

Telegraphie des Deutschen Reichs.

Telegramm

Sr. Majestät des Kaisers und Königs.

informed me he understood, British Government would guarantee neutrality of France, & wished to know whether Germany would refrain from attack. I telegraphed to H.M. the King personally that, mobilization being already carried out could not be stopped, but if H.M. could guarantee with his armed forces the neutrality of France, I would refrain from attacking her, leave her alone, & employ my troops elsewhere. H.M. answered that he thought my offer was based on a misunderstanding; & as far as I can make out Sir E. Grey never took my offer into serious consideration. He never answered it. Instead he declared England had to defend Belgian neutrality, which had to be violated by

Telegraphie des Deutschen Reichs.

Telegramm

Sr. Majestät des Kaisers und Königs.

Germany on strategical grounds, ~~knowledge~~ news having been recieved that France was already preparing to enter Belgium & the King of the Belgians having refused my petition for a free passage under guarantee of his country's freedom. I am most grateful for the Presidents message.

William I.R.

10th AUGUST 1914

James Gerard
U.S Ambassador to Germany 1913-17

Brand Whitlock
U.S. Ambassador to Belgium

Hugh Gibson
U.S. Secrétaire de la Légation Belgium

BERLIN. Kaiser Wilhelm asks Ambassador Gerard to cable his letter to President Wilson; then to the U.S. State Department, and simultaneously pass it to the press. Gerard dispatches the documents. Kaiser Wilhelm continues to talk despondently to James Gerard, about the tragedy of the Great War. *The British change the whole situation ~ an obstinate nation. They will keep up the war. It cannot end soon.* After lunch, a man holding high position in Germany (who never reveals his name) sends for Gerard. He said: *You had an interview today with the Kaiser… What happened?*
Gerard: *The message given to me for the President, was intended for publication by the Kaiser.*
He said: *I think you ought to show that message to me; you know the Kaiser is a constitutional Monarch and there was once a great row about such a message.* Ambassador Gerard shows him the Kaiser's message. He reads it and advises: *I think it would be inadvisable for us to have this message published, and in the interest of good feeling between Germany and America. If you cable it, please ask that publication be withheld.* Gerard complies with this request, and President Woodrow Wilson agrees to suspend its publication.

PARIS. France formally declares war on Austria-Hungary in the evening, following the failure of Vienna to provide any satisfactory explanation of the movements of A-H troops from the Tyrol who have now crossed Lake Constance and are now camped at Basle on the Swiss border, in preparation to attack French forces operating in Alsace. A-H Ambassador Count Nilolaus Szécsen von Temerin is sent for and asked to give some definition of the Austrian position. As this is not forthcoming, the Quai d'Orsay officials therefore hand the A-H Ambassador his passports, and war is duly declared. Count Nikolaus Szécsen von Temerin with his staff, depart for Vienna that night.

BRUSSELS. *On my way back to the Legation, I stopped at the British Legation to see Sir Francis Villiers. He told me of Germany's request for an armistice of twenty-four hours at Liége, to bury their dead ~ piled up, Sir Francis said, in windrows on the glacis of the forts. The armistice has been refused. Sir Francis, of course, was delighted. Indeed, the heroic resistance of the little Belgian Army in the forts along the Meuse ~ the forts that Leman, who now commands them, helped to construct ~ has produced an extraordinary enthusiasm that vibrates nervously in the sparkling sunlight; a kind of contagious exhilaration, intoxication, when men meet each other on the streets and say: The forts still hold out! When I came back to the Legation I was met at the door by the big officer of the civil guard, who greeted me in his deep bass voice; Your Excellency, have you any news? I told him of the heroism of his comrades in arms, of the request for an armistice, and his face lit up. It is glorious! It is a day that will go down in history. And he went to tell his men.* Brand Whitlock

BELGIAN FRONT. The British Royal Flying Corps fly across the English Channel to France.
All Belgian forts to the east and north of Liége have fallen, and are now surrendering to the Germans.

BERLIN 14th August. Kaiser Wilhelm leaves Berlin for the forward German GHQ at Coblenz.
Thereafter he practically abandons Berlin, and lives either at Pless or places near to the Western front.

The A-H Embassy in Paris ~ Ambassador Count Nikolaus Szécsen de Temerin with his staff.

Minerva armoured touring cars of the Belgian Army.

BELGIUM 15th August. *The Belgian troops have so far had to dam the flood of Germans with little or no help from the allies. The Kaiser expected to sweep through Belgium with little opposition and be fighting in France in three days! The Belgians have knocked his schedule out by twelve days already! My military advisers tell me of the German's great necessity for a quick campaign in France, to get their army back in time to head off the Russian flood when it begins to pour over the northern frontier ~ the loss of this much time is equivalent to their loss of a great battle. The moral effect is also tremendous.* Hugh Gibson, Secrétaire de la Légation USA

As the vast German Army storms westward to assault Belgium, King Albert's heroic forces cover their homeland with a screen of cavalry, supported by machine-gun units and horse artillery. Inspired to shake the aggressors, Lieutenant Charles Henkart directs the engineers of Cockerill & Co. Antwerp to attach some protective sheets of steel plate to the sides of his two Minerva touring cars while leaving their tops open for a soldier to operate a lethal Hotchkiss machine gun. Henkart then places these armoured cars at the complete disposal of the Belgian Army. These simple armoured vehicles are followed by more formal designs from the Minerva Motor Car Company of Antwerp, whose workers fit all Minerva armoured cars with spotlights, for evening raids against the Kaiser's forces. Lieutenant Charles Henkart's squadron of armoured cars, driven by the most capable racing drivers in Belgium make repeated forays against the German Army, reconnoitering to acquire information, raiding to capture prisoners, and providing effective fire support to numerous Belgian counterattacks. Operating out of Antwerp, the Minerva crews carry out many raids against an enemy still marching in open order across fields or along country roads, where a single machine gun can often create havoc.

I witnessed an example of the cool daring of these Belgian drivers during the fighting around Malines. I was watching the withdrawal under heavy fire of some Belgian troops, when a Minerva armoured car, the lean muzzle of its machine-gun peering from its turret, dashed past toward the Germans at fifty miles per hour, spitting a murderous spray of lead as it bore down on the advancing Germans. Bullets were kicking up spurts of dust in the road and ping-panging against the turret. By this time, the German artillery had acquired their range and shrapnel burst all about them. As the car straightened out for its retreat, the Belgians gave the Germans a jeering screech from their horn, a parting blast of lead from their Hotchkiss machine-gun, and went racing off towards Antwerp. Alexander Powell ~ *New York World*

Every man in France is envying the young dragoon officer, Lieutenant Bruyant, who has been given the first Cross of the Legion of Honour in the war. The lieutenant with six men was scouting near the frontier, when suddenly he saw a number of horsemen moving a good way off, and made them out to be a patrol of 27 Uhlans. Shots were exchanged and a German fell. Then the Uhlans cantered away. They were four to one, but did not care to fight. The French followed up resolutely, but the Germans kept their distance. When the dragoons trotted, the Uhlans trotted too. Now the former would gallop across a bit of open country, and the Germans would gallop away just as quickly. Evidently, they were making for shelter. Soon Lieutenant Bruyant saw that they were trying to reach a wood, where they could take cover. No time was to be lost. He knew that if they got there they would escape him. Now was the moment to unchain the ardor of his men. He orders ~ Draw swords! Charge! The seven spurred their horses and fell upon the 27 with shouts of defiance. The shock demoralized the Germans, who made no stand at all. One was killed by a lance thrust. The officer in command was drawing his revolver when Lieutenant Bruyant cut him down with his sabre. Six more were wounded and knocked off their horses. The rest fled in disorder.

Charles Inman Barnard ~ Paris correspondent, *New York Tribune*

19th AUGUST 1914

French artillery passing by Chateau du Versailles

Eighteenth day of mobilization.
Fine summer weather, with a light northerly wind.
Absolute silence concerning military movements in Belgium.
No official communication was made today at the Ministry of War.
Parisians feel that momentous events are about to take place but look forward with calm confidence.

I called upon my old friend, René Baschet, manager of l'Illustration. I hastened to tell Baschet that I had just received a private telegram from Rome announcing that the Pope was so ill that his physicians did not think that His Holiness could live through the night. Baschet paid tribute to Lord Horatio Kitchener's instructions: To every soldier of the British Expeditionary Forces ~ You may find temptations, both in wine and women. Resist both temptations, and while treating all women with perfect courtesy, you should avoid any intimacy.

At the Ministry of Foreign Affairs, I met Jules Cambon, French Ambassador at Berlin, who after being treated discourteously by the Germans and dealt with practically as a prisoner, reached Paris by way of Denmark and England. It would have been indiscreet to ask Jules Cambon to disclose diplomatic secrets, he drew an important distinction between German diplomacy, and the German military clique. The former were willing only to go so far as risking a war, while the latter seized the opportunity to bring on the war and attack France. The discussion lasted two days, the military caste became omnipotent, and from that moment war was inevitable. In regard to France, Germany constantly repeated the formula : Put strong pressure upon Russia, your ally, to prevent her from helping the Servians! To this France replied : Very good, but you yourself should put strong pressure upon Austria, your ally, to prevent her from provoking a catastrophe! To this Germany rejoined : Ah! But that is not the same thing! Thus it was in this cercle vicieux that the diplomatic conversation continued, which under the circumstances could end in nothing else but war.

The curator of the Louvre Museum has taken every possible precaution to ensure the safety of the works of art under his care. The Venus of Milo has been placed in a strongroom lined with steel plates ~ a sort of gigantic safe ~ and stands in absolute security from any stray Zeppelin bombs. The Winged Victory of Samothrace is also protected by armor plates. Mona Lisa once more smiles in darkness. The Salle Greque, containing masterpieces of Phidias is protected by sand bags. Many unique treasures of statuary and painting are placed in the cellars. The upper stories of the Louvre, which are roofed in glass, are being converted into hospital wards, and thus the collections of the National Museum, which belong to all time, enjoy the protection of the Red Cross flag.

I made a brief trip to Versailles, which has been transformed into an arsenal and a vast supply depot for food and forage. Troops of the military commissariat train are cantoned in the parks and shooting preserves of Prince Murat. Versailles is the mobilization center or assembly for the S.W. military regions, and over 50,000 men have been equipped here and sent on to their destinations at the front. Herds of cattle and flocks of sheep are grazing on the meadows of the chateau.

Charles Inman Barnard, *New York Tribune*

BRUSSELS. *A day of dreadful excitement; the German Army has entered Brussels and we have witnessed the humiliation of a proud, old city, the finest in the world. How to describe it!*

We went to bed last night not knowing what the night would bring, and yet to my surprise thinking it over, I fell into a heavy sleep and was awakened very early this morning by loud knocks on my door. Slipping into my dressing gown, I opened the door, and there stood poor Gustave, weary, haggard, and frightened, looking as though he had not been to bed at all ~ as indeed he had not. It had been decided by the authorities after all, not to offer any resistance; the civil guard had been withdrawn and disbanded, and this had been done on orders from the King at Antwerp. The German Army was to enter the city during the day.

And so we waited for eleven o'clock, straining our ears. But we saw no Germans and we had a kind of perverted disappointment, as though we were waiting for a circus parade. Excited crowds were swaying this way and that, running in the direction of the hoof beats. And there, between the silent crowds packed along the sidewalks, and turning into the rue de la Montagne, which twisted away to our left, riding in columns of twos, in grey uniforms, the black and white pennants fluttering from their lances, was a squadron of German hussars, and as they rode they chanted in chorus, Heil Dir im Siegeskranz. The crowds sullen and silent in the glitter of the sunlight, the horses' hoofs clattering on the stones of the uneven pavement, the lances swaying, the pennants fluttering, and that deep-throated chant was all weird, unreal. The scene had the allure of medievalism; something terrible too, that almost savage chant, and those grey hosts pouring down out of the Middle Ages into modern civilization. The column had halted, the chanting ceased; the last two troopers promptly turned their horses around ~ no rear attacks! The column moved again, taking up their savage hymn, and rode out of sight.

We thought we had seen it all. All we had seen was a mere outpost; vedettes as it were, a little advance guard, for there up and down the boulevard, as far as we could see under the trees, were the lumbering guns, nervous horses, and the glint of fields of bayonets and those grey, grim hordes. We stopped, and sat there in the motor, and gazed. Heavy guns, their vicious mouths of steel lowered, drawn by superb horses, are lumbering by, the officers erect, either very thin of the Prussian type, with hard and cruel faces, scarred by dueling, or the heavier type, with rolls of fat at the back of their necks, and with heavy red brutal faces smoking cigarettes, looking about over the heads of the silent, awed, saddened crowd, with arrogant, insolent, contemptuous glances. Finally, I fancied that each regiment of infantry was supported by a troop of cavalry, and followed by a battery.

The uniforms seemed to be all alike; all of that rather ugly grey, helmets and every bit of metal hidden by grey covers, though the officers' uniforms of course fitted them and the men's did not always. Each officer carried his maps in a leather case to protect it and a little electric lamp hung about his neck. I could imagine them studying those maps at night in the field in the glow of those little electric lamps; and this efficiency was everywhere marked, even down to the extra pair of boots each soldier carried. Then came a long line of steel pontoons, the mud of the Meuse still on their bottoms. Then the commissariat, cookstoves with fires burning and smoke coming from the long stacks, soup simmering in their great kettles.

Then, suddenly, far down the boulevard, the crack of the music of a military band, with the high shrill, fierce notes, the horrid clang of mammoth brass cymbals, music calculated like that of savages to strike terror. A great band passed by, and then officers and again the grey hordes, pouring down over the peaceful earth, bringing ruin and destruction, and the grey hordes filed on, like grey ghosts in a grey twilight.

We drove on to the Grande Place ~ that square of golden beauty ~ and there, artillery was parked and cookstoves were steaming. There were mounted sentinels at the entrances and the soldiers were comfortably settling themselves; the horses were munching their provender. The sentinels saluted us as we entered. The Germans demanded hostages and a contribution de guerre of 50,000,000 francs, and enormous quantities of food and forage. Max refused the hostages, held out, and gained his point. But the levy must be paid. For three days the Germans will pass through Brussels. The troops that came through comprise the 4th Army Corps. All night long and all through the day the grey hosts have been passing through the city, cavalry, infantry and artillery, ammunition carts and supply wagons, a never-ending stream of over 200,000 men. The population is calm, the calm of sorrow and humiliation. The walls of Brussels bear placards in French and German signed by General Sixt von Arnim warning of reprisals if any overt act of hostility occurs. The Belgian flags have been taken down. Around the town, German officers race in automobiles; in each motor one or two soldiers with guns across their knees ready to fire at the slightest provocation. An appalling sight. Tonight, the grey ghosts still pour in by the Chaussée de Louvain and file through all the streets---a horde like Attila, the scourge of God. Poor little Belgium, and poor lovely Brussels! It is awful to witness the humiliation of a proud city. Ambassador Brand Whitlock

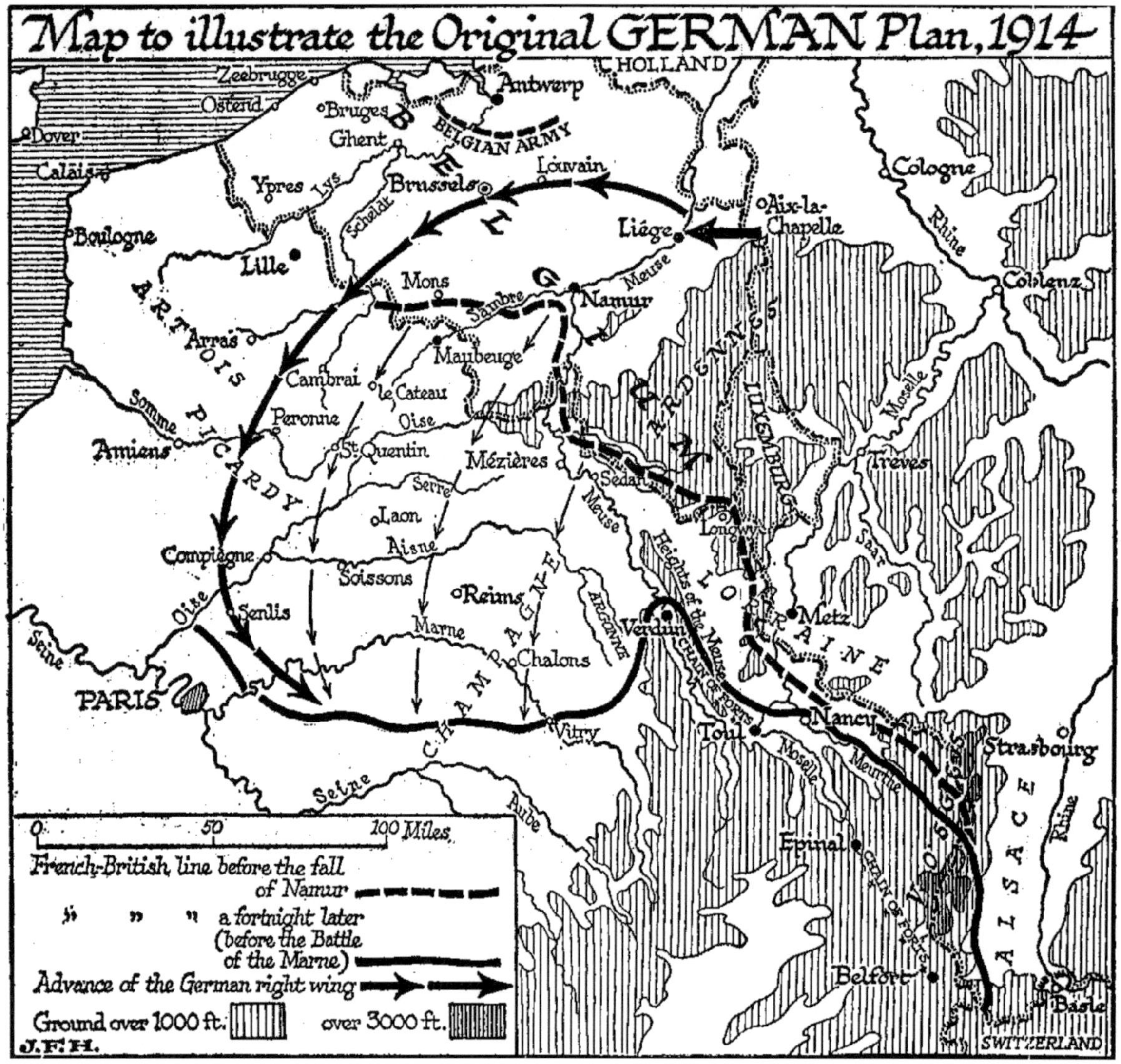

Only via the Kingdom of Belgium into northern France are there level fields of vast breadth for most of the distance, clear of obstacles to offer every logistic facility for the rapid deployment and continuing supply of a large military force. Highways, rail lines and even useful canals all stream in the decisive direction. Metz, as the German city nearest the French border is 180 air miles from Paris. Via the Belgian plain, the German Army has to march 250 miles, yet the route provides smooth going. Once the invaders win through Liége, they spread westwards around the northern Ardennes Forest, through Louvain and Brussels, and on an expanding front to then continue southwest past Mons, Valenciennes, Cambrai, Saint Quentin, and on to Paris…

During much of September 1914, the Belgian Army stings the Germans by occasional raids from Malines and Antwerp. On the 27th the Germans capture Malines, and the next day began shelling the forts defending Antwerp. The heavy German siege guns prove irresistible, and on 8th October 1914, Belgian and British forces evacuate the city. Following the fall of Antwerp, the suffering in Belgium becomes even more widespread and intense. Numerous towns and villages are put to the torch, many of their inhabitants shot, and much Belgian property is stolen. Heavy money requisitions are laid upon the conquered Belgian cities. Members of the Belgian Cabinet visit the U.S.A. and Great Britain to appeal for aid for the Belgium Relief Fund, and their cooperation averts a catastrophe in Belgium.

Henry Morgenthau
U.S. Ambassador to the Ottoman Empire

Baron Hans von Wangenheim
German Ambassador to the Ottoman Empire

Sir Louis du Pan Mallet
British Ambassador to the Ottoman Empire

Cardinal Désiré-Joseph Mercier of Mechelen

Princess Elena of Montenegro
Queen Elena of Italy

Vittorio Emanuele III
King of Italy

Crown Prince Alexander of Serbia

Admiral Andrei Eberhart
CO Black Sea Fleet 1911-17

Admiral Wilhelm Souchon

In August of 1914, Sir Francis Bertie held the plum posting in the British Diplomatic Service ~ the ambassadorship to Paris. Not only did it carry a yearly salary of £11,500 (when Foreign Secretary Grey was paid just £5,000) but the Ambassador also got to live at the Hôtel de Charost, the grand mansion bought by the Duke of Wellington in 1814 from Napoleon Bonaparte's own sister, Pauline Borghese, to serve as the British Embassy. Bertie had been due to retire in 1914, but with the outbreak of war he was confirmed in his post for the duration of hostilities. During these years, Sir Francis kept a diary. With the German Army heading towards the French capital, Bertie wrote the following observations ;

29th August
The Germans are at Abbeville. There will be a great battle tomorrow to prevent their breakthrough.
The French Government are preparing to move south and the embassy will follow.

30th August
Fighting is going on at Compiegne ~ the Germans in overwhelming numbers may be before Paris in a few days. I am burning confidential papers, and preparing to leave at short notice when the Germans seem sure to succeed in occupying Paris. They are quite capable of looting everything here, in which case I shall lose all I possess.

31st August
The military position is better today than yesterday; I fancy that the flight to the south will be on Wednesday. We have suffered terrible losses, but the Germans have lost many more men. That they can afford better than we.

2nd September
We are very busy packing up. We are off to Bordeaux tonight.
This may be the last time that I shall ever write from this house, for the war is likely to be long.

3rd September
BORDEAUX. We arrived between 1 & 2 PM today.
Endless cases containing archives filled a big and small furniture van. Such a conglomeration! We came straight to the house allotted to us ~ full of fine things. The drawing rooms have Louis XVI chairs, statues and tapestries. The country round here is like some parts of England, barring the vines ~ hedgerows and meadows.

4th September
There are none of the horrors of war here, no wounded and no refugees from the invaded provinces.
How I wish that a victory such as that of the Russians over the Austrians could be announced on our Front.

6th September
I am very anxious as to what the next three days may produce.
I have had a line from Declassé saying that the offensive is imminent.
Russian troops have advanced into East Prussia. To the south another Russian Army has knocked out a large Austrian force and occupied Galicia. The Germans are forced to withdraw vital troops from the Western Front. The Tsar has honoured his commitment to France ~ Fancy Russia the saviour of civilization!

7th September
The military situation is a little better today. We are attacking in the Marne.
We breakfast and dine every day at Le Chapon ~ unless you have reserved a table, the door is shut in your face.

10th September
Matters military look better.
We have driven the Germans back some 40km.

11th September
Thank goodness, the war news is better today. If the Belgians had not resisted and we had not come to the rescue with our troops and the fleet, the Germans would be in Paris now.

13th September
At last there is really good news.
With the recent victories, the spirits of the French are reviving.
The Germans are on the run, and they are, it is reported very tired and demoralized.
Yet, there are many people here already in mourning, and the war has only just begun.

16th September
At Senator Charles Jonnart's house the officers of Indian regiments drank all his wine and all his old brandy, He had a very fine cellar. I dined with Madame de G. last night; we were nine, with a dirty tablecloth.

17th September
We await the result of the battle that has been going on for the past three days.
This place is full of soldiers of all armies and of all colours. The Germans have entrenched themselves strongly.
Things here go on much as usual ~ with a conviction that in the end France, Russia, and England must win.

20th September
Went to Château Lafite of the Rothchilds family.
It is most unkempt and full of terrible Louis Phillipe furniture.
We saw the cellars in which there are 80,000 bottles of wine.
The old couple drank chocolate, and when their cups were half drunk they filled them up with tea.

24th September
Still awaiting the result of the battle of the Aisne.
Here they are too confident that the Germans will soon be driven out of France.
I hear that the since the war began French casualties have been 40,000 killed and 140,000 wounded.

29th September
Every day we are told that a great and deciding battle is to be brought to a finish, but every day it is ~ demain…
Everyone is confident that the Germans cannot break thru the Anglo-French lines. God grant that it will be so.

9th October
We are trembling for Antwerp.
There are reports of it having already surrendered after bombardment. Its fall would be serious for us all.

11th October
Antwerp has fallen!
All that can be said in mitigation is that it might have been worse: the Belgian Army might have been shut in.
The Queen of the Belgians walked out of Antwerp with the King, a long trudge.
It is such conduct that justifies sovereignty to a people.

12th October
Our plan of campaign will be revised. A good many French Generals have been removed from their commands.
Reason for Joffre's delay ~ now it is said that he is waiting for rain as the German heavy pieces stick in the mud.
The red trousers of the French soldiers make good targets for German bullets.
It is suggested that workmen's blue cotton trousers be worn over them.

5th November
Turkish ambassador left with an unpaid hotel bill of £480. He said he would send a cheque from Constantinople!

10th November
The military people seem confident of being able to prevent the Germans from piercing our lines, and getting to Boulogne, Calais or Dunkirk. The Germans have made their great effort and failed. Their plan of campaign that depended entirely on a rapid arrival at Paris has to be recast. Manoury and Galiieni requisitioned all the autotaxis and automobiles in Paris to convey their infantry to the north. Advancing on Paris, von Kluck found French troops from Paris on his right and fearing for his flank, he bore off to the east.

Grey's hens are coming home to roost,
He must feel very embarrassed about promising Constantinople to Russia.
Isvolsky is against the Greeks taking any part in the Dardanelles business…
He would like France and England to have all the hard knocks for the benefit of Russia.

The French have set their hearts on a Japanese army of 400,000 in France to help clear out the Germans.
Will the Japanese be willing, and if so, at what price?
French Indo-China is thought to be suitable bait…

The next move of an interested Empire with cause to enter the Great War comes in the Far East. Japan and Britain having entered into an alliance in 1902, Japan proposes to the British Foreign Office that Japan will enter the Éntente to secure (and acquire) some of Germany's nearest Pacific territories. On 7th August, Great Britain requests Japan to seek and destroy all German warships presently in and around Chinese waters. Tokyo thus drafts an ultimatum to Berlin, which is conveyed on 16th August;

TOKYO 16th August 1914. We consider it highly important and necessary to take measures to remove the causes of all disturbances of the peace in the Far East, and to safeguard the general interests as contemplated by the agreement of alliance between Japan and Great Britain. In order to secure an enduring peace in Eastern Asia, the Imperial Japanese Government gives advice to the Imperial German Government to carry out the following. First. To withdraw immediately from Japanese and Chinese waters German armed vessels of all types, and to disarm at once those which cannot be so withdrawn. Second. To deliver on a date not later than 15th September to the Imperial Japanese authorities, without consideration or compensation, the entire territory of Kiao-chau. The Imperial Japanese Government announces, that in the event of its not receiving by noon on 23rd August 1914, an answer from the Imperial German Government, signifying its unconditional acceptance of the above advice, Japan will be compelled to take such action, as she may deem necessary to meet the situation.

When the time limit set for Berlin's expected reply came and passed with no official communiqué, Japan, faithful to her alliance with Britain consequently declares war against Germany on 23rd August:

Issued at Tokyo, 23rd August 1914 at 6 PM. We, by the Grace of Heaven, Emperor of Japan, seated on the throne occupied by the same dynasty from time immemorial, make the following proclamation to our loyal and brave subjects. We hereby declare war against Germany, and we command our army and navy to carry on hostilities against that empire with all strength, and we also command our competent authorities to make every effort, in pursuance of their duties, to attain the national aim by all means within the limits of the law of nations.

Since the outbreak of war in Europe, the effect which we view with grave concern, we have entertained hopes of preserving the peace in the Far East by the maintenance of strict neutrality, but the actions of Germany has at length compelled our ally Great Britain to open hostilities against that country, and Germany is at Kiao-chau, its leased territory in China, busy with warlike preparations, while its armed vessels cruising the seas of eastern Asia threaten our commerce and that of our ally. Peace of the Far East is thus in jeopardy. Accordingly, our Government and that of his Britannic Majesty, after full and frank communication with each other, agreed to take such measures as may be necessary for the protection of general interests contemplated in the 1902 Agreement of Alliance. We on our part, being desirous to attain that object by peaceful means, commanded our Government to offer with sincerity an advice to the Imperial German Government. But on the last day appointed for that purpose, however, our Government failed to receive an answer accepting their advice. It is with profound regret that we, in spite of our ardent devotion to the cause of peace, are thus now compelled to declare war, especially at this early period of our reign. It is our earnest wish that by the loyalty and valour of our faithful subjects peace may soon be restored, and the glory of the empire be enhanced.

Berlin, as before, ignores the Japanese. German GHQ instead directs the garrison at Kiao-chau to defend itself against any attacks made by the Japanese, and when the CO there hears of the Japanese declaration, he issues a statement to invite the eager enemy ~ *If you want the place, come and fight for it!* So much for the theory long cherished in Berlin's military circles that because Prussian officers had trained the Japanese Army, Japan would join Germany. In reality, while Japanese diplomats are packing their trunks and preparing to leave their Embassy in Berlin for home, an uninformed German crowd with flags assemble out front to cheer Japan ~ unaware they are the latest allies of the Éntente! As Vienna refuses to recall their cruiser SMS *Kaiserin Elisabeth* from Tsingtao, Japan declares war on Austria-Hungary on 25th August, and two days later, Austria-Hungary declares war against Japan, and the following day on Belgium. Within the week, Japanese and British naval forces commence their blockade of Tsingtao, the principal German city of her Pacific territories. Troops from Japan, India, and elsewhere land to besiege German-held Kiao-chau. After a six-week siege, and a preliminary bombardment preparatory to launching a final assault, the garrison capitulates on 7th November 1914.

GERMAN COLONIAL POSSESSIONS IN THE NORTH PACIFIC

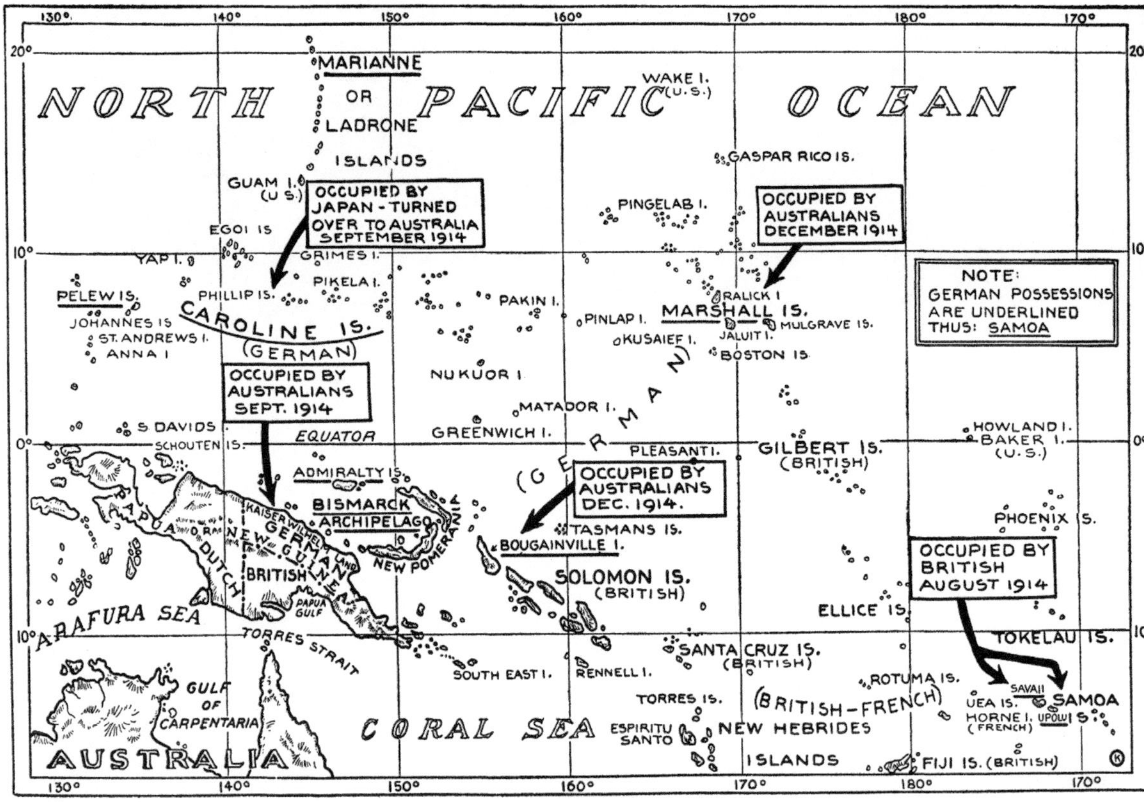

Before the war, Australia knew Imperial Germany from two facets of distinct first-hand evidence. One was the *peaceful penetration* of Australian industry by energetic German firms and businesses, especially in the purchase and refining of Australian minerals. The other was the presence of German New Guinea, which bordered Dutch Papua and the Australian territory of British New Guinea.

Australia observed the presence of Imperial Germany in the South Pacific as far back as 1883, when the colony of Queensland grew fearful of growing foreign influences ~ principally German, when Germany annexed the S.E. part of New Guinea against the wishes of the British Government. German interest extended to the remaining third of the island and on 3rd November 1884, the German flag was raised over *Kaiser-Wilhelmsland,* the Bismarck Archipelago, and the German Solomon Islands. On 17th May 1885, the Kaiser granted an Imperial charter to the newly founded *Neuguinea-Kompanie.* Only then, did Britain proclaim a protectorate over S.E. New Guinea and groups of nearby islands.

The thrusting German programme of investment and organization throughout the Pacific saw them establish a network of wireless stations on all their colonial acquisitions, and these powerful radio stations that kept in touch with the German Naval HQ on the China coast were located in secret locations and the power output of these transmitters was kept as secret as possible by the deliberate use of low-powered transmissions to lull listeners into under-estimating their potential radio strength. Two days after the outbreak of war, Britain asked the Australian and New Zealand Governments to seek out and destroy German wireless stations in the Pacific, and occupy all territories seized. Following the British request, the New Zealand expedition asked London for details of what defences to expect. London replies ~ *For information regarding the defences of Samoa, see Whitaker's Almanac.*

Of the expedition to Samoa, *The Sydney Bulletin* reports ~

HMAS Australia and HMAS Melbourne called in at Wellington for a force of New Zealanders. The vessels then steamed to Noumea to pick up the French cruiser Montcalm. Noumea had been worried since the war began the German fleet from China would come and bombard the place. Notices went up to the effect that five shots would signify the arrival of the Germans, and that every inhabitant was to grab rations and then to evacuate. The welcome the French handed to us would have stirred the blood of a jellyfish. The night before we reached Samoa, an English military officer on board told me that it was remarkable, and highly creditable, the rapidity with which the men had adapted themselves to their changed circumstances. They reached Apia on Samoa on the 28th and without a blow being struck, Samoa proved a walk over. Schoolboys with Shanghais and a hatful of rocks could have taken it. The German fleet hadn't been around for eleven months. *The Sydney Bulletin*

With Samoa secured, Australian forces prepare to seize the remaining German territories in the Pacific from the Caroline and Marshall Islands in the north, to New Britain and German New Guinea in the south, to deny the enemy of their intelligence communications network and logistical facilities on these islands capable of supporting warships of the Imperial German Navy based in China.

The naval group led by Admiral Sir George Patey, commander of the Royal Australian Navy onboard his flagship HMAS *Australia* leaves Sydney Heads, and the ships turn north. In the early hours of 11th September, as the naval force arrives off Kabakaul Bay some 80 kilometres east of Rabaul for what promises to be the first Australian action of the Great War, the morning is crisp and clear. Their objective is to locate and destroy the newly built wireless station, suspected to be powerful enough to communicate with Berlin directly. Crates seen at the shoreline and wheel marks in the road indicate this is the route leading to the wireless station.

As the Australians begin their advance inland through jungle and plantations to seize the two wireless stations thought to be inland; two scouts Petty Officer G.R. Palmer and Able Seaman Eastman deviate from the main road and push into thick jungle a short way away and almost parallel to the road they have just left. A few minutes later, the two scouts near a clearing and see a trench to the rear positioned across the road ahead and on a blind corner. Three Germans supervising 30 native troops, now waiting to ambush the Australians, are occupying this trench as the German first line of defence. Petty Officer Palmer as the lead scout, raises his rifle and fires at the German NCO in charge, wounding him in the right hand. In the following exchange of fire, several New Guinean troops are killed and the others, including the wounded German are taken prisoner after a short skirmish.

The trench surrendered with two Germans and 30 natives. Here an incident worth noting occurred, as it no doubt saved a big disaster to us. Captain Harkin, of the machine gun section, who was then present, asked one of the Germans if there were any rifles, ammunition, or other explosives on the spot. The German replied that there was nothing of the kind about to his knowledge. Being suspicious, however, we searched the surrounding bush, and found a large case of ammunition. Captain Harkin placed his hand menacingly on the butt of his revolver, and coolly approached the German ~ I understood you to say there was nothing of this kind hereabouts? The German stammered a reply to the effect, that he had overlooked the ammunition ~ it had slipped his memory. Well, replied Captain Harkin calmly, tapping his revolver ~ Take care that your memory doesn't fail you again!

Lieutenant Bond, Captain Travers, Corporal Conrad Eitel, an interpreter from the machine gun section, and a German prisoner Kempf, continue towards the wireless station. Arriving at a police barracks one kilometre from the wireless station, eight Germans and 20 New Guineans then appear. The Germans are well armed with magazine pistols and the latter with Mauser rifles. Through Kempf they are ordered to surrender ~ they all refuse. Bond in a feat of daring hands his pistol to Travers and warns him to provide cover. Bond walks towards the Germans, keeping them between himself and the New Guineans. While the Germans are busy conferring, the unarmed Bond turns towards the enemy who are so surprised by Bond's sudden daring action, they are unable to defend themselves. Pulling their pistols from their holsters, Bond swiftly manages to disarm all eight Germans in the presence of the 20 New Guinean troops drawn up under arms. The surrender of the New Guineans follows and the prisoners are marched off. The wireless station is reached at 7 PM and found to be abandoned. The mast has been dismantled, although the instruments and machinery remain intact. The surviving German prisoners are secured on the second floor of the radio station building, with the New Guineans in a nearby building. The fighting is now over. For his courage and quick-thinking Lieutenant Thomas Bond becomes the first Australian to be decorated in the war, receiving a D.S.O.

News of the capture of the wireless station reaches Admiral Patey at 1 AM on 12th September. The Imperial German flag is hauled down and replaced with the Union Jack of the British Empire. At 3 PM on 13th September, the British flag is formally hoisted at Rabaul in a ceremony held in an open space overlooking the harbour where the Australian naval force riding at anchor, give a 21-gun salute. Within weeks, most German territories in the area, including Bougainville and the Admiralty Islands are occupied without further opposition. The Australians permit generous terms to the few hundred German citizens in their settlements and surrounding plantations. Most Germans choose to remain after swearing an oath of neutrality to take no further part in the war. Later a cache of arms and ammunition is found in a plantation owner's house, where a woman is resting supposedly sick in bed. A subsequent search discloses that the mattress she is lying on is stuffed with rifles.

On Monday 14th September, as a result of operations of the Australian Expeditionary Force, Rabaul, the seat of government in German New Guinea has been occupied. The Australian Naval Reserve captured the essential wireless station at Port Herbertshöhe at the northern end of the island at 1 PM on 12th September, after 18 hours bush fighting over about six miles. An advanced landing party was established at dawn before the enemy were aware of our intention. From within a few hundred yards of landing there was a bush fight for four miles. The roads had rifle pits dug, and snipers were hidden in the trees. Have landed the 12-pounder guns ~ if the station does not surrender we intend shelling. Our force reconnoitered enemy strength holding the station. The troops of the Royal Australian Navy force met with vigorous opposition. The Australians continued to fight their way, losing six officers, and four men, and captured the wireless station. The German flag was hauled down. The British flag was hoisted over the town, and it was saluted. With the Union Jack flying triumphantly in its stead, thus the Bismarck Archipelago was lost to the Kaiser. *The Australasian of Melbourne*

Japan also captures Germany's Micronesian colonies of the Marianne Islands (without Guam). The Caroline Islands surrendering to a Japanese squadron are initially occupied by the Japanese to be promptly handed over to an Australian and New Zealand force. The Marshall and Solomon Islands are likewise occupied in December, to complete the tale of German colonial possessions in the Pacific. While there remain large areas in East Africa and the Kamerun still under German control, in both cases the coastline is in the secure possession of various Éntente forces.

While the Imperial German Navy began the war with three principal codes, within four months, the British Admiralty had secured copies of all three. To consolidate their useful triumph, the former Director of Naval Education, Sir Alfred Ewing who was appointed to head the newly created Admiralty code-breaking department, made several visits to the library of the British Museum to examine its collection of historic codebooks. He then gathered around him a small detachment of eminent cryptographers as well as German scholars to establish a secret department in Room 40 in the Old Building of the Admiralty, on the same corridor as First Sea Lord Winston Churchill's office. Within months of the outbreak of the war, Room 40 flourished from immediate access to the German Imperial Navy's wireless traffic, and was able to decipher many German signals that enabled the Naval War Staff to respond to several German naval maneuvers. In March of 1915, they were also able to capture an important German diplomatic code and thus decrypt inter-communications to America. This overwhelming intelligence advantage, was one from which the unaware Imperial German Navy was never to recover. Any wireless signal made by the German Navy (which could be intercepted) became available to Winston Churchill, Admiral *Jacky* Fisher (subsequent 1st Sea Lord), and their immediate subordinates. Conscious of the importance of this secret, very few were in on the secret. The following briefly describes how Britain came into possession of these crucial German codebooks.

Handelsverkehsbuch or *HVB* ~ MELBOURNE 11th August. The German-Australian steamship *Hobart* and her crew, unaware that war has been declared, are seized off Port Philip heads, Melbourne. A boarding party, under Captain J.T. Richardson RAN in civilian clothes under the pretense of a quarantine inspection party give the German master of the *Hobart* sufficient leeway, so he will feel unobserved, and attempt to destroy his classified papers ~ which will reveal their whereabouts. Richardson's ruse succeeds brilliantly. In the early hours of the morning, the German master creeps from his bunk, slides back a secret panel in his cabin, and withdraws the precious German documents. Richardson however, who is waiting for such a moment is now behind him, and at pistol point, forces the duped German to hand them over. The cache of documents includes *HVB* code used by German warships for intercommunication within the High Seas Fleet, their merchant vessels, naval shore commands, coastal stations, U-boats and Zeppelins. A copy of this code is made by the Royal Australian Navy and shipped to England on a fast steamer, which arrives toward the end of October.

Verkehrsbuch or *VB* ~ Off the Dutch coast, 17th October 1914. Captain Cecil Fox on the light cruiser *Undaunted,* with the four destroyers, *Lance, Lennox, Legion* and *Loyal,* heads out to sea to carry out a quick sweep up the Dutch coast. As Captain Fox nears the Texel, he identifies four elderly German destroyers steaming south in line abreast. As the Germans turn away, two are hit and began to sink. Soon Captain Fox's squadron has sent all four to the sea floor. Further loss comes for the Germans after an officer on the vessel *S-119,* properly jettisons his warship's secret documents in a lead-lined sea chest to the seabed, and thinks little more of it. Some weeks later on 30th November, a British trawler fishing off the Texel drags up the German lead-lined chest and its cache of secret papers. When opened, it contains a secret grid chart of the North Sea used to plot the location of friendly and enemy warships ~ plus a copy of the *Verkehrsbuch* or *VB,* which is an odd find, because this code was rarely used at sea ~ only by Flag officers or to signal units of the German Army ashore.

Wreck of the Magdeburg. BALTIC SEA morning 26th August. In thick fog, the German light-cruiser *Magdeburg* runs aground on a stretch of rocky coast, 400 yards off the northwestern tip of Odensholm, overlooked by the Odensholm lighthouse off the coast of Russian Estonia at the entrance to the Gulf of Finland. Her commanding officer, *Korvetten-kapitan* Habenicht tries to get the *Magdeburg* underway by lightening her and by towing, but failing, he gives the order to detonate her forward magazine which destroys the stranded warship as far aft as her second funnel. A German destroyer rescues most crew, but her captain with his officers and five men, remain behind. As the fog begins to lift, two Russian light cruisers, *Pallada* and *Bogatyr* under the command of Captain Nepenin from the communications and intelligence section of the Russian Baltic Fleet, steam through the mist and promptly open fire. (Previously, the keeper of the Odensholm lighthouse had reported to Captain Nepenin that he had heard a number of men talking German through dense fog, then too thick for anything to be seen). When Captain Nepenin locates the wreck of the *Magdeburg* he takes a ship's boat and approaches the enemy. Boarding the wreck, he is surprised by *Korvetten-kapitan* Habenicht who waves his sword at him... Nepenin orders the German to replace the Kaiser's naval ensign with his Russian naval flag, the blue cross of St Andrew, and the Germans surrender. When it is reported to Kaiser Wilhelm that one of his light cruisers has been destroyed, he responds ~ *Well, splinters are bound to fly at a time like this!*

When the German light cruiser Magdeburg was wrecked in the Baltic, the body of a drowned German under-officer was picked up by the Russians a few hours later, and clasped in his bosom by arms rigid in death, were the cipher and signal books of the Imperial German Navy. On 16th September, the Russian Naval Attaché came to see me. He had received a message from Petrograd that the Russians felt that as the leading naval Power, the British Admiralty ought to have these books and charts. If we would send a vessel, the Russian officers in charge of the books would bring them to England. First Sea Lord Winston Churchill

One fine Sunday morning, with the Poltava at anchor off Reval, I was keeping watch till noon, pacing the quarterdeck and listening to the chants of our fine choir singing mass on the maindeck below, when a yeoman handed me a signal just having come through from the flagship: I was to report immediately to the Flag Captain. In a few minutes I was on my way to the Petropauolsk; amazed and delighted to receive the staggering news ~ to proceed to St Petersburg to report to the Admiralty for instructions for a confidential mission to London.

The Flag Captain kindly let me have the gist of it then and there. Sometime before the German light cruiser Magdeburg, hard-pressed by our forces, had been driven ashore on the island of Oesel, and abandoned by its crew. In so doing the German navigator, incredible as it may seem, failed to take with him or at least destroy, three copies of the main signal book in use by the German Navy since the beginning of the war. It was discovered by our people in its customary place in the charthouse of the Magdeburg plus a secret gridded chart of the Baltic. The secret of this discovery was so well kept that the Germans continued to use the same code throughout the war, with incalculable advantage to the Allies. I was to carry to the British Admiralty a copy of this book.

Ten days later I was on my way by rail to Archangel, in civilian clothes, all complete with a satchel, worn slung over my shoulder on a strap, containing the code, with a large piece of lead sewn in to enable me to jettison it in case of emergency. An extremely inconvenient way of carrying, not to speak of hiding, the fat little volume. At that time, Archangel was the only port from which the West could be reached at all, with the land front stretching right across from sea to sea, and the entry of Turkey into the war closing the Bosphorus. On arrival, I had to wait a few days while HMS Theseus was got ready for its journey to Hull. As we approached England, knowing we would make port late in the day, a night train would convey me to the Russian Embassy at dawn. With our attaché, we proceeded to the Admiralty, to be greeted with much enthusiasm on delivery of my satchel.

A few days would have sufficed to get all the information available, but my stay was prolonged by a special circumstance ~ I had to wait for my return shipment. British Naval Intelligence had somehow got hold of the German re-coding tables, which together with the signal code I had brought completed the Allied knowledge of the German signaling system. These tables, reproduced on onion paper, I was to take back to Petrograd, but as the matter was urgent, this time via Norway and Sweden to the Russo-Finnish frontier, as the whole journey by that route would take not more than three days. But of course these countries were neutral and what is more, Sweden openly favouring Germany : therefore I had to go as a civilian, with the "war material", under the threat of confiscation in neutral territory, sewn into the lining of my waistcoat. I was given a diplomatic pass indicating that I was returning naturally from a visit to my father, the Russian Ambassador to Great Britain. But still the position was delicate ~ if discovered I would have been interned, everything confiscated, and what was more important, the Germans would have got to know about it and changed their code. Returning to Russia when our train steamed through Sweden, and when safely across the Russo-Finnish border, I sent a message : All our belongings are intact. Captain Count Constantine Benkendorff
Imperial Russian Navy

British advantage thereafter rested on German ignorance of the range at which their short-wave messages could be received. Fortunately for the Admiralty, the Germans used wireless a great deal. The Royal Navy on the other hand, maintained the greatest possible wireless silence. Many important German orders were transmitted from one warship to another ~ some while even in the same harbour, in the firm belief that these signals could not be detected more than 15 miles away. Most of these encoded secret messages were clearly received in England, and soon after Room 40 in London was able to decipher and read secret German messages. This solved the main strategic problem for the next four years, and it quickly proved its value. Warning was given of the German raids on Scarborough and Hartlepool in time for British warships to steam to intercepting positions. Without Room 40, the naval actions of Dogger Bank and Jutland would not have been fought, the Irish Easter rebellion may have succeeded, and the United States of America might not have been drawn into the war during April 1917. The excellence of British intelligence was the foundation of the Allied victory.

THE BELGIAN FRONT. In their first mass execution of civilians, German troops burn the town of Aarschot and shoot 156 hostages. Next are Andenne (110 dead), Tamines (383 dead), and Dinant (674). Under the pretext that German troops have been fired on by civilians (or under no pretext at all) entire Belgian villages are razed to the ground. Unique centres of European culture are not spared. Monuments respected during centuries of wars involving Belgium are now deliberately destroyed. Open cities are bombarded, and then steep taxation is extorted from such conquered towns, and if the inhabitants are unable to pay the taxes, a large number of their houses are thereupon put to the torch. This happens to Wavre, whose 8,500 inhabitants are unable to pay. Termonde with 10,000 inhabitants is torched on 15th September, when only 282 houses are left standing from Termonde's original 1,400. German troops completely destroy the occupied Belgian town of Visé located near Liège, executing, deporting or expelling the entire population. The Germans then begin a policy of mass deportations.

Most distressing news has come through from Tamines. I had a long talk today with a trustworthy man from there, and his story was enough to make one's blood run cold. He says that ~ One evening the Germans entered the village after a brush with French troops, which were still in the neighbourhood. Infuriated by the resistance offered to their advance, they proceeded to vent their rage on the town. They shot down a lot of villagers, and arrested many more. A great many escaped to the country. A lot of houses were first sacked, and then burned. The orgy continued during the night, and through the next day. Over 400 men were collected near the church and lined up to be shot. The work was done for a time by a firing squad, which fired into the crowd. But this was too slow, and finally a rapid-fire gun was brought out and turned loose. A great many were not killed outright and lay groaning among the dead. Now and then a German would put one out of his misery by a bayonet thrust. Others settled their own troubles by rolling themselves into the nearby river. Altogether over 600 people were shot down. After the shooting was over, civilians were brought and compelled to bury the dead. My informant says that some of the scenes attending this duty were quite as poignant as the shooting itself, for some buried their own fathers and brothers. One man about to be thrown into the trench was found to be still alive, but the German doctor, after a cursory examination, ordered him buried with the rest. The man had enough life left in him to raise his hand in appeal but the doctor shrugged his shoulder and repeated his order. There were many incidents, most of them horrible. The man who told the story seemed still dazed and spoke quietly, with few adjectives and little emphasis on anything he said. It was a bare recital of facts, and far more moving than if he had striven for effect.

There is bad news from Louvain. The reports we have received agree that there was some sort of trouble in the square before the Hôtel de Ville a day or two ago. Beyond that, no two reports are alike. The Germans say that the son of the Burgomaster shot down some staff officers who were talking together at dusk before the Hôtel de Ville. The only flaw in that story is that the Burgomaster has no son. Some Belgians say that two groups of Germans who were drunk met in the dusk; that one formation mistook the other for French, and opened fire. Other people tell with convincing detail that the trouble was planned and started by the Germans in cold blood. However that may be, the affair ended in the town being set on fire, and civilians shot down in the streets as they tried to escape. According to the Germans themselves, the town is being wiped out of existence. The Cathedral of St. Pierre, a marvelous example of Gothic art, the University, the Library with more than 70,000 volumes, of which a large number were ancient manuscripts, and nearly all the houses of the town were deliberately burned. The town of Louvain, with its 42,000 inhabitants, was one of the finest centres of Belgian culture ~ now nothing more than heaps of ashes. People have been shot by hundreds and those not killed are being driven from the town. They are coming to Brussels by the thousands, and the end is not yet. Two old priests have staggered into the Legation more dead than alive after having been compelled to walk ahead of the German troops for miles as a sort of protecting screen. One of them is ill, and it is said that he may die as a result of what he has gone through.

Subsequent visits to Louvain, and conversations with people who were there when the trouble began, have only served to strengthen the impression that the whole affair was part of a cold-blooded and calculated plan to terrorize the civilian population. While we were there, it was frankly stated that the town was being wiped out; that its destruction was being carried out under definite orders. When the German Government realized the horror and loathing with which the civilized world learned of the fate of Louvain, the orders were cancelled and the story sent out that the German forces had tried to prevent the destruction, had fought the fire, and by good fortune had been able to save the Hôtel de Ville. Never has a government lied more brazenly. When we arrived, the destruction of the town was being carried on in an orderly and systematic way that showed preparation. The only thing that saved the Hôtel de Ville was the fact that the German troops had not progressed that far with their work when the orders were countermanded from Berlin.

Hugh Gibson
Secrétaire de la Légation USA

ORDRE
A LA POPULATION LIÉGEOISE

La population d'Andenne, après avoir témoigné des intentions pacifiques à l'égard de nos troupes, les a attaquées de la façon la plus traitresse. Avec mon autorisation, le général qui commandait ces troupes a mis la ville en cendres et a fait fusiller 110 personnes.

Je porte ce fait à la connaissance de la Villé de Liége pour que ses habitants sachent à quel sort ils peuvent s'attendre s'ils prennent une attitude semblable.

Liége, le 22 Aout 1914.

Général von BULOW.

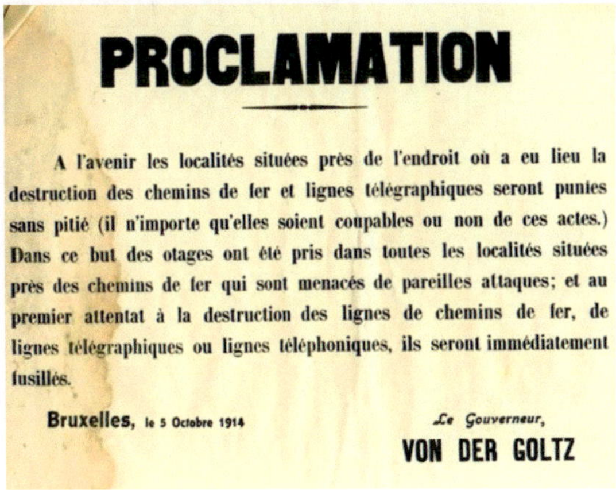

PROCLAMATION

A l'avenir les localités situées près de l'endroit où a eu lieu la destruction des chemins de fer et lignes télégraphiques seront punies sans pitié (il n'importe qu'elles soient coupables ou non de ces actes.) Dans ce but des otages ont été pris dans toutes les localités situées près des chemins de fer qui sont menacés de pareilles attaques; et au premier attentat à la destruction des lignes de chemins de fer, de lignes télégraphiques ou lignes téléphoniques, ils seront immédiatement fusillés.

Bruxelles, le 5 Octobre 1914

Le Gouverneur,

VON DER GOLTZ

The population of d'Andenne, after manifesting peaceful intentions toward our troops, attacked them in the most treacherous manner. With my authorization, the general who commanded these troops has reduced the town to ashes and has shot 110 persons. I bring this fact to the knowledge of the City of Liége so that its people may understand the fate which awaits them if they assume a like attitude. General von Bulow

22nd August 1914

In future, villages in the vicinity of places where railway and telegraph lines are destroyed will be punished without pity (whether they are guilty or not of the acts in question). With this in view hostages have been taken in all villages near the railway lines, which are threatened by such attacks. Upon the first attempt to destroy lines of railway, telegraph, or telephone, they will be immediately shot. Governor Count von der Goltz

5th October 1914

The old Reich knew how to act with firmness in the occupied areas. That's how sabotage to the railways in Belgium was punished by Count von der Goltz. He had all the villages burnt within a radius of several kilometres, after having had all the mayors shot, the men imprisoned, and the women and children evacuated.

Adolf Hitler

In this war, German occupation of territory is systematically followed by (and is at times preceded by and accompanied by) acts of violence against the civil population, contrary to the conventions of war and principles of humanity. The procedure of the Germans is everywhere the same. They advance along the roads, shooting inoffensive passersby, particularly cyclists, and even peasants occupied in the fields which the Germans traverse.

In the towns and villages where they stop, the Germans first of all requisition victuals and drinks which they consume to the point of drunkenness. Then they begin to shoot wildly, sometimes from the interior of empty houses, declaring that the inhabitants have fired the shots. It is then that the firing scenes begin, and murder and especially pillage accompanied by acts of cold cruelty that respect neither sex nor age. Even where they claim to know the perpetrator of the deeds which they allege, they take advantage of the occasions to decimate the population, to pillage the inhabitants and set fire to them. After a first massacre, they shut the men into the church of the town and order all women to go back to their houses ~ and leave their doors open during the night.

In several localities, the populace has been sent to Germany, to labor in the fields, as in the slave days of old. Numerous cases are known where the inhabitants were forced to make trenches for the Germans. Numerous depositions reveal that in their march, and even in their attacks, the Germans put before them civilians, men and women, in order to prevent our soldiers from firing. Other testimony proves that German detachments do not hesitate to fly either a white flag or a Red Cross flag, so as to approach our troops without being suspected. On the other hand, they fire on our ambulances and ill-treat our ambulance nurses. They ill-treat and even kill our wounded. Clergymen seem to be particularly the object of their attacks. Last, but not least, we have in our possession explosive bullets left behind by the enemy at Wechter, and we are also in receipt of medical certificates testifying that such bullets must have inflicted the wounds. Count Goblet d'Alviella

BRUSSELS. *Had a call from Dr. Barnich, who has been in Luxembourg since the war began. He had seen the Kaiser who descended from the German Legation in Luxembourg heavily guarded; miltraileueses all about, and a searchlight sweeping the sky at night over the palace where the royal head reposed. He said the Kaiser looked 15 years older, grey and haggard with dark circles under his eyes ~ He is not lighthearted, he isn't happy. Many tales of distress down that way. Drunken officers. Children and peasants stood up against walls and shot, women and girls ravished. Said he would never get the cries of the children and out of his ears.* Hugh Gibson

Védrines, flying a Bleriot 160hp Blindé known as *la vache* (French : the cow) claimed the downing of a Taube over Suippe on 2nd September. *At 0750 a Taube dropped a bomb 60m from our camp, Védrines took off ~ the Taube was shot down and landed 10km from our camp, radiator holed. Pilot and observer were able to get back to German lines after two days of pursuit in woods.* Védrines' forward gunner René Vicaire said ~ *There is no doubt my shots downed the Taube!*

10th October. Following the fall of Antwerp, King Albert's Belgian Army who unable to resist, withdraw to the Yser River and canal ~ the waterways on which English and Australian fine wool are carried inland from the sea below Nieuport to the mills of Ypres. There, King Albert orders their halt. The dykes at Nieuport and Diksmuide are opened, and the Yser plain is deliberately flooded in late October to provide an obstacle to the enemy and keep western Belgium free from German occupation. Following this, the water level is raised with seawater through the hydrologic complex of Nieuport. This inundation prevents a further German breakthrough and saves the exhausted Belgian Army. Behind the mile-wide reaches of the Yser, the defenders make their most defiant and effective stand. Belgian troops begin to dig holes connected to each other, and these first trenches of the war on the Yser Line are typically built above ground, because the water level makes digging impracticable. Sectors are fortified with successive lines of resistance. Behind the main line of resistance there is a second line with the same length. Where necessary, other lines of resistance are added. Outposts and listening posts are set up in front of their main line of resistance to watch the enemy. The trenches, which in time become the symbol of the stalemate on the Western Front are improved and extended. Tens of thousands of sandbags and endless rows of barbed wire are required to achieve these field works. Several bunkers and shelters are also built. In time, the entire Western Front will become a scene of hellish trench warfare, as it first appears in the defence of Belgium.

10th November. There is further fighting around Nieuport, which will remain in Belgium hands. With French support on the Yser Line and around Diksmuide, the German invasion is blocked and Belgium wins the respect of the world. A triangular portion of Belgium remains free to the defenders who establish their GHQ at La Panne, where King Albert will base himself until the close of the war. France meanwhile accommodates the Belgian Government and the diplomats accredited to it, within the coastal city of Le Havre, in the hotels and villas of the summer resort suburb of St Addresse. Le Havre soon becomes an international city, and its vast harbour an imposing French naval station.

After the summer of 1914, the Western Front becomes a tale of woe. Quick-firing field guns, machine guns, trenches, and coiled barbed wire entanglements all combine in a revolutionary quartet to degrade the experience of warfare beyond all recognition. The generals, most of whom expected a war of movement, try to break the deadlock by bombarding the enemy trenches and then sending their infantry forward to create a breach, in the forlorn hope of using their cavalry to exploit it so mobility might be restored to the battlefield ~ but these tactics fail repeatedly, with severe casualties.

THE DEFENCE OF THE YSER LINE

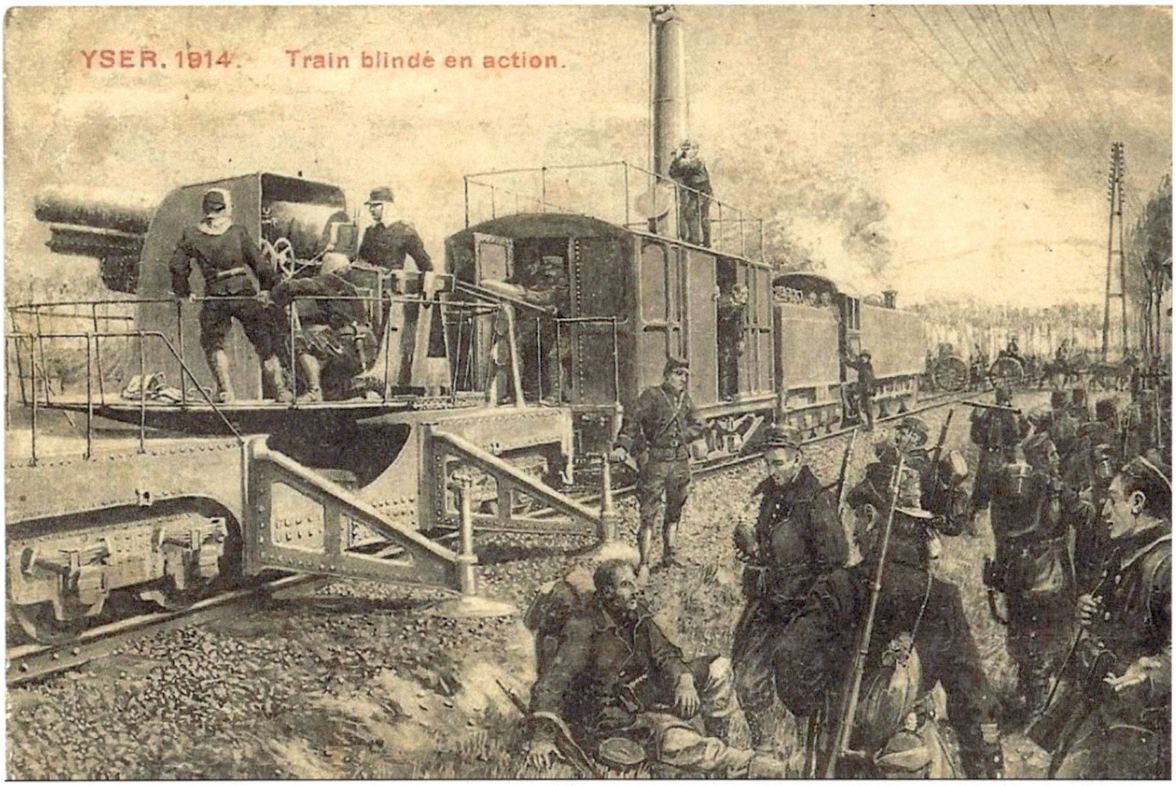

Cardinal Mercier of Mechelen while returning from the conclave of Rome passes thru Le Havre where he visits many wounded Belgian, British, and French troops. Once back in his archdiocese he sees the ruins of Mechelen Cathedral and learns that 13 priests from his diocese have been murdered. In response, Mercier writes the pastoral letter *Patriotism and Endurance* to be read in Belgian churches. The Belgian people take his passionate and unflinching words of resistance to the invaders to heart. Pope Benedict XV sends his portrait with a letter of support to Mercier ~ *You saved the Church.*

My very dear Brethren,

It was in Rome itself that I received the tidings ~ stroke after stroke ~ of the partial destruction of the Cathedral church of Louvain, next of the burning of the Library and of the scientific installations of our great University and of the devastation of the city, and next of the wholesale shooting of citizens, and tortures inflicted upon women and children, and upon unarmed and undefended men. And while I was still under the shock of these calamities the telegraph brought us news of the bombardment of our beautiful cathedral, and a great part of our city of Malines. A disaster has visited the world, and our beloved Belgium is the first sufferer. She bleeds. Her sons are stricken down within her fortresses and upon her fields, in defense of her rights and of her territory. When on my return from Rome, I went to Le Havre to greet our many Belgian, French and English wounded; when it was given to me to take the hands of those brave men who carried a bullet in their flesh or a wound on their forehead, because they had marched to attack the enemy, or borne the shock of his onslaught, it was a word of gratitude to them that rose to my lips; O valiant friends, it was for us, it was for each one of us, it was for me, that you risked your lives and are now in pain. I am moved to tell you of my respect, of my thankfulness, and to assure you that the nation knows how much she is in debt to you ~ For in truth our soldiers are our saviors.

A Prayer to the Holy Spirit ~ I am going to reveal to you the secret of sanctity and happiness. Every day for five minutes control your imagination and close your eyes to all the noises of the world to enter into yourself. Then, in the sanctuary of your baptized soul (which is the temple of the Holy Spirit) speak to that Divine Spirit; O Holy Spirit, beloved of my soul, I adore You. Enlighten me, guide me, strengthen me, and console me. Tell me what I should do; give me Your orders. I promise to submit myself to all that You desire of me and to accept all that You permit to happen to me. Let me only know Your Will. ~ If you do this, your life will flow along happily, serenely, and full of consolation, even in the midst of trials. Grace will be proportioned to the trial, giving you the strength to carry it and you will arrive at the Gate of Paradise, laden with merit.
This submission to the Holy Spirit is the secret of sanctity.

Désiré-Joseph Mercier
Cardinal-Archbishop of Mechelen

GERMAN FRONTLINE DEFENCES

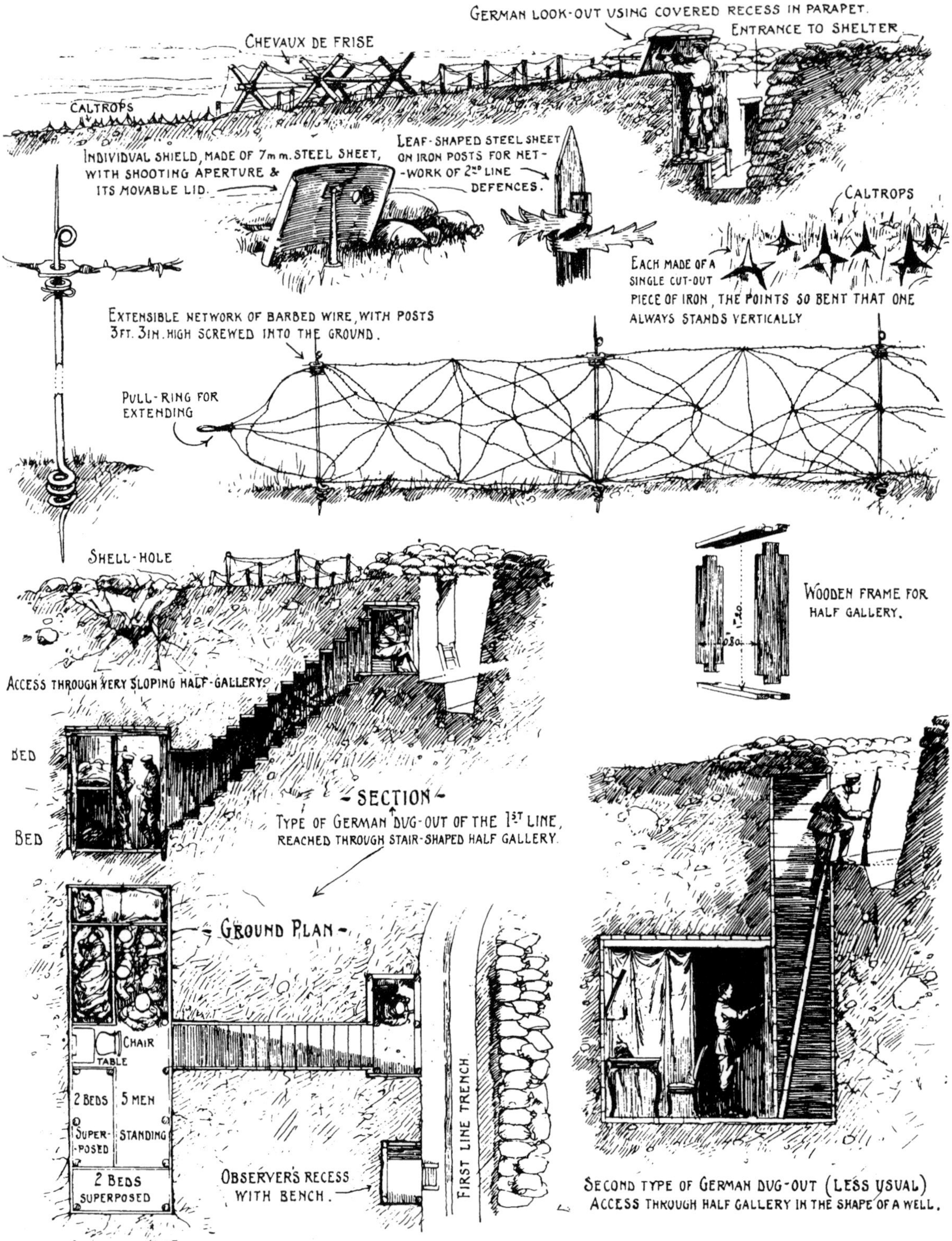

All along the trench lines of the Western Front now covered with wire entanglements and other obstacles, all the while being strafed by enemy machine-gun and artillery crossfire, the soldiers dated field tactics flounder in futile immobility. As they hide within the cover of their defensive earthworks to endure slashing scythes of fire so severe that attacking infantry find it nearly impossible to advance, surprise seems unattainable ~ since the length of time needed to break the enemy wire by artillery bombardment signals the direction and timing of an imminent attack, and after such a preparation, maps soon become useless, for everything soon has the appearance of the surface of the moon above. The bewildered generals cling to their one timeless principle ~ offensive action ~ and a good example of the futility of advancing in the open against modern weapons is provided in the first A-H offensive.

The Austro-Hungarian Empire soon began to taste the bitter fruits of Colonel Redl's betrayal. Conrad of Vienna's GHQ would have preferred to interrogate Redl on the exact extent of his betrayal, but his *Evidenzbureau* operatives insufficiently enquired on the night when Major Max Ronge had preferred to see Redl dead. Subsequently no major changes were made to *Plan III* until early 1914. Many ascribed several A-H Army defeats in the opening battles against Serbia and Russia to Redl's treachery, and no doubt, tens of thousands of A-H troops died because of the treason of Alfred Redl.

From the military secrets Redl had sold, Russian forces first blocked the A-H Army thrust into Poland, then counter-attacked to capture several A-H frontier fortresses and occupy eastern Hungary. Alfred Redl's espionage enabled the Russian Army to reply to all movements of the A-H Army, and Russian military countermoves often displayed their anticipation of pre-planned Austrian responses.

Russian Armies flowed west through Galicia and approached Tarnow. Przemysl was isolated; tens of thousands of prisoners, scores of guns and quantities of stores fell to the victorious Russians. Whereas it was widely thought that Russia did not want a war in 1914, as their army would have been in a stronger position to fight in 1916 (having completed its armaments modernization programme) their winning edge provided by Redl's reports, saw Russia in fine fighting trim for the opening phase of the war and they fared well, until their tapering munitions supply ruined their realm.

On the Serbian Front, A-H Commanding General, Conrad von Hötzendorff is unwisely hoping that his *Plan III* has remained secret, as he hurls the Austrian Army across the border into Serbia. At every point of attack, the Serbians under Marshal Radomir Putnik are prepared to repel them; Serbian prior knowledge of the Austrian plans having already been passed to them by the *Okhrana*. The A-H plan might be altered, but its main features remain the same, and the Serbian General Staff, have an insight into the mind of the A-H General Staff. For months, their competent Marshal Putnik has been pondering over *Plan III*. He has known it by heart for months, and can easily see what changes will occur. What good use the Serbian commander makes of his knowledge is clearly seen when three times Conrad tries variants of *Plan III* ~ and three times his troops are defeated.

A-H General Oskar Potiorek, who as the 8th Governor-General of Bosnia and Herzegovina and who was a passenger in the car carrying the Archduke Franz Ferdinand and his wife Duchess Sophie when they were assassinated in Sarajevo, now commands all A-H forces in the Balkan campaign. *Survivor's guilt* leads Potiorek to take active command of the A-H Army and lead the first mission to punish Serbia. Zealous in action, he often says ~ *I was spared at Sarajevo, so that I may die avenging it!*

General Oskar Potiorek rushes the first offensive against Serbia in his desire to win a victory before Franz Josef's birthday, and in trying to knock Serbia out as soon as possible he makes two strategic errors. He first chooses to attack uphill on the difficult hilly terrain of western Serbia instead of the open plains of the north, and then sends in the opening A-H assault with half of its strength.

After the bombardment of Belgrade in response to the destruction of the Semlin-Belgrade bridge, and after many ineffectual attempts to cross the rivers Danube, Sava, and Drina, where the A-H Army are repulsed with great losses, the Austrians manage to secure a passage across the Drina near Loznica, by throwing a bridge across the river ; they also secure a crossing of the Sava at Šhabatz. The Serbians retire to the hills above the Šhabatz to inflict a smashing defeat on the A-H Army which surprisingly still attempt to advance up through the Valley of Šhabatz. On 19th August, the A-H morale collapses and thousands of soldiers retreat back into A-H, with many drowning in the Drina.

The defeat of the A-H invaders involves their serious loss of much artillery and war material. On 24th August, the Serbs re-enter Šhabatz to conclude the battle. After the fierce four-day battle, the casualties number over 23,000 for A-H (of whom 4,500 surrendered), and at least 14,000 for Serbia. After the defeat of his mishandled A-H troops, General Oskar Potiorek is relieved of his command. When he is seen to, and reported to have lost his reason, he is placed in a lunatic asylum.

Having won, and yet weakened by the loss of too many men, Marshal Putnik orders a retreat into the hills and the front degrades into trench warfare, which is extremely unfavourable to the Serbs, due to their lack of industry, deficiency in heavy artillery, ammunition stocks, and poor shell production having only a single factory producing less than 100 shells a day. In time, the Serbian guns fall silent... *There are only 120,000 rifles available. The Schneider-Creusot 75's have been subjected to severe strain during the Balkan wars and many are worn out ~ the deficiency is by no means made up by the Krupp guns captured from the Turks. There are only 12 seige guns, and no heavy artillery. Munitions are very short, there are no tents, and our transport services are poor. The army is, so to speak, naked and barefoot.* General Z. Pavlovic

13th OCTOBER 1914 ~ EVIDENCE HEARD AT SARAJEVO

The Sarajevo trial in progress of those involved in Archduke Franz Ferdinand's assassination (Sarajevski Atentat)
Front row ~ left to right : Trifun Grabez, Nedjeliko Cabrinovic, Gavrilo Princip & Veljko Cubrilovic (also above)
~ who recruited by Danilo Ilic, survives the Great War to become a Professor of History at Belgrade University.

SARAJEVO. At their trial, the Serbs stare indifferently at their inquisitors as Gavrilo Princip states under cross-examination ; *I am a Yugoslav nationalist, and I believe in unification of all South Slavs in whatever form of state, and that it be free of Austria.* When asked how he intends to realize his goal, Princip replies ~ *By means of terror.* The judge summarizes: *The court regards it as proved by the evidence, that both the Narodna Odbrana and military cliques in the Kingdom of Serbia in charge of the espionage service, collaborated in the outrage.* At the end of the trial, the presiding judge asks all of the accused to stand up. *Do any of you feel sorry for your acts?* Nedelijko Cabrinovic expresses his regrets for the murders and post-sentencing receives a letter of forgiveness from the three children, the assassins have orphaned. All rise except Princip. *Why do you not rise?* Princip, *I am sorry for the children who have lost their parents, and I am especially sorry for killing Duchess Sophie. The bullet that hit her had been for General Oskar Potiorek.*

When one week later the judges return with their verdicts, Gavrilo Princip who is still two weeks short of his 20th birthday is too young to be executed as A-H law can only apply the death sentence to criminals aged 20 or more. Instead, he is sentenced to 20 years solitary confinement with the proviso that every 28th June, his cell will be darkened, his food denied to him, and his mattress removed.

Gavrilo Princip was taken to Theresienstadt, an old Bohemian fortress north of Prague, a military prison. Chained with shackles weighing 22 pounds, the 145-pound prisoner was kept in solitary confinement in an unheated cell. Tuberculosis consumed him, and on 28th April 1918, Gavrilo Princip died weighing 88 pounds. Fearing his bones might become relics; Princip's Austrian jailers took the body in secret to an unmarked grave. In 1920 Princip and the others were disinterred and brought to Sarajevo's St. Mark's Cemetery, where they were buried together beneath a chapel. Visitors are discouraged by a locked gate, weed-choked pathways, and packs of dogs that feed on the heads of sheep thrown at the chapel by butchers at a nearby market. David de Voss

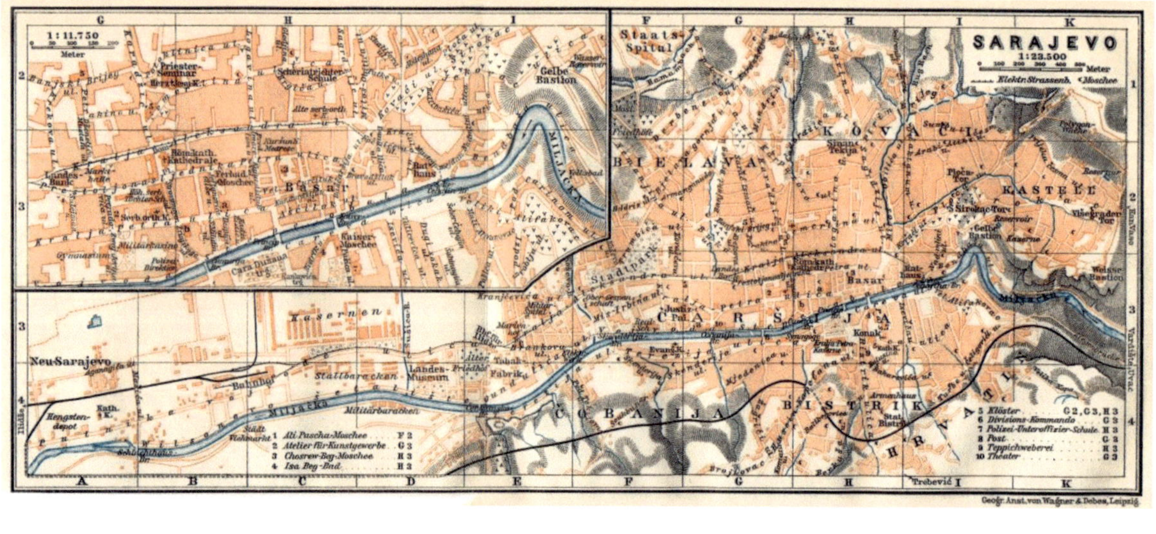

We need a strong leader ruling over Constantinople, and since we cannot let any other power assume this role, we must take Constantinople for ourselves. For us to accomplish this without waging war on the Ottoman Turkish Empire would be impossible. The time has come for Russia to settle accounts with her ancient enemy and to liquidate the Straits Question once and for all. Russian Ambassador M.N. Giers, Constantinople

In 1914, the Turks still fear Russia, as they have feared her since the days of Peter the Great. Russia is the historic enemy, the nation that has delivered freedom to Bulgaria and Romania, and which has been most active in dismembering the Ottoman Empire, and which still presumes herself to be the power to ultimately possess Constantinople. This fear of Russia is the one factor above everything else that induces Turkey into the arms of Germany. For more than half a century Turkey has regarded the British Empire as her surest safeguard against Russian aggression, but ever since Britain has become Russia's Éntente partner, there is a general belief that the Turkish élite share that Britain is willing for Russia to inherit Constantinople and the Dardanelles, as Tsar Nicholas proclaims:

The acquisition of the Straits is of vital interest to the Russian Empire; far more important than all the territorial advantages Russia may obtain at the expense of Germany or Austria-Hungary. The neutralization of the Bosphorus and the Dardanelles would be an imperfect, hybrid compromise, fraught with peril for the future. Constantinople must be a Russian city! The Black Sea must become a Russian lake. My views have not changed but there is one point on which to be more precise ~ Constantinople. The question of the Straits is preoccupying public opinion in Russia to a highest degree. It is a current that flows more strongly every day. I could not admit my right to impose on my people the terrible sacrifices of this war if I did not reward them with the realization of their time-honoured ambition. My mind is now therefore made up. Constantinople and southern Thrace must be incorporated into my Empire. Of course, I should be prepared to allow the city to be administered on principles designed to safeguard foreign interests. England has already expressed her approval. King George told my Ambassador recently : Constantinople must be yours. That pronouncement is a guarantee of goodwill.

However, the plans of Kaiser Wilhelm II for influencing Turkey that he has carefully pursued for a quarter of a century, are set to achieve success. German schemes in the Ottoman Turkish Empire and Near East have minimized Russian interests in the region to transform the Baltic and Black Seas into German waters, and have seen them construct several promising rail routes by way of the Caucasus. For the time being, Germany and Austria-Hungary dominate Serbia, Bulgaria, Romania, and Turkey, and regard their aspirations for a Teutonic Empire extending from the North Sea to the Persian Gulf, as practically realized. The world (rightly or wrongly) sees that Berlin's position has precipitated the July Crisis to destroy Serbia, to seize control of the Balkans and transform the wider Ottoman Empire into a vassal entity, pursuant to attaining a vast oriental empire to form a basis for world dominion.

For the work of persuasion in Turkey, and transforming its army and territory into instruments of Germany, the Kaiser has sent to Constantinople an ambassador who is ideally fitted for the mission. The mere fact that the Kaiser has personally chosen Baron Hans von Wangenheim for this post shows that he has accurately gauged the human qualities required for this enormous diplomatic enterprise. Kaiser Wilhelm detected early in Wangenheim an instrument ideally qualified for oriental intrigue, and has several times summoned him to Achilleion Palace on Corfu for his summer vacations, and here the two pass many days discussing German ambitions in the Near East. Wangenheim delights in describing the Kaiser's caste organization of German estates, how he has made them non-transferable, and arranged it so the possessors or prospective possessors cannot marry without Imperial consent. *In this way, we keep our governing classes pure, unmixed of blood.* Like all of his social order, Wangenheim admires the Prussian military system. His bearing shows that he has himself served well in the army, and that he now regards practically every situation in life from a military standpoint.

Upon Wangenheim, depends the success of Kaiser Wilhelm's many ambitions in the Near East. Unless Germany can obtain the Ottoman Empire as its ally, there is little chance that they can succeed in a wider European war. Once France achieved her alliance with Russia, the manpower of 170,000,000 people was arrayed on her side in the event of a war with Germany. For more than 20 years Germany has sought to diplomatically detach Russia from this French alliance ~ but has failed. There is only one way in which Germany can neuter the Franco-Russian Alliance ~ by obtaining the Turks as an ally. With Turkey on her side, Germany can sever the Straits, the only practical line of communication between Russia and her western allies. This simple act will deprive the Tsar's army of war munitions, destroy Russia economically by stopping her grain exports ~ her greatest source of wealth ~ and thus detach Russia from her partners in an escalating war. Therefore, Baron Hans von Wangenheim's mission is to assure that the Ottoman Turkish Empire shall align herself with Imperial Germany.

Wangenheim had in combination the jovial enthusiasm of a college student, the rapacity of a Prussian official, and the happy-go-lucky qualities of a man of the world. This huge figure of a man, sitting at the piano, improvising on some beautiful classic theme ~ and then suddenly starting to pound out uproarious German drinking songs or popular melodies. I still see him jumping on his horse at the polo grounds, spurring the splendid animal to its speediest efforts, but never making sufficient speed to satisfy the ambitious sportsman. Indeed, in all his activities, grave or gay, Wangenheim displayed this same restless spirit of the chase bending Turkish officials to his will in the interest of Germany. All life was to him a game, which was to be played more or less recklessly, and in which the chances favoured the man who was bold and willing to pin success or failure on a single throw. And this greatest game of all, as Bernhardi has expressed it ~ World empire or downfall.

Wangenheim did not play languidly. He was fire and flame for it. He had the consciousness that he was a strong man selected to perform a mighty task. All the time that like the Imperial German Government which he served so loyally, he was fundamentally ruthless, shameless, and cruel. He saw only a single goal, and with realism and logic he would brush aside all feelings of humanity and decency that might interfere with success. He accepted Bismarck's famous dictum that a German must be ready to sacrifice for Kaiser and Fatherland not only his life but his honour as well. U.S. Ambassador Henry Morgenthau

On 9th August 1914, Russian FM Sazonoff instructs Ambassador Giers in Constantinople one day prior to the German-made heavy-cruiser *Goeben* and light-cruiser *Breslau* steaming through the Straits: *You possess the authority to mobilize the Black Sea Fleet in case the Goeben and Breslau come through the Dardanelles ~ flying the German flag, we authorize Black Sea Fleet commander Admiral Andrei Eberhardt to pursue all measures in his power to prevent them from entering the Black Sea and to annihilate them, even if this means we must violate Turkish territorial waters. Such measures, be warned could easily provoke Turkey into declaring war, a possibility we wish you as our ambassador to do your utmost to prevent, as we believe that a war with Turkey will not be advantageous for us Russia at the present time.*

On 11th August, Henry Morgenthau visits the German Embassy, where Wangenheim has little interest in routine matters. He cannot rest in his chair more than a few minutes at a time, for he is constantly jumping up, rushing to the window and looking anxiously out toward the Bosphorus, where the *Corcovado,* is moored one kilometre away. Wangenheim uses the radio on board the German steamer *Corcovado,* moored opposite his Embassy in the Bosphorus, for all practical purposes, as a constant telephone connection with Berlin. Wangenheim's eyes are shining as he stalks to the window again for another look at the *Corcovado.* Henry Morgenthau suggests ; *Something is seriously distracting you, I will go and come again another time.* Wangenheim raises his voice; *No, I want you to stay here... This will be a great day for Germany! If you will only remain for a few minutes you will hear a great piece of news ~ something that has the utmost bearing upon Turkey's relation to the war.* Then he rushes out on the portico and leans over the balustrade. At the same moment, a launch puts out from the *Corcovado* toward the Ambassador's dock. Baron Hans von Wangenheim hurries down, seizes an envelope from one of the sailors, and a moment afterward bursts into the room and shouts ~
We've got them! Got what? Asks Ambassador Henry Morgenthau.
The Goeben and the Breslau have passed through the Dardanelles!

Wangenheim waves the wireless message, momentarily checking his enthusiasm, humorously shakes his forefinger, lifts his eyebrows, and tells Morgenthau. *Of course, you do understand that we have sold those ships to Turkey!* He adds with a wink ~ *Admiral von Souchon, will now enter the Sultan's service!* The arrival of these ships is the greatest day in Wangenheim's diplomatic career. The voyage of the *Goeben* and the *Breslau* is his personal triumph, having arranged with the Turkish Cabinet for their passage through the Straits and having directed their movements by wireless, in the Mediterranean. By safely guiding the *Goeben* and the *Breslau* into Constantinople, Wangenheim has definitely clinched Turkey as Germany's ally. All his intrigues and plotting for three years have now finally succeeded. From the *Corcovado* wireless station in the Bosphorus, Wangenheim sends word to Admiral Souchon, telling him to hoist the Turkish flag for the German cruisers have become vessels of the Turkish Navy.

No longer the *Goeben* and *Breslau,* Wangenheim has changed them to the *Sultan Selim* and *Medilli.* These warships spent the months preceding the war cruising in the Mediterranean, and when the war declaration came, these two vessels were anchored near Turkey. The selection of the *Goeben* was apt, as twice before she has visited Constantinople and her officers and crew know the Dardanelles well. Wangenheim boasts that Germany has 174 German gunners at the Dardanelles, and that the Straits can be closed within 30 minutes. Admiral Souchon informs him ~ *The Straits are now impregnable!*

Wangenheim's manœuvre has accomplished two purposes. It places Germany before the people as Turkey's friend, and having provided a subterfuge for getting the ships through the Dardanelles, now enables them to remain in Turkish waters. Cemal issues a communiqué to announce the purchase of the *Goeben* and *Breslau* by the Ottoman Government for 80 million marks and for the warships to replace the dreadnoughts *Sultan Osman* and *Resadiye* previously requisitioned by the Royal Navy. The *Goeben* and *Breslau* change their names, and the German sailors adorn themselves with Turkish fezzes.

But the sale is a sham. A few Turkish sailors are placed on board for appearance sake ~ but their German officers and German crews still retain active charge. As the *Goeben* and *Breslau* steam to their new moorings in the Bosphorus, the Ottoman Minister of Finance Djavid Bey happens to meet a Belgian visiting Constantinople. The most supportive Turkish minister ~ *I have terrible news for you. The Germans have captured Brussels!* The Belgian, puts his arm upon the shoulder of the Turk and points to the *Goeben* and *Breslau*... *I have even more terrible news for you. The Germans have captured Turkey!*

Constantinople's policy is to confuse the Éntente, so that they never know where Turkey stands. Whether she will remain neutral ~ or enter the war on Germany's side. With Constantinople's inclination so uncertain, Russia is compelled to keep large forces in the Caucasus; Britain is obliged to strengthen her forces in Egypt and India, and to also maintain a fleet at the mouth of the Dardanelles. This all works in harmony with Berlin's plans, for these detached forces weaken Britain and Russia in Europe at a time when Germany expects to defeat France and Russia with the aid of her ally A-H, and obtain a victory that will enable her to dictate the future of Europe. Even if Turkey were engaged in military operations, she can do no more toward bringing about this victory than she is doing now, by keeping Russian and British forces away from more important fronts.

German plans have no interest in having Turkey as an active ally ~ except in the event that German forces cannot speedily win an anticipated triumph. But if Russia ever makes viable progress against Austria-Hungary, then the active alliance of the Ottoman Turkish Empire will have great value (especially if her entry can be timed to bring in Bulgaria and Romania as additional allies).

Meanwhile, Baron Wangenheim plays a waiting game shaping Turkey as a potential German ally, strengthening her army and her navy, and preparing to use her, whenever the moment arrives for using her to the best advantage. If Germany cannot win the war without Turkey's aid, Germany is prepared to include her as an ally. As the Turkish Army and Navy prepare for various eventualities, it becomes apparent that Germany cannot triumph over France and Russia in their opening offensive. In September 1914, the French repel the invasion and drive the German Army into entrenchments along the Aisne. The Russians sweep triumphantly through Galicia, and it seems probable they will soon cross the Carpathians into Austria-Hungary. Now the time has come, when Imperial Germany needs the Ottoman Turkish Empire, and some situational pretext is sought to respond to.

26th September. A Turkish torpedo boat is steaming through the Dardanelles, to enter the Ægean. A British warship stationed outside hails the ship to examine it, and finds German sailors on board. As soon as the English Admiral orders the vessel to go back, Weber Pasha, the German general in charge of the Dardanelles fortifications, does not consult the Turks, but immediately gives orders to *Close the straits!* ~ and in less than 30 minutes it is done. Down into the water plunge the mines and nets, lights in lighthouses are extinguished and signals notify near ships of the deed ~ *No thoroughfare!* Soon after, the British Ambassador in Constantinople Sir Louis Mallet enters the U.S. Legation to announce to Henry Morgenthau, *I have something important to tell you. They have closed the Dardanelles!*

St PETERSBURG 1st October. The French Ambassador to Russia Maurice Paléologue announces:
The Turkish Government has closed the Straits on the pretext of the presence of an Anglo-French squadron off the entrance to the Dardanelles. This action does incalculable harm to Russia, which is left without maritime communications except by Vladivostok and Archangel. Now it must be remembered that Vladivostok is 10,500 kilometres from Petersburg and that the port of Archangel may be closed by ice at any time until the end of May. For some time, I have been receiving reports from Moscow and Kiev that the old Byzantine dream is reviving. This war will have no meaning for us unless it brings us Constantinople and the Straits. Our historic mission and our Holy duty is to set the cross of the Orthodox Faith on the dome of Saint Sophia once more. That is what is being said and spread in political, religious, and university circles and in the depths of the Russian conscience.

Few people realize, even today, what an overwhelming influence this act wields upon subsequent naval and military operations. Germany has separated Russia from her allies, and in less than a year this will lead to Russian military defeat and collapse, which in turn facilitates the Russian revolution. The enormous lands of Imperial Russia have just four ways of reaching the seas. One : by way of the Baltic, but this, the German Navy has already closed. Two : Archangel a port on the Arctic Ocean, frozen over for several months in the year connecting with Russia by a long, single-track railroad. Three : the Pacific port of Vladivostok; ice bound for three months that connects Russia only by the thin line of the 5,772 miles long Siberian railway. The fourth passage is that of the Dardanelles Straits; the only practicable one ~ the narrow channel through which Russian exports reach the world with 90% of Russian trade being shipped this way. As Russia's principal Black Sea export was grain with for example, over 20 million tons shipped through the Straits to world markets in both 1911 and 1912, the security of her economy depended on access to the Straits.

By suddenly closing it, Germany has disabled Russia, both as an economic and a military power. By shutting off the exports of Russian grain, she deprives Russia of the financial power essential to pursue successful warfare. What is perhaps even more fatal, she prevents Great Britain and France from getting munitions to the Russian battlefront in sufficient quantity to stem the German onslaught. As soon as the Dardanelles is closed, Russia has to fall back on Archangel and Vladivostok for such supplies as she can get from these ports. The cause of the 1915 military collapse of Russia will be that her soldiers simply have no ammunition with which to fight. The task of undoing the Franco-Russian treaty, and driving such a barrier between Russia and her western allies is proving an easy one. Simply a matter, of closing a little strip of water about 20 miles long, and just two or three miles wide. It is one of the most dramatic military triumphs of the war, and it is all the work of sharp German penetration and diplomacy. This bottling up of Russia sees mountains of useless ammunition pile up in distant Arctic and Pacific ports, with never enough rail capacity to deliver it to the field of battle. When Ambassador Paléologue learns that the Russian artillery are short of shells and that their infantry are short of rifles, he meets with Russian Army Chief of Staff, General Bielaiev who explains:

Our losses in men have been colossal, though if it were merely a matter of replacing wastage we could soon do so as we have more than 800,000 men in our depots. But we're now short of rifles to arm and train these men. Our original reserve was almost 5,600,000 rifles ~ at least we thought so. Grand Duke Nicholas thought so, and I thought so myself. We have been criminally deceived : our magazines are nearly empty. Forgive me for giving no further explanation of a very painful matter. To make good the deficit we are about to purchase a million rifles in Japan and America and we hope to arrive at an output of one hundred thousand a month in our own factories.

Perhaps France and England could also let us have a few hundred thousand. The position is hardly less difficult as regards artillery ammunition. The consumption has surpassed all our calculations and anticipations. At the beginning of the war we had 5,200,000 rounds of 76mm shrapnel in our arsenal. Our entire reserve is exhausted. The armies need 45,000 rounds per day. Our maximum daily output is 13,000. We hope it will reach 20,000 by February 1915. Until that date, the situation of our armies will not only be difficult ~ but dangerous. I have kept nothing from you. General Mikhail Bielaiev

Berlin deciding to intrigue to produce a *casus belli* directs Admiral Souchon to recurrently send the *Goeben* and the *Breslau* to manœuvre through the Black Sea, hoping Russian warships will strike. Three Turkish torpedo boats enter the harbour of Odessa; sink a Russian gunboat, fire on the French mail boat *Portugal,* then turn their guns on the Black Sea port of Odessa to destroy a sugar refinery. They then steam away at top speed pursued by a Russian destroyer. German officers commanding these three Turkish vessels have few Turks on board as the Turkish contingent of their crews have been given a holiday for the religious festival of *Bairam.* The German officers have raided Odessa deliberately to make war inevitable, after constantly threatening to commit some act if Turkey did not, and now they have done it. The *Breslau* next fires off 200 shells into Novorossiysk sinking 14 Russian grain ships and blowing up another 50 petroleum tanks, and next visits Theodosia, Kerch, and Yalta.

When this news reaches Constantinople, Minister of Marine Ahmed Cemal is playing cards at Cercle d'Orient. As an official act of Turkey, it supposedly could only have been made on his orders. And yet, when someone calls him from the card table to tell him of the news ~ *I know nothing about it! It has not been done by my orders.* On the evening of 29th October, Ambassador Morganthau talks with Talaat, *I knew nothing of this attack beforehand. Responsibility rests with German Admiral Wilhelm Souchon.* By drawing Turkey into the war on the side of Germany, Admiral Souchon has achieved strategic results out of all proportion to the forces available to him, to shape the course of the war in his theatre.

CONSTANTINOPLE 30th October. Ambassador Henry Morgenthau visits the British Embassy.

One final picture I have of these exciting days. On the evening of the 30th I called at the British Embassy. British residents were already streaming in large numbers to my office for protection and fears of ill treatment, even the massacre of foreigners filled everybody's mind. Amid all this tension I found one imperturbable figure. Sir Louis du Pan Mallet was sitting in the chancery, before a huge fireplace, with large piles of documents heaped about him in a semi-circle. Secretaries and clerks were constantly entering, their arms full of papers, which they added to the accumulations already surrounding the Ambassador. Sir Louis would take up document after document, glance through it and almost invariably drop it into the fire. These papers contained the embassy records for probably a hundred years. The records of other great British ambassadors now went, one by one, into Sir Louis Mallet's fire. The long story of British ascendency in Turkey had reached its close. The 20-years campaign of the Kaiser to destroy England's influence and to become England's successor had finally triumphed, and the blaze in Sir Louis's chancery was really the funeral pyre of England's vanished power in Turkey. As I looked upon this dignified and yet pensive diplomat, sitting there amid all the splendours of the British Embassy, the British Ambassador, as is usually the case with all British diplomatic figures, was quiet and self-possessed. We sat there before his fire and discussed the details of his departure. He gave me a list of English residents who were to leave, and those who were to stay. I made arrangements with Sir Louis for taking over British interests.

St PETERSBURG. West of the Vistula, the Russian Army is still advancing along the whole front. Serge Sazonoff receives the news calmly, and arranges to see Tsar Nicholas, *I have decided that not a man shall be withdrawn from the German Front. Our first concern is to beat Germany. The defeat of Germany will necessarily involve the ruin of Turkey. We shall keep down to a minimum the forces required against the Turkish fleet and army.* Sazonoff: *We shall make Turkey pay for her mistakes. We must have tangible guarantees on the Bosphorus. As regards Constantinople, personally I don't want the Turks to be cleared out. I'd gladly leave them the old Byzantine city with a good-sized kitchen garden all round. But no more!*

2nd November. Tsar Nicholas II issues his Proclamation of War in leaflet form to his people;

It with complete serenity and faith in the assistance of God, that Russia takes on the appearance of this new enemy, this ancient oppressor of the Christian faith and of all Slavic nations. It will not be the first time that Russia's valiant arms overcome the Turkish hordes and chastise this insolent enemy of our motherland. Together with all Russian nations we believe without fail that Turkey's reckless intervention in the present conflict will only accelerate her submission to fate and open up Russia's path towards the realization of the historic task of her ancestors along the shore of the Black Sea.

Under German command the Turkish feet has had the treacherous effrontery to attack our Black Sea coasts. We share with all the peoples of Russia the unshakable conviction that the rash intervention of Turkey will only hasten that country's downfall and open Russia's path towards the solution of the historic problem, which our ancestors have bequeathed to us on the shores of the Black Sea. With this declaration, many Russians are inspired to re-read the old poems of Fyodor Tiutchev, the poet of *slavianophilstvo*, in particular his verses of *Russian Geography,*

Moscow, the City of Peter, and the City of Constantine, are the three sacred capitals of the Russian Empire. But where are its frontiers on the north and the east, the south and the west? Destiny will show us in the future. Seven inland seas and seven great rivers; from the Nile to the Neva, the Elbe to China, the Volga to the Euphrates, the Ganges to the Danube, there is the Russian Empire, and it will last throughout the centuries!
The spirit has predicted this, and Daniel has prophesied it. Soon the day will be fulfilled, the hour will sound!
And in Byzantium, born again, the ancient vault of Saint Sophia will once more shelter the altar of Christ.
Kneel before that altar, O Tsar of Russia, and rise, Tsar of all the Slavs!

CONSTANTINOPLE. Russian Ambassador Mikhail Giers: *Russia, herself, will guarantee the Turkish Fleet does not again come into the Black Sea.* Talaat calls on Morgenthau during the afternoon to advise, *I have just had lunch with Baron Wangenheim. The Cabinet has the Russian reply under consideration. The Grand Vizier wishes to have Giers's terms put in writing ~ will you attempt to get it?* But by this time, Italian Ambassador Marquis Garroni has taken charge of Russian affairs, so Talaat tells Morgenthau: *Such negotiations are all out of my hands and that further negotiations must now be conducted through him.* Morgenthau: *Why don't you drop your mask as messenger boy of the Grand Vizier and talk to me as Talaat?* Mehmet Talaat laughs : *Well, Wangenheim, Enver, and I prefer that war shall come now!* Governor of Syria, Dejmal Pasha concurs ~ *I consider this new crisis that has emerged to be a blessing. I believe that it is the Turk's ultimate duty ~ either to live like a honourable nation or to exit the grand stage of history gloriously.*

Baron Max von Oppenheim is appointed head of Germany's new *Nachrichtenstelle für den Orient Intelligence Bureau for the Orient* in Berlin to gather intelligence and to spread Pan-Islamist ideas among Muslims to encourage participation in a *Jihad* against Great Britain, France, and Russia ~ to convince Muslim opinion that Germany is the true friend of Islam, and to report German military successes. With sections devoted to Arabia, Persia, Turkey, India, and Russia, the *Nachrichtenstelle* employs German academics specializing in various branches of *Oriental Studies* and a number of Muslim associates from Egypt and North Africa. In addition to placing pro-German articles in newspapers in Constantinople and elsewhere, and producing Oppenheim's own bi-monthly Arabic language paper, *El-Dschihad*, the Bureau prints and distributes leaflets and pamphlets. Oppenheim wrote this piece:

Exposé Concerning the Fomenting of Revolutions in the Islamic Territories of our Enemies: Our consuls in Turkey and India, and our agents, must incite the Mohammedan world to wild revolt against this unscrupulous nation of grocers. Should we have to haemorrhage, England should at least lose India. The aim is to stress the religious element, especially through the Al-Azhar Mosque and the brotherhoods. Small-scale coups and terrorist attacks could also contribute to making the British in Egypt more headless than they apparently already are. For England especially, the intervention of Islam in the war is a fearful blow. Let us do everything in our power; let us use all possible means to make it a fatal one! As soon as Turkey strikes, the call to Holy War and emancipation from foreign rule must immediately be sounded.

What von Oppenheim has in mind, is not simply just to instigate religious revolutions in Russian Central Asia, French North Africa, and British India. Not just a shattering military blow to knock out enemy fighting capabilities, but the larger geo-political strategy to wreck the British Empire and for Germany to supersede Britain as a world power. Baron Oppenheim assures Chancellor Bethmann: *Even if religious rebellion fails to eventuate, the mere threat of a Moslem uprising in India will force England to agree to peace terms favourable to us.* Chief of German General Staff, von Moltke agrees; *It is of the greatest importance to start insurrections in India and Egypt, also in the Caucasus. By means of the treaty with Turkey, the Foreign Office will be in a position to bring this idea to realization and to excite the fanaticism of Islam.*

Wangenheim explained to me one of Germany's main purposes in forcing Turkey into the conflict. He made his explanation quietly and nonchalantly, as though it had been quite the most ordinary matter in the world. Sitting in his office, puffing away at his big black German cigar, he unfolded Germany's scheme to arouse the whole fanatical Moslem world ~ Germany has planned a Jihad or Holy War as one means of destroying English and French influence in the world. Turkey herself is not the really important matter. Her army is a small one, and we do not expect it to do very much. But the big thing here is the Moslem world. If we can stir the Mohammedans up against the English and the Russians, we can force them to make peace. Henry Morgenthau

11th November. Nine days after Russia responds to the Turkish naval attacks by declaring war on the Ottoman Empire, and six days after Britain and France follow, Turkey declares war on Britain, France and all of their allies. The Young Turks inform Berlin; *If victorious we desire territorial gains at Russia's expense and for Germany to secure for Turkey the correction of her eastern border to place Turkey in direct contact with the Moslems of Russia.* At a ceremony in the mosque of Mehmet II the Conqueror, the Banner of the Prophet is unfurled and Sheikh Khairi girds the Sultan with the Sword of the Prophet. The Sultan delivers a fiery speech calling upon the people to join the struggle against the foes of Islam.

The pursuit of Jihad against the Éntente is the holy obligation of every individual Muslim, including Muslims living under the rule of any one of those powers and thus subject to the harshest reprisals from their rulers, such as death for themselves and the destruction of their families. Any Muslim failing in this duty, will incur the wrath of God; any Muslim fighting in the British, French, and Russian armies who engages in combat against the soldiers of Islam will merit the fires of hell.

14th November. The formal promulgation of the *fatwas* calling for *Jihad* is marked by a spectacle carefully planned and directed by the German authorities. A crowd of demonstrators with a band gather in front of the German Embassy where they are greeted from the balcony by von Wangenheim and a group of Moroccan, Tunisian, and Algerian POWs, specially brought in from camps in Germany to create a vivid image of the solidarity of all the world's Muslims. After the crowd respond with cheers for the Kaiser and Germany as Islam's ally, the demonstration moves on to the A-H Embassy and then to the inner city, where one of the accompanying policemen, fired up by patriotic enthusiasm, enters a hotel, takes out his revolver and fires at an English grandfather-clock in the foyer.

The Sheik-ul-Islam published his proclamation, summoning the whole Moslem world to arise and massacre their Christian oppressors ~ Oh, Moslems, Ye who are smitten with happiness and are on the verge of sacrificing your life and your goods for the cause of right, gather now around the Imperial throne, obey the commands of the Almighty, who in the Koran, promises us bliss in this and in the next world; embrace ye the foot of the Caliph's throne and know ye that the state is at war with Russia, England, France, and their Allies, and that these are the enemies of Islam. The Chief of the believers, the Caliph, invites you all as Moslems to join in the Holy War!

The religious leaders read this proclamation to their assembled congregations in the mosques; all the newspapers printed it and it was broadcast in all the countries which had a large Mohammedan population ~ India, China, Persia, Egypt, Algiers, Tripoli, Morocco, and in all these places it was read to the multitudes and the populace was exhorted to obey the mandate. The (German owned) Turkish newspaper Iqdām then incited ~ The deeds of our enemies has brought down the wrath of God. A gleam of hope has appeared. All Mohammedans, young and old, men, women and children, must fulfil their duty so that the gleam may not fade away, but give light to us for ever. How many great things can be accomplished by the arms of vigorous men, by the aid of others, of women and children! The time for action has come. We shall all have to fight with all our strength, with all our soul, with teeth and nails, with all the sinews of our bodies and of our spirits. If we do it, the deliverance of the subjected Mohammedan kingdoms is assured. Allah is our aid and the Prophet is our support.

The Sultan's proclamation was an official public document, and dealt with the proposed Holy War only in a general way, but about this time a secret pamphlet appeared which gave instructions to the faithful in more specific terms. It was printed in Arabic, the language of the Koran. It was a lengthy document full of quotations from the Koran, and its style was frenzied in its appeal to racial and religious hatred. It described a detailed plan of operations for the assassination and extermination of all Christians ~ except those of German nationality.

O people of the faith and beloved Moslems, consider for a moment the present condition of the Islamic world. For if you consider this you will behold a bewildering state of affairs, which will cause tears and grief. You see the great country of India, which contains hundreds of millions of Moslems, fallen, because of religious divisions and weaknesses into the grasp of the enemies of God, the infidel English. You see forty millions of Moslems in Java shackled by the chains of captivity and of affliction under the rule of the Dutch. You see Egypt, Morocco, Tunis, Algeria, and the Sudan groaning in the grasp of the enemies of God and his apostle. Wherever you look you see that the enemies of the true religion, particularly the English, the Russians, and the French, have oppressed Islam and invaded its rights in every possible way. We cannot enumerate the insults we have received at the hands of these nations who desire totally to destroy Islam and drive all Mohammedans off the face of the earth. This tyranny has passed all endurable limits; the cup of our oppression is full to overflowing. In brief, the Moslems work and infidels eat; the Moslems are hungry and suffer and infidels gorge themselves and live in luxury. The world of Islam sinks down and goes backward, and the Christian world goes forward and is more and more exalted. The Moslems are enslaved and the infidels are the great rulers. This is all because the Moslems have abandoned the plan set forth in the Koran and ignored the Holy War, which it commands. Time has now come for the Holy War. Know that the blood of infidels in Islamic lands may be shed with impunity ~ except those to whom the Moslem power has promised security and who are allied with it. The killing of infidels who rule over Islam has become a sacred duty, whether you do it secretly or openly, as the Koran has decreed. Take them and kill them where you find them. He, who kills even one unbeliever of those who rule over us, whether he does it secretly or openly, shall be rewarded by God. Let every Moslem know that his reward for doing so shall be doubled by the God who created heaven and earth. Rise up as the rising of one man, in one hand a sword, in the other a gun, in his pockets balls of fire and death-dealing missiles, and in his heart the light of the faith, and that we should lift up our voices, saying ~ India for the Indian Moslems, Java for the Javanese Moslems, Algeria for the Algerian Moslems, Morocco for the Moroccan Moslems, Bokhara for the Bokharan Moslems, Egypt for the Egyptian Moslems ~ and the Ottoman Empire for the Turks and Arabs.

Specific instructions for carrying out this Holy purpose follow. There shall be a Heart War ~ every follower of the Prophet shall constantly nourish in his spirit a hatred of the infidel; a Speech War ~ with tongue and pen every Moslem shall spread this same hatred wherever Mohammedans live; and a War of Deed ~ fighting and killing the infidel wherever he shows his head. The Holy War will be of three forms. First, the individual war, which consists of the individual personal deed. This may be carried on with cutting, killing instruments, like the slaying of the English chief of police in India, and like the killing of the official arriving in Mecca by Abi Busir (may God be pleased with him). Second, believers are to organize bands, and to go forth and slay Christians. The most useful are those organized and operating in secret. It is to be hoped that the Islamic world will profit very greatly from such secret bands. The third method is by organized campaigns, by trained armies.

With the proclamation of *Jihad*, special agents are sent to the consulate at Tripoli with orders to subvert French rule in Morocco, Algeria, and Tunisia with the aim of persuading the rulers of those countries to engage in military actions directed against India and British interests in the Persian Gulf. Chancellor Bethmann supports Oppenheim's intriguing with the revolutionary Indian Ghader Party to encourage terrorist action in India and mutinies among the Hindu troops in the British Army while the Kaiser reassures his people regarding their alliance with a Muslim nation against other Christian, European nations when he asks a leading oriental academic to explain;

Only when Islam is threatened can the Caliph, by proclaiming Holy War, call upon all Muslims to unite under him and serve him with all their might. The outbreak of war has roused the feeling of community among all Islamic peoples, irrespective of origin and race to fever pitch, so that we can now observe the emergence of a powerful movement directed against those nations that in the course of time has overrun Muslim lands and subjected their populations to unrestrained domination and arbitrary rule ~ i.e. Britain, France, and Russia ~ but not Germany or Austria-Hungary. Many non-Muslims have it appears, misinterpreted the recent Jihad as a movement to impose the Muslim religion on the entire world.

When the *fatwas* are distributed, in Arabic, Persian, and other languages, in addition to Turkish, the Éntente simply obtained *fatwas* from local legal experts declaring obedience to colonial powers as lawful and binding on the Muslims in their jurisdiction. Oppenheim hoped through his personal contacts with Prince Faisal, one of the sons of Sharif Hussein of Mecca and guardian of Muslim holy places, to win the support of the Arabs for his *Jihad*, but his British counterpart Captain T.E. Lawrence outmanoeuvres him. After playing a double game for a while, Sharif Hussein of Mecca, (encouraged by the British) campaigns to have himself recognized as the legitimate Caliph rather than the Ottoman Sultan whose right to the title of Caliph is not recognized by all Muslims. Hussein raises his standard against Constantinople to confront ; *Weak government, anti-religious legislation, and unjust punishment, including the hanging of people of rank.* Ernst Herzfeld from Tell Halaf declares that ~ *The Jihad is a farce, and the war a crime. Why would the Ottoman Empire's subjects fight for a hated, corrupt, and deceitful regime?* When von Schowingen heads the *Nachrichtenstelle* with a monthly allowance of 5,000 marks, he writes; *The Odol-Toothpaste Company invests more in advertising than Imperial Germany does in war propaganda!* Harry Stuermer as the Constantinople correspondent of newspaper *Kölnische Zeitung* likewise writes;

The artificial character of the Jihad, which was only set in motion against a portion of the unbelievers, while the others became more and more the ruling body in Turkey, is the proof of the untenability of this theory. Practically all its hopes in that direction have been shattered. Its costly efforts against European civilization in Mohammedan countries have resulted in the terrific counter-stroke of the defection of the Arabs and the foundation of a purely Arabian Caliphate under English protection. The so-called Holy War, if it had succeeded, would have been one of the greatest crimes against human civilization that even Germany has on her conscience. But the attempt against colonial civilization has not succeeded. The Jihad, proclaimed as it was by the Turanian pseudo-Caliph and violently anti-Éntente, was doomed to failure from the very start from its obvious artificiality. It was a miserable tragicomedy, the present ending of which, the defection of the Arabian Caliphate, is the direct contrary of what had been aimed at with such fanatical urgency and the use of immoral propaganda. The attempt to unloose Holy War was due to absurd illusions. It would seem that in Germany, the home of so many eminent doctors of research, even the scholars have been dazzled by wild political illusions, or surely, knowing the countries of Islam as they must, they would have raised their voices against such arrant folly.

Of the shady characters employed by Oppenheim's Nachrichtenstelle to spread propaganda in support of the Jihad ~ many of the individuals claiming to be devout Muslims, offered their services and received funding from the Nachrichtenstelle, were simply crooks who milked the German Government ~ and Oppenheim himself, since he was helping to finance the program with his own money ~ for all they were worth. Numerous examples might be cited of the German Embassy being made the dupe of greedy adventurers who treated them as an inexhaustible source of gold. First Oskar Niedermayer appeared on the scene to announce himself as the one to cope with Afghanistan, and then Wilhelm Wassmuss came along to play the great man on a special mission for a time in Persia while money belonging to the German Empire found its way into all sorts of low haunts. Even a bona fide connoisseur of the East like Baron von Oppenheim, who had already made tours of considerable value for research purposes right across the Arabian Peninsula, and so should have known better than to share these false illusions, doled out thousands of marks from his own pocket ~ and millions from the Treasury! ~ to stir up the tribes to take part in the Jihad.

* * *

WESTERN FRONT. During the month of December 1914, a distinct softening of feeling took place in the German lines, showing the deep attachment of the Teutonic mind, to the candle and the fir tree.

A week ago, at Armentiéres an officer told me that the Germans threw into the British trenches a splendid cake of chocolate with a note tied to it which read ~ We propose having a concert tonight as it is our captain's birthday, and we cordially invite you to attend ~ provided you will give us your word of honour as guests that you agree to cease all hostilities between 7.30 and 8.30. If you will honour us with your company, please write back saying so, and wrap the note round all the tobacco you can spare, as we haven't got a pipe full left. We shall return it tomorrow. When you see us light the candles for footlights on the edge of our trench at 7.30 sharp you can safely put your heads above your trenches, and we shall do the same, and begin the concert.

Word was passed up and down the long trenches of both sides to avoid accidents. The invitation was accepted in tobacco, the long row of footlights was lit to time, and on the minute a double quartette of whiskered heads shot up above the trench like Christy Minstrels in a row and sang. Our men applauded like the rattle of machine guns loaded blank. Before the concert finished one big voice yelled ; Please com mit us into der chorus! One of our boys, before we could stop him, shouted back : We'd rather die than sing German! And the big voice came back ; It would kill us if you did! Splendidly sung ~ Die Wacht am Rhein ~ closed the truce.

One by one the battlefield footlights went out, and as heavy clouds began hiding the sentinel stars two long rows of human beings built in God's own image sank with moistened eyes into the living graves where men kill men. As a few bullets aimed high sang in mockery through the golden reflection of the trench fires the soldiers Good Nights passed each other across the space. *Daily Express*

General Walter Congreve who had won the Victoria Cross in the Boer War, and had since became a divisional commander on the Western Front, wrote to his wife on Christmas Day to describe the *Christmas Truce* ~ a unique experience that arose spontaneously in several parts of the Western Front.

Christmas Day,

Darling dear,
As I cannot be with you all, the next best thing is to write to you so I get closer.

We have had a seasonable weather day ~ which means sharp frost and fog ~ and never a spot of sun. I went to church with two of my battalions in a factory room, and after lunch took down to the North Staffords in my old trenches at Rue du Bois. (Mother's gifts of toffee, cigarettes, pencils, and writing paper have arrived)

There I found an extraordinary state of affairs. This morning a German shouted out that they wanted a day's truce, and would come out if we did. So very cautiously, one of our men lifted himself above the parapet and saw a German doing the same. Both got out, then more, and finally all day long in that particular place they have been walking about together all day giving each other cigars and singing songs. Officers as well as men were out, and the German colonel himself was talking to one of our captains.

My informant, one of the men, said he had had a fine day of it and had smoked a cigar with the best shot in the German Army, who was not more than 18. They say he's killed more of our men than any other 12 together, but I know now where he shoots from and I hope we down him tomorrow. I hope devoutly they will.

Next door the two battalions opposite each other were shooting all day. And so, I hear it was further north, 1st RB playing football with the Germans opposite them, the next regiments shooting each other. I was invited to go and see the Germans myself, but refrained as I thought they might not be able to resist shooting a general…
~ Walter

THE CAUSE OF THE GREAT ARAB REVOLT

Captain T.E. Lawrence

Lawrence of Arabia

Auda abu-Tayi of the Howeitat

Following the dramatic coup of the Young Turks in Constantinople, they astonish the world when they abandon the traditional stability of the old Ottoman pluralistic and pan-Islamic policies, instead to pursue a policy of secular Turkish nationalism in which the formerly cosmopolitan and tolerant Ottoman Empire now decide to discriminate against many of their non-Turkish inhabitants. The Arabs in particular are faced with political, cultural and linguistic persecution when the new Ottoman régime upholds the ban on the official use of the Arabic language and it's teaching in schools and arrests many Arab nationalist figures in Damascus and Beirut. When the Arabs are further threatened by the construction of the Hejaz Railway to connect Damascus-Mecca, with the potential to facilitate the mobility of Turkish troops into the Arab heartland, Arab nationalist groups in Syria, Persia and Arabia begin to rally behind the Hashemite banner of Emirs Abdullah and Faisal, the sons of Sharif Hussein bin Ali. Hussein, Sharif of Mecca who claims direct descent from Muhammad's daughter Fatima thru the Hashemite family; is Guardian of the Holy Cities of Islam, and whilst the Bedouin of the Hejaz receive income from tolls charged during the annual pilgrimages to Medina and Mecca, they too oppose the Turkish extension of the railway from Medina-Mecca. When the Ottomans use troops to force the rail extension, Hussein's second son Abdullah travels to Cairo to meet with Lord Horatio Kitchener to secure the support of the British Empire. Lord H.H. Kitchener confirms ; *I promise to give the Arabs assistance against foreign aggression in return for Arab assistance against the Turks.*

Consequently, as head of the Arab nationalists and now in alliance with Great Britain and France, Sharif Hussein bin Ali initiates the Great Arab Revolt against Ottoman rule early one morning as the call to prayer trails off over the rooftops of Mecca. Sharif Hussein points his rifle from a window of his home and fires, unfurls the flag of his influential Hashemite clan, and the Great Arab Revolt begins. Sharif Hussein's objective in undertaking the Great Arab Revolt is to establish a single independent and unified Arab state stretching from Aleppo in Syria to Aden in Yemen, based on the ancient traditions and culture of the Arab people, the upholding of Islamic ideals and the full protection and inclusion of ethnic and religious minorities. Arab nationalists in the Fertile Crescent and the Arabian Peninsula find in the Hashemite commanders of the Great Arab Revolt the leadership that can realize their aspirations, and thus coalesce around them. His sons, the Emirs Abdullah and Faisal, lead their Arab forces on a well-planned, surprise attack that forces the Ottoman troops in Mecca to surrender. After they are savagely repulsed by artillery fire at Medina, Emir Faisal meets the enigmatic and now legendary Captain T.E. Lawrence for the first time. Together they agree to deploy mobile Arab forces against static Turkish defences to sever Ottoman communications and rail lines to isolate the Turkish garrison within Medina. Lawrence also meets with the powerful Howietat leader Auda abu-Tayi to inspire his followers to successfully seize the Turkish-held seaport of Aquaba to finish the fight in the Hejaz. The Arab victory of Aquaba concludes the war in the Hejaz, and thereafter, both Arab and British forces have access to a strategic port for their subsequent Palestine Campaign.

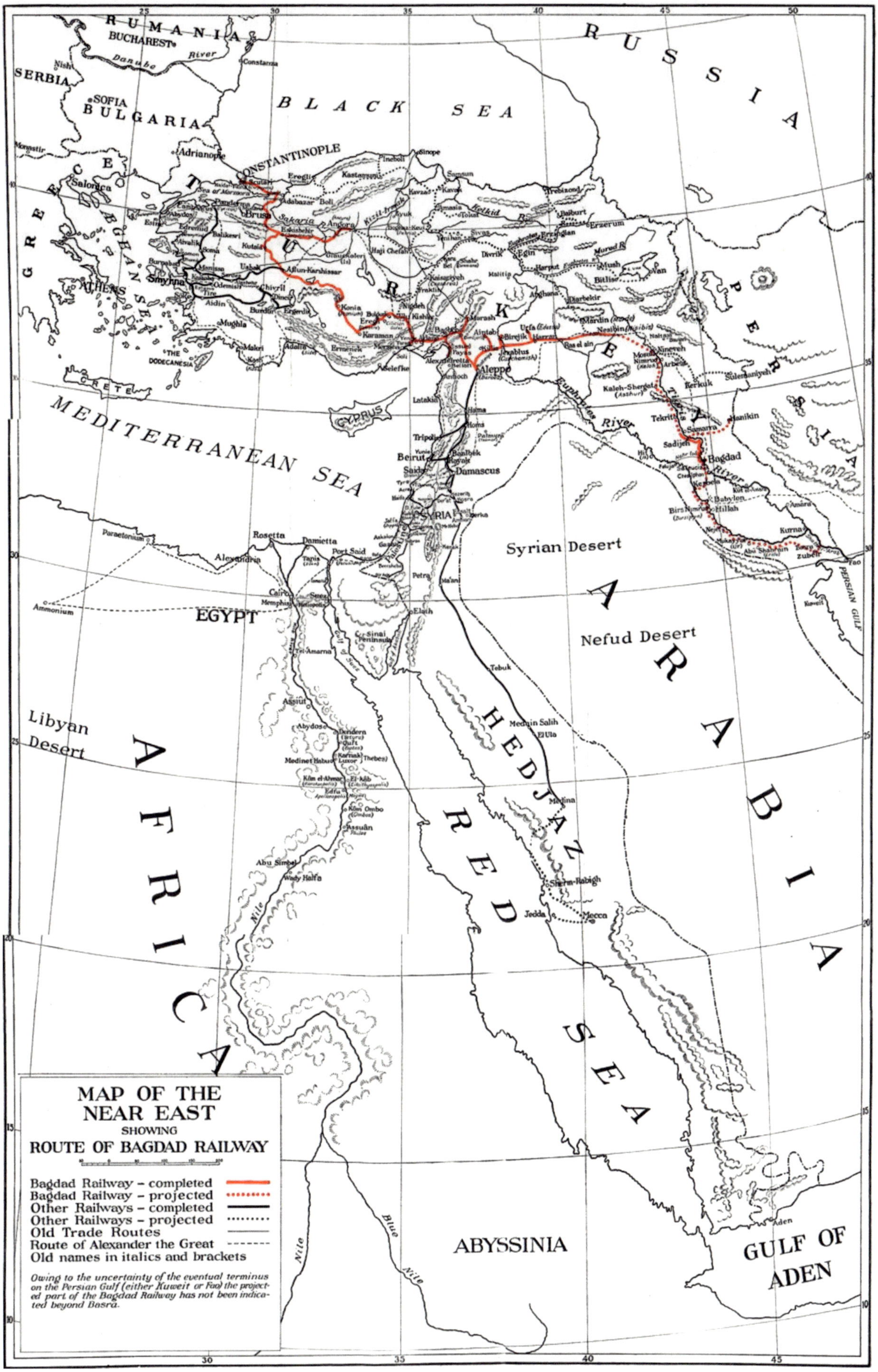

Construction of the *Berlin to Bagdad Railway* is well advanced with a new branch line to the port of Alexandretta on the Mediterranean ~ but difficulties in the Taurus Mountains mean that the rail track is 480km short of completion, limiting its practical use in the war, while Arab forces loyal to Colonel Lawrence continually sever the Hejaz Line. What is very useful to Germany and the Ottoman Turkish Empire ~ is their Balkan Express.

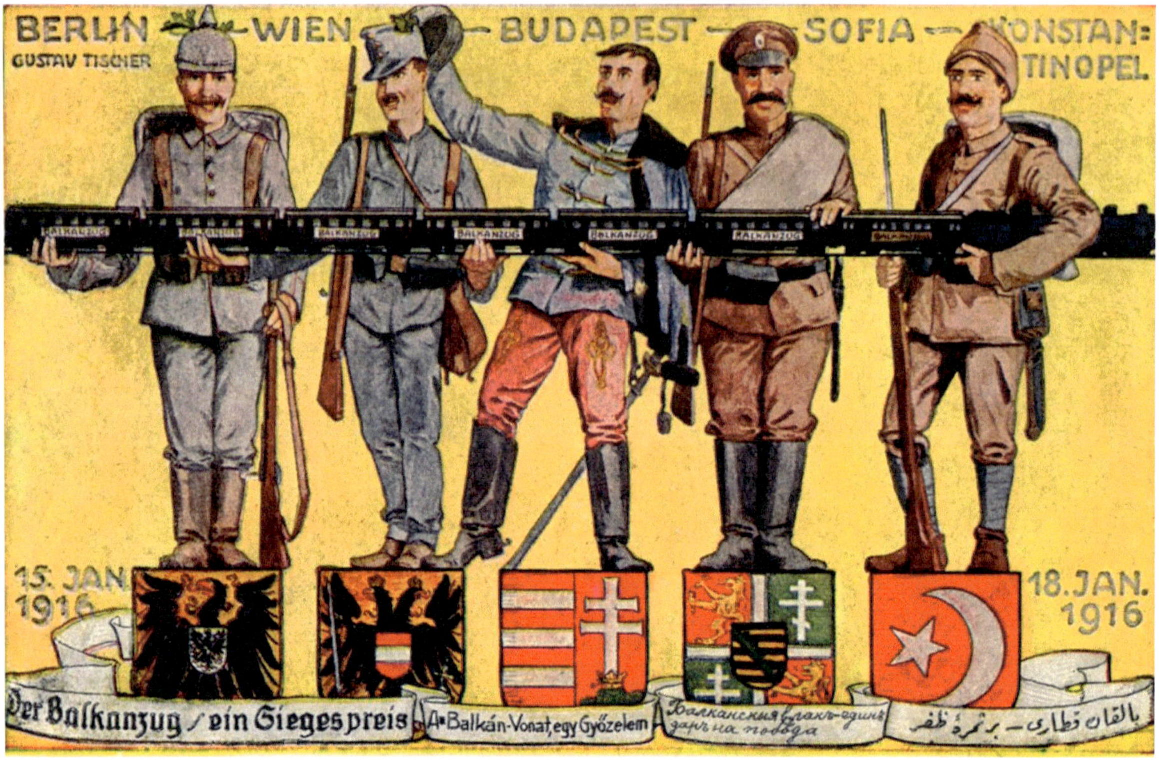

First of all, the Balkan Express does exist, as I have travelled by it myself. It is one of the most perfectly organized railway services I have ever seen, and I have seen many. This service enables the Germans to transfer all sorts of materiel to and from Berlin to Constantinople, and by its aid German troops can be rushed to Constantinople within 56 hours, and from there transferred to whatever front most needs them. The Germans have great ambitions as regards the Near East and Far East also, and much of their energy is concentrated on the realization of ambitions in Turkey, the Balkans, and Asia Minor. To Egypt! is more than a mere political cry.

The Balkan Express is the show train of the world. It is probably the handsomest train in Europe, and beyond doubt has been designed by the Germans with the object of impressing the thousands of people of various nationalities who gaze on it in wonder twice a week on its way from Berlin to Constantinople and twice a week from Constantinople to Berlin. The Balkan-Zug, as it is called in the Central Empires, is a source of unqualified delight to Germans, Austrians, Hungarians, Bulgarians, and the rest of the people who see it on its journey. Its name is blazoned in three-foot letters on each wagon. Engine and carriages are decorated with flags and flowers, and every passenger wears in his buttonhole a German flag on which appear the words Balkan-Zug and the date. On the Balkan-Zug the quality of food is infinitely better than can be obtained in Constantinople or in Vienna, bread-tickets are unknown, and for a mark I had an early breakfast of coffee, rolls, butter and marmalade.

It was ten o'clock at night when we reached Belgrade, which I was anxious to see. On inquiry, I found that the Balkan Express was to remain there for an hour, and determined not to be disappointed, I left the station to stroll round the ruins of the town. Some idea of the accuracy of modern artillery fire may be gleaned from the fact that the besieging Austrian gunners were able to aim with such precision that not one shell had fallen on the railway station. It must be remembered that they were firing from the other side of the Danube at its widest part. The Austrian Staff had obviously realized that their advancing army would have need of the railway as soon as the Serbians had been forced back, and doubtless the artillery had been instructed to spare this important point. The remarkable thing however, is that houses within yards of the station itself have been absolutely destroyed, yet there was not so much as a mark that I was able to see upon any of the station buildings themselves.

Already the river was spanned by a wonderful new wooden bridge, such as could not be constructed in a few weeks, or months for that matter. In all probability, this and many other bridges had been built years back in preparation for the great struggle that Germany and Austria alone knew was impending. This was no temporary makeshift, but as good as the fine American trestle-bridges in use on the best American railways. As I returned to the station, having just seen the terrible fate that had overtaken the Serbian capital, I could not help wondering why it is that England seems incapable of appreciating her danger. I refer, of course, to the population in general, for many of those in high places, I am convinced, have no illusions as to the political and strategical situation. Belgrade is now the Austrian main headquarters on that front, and is essentially a military town.

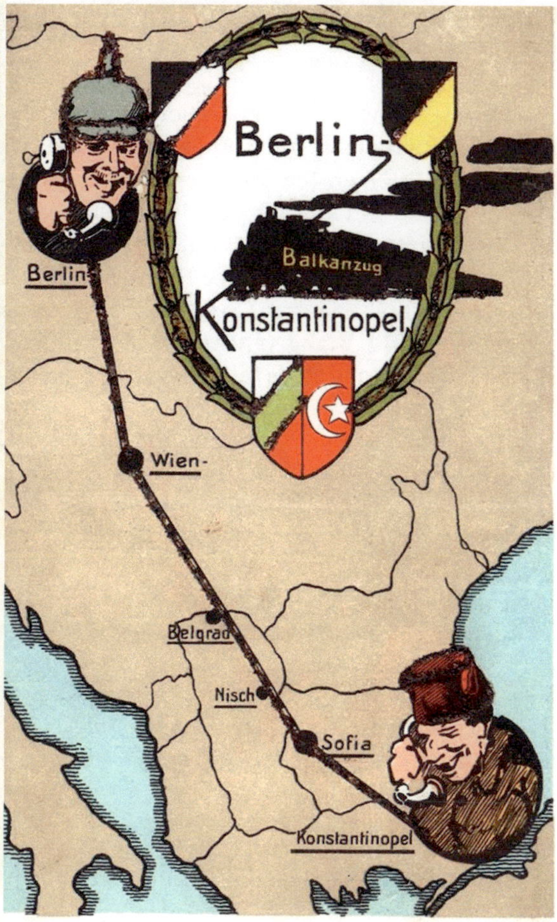

BALKANZUG ~ THE BALKAN EXPRESS ~ Berlin to Constantinople in 50 hours.
As the German replacement of the Orient Express, the Balkanzug connects Germany to its allies A-H, Bulgaria and Turkey.

We drew out of the station shortly before midnight, and arrived at Buda Pesth at ten o'clock the next morning. In the Hungarian capital, the train received a tremendous reception ~ ovation would be a better word. At the Nord Bahnhof there was an enormous crowd, the greatest I have ever seen at a railway station. The excitable Hungarians tumbled over each other in their anxiety to get near the Zug. Wine was brought for the engine driver and fireman, and the passengers, with their little Balkan-Zug flags in their coat buttonholes, were promptly lionized, and for once in their lives experienced the sensation of being popular heroes. The crowd patted them on the back, insisted on shaking hands with them, cooed over them, crowed over them, and laughed with hysterical joy. At Buda Pesth the Balkan-Zug was tidied and made presentable. Windows were cleaned and the compartments and corridors swept. At Buda Pesth the train divided itself into two parts. Another beflagged locomotive appeared. One half of the train goes to Berlin and the other half to Vienna. Those of us who had come from Constantinople were looking forward to improved food obtained from Vienna. The line to Constantinople has had time to take both the Balkan-Zug, and German military trains carrying supplies, men, and munitions.

During the journey to Vienna I talked with a Turkish gentleman and his wife and daughter. I was greatly amused to hear that, although the women had left Constantinople veiled and dressed in Eastern costume, as soon as they crossed the border both put on European clothes and dropped the veil. They expressed the opinion that now Germany had opened up Turkey with railways, the state of semi-starvation in Constantinople would cease. The old Turkish gentleman was shrewd and kindly, and he expressed his regret at the closing of all the French schools in Constantinople. He volunteered the information that, in order that his son should not absorb the principles of German militarism, he had sent him to be educated at a school in Switzerland.

Vienna gave the train what the newspapers call a rousing reception. Even the official mind gave way before it, and the Custom House officers and other functionaries spared us the usual examination and interrogation. Not even our passports were examined on the efficient Balkan-Zug running from Constantinople to Vienna. Knowing, however, the ways of the military authorities, and that later on I should be obliged to prove my arrival, I insisted on having my papers stamped by the military authorities at the railway station. At Vienna tickets were collected from the passengers as they left the station. While my journey from Vienna to Constantinople by way of Bucharest had occupied five days, the Balkan-Zug reduces this to two nights and two days, 50 hours exactly. Even now the train arrives at the various stations with remarkable punctuality, always within five minutes of the scheduled time, which in itself is a triumph for German organization.

Anon.

Despite the warlike preparations under way in Constantinople during late 1914, and its external characteristics that suggest nothing but war, it unexpectedly now becomes a focus of peace. While the Royal Navy threaten the Dardanelles, and Turkish troops march through the streets, these activities do not first and foremost engage the attention of the German Embassy, for Baron von Wangenheim is thinking of one thing only ~ this hawk has apparently become a Christmas dove of peace, who realizes that the greatest service that a German Ambassador can render unto his Kaiser is to end the war on terms that will save Germany from exhaustion ~ and to someway obtain a peaceful settlement. In August of 1914, Wangenheim told Morgenthau ~ *If we don't get Paris in thirty days, we are beaten!* Having failed to take Paris by November, he reopens this theme with Henry Morgenthau...

It is part of Imperial Germany's system, not only to be completely prepared for war ~ but also for peace. A wise general, when he begins his campaign, always has at hand his plans for a retreat, in case he is defeated. This principle applies just the same to a nation beginning war. There is only one certainty about war, and that is that it must end sometime. So, when we plan war, we must consider also a campaign for peace.

Germany had only prepared for a short war because we expected to crush France and Russia in two brief campaigns, lasting no longer than six months. The great rush campaigns have collapsed, all Germany can now look forward to is a tedious, exhausting war, and that all we can obtain from the existing situation will be a drawn battle. Clearly this plan has now failed, and there is little likelihood that Germany will win the war. Germany, will make a great mistake if she persists in fighting the war to exhaustion, for such a fight will mean the permanent loss of her colonies, her mercantile marine, and whole status. We have made a mistake this time, in not laying in supplies for a protracted struggle. It was an error that we shall not repeat. Next time, we shall store up enough copper and cotton to last for five years. Baron Hans von Wangenheim

After severe losses on the Western Front, many Germans by December 1914 are anxiously looking for peace and turn to Constantinople as the most promising place where peace negotiations may start. The Germans hope that President Woodrow Wilson will be the world peacemaker, but Germany with no access to American Ambassadors in enemy capitals turns to the U.S. Embassy in Constantinople. Wangenheim informs Ambassador Morgenthau. *January 1915 will be an excellent time to end the war. Italy has not yet entered, though there is every reason to believe that she will do so by spring. Bulgaria and Romania are still holding aloof, though no one expects that their waiting attitude will last forever. France and England are preparing for their first spring offensives, and their warships are gathering at the Dardanelles.*

Henry Morgenthau: *Yet, badly as Wangenheim wants peace in January 1915, it is quite apparent that he is not thinking of a permanent peace. The greatest obstacle to peace is the fact that Germany shows no signs that she regrets her crimes, and there is not the slightest evidence in Wangenheim's attitude now. Germany has made a bad guess ~ that is all. What Wangenheim and the other Germans see in the situation is that their stock of wheat, cotton, and copper was inadequate for a protracted struggle. In my conversations with Wangenheim, I find him frequently using such phrases as ~ the next war, the next time, and in confidently looking forward to another greater world cataclysm than the present, he merely reflects the attitude of the dominant military class. The Germans apparently want a reconciliation ~ a kind of an armistice that will give their generals and industrial leaders time to prepare for the next conflict.*

Wangenheim's plan is that representatives of the warring Powers should gather around a table and settle things on the principle of ~ give and take. From Wangenheim's conversation I obtain a slight inkling of Germany's terms. Wangenheim is quite insistent that Germany must have permanent naval bases in Belgium with which her navy could at all times threaten England with blockade, and so ensure ~ the freedom of the seas. Germany wants coaling rights everywhere. She might give France a piece of Lorraine and a part of Belgium in return for the payment of an indemnity.

Wangenheim requests that I should place Germany's case before the American Government. My letter to Washington dated January 1915, ventures fully into the internal situation which prevails and gives the reasons why Germany and Turkey desire peace. An interesting part of this incident is that Germany apparently ignores Austria. The Austrian Ambassador Pallavicini knew nothing of the negotiations until I myself informed him. Nothing came of this proposal. Our Government declined to take action, not regarding the time as opportune. Both Germany and Turkey recur to this. Henry Morgenthau
U.S. Ambassador in Constantinople

" TSARGRAD."

The fair city of Constantinople has been blotted out by the Tsar with one stroke of the pen. *We have commanded our ambassador at Tsargrad to leave the precincts of Turkey forthwith,* says the Tsar in declaring war on Turkey; and so, Constantinople henceforth, is to be Tsargrad ~ the City of the Tsars. The Moscow Town Council strikes these Byzantine notes in its address to the Tsar :

Many a time have the conquering hosts of Russia, under the Imperial leadership of your ancestors, knocked at the gates of Tsargrad. Now hath the Lord placed the sword into the hands of your Majesty that the wishes of the Russian people may be fulfilled.

The reactionary representatives of the landed interests in Russia ~ the power behind the throne ~ had not their heart, even from the outset, in the fight against their spiritual kith and kin, the Junkers of Germany. It is therefore not surprising that now, when the *Beauty of the Orient* has kindled their imagination, they express freely what they think about the war with Germany. The *Russkoe Znamia* (*Russian Banner*) of the ultra-nationalist *Black Hundreds* (the Tsar being their chairman) plainly says : *We have lost all moral interest in the fight with the callous, dishonest German.*

The benefit of Russia's victory over Germany will be entirely reaped by England and France, and the only satisfactory compensation acceptable to Russia must be at the expense of Turkey. Why, should the German heritage in Turkey ~ the Persian Gulf, Mesopotamia, the region of the Berlin-Baghdad Railway ~ go to England, which is impotent against Imperial Germany, and not to Russia, who is destined to crush the German might? The real compensation for our efforts in Germany can only be Germany's inheritance in Turkey. For the future generation, for the future of Russia, we must have an outlet in the south to the warm seas. *Novy Ekonomist*

It must not be assumed however, that the grabbing propensities developed in this war are confined to the reactionary elements of Russia alone. In the matter of grabbing they are backed up by the whole Russian Press. The Government has taken good care there should be no discordant note in the chorus, and from the beginning of the war it suppressed Socialist papers, dissolved Trade Unions, and clapped into prison all prominent leaders of Labour movements and editors of the Socialist Press. The clamour to get Constantinople and the Straits ~ *the keys of her own house* ~ are seen as appetizers by other Russian leaders. The real dish remains Turkey, but they are not satisfied with Armenia, Constantinople, and acquiring the Turkish Straits ~ they crave for Palestine and the Holy Sepulchre.

Russia should take not only the Dardanelles and Constantinople ~ but also Asia Minor! It is beyond doubt that in the present war Russia plays the leading role. We are not in need of others, but others are in need of us! When the Austro-German hordes have been crushed on the banks of the Vistula, and Galicia has been occupied by Russia: when Russia has been cleared of the enemy and positions have been secured making it impossible for the enemy to attempt further invasion. Russia can afford to stop its conquering march and calmly await the natural exhaustion of Germany. Such a waiting policy would hardly prove safe for the Anglo-French armies.

Professor P. Migulin, editor of the *Novy Ekonomist*
Chairman Special Commission, Russian Ministry of Finance

This high-level Russian official and prominent voice for acquiring the Straits and Constantinople, speaks in a *Freudian Slip* of an Imperial Russian *waiting policy* to prove unsafe for Anglo-French forces.

The conquest of Constantinople was felt so imminent in Russia when war started in mid-1914, that cheerful artworks were published, and Fyodor Uspenskii, Director of the Russian Archaeological Institute in Constantinople, called for the establishment of a commission to oversee the preservation of all Byzantine monuments in Constantinople ~ once victorious Russian troops could enter the city. The commission is chaired by President of the Imperial Academy of Sciences, Grand Duke Konstantin. An early draft of the Constantinople Agreement proposes that old Constantinople within the historic city walls containing a number of byzantine monuments will be given to Russia along with the Straits.

On 27th November 1914, the Russian Government gave a denial to a revealing statement which was published in the *Birzbevia Yedomosti* on the previous day, alleging that an agreement had been arrived at between Russia, France, and Britain concerning the Turkish Straits and Constantinople.

Le Vin ~ Scissors & Paste ~ Dublin 20th January 1915

The Constantinople Agreement was swathe of secret assurances made by the Éntente in which France and Great Britain agreed to give Constantinople and the Dardanelles to Russia… if victorious.

The Tsar speaking to French Ambassador Paléologue in September 1914 ~ *Annexing the Straits, European Thrace and Turkish Armenia ~ we will need substantial British assistance to breach the Straits.* FM Serge Sazonoff knows that Russia should at least offer some forces to any plan to force the Straits, in order to justify Russia's claim on them. He asks the generals repeatedly and is repeatedly told ~ *No!*

From 31st October 1914, once the Ottoman Turkish Empire declares war and closes the Dardanelles supply route to Russia, thereafter wheat from which the Tsar's empire derives most of its foreign income cannot be exported ~ and worse, British military supplies can only be delivered to Russia through the northern port of Archangel. The next development comes in December, when a message arrives via British Ambassador Sir George Buchanan from Grand Duke Nicholas asking ; *Could Britain arrange a demonstration against the Turks to draw away part of their forces attacking Russian positions in the Caucasus?* Grand Duke Nicholas who repeatedly overlooks French requests to prioritize the German front over the A-H ~ no matter how many times French asks, soon receives reports of the Turkish IIIrd Army having inadequate kit for their winter offensive, of them abandoning guns that cannot be dragged through deep snow, and of three Turkish divisions encircled by his forces. Evidence enough, that any *supposed* threat from the Turkish IIIrd Army in the Caucasus has passed… but the Grand Duke does not retract his request for Allied action. Russia wants the Straits and more.

A naval action is proposed to pierce the barriers that separate Imperial Russia from her allies. Lord Horatio Herbert Kitchener states that a naval action is vitally important ~ the great aspect of this form of offensive action is that ~ *If satisfactory progress cannot be made, the attack can be easily broken off.* First Sea Lord *Jacky* Fisher declares ~ *If necessary, a fleet of old pre-dreadnought battleships could force through the Dardanelles into the Sea of Marmara, then press on to shell Constantinople by naval gunfire to eliminate Turkey from the war!* Lord Kitchener agrees ~ *This idea has merit!*

As practically demonstrated by the advancing German Army how easily old fortifications of the Belgians and French could be demolished 1914, and as the Turkish coastal forts are so much older, the Admiralty therefore believes that soon after the methodical destruction of the Turkish Straits forts, the minefields can be cleared so that an Allied fleet can steam up the Dardanelles to Constantinople. If the Ottoman Empire will not surrender, then the battleships will target the Turk's few munitions factories all located within Constantinople, then shell their rail link to Europe and Anatolia situated along the coast of the Marmara to cut their military supply-line from Germany and from A-H. To compel the Turks to surrender, Constantinople consisting of many old wooden buildings could be shelled to create a blazing inferno, and this proposed operation could knock Turkey out of the war, reopen the shipping-route to Russia thru the Straits, and secure the Tsar's troubled Empire.

Lord *Jacky* Fisher, signals to Vice-Admiral Sackville Carden, CO of the Royal Navy squadron in the Ægean ; *Do you consider the forcing of the Dardanelles by our ships alone to be a practicable operation? It is assumed older battleships fitted with mine-bumpers will be used, preceded by colliers, or other merchant craft as mine-bumpers and sweepers. Importance of results would justify severe loss. Let me know your views.* Carden responds favourably proposing a three-stage operation : bombardment of the entrance forts, the clearing of the minefields, and steaming up the Straits thru the Sea of Marmara to Constantinople.

To support the naval operation, Lord Kitchener appoints his friend General Sir Ian Hamilton to command a force of British soldiers, French North Africans and ANZACS ; *to help the navy to reap the fruits of success.* Hamilton : *I bade him good morning and walked up to his desk where he went on writing like a graven image. After a moment, he looked up and said in a matter-of-fact tone ~ We are sending a military force to support the fleet now at the Dardanelles and you are to have command.* Kitchener resumes writing while Hamilton remains silent in front of his desk. When Kitchener looks up, *Well? Once the fleet gets through, Constantinople cannot hold out. The fleet with their guns will dominate the place and if necessary burn the place to ashes. There will be a revolution at the mere sight of the smoke from the great funnels of our warships. The Turks are busy elsewhere. I hope that you will not have to land at all. If you do have to land on the shores of the Bosphorus to occupy Constantinople, then the powerful fleet at your back will be the prime factor in your choice of time and place. But we must not land troops on the Asian side of the Dardanelles. Once we begin marching about continents, situations calling for heavy reinforcements will be created. Undertake military operations only in the event of the fleet failing to get through after every naval effort has been exhausted.* Hamilton is given few useful details : an out-of-date map of the Dardanelles defences, a 1903 report on the Turkish Army, plus a phrase book and some tourist guides for sightseers in Constantinople.

By now, Winston Churchill who has reasonably expected the Dardanelles Campaign to become a joint Allied operation in which Russia will play some effective contributory role is feeling much amiss in the intent of Russia's war effort ~ as evidenced in his 20th January 1915 communiqué to Sazonoff : *Russia must contribute both warships and amphibious landing forces to the campaign. Your Black Sea Fleet is to begin shelling the upper Bosphorus defences as soon as the outer Dardanelles forts are destroyed by the Allies.* Russia replies : *No more than 40,000 troops can be made available this year.* Russia fails to contribute any troops that France and Britain expect, and their Black Sea Fleet remains inactive to Churchill's request.

~ Russian soldier having a chat at Tsargrad ~

IN THE SHELL-SWEPT WATERWAY OF THE DARDANELLES

THE BRITISH AND FRENCH NAVAL ATTACK ON THE DARDANELLES IN MARCH 1915 WAS A MOVEMENT DARING IN ITS CONCEPTION AND STRONG IN ITS EFFECT ~ THAT IT WAS NOT IMMEDIATELY SUCCESSFUL IS NO REFLECTION UPON THE WISDOM OF THE DECISION THAT UNDERTOOK IT. THE PRIZE THAT WOULD HAVE FALLEN ON THE STROKE OF SUCCESS WAS SOMETHING WELL-WORTH RISKING A GREAT DEAL FOR : NOTHING LESS THAN THE FULL PARALYSIS OF TURKEY FOR FURTHER OFFENSE, THE OPENING OF THE TRADE ROUTE THAT WOULD HAVE ADMITTED A STREAM OF MUCH-NEEDED MUNITIONS AND MILITARY SUPPLIES TO RUSSIA'S SOUTHERN PORTS, AND COULD HAVE FREED HER STORES OF GRAIN TO FEED HER ALLIES IN THE WEST, AND A LESSON ON THE STRENGTH OF THE ALLIES TO UNDECIDED BALKAN NEUTRALS.

THE ILLUSTRATION ABOVE IS A BIRDS-EYE VIEW OF H.M.S. TRIUMPH, WHICH DESTROYED THE FORTS AT SMYRNA, WITH ANOTHER BATTLESHIP AHEAD, MOVING UP THE STRAITS BEHIND THE MINESWEEPERS AND UNDER HEAVY SHELLFIRE FROM THE TURKISH LAND FORTS. THE NEWS OF THE NAVAL ACTION IN THE DARDANELLES CONVEYED TWO SURPRISES TO THE WORLD ~ THAT THE H.M.S. QUEEN ELIZABETH WITH HER HUGE 15-INCH GUNS WAS IN COMMISSION AND THUNDERING AT THE ENTRANCE TO THE WATERWAY TO CONSTANTINOPLE, AND THAT THE ARK ROYAL, THE FIRST SHIP TO ACCOMMODATE AEROPLANES WAS ALSO DOING HER PART IN NAVAL WARFARE. THE GREAT RANGE OF THE QUEEN ELIZABETH'S POWERFUL GUNS ENABLED HER TO FIRE HER MAMMOTH SHELLS FROM THE SAROS GULF OVER THE INTERVENING HILLS INTO THE FORTS OF THE NARROWS.

FIRST STAGE : SHELLING THE DARDANELLES FORTS

AUX DARDANELLES

Un fort détruit par l'artillerie des navires alliés

A FORT DESTROYED BY THE ARTILLERY OF THE ALLIED NAVIES

The view of German commander General Liman von Sanders is ; *If the Allies are determined to force a passage through the Straits, they will probably try to land troops on Turkey's Asiatic shore near Besika Bay. From there an advance over the Mendere plain across open country with few natural obstacles to movement during the spring and summer months, and an outflanking operation designed to take the Turkish defences in the rear offers a real prospect of success. I view the area around Bulair with apprehension, for if the Allies can cut the isthmus, all communication with Constantinople will be severed and the Dardanelles are as good as lost.*

By mid-February Allied naval bombardments of Turkish coastal forts are a daily event, as are minesweeping operations. At this moment, an Allied naval assault could deliver a fatal blow to the Turkish defence ~ but as Liman von Sanders writes ; *The English allowed us four good weeks of respite. This respite just sufficed for the most indispensable measures to be taken.*

On 19th February, Admiral Carden attacks the Dardanelles forts with most of his obsolete fleet. His flagship, *Queen Elizabeth, Agamemnon,* five other British warships and four ships from the French squadron take part with a slow deliberate bombardment all day, first at a distance, and then a more intense fire gradually closer in. The guns of the outer defensive forts at Sed-el-Bahr and Kum Kalessi are both silenced. Parties of marines and bluejackets are put ashore to finish the destruction of the abandoned batteries of the outer forts. In their first landings, they meet with little serious resistance, and 50 guns are destroyed for trifling casualties. Then the minesweepers steam in to clear the way up the Straits for six miles so that the battleships can close in to destroy the Kilid Bahr and Chanak forts guarding the Narrows with observed fire. The minesweepers have difficulty against the five-knot current sweeping down through the Dardanelles, but they do penetrate six miles inside the Straits and they do clear their first designated area of mines. But, when the minesweepers steam forward to clear the next section of the channel, they cannot continue to face the more accurate intense gun fire coming from the shore batteries ~ the minesweepers cannot continue, until those batteries are silenced…

The Allied fleet prepares for two weeks and steams in to attack on 18th March in their attempt to cross the Dardanelles. As the Allied fleet, now under the command of Admiral Sir John de Robeck, tries to force the Straits, this renewed attack by 16 Allied battleships turns out to be a dramatic failure, mainly due to unsuspected mines laid ten days previous by the Ottoman Navy mine-layer *Nusrat.* The diary entry of *Nusrat* for 18th March 1915 : *Upon the order received, at 05:30 in the morning Nusrat established a line of 26 carbonic mines between Paleokastro and Erenköy coast and safely returned home. No enemy was sighted. Mines are laid at a depth of 4.5 meters with intervals of 100-150 meters. There has been light bombardment from enemy fortifications. Nusrat* has effectively changed the course of the Great War. Five Allied warships are sunk or disabled by mines : the British *Inflexible, Irresistible* and *Ocean* and the French *Bouvet* and *Gaulois.* Furthermore, the Allies lose two destroyers and seven mine-sweepers. The Turkish report states ~ *The forts and batteries were damaged, but in the end the Allied fleet withdrew.*

Sir Ian Hamilton and his staff arrive in the cruiser *Phaeton* to establish a HQ on the liner *Franconia.* On 22nd March, he steams across to Admiral de Robeck's flagship *Queen Elizabeth,* and with de Robeck makes the terrible decision, that ; *Only an attack with the combined naval and military forces can capture the Gallipoli Peninsula to command the Dardanelles.* Very many British and French naval officers think differently ~ and remain forever convinced that a second naval assault would have been successful. How unfortunate they were overruled. The British Navy after having made plans and being prepared to lose a certain number of ships as the cost of conquest, and now having lost almost that number, have just withdrawn at the very moment the Turks are ready to quit ~ for the Ottoman hierarchy are actually fleeing with their national treasure from Constantinople into Anatolia. Had the Allies pressed home their naval attacks against the Turks, they could have by this time, been in Constantinople ; but now the Turks are determined to defend the Dardanelles, and they intend to fight the Allies hard. Then having given warning, the British commanders hold back their landing forces just long enough to enable the waxing Cresent to mobilize enough troops to render them impotent.

Throughout the Greek islands, and on the mainland German agents are scattered in profusion. Even if King Constantine of Greece who married to the Kaiser's younger sister Princess Sofia, favours a policy of neutrality, every move made by the British forces preparing in Mundros harbour on the island of Lemnos for the arrival of their fleet is known to their opponents, almost before it is made. Then, Greek PM Eleutherios Venizelos, having obtained the permission of King Konstantine places at the disposal of Great Britain 250,000 Greek troops and the entire Greek Navy. Sir Winston Churchill : *The Greek army and navy offered a means of settling the difficulties of the Dardanelles in a most prompt and effective manner. The Greek general staff were known to be ready with well thought out plans for its seizure…*

St PETERSBURG 4th March 1915. Russian FM Serge Sazonoff to British and French Ambassadors;
The course of recent events leads His Majesty Tsar Nicholas II to think that the question of Constantinople and of the Straits must be definitely solved, according to the time-honoured aspirations of Russia. Every solution will be inadequate and precarious if Constantinople, the western bank of the Bosphorus, of the Sea of Marmara and of the Dardanelles, as well as southern Thrace to the Enez-Midye line, should henceforth not be incorporated into the Russian Empire. Similarly, and by strategic necessity, the Asiatic shore that lies between the Bosphorus, the Sakarya River and a point to be determined on the Gulf of Izmit, and the islands of the Sea of Marmara, the Imbros Islands and the Tenedos Islands must be permanently incorporated into the Russian Empire. The Imperial Government entertains the hope that the above consideration will be sympathetically received by the two Allied Governments. The Allied Governments are assured similar understandings on the part of the Imperial Government for the realization of plans which they may frame in other regions of the Ottoman Empire. As the Russian generals and Sazonoff plunged Europe into war in 1914 to secure Constantinople and the Straits ~ so they are now directly enlisting British and French forces for that same purpose. Curiously, after the Greek Government offers to contribute their entire forces ~ Russia vetoes the idea.

LONDON 12th March 1915. After Great Britain agrees to Tsar Nicholas and Sazonoff's request to incorporate both Constantinople and the Straits, France demands the right to annex Syria for its acquiescence. After a British memorandum stating ~ *Russia alone, will if the war is successful gather the fruits of these operations* ~ Sir E. Grey passes a secret *aide-mémoire* to Russian Ambassador Benkendorff;
Subject to the war being carried on and brought to a successful conclusion, and to desiderata of Great Britain and France in the Ottoman Empire and elsewhere being realized, as indicated in the Russian communication herein referred to, H.M. Government will agree to the Russian Government's aide-mémoire *relative to Constantinople and the Straits, the text of which was communicated to H.M. Ambassador by Sazonoff.*

St PETERSBURG 14th March 1915. French Ambassador Maurice Paléologue delivers the recent reply from Paris that France accepts Russia's demands to Sazonoff. In return France wants to annex Syria and more ~ as Syria is not a clearly defined region, it becomes apparent to the Russians that the French are attempting to lay claim to Palestine and the Holy Places. The French diplomatic note states;
I should be grateful to Your Excellency for informing His Imperial Majesty that the Government of the French Republic, having studied the conditions of the peace to be imposed on Turkey, would like to annex Syria together with the region of the Gulf of Alexandretta and Cilicia up to the Taurus mountain range. I should be happy to inform my Government, of the Imperial Government's consent. Russian Assistant FM advises Sazonoff;
The French Ambassador has told me that it is his impression that Syria ~ includes Palestine.
I deemed it useful to remind him that there is in Jerusalem an independent governor.

PARIS 16th March 1915. Russian Ambassador Isvolsky receives wire from Sazonoff;
After arrival at General Headquarters, the French Ambassador informed me of the contents of Delcasse's telegram which asks for consent by Russia to the annexation of Syria and Cilicia by France. Paléologue explains that in his opinion the French Government refers also to Palestine when speaking of Syria. However, since in this telegram there is no question of Palestine, it would be desirable to elucidate whether the explanation of the Ambassador really corresponds to the view of the French Government. This question appears important to us; for, if the Imperial Government should be prepared largely to satisfy France's desires concerning Syria and Cilicia proper, it is indispensable to study the question with closer attention, if the Holy Places are involved.

St PETERSBURG 18th March 1915. British Ambassador Sir George Buchanan delivers the British reply to Tsar Nicholas himself with Sazonoff in attendance that Britain agrees to Russian occupation of the Straits, provided the free passage thru them will be provided ~ in return for an increased British influence in Persia. Sazonoff wires to Russian Ambassador Isvolsky in Paris : *Now today, the British Government has expressed to us in writing its full accord in the matter of the annexation by Russia of the Straits and Constantinople within the boundaries fixed by us.* Sazonoff to the Russian Ambassador in London ; *Referring to the memorandum of the British Embassy of 12th March, express to Sir Edward Grey the profound gratitude of the Imperial Government for the complete and definitive approval of Great Britain to a solution of the question of the Straits and Constantinople that satisfies Russia.* From January-April in the Dardanelles, Russia contributes nothing… then CO of Russia's Black Sea Fleet, Admiral Eberhardt informs London;
Only after the annihilation of the Turkish Fleet and the union of the Allied fleets in the Black Sea will Russia be ready to send amphibious forces to secure the Bosphorus coastline. I am assembling an expeditionary force. British and French commanders should understand that this is only a symbolic force put together for the final occupation of Constantinople, and that it can only possibly land once the Allies have arrived in Constantinople and the Turkish Fleet has been destroyed. This moment will not likely appear in the near future…

(The *Constantinople Agreement* remains unfulfilled, due to the failure of the Dardanelles Campaign)

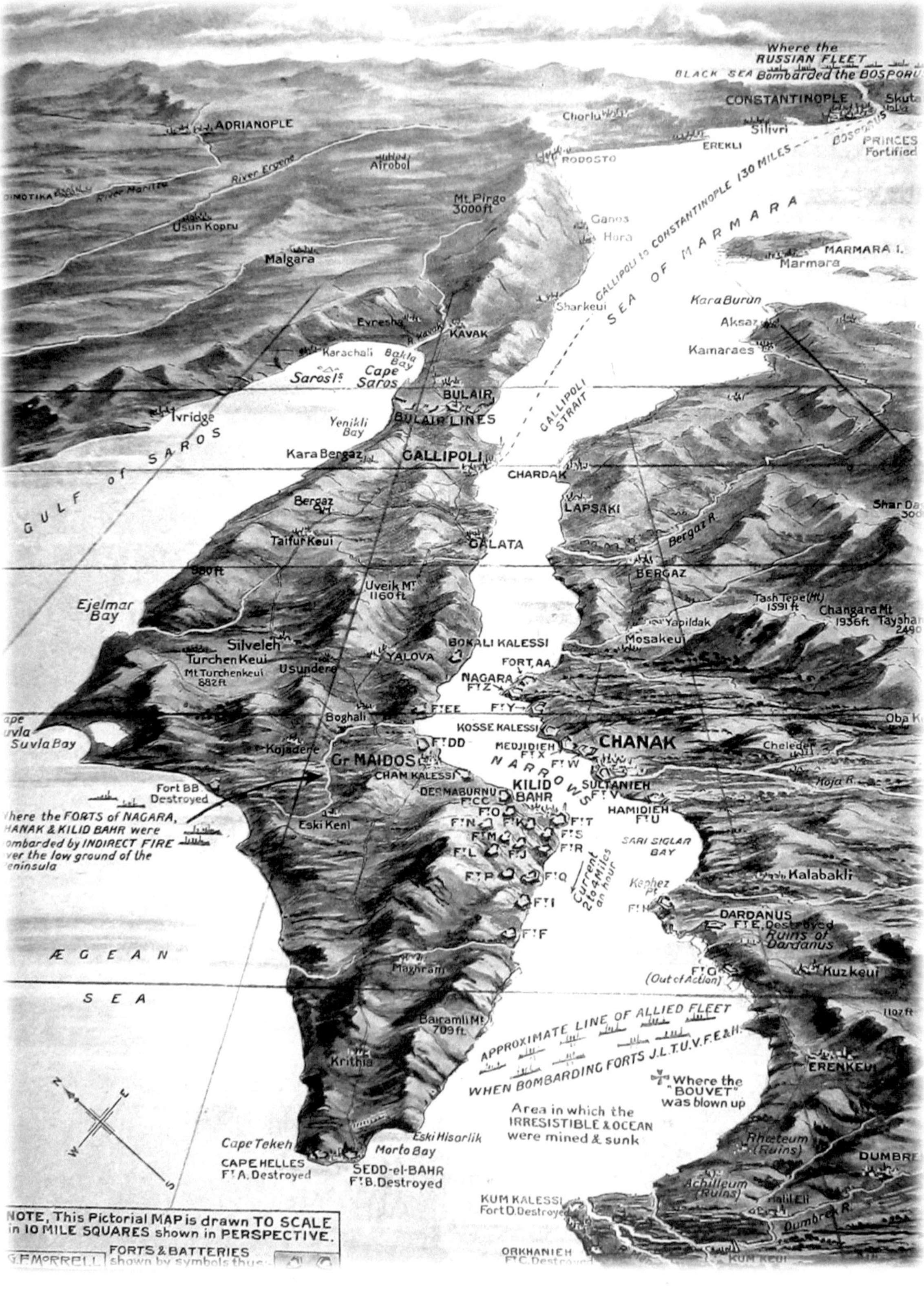

PROGRESS OF THE ALLIED FLEET IN THE DARDANELLES ~ AFTER A TEMPORARY CESSATION OF HOSTILITIES DUE TO WEATHER, THE ATTACK ON THE DARDANELLES FORTS HAS BEEN RENEWED WITH DOUBLE VIGOUR BOTH FROM THE GULF OF SAROS AND FROM WITHIN THE STRAITS AS INDICATED ABOVE, IN SPITE OF THE PERIL OF GUNFIRE AND DRIFTING MINES.

When the Black Sea Fleet finally appears on 25th April 1915, the day of the Allied landings on the Gallipoli Peninsula, it makes no impact on the Dardanelles campaign. Admiral Eberhardt reports : *Projectiles from our ships reached as far as Sariyer and Beykos. The shells fired did no visible damage to any populated areas aside from Akbaba, a small unfortified town on the Black Sea coast.* Eberhardt 3rd May 1915 : *In a naval attack lasting several hours, few hits were scored out of 160 six-inch shells fired, with at least one visible explosion scored at the fortress of Elmaz ~ a fantastic result. However, most shells fired at the Bosphorus shoreline, detonated without result.* The Russian BSF contributes little else to the Dardanelles campaign.

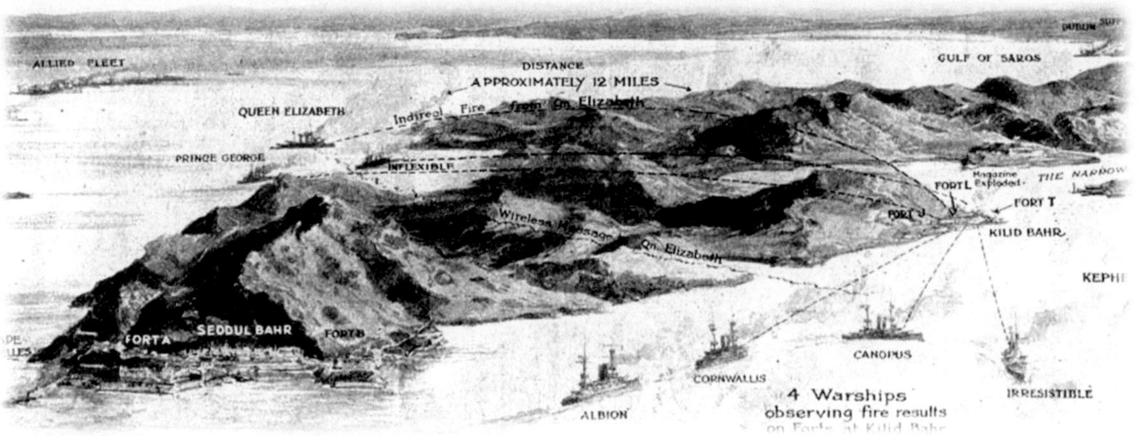

After the British, French, Indian, and ANZAC brothers-in-arms are deceived into fighting and suffering on the Gallipoli Peninsula ~ not for Britain and France ~ but for the despotic Tsarist régime, and after 97,000 Allied and Ottoman Empire lives are wasted fighting for the Turkish Straits route (and for Russia's quieter quest and now centuries old ambition for the re-establishment of a new Byzantine Empire ruled by Tsar Nicholas on the new Christian throne in *Tsargrad* on the Bosphorus) on 3rd November 1915, Lord H.H. Kitchener sends a message to Birdwood, commanding at Gallipoli :

Most secret. Decipher yourself. I leave here tomorrow night to come out to you. Have seen Commodore Keyes and the Admiralty will, I believe, agree to another naval attempt to force the Straits. We must do what we can to help them, and as soon as the ships are in the Marmara, we should seize and hold the Bulair Isthmus. The admiral will probably be changed and Wemyss given command to carry through the naval part of the work. We must do it right this time. I absolutely refuse to sign orders for evacuation, which I think would condemn a large percentage of our men to death or imprisonment.

But when the British Admiralty decide to evacuate Anzac Cove and Suvla, Vice-Admiral Wemyss calls the decision ~ *A disastrous mistake! That naval action would have involved heavy losses is probable, but the sacrifice would have been no greater than those offered up almost daily on the Western Front with less chance of success. The results of success could have been far more reaching than in any other theatre of war. Once through the Narrows : Turkey would become a negligible factor ; Russia would be rejoined to the Allies, Egypt would be saved, and the end of the war brought within measurable distance.*

When Winston Churchill becomes British Prime Minister on 10th May 1940, commenting on the naval aspect of the 1915 Dardanelles Campaign he states ~ *Searching my heart, I cannot regret the effort. It was good to go as far as we did. Not to persevere ~ that was the crime.*

Churchill's promising enterprise could have succeeded in breaking thru the Dardanelles and led to the subsequent fall of Constantinople. This is evidenced by Turkish Government contingency plans; the words of the military commanders of the forts to their subordinates on the night of 18th March; the local German commander, the Turkish PM ~ as summarized by U.S. Ambassador H. Morgenthau ;

Had the Allied commanders but endured their heavy losses a day longer, the entire issue of the Great War might have been changed. Its three last terrible years might have been escaped. Should it have come back? Could the Allied ships really have captured Constantinople? I wish it understood that I am giving not my own views, but merely those of the officials then in Turkey who were best qualified to judge. Had the allies returned on the morning of the 19th what would have happened? Inspector-General of Coast Defences, Johannes Mertens at the Straits stated that they may be able to hold out for only a few hours, and that he should get up at 6:00 in the morning and take to the Anatolian Hills. Mertens said that was what the soldiers were going to do. The troops at all the fortifications had orders to man the guns until the last shell had been fired and then to abandon the forts.

Thus, the Allied fleet would have appeared before Constantinople on the morning of the 20th. What would have happened then? Talaat Bey then controlling the Turkish Government, stated that, far from opposing the arrival of the Allied fleet, his government would have fled. Talaat had obtained scores of automobiles, which he had loaded with extra tires, gasoline, and all the other essentials of a protracted journey. Talaat stationed these automobiles on the Asiatic side of the city with chauffeurs constantly at hand. Everything was prepared to leave for the interior of Asia Minor at a moment's notice. But the Great Allied Armada never returned to the attack. And so, the Turks and the Germans prepared to abandon Constantinople. Their forts had no remaining shells, and the few remaining mines were so widely spaced that they could not have held up the Allied fleet for long. But the combined Allied fleet withdrew one day before total victory was theirs…

* * *

Imperial Russian warships in the Black Sea led by the Ioann Zlatoust & the Panteleimon

So weak was Russia's effort, that FM Serge Sazonoff reprimanded the Imperial Russian Navy; *For not coming close to providing the diversionary help from the Bosphorus that our Allies had counted on. The sufferings of Russia's Allies on the Gallipoli Peninsula are greatly exacerbated, because the Turks are able to concentrate against them all of their strength.* And so, the alarming question is raised. Was the Russian BSF capable of dominating the waters of the Black Sea pursuant to contributing to the Allied plan to force the Turkish Straits and to capture Constantinople in which they were expected to take part? Did any aspects of Russian BSF operations during 1914-15 provide evidence that they could do more?

The Bosnian Crisis of 1908 forced a strategic reconsideration for Russia ; new battleships, cruisers, and destroyers were ordered with significant foreign participation. But, after the outbreak of war in 1914, all equipment being built in Germany was confiscated and other equipment on order from Britain being unable to reach the Black Sea Fleet was promptly diverted to the Allies' own war effort. The Black Sea was the domain of Imperial Russia and the Ottoman Turkish Empire, but the Russian Black Sea Fleet was expected to easily dominate the region, as the Ottoman Turkish Navy consisted of three old battleships, two light-cruisers *Hamadie* and *Midilli*, two other small light-cruisers of antique vintage, and eight torpedo boats ~ but in stark comparism to this small force, Turkey's old enemy of Tsarist Russia maintained their BSF at Sebastopol ~ six battleships, two cruisers, 26 destroyers, 17 torpedo boats, and eight submarines ~ plus three powerful dreadnoughts being built at Nikolayev.

The naval war in the Black Sea started when the Ottoman Turkish Fleet bombarded several Russian ports in late October of 1914 by the most advanced ships in the Ottoman Turkish Fleet, the two German ships: the battle-cruiser SMS *Goeben* and the light-cruiser SMS *Breslau,* commanded by Admiral Wilhelm Souchon. The Ottomans informed the Germans that they would go to war as soon as they received the equivalent of two million *liras* in gold ~ money the Ottoman Turkish Empire would need to fund a war. The money shipped thru neutral Romania all arrived by 21st October. Informants working for the Russian Ambassador in Constantinople Mikhail Griers, forwarded the information about the payments to Russian FM Serge Sazonoff. Sazonoff suspecting Ottoman and German intentions, warned BSF command in Sebastopol to be prepared for an attack. On 21st October, Admiral Kazimir Ketlinski assured Sazonoff ~ *The Black Sea Fleet is completely ready for action.*

On 25th October, Ambassador Giers forwarded his informant's predictions to FM Serge Sazonoff: *The attack will take place on 29 October.* Russian BSF naval officers were then placed under specific instruction not to fire first on the Ottomans in the event of a confrontation. The Russian Government waited… wanting to make it clear to all, that the Ottomans would be the ones to instigate hostilities. On the afternoon following the raids on Odessa, Sebastopol, Feodosia, Yalta, and Novorossiyisk Admiral Souchon radioed Constantinople that : *Russian ships have shadowed all movements of the Turkish Fleet and systematically disrupted all exercises, and as such have opened hostilities.* The Russians attempted to but were *unable to pursue the Ottoman fleet.* German officers were disappointed by the limited extent of the attack, in which Russia's BSF had not been seriously damaged by the raid. Russians declared war on the Ottoman Empire on 2nd November 1914. Admiral Andrei Eberhardt ordered the BSF to retaliate against the Ottoman Empire and on 4th November, a BSF force bombarded Zonguldak.

The German heavy-cruiser SMS Goeben with the much smaller light-cruiser SMS Breslau in 1914.
Goeben would be damaged on at least four different occasions and usually ~ but not always, disengaged from the numerically superior BSF.

Battle of Cape Sarych. Morning, 17th November 1914. BSF force of five pre-dreadnought battleships, three cruisers, and 13 destroyers under Vice–Admiral Eberhardt attack Ottoman port of Trebizond.

German Admiral Wilhelm Souchon decides to intercept the Russian BSF returning to their port. At 13:00 the battle-cruiser *Goeben* and the light cruiser *Breslau* steam from the Bosphorus for the BSF base of Sebastopol at high speed. Admiral Andrei Eberhardt commanding the BSF becomes aware of the two German cruisers departure and so alerted, prepares for action. Having the superior force the Russians have a rare opportunity to destroy the potent German cruisers of the Ottoman Navy.

The following morning, the two German-made warships in Ottoman Navy service begin patrolling the Crimean coast in spite of foggy conditions. At 12:05 the *Breslau* sights a Russian cruiser to starboard. *Goeben* turns to face the cruiser at full speed. Other Russian ships are soon sighted to the starboard, and soon five BSF pre-dreadnoughts are identified. Admiral Wilhelm Souchon orders the light-cruiser *Breslau* to keep out of range, preferring to engage the BSF with his heavy-cruiser *Goeben.*

At 12:02 The German directed *Goeben's* first artillery officer first spots the BSF warships through the fog and orders his gun crews to target *Evstafi*. Their first salvo overshoots the Russian warship, though one shell pierces *Evstafi's* centre funnel and causes splinter damage to the wireless antennae. The Ottoman's third salvo scores two hits on the *Evstafi*. The first one disables the forward starboard 6-inch casemated gun and causes casualties among its crew. The second shell penetrates the casemate, setting fire to some 6-inch cartridges that detonate beyond into the officer's galley, causing severe damage while splinters penetrate the deck and impact partially into the boiler room below.

At 12:20 *Evstafi*, with a clear view of the *Goeben and Breslau*, quickly opens fire on the *Goeben*, striking and disabling one of her 15-cm guns with its first salvo. The BSF has proved itself in action.

At 12:24 the larger number of Russian warships now oddly turn away losing heart for the fight, and as *Goeben* now begins to lose sight of the faster retreating BSF group, while always outnumbered, and continuously firing on *Evstafi* with its aft guns, Admiral Wilhelm Souchon decides to withdraw. The *Goeben* ceases firing at 12:32 and sets course for Cape Sinop, arriving at 08:00 on 19th November.

The Imperial Russian Battleship Prince Potemkin (1906) ~ she became famous when her crew rebelled against her officers in 1905 (during that year's attempted revolution, now viewed as an initial step towards the Russian Revolution of 1917. After the mutineers sought asylum in Costanzia, Romania, and after the Russians recovered the ship, her name was changed to Panteleimon. She covered several bombardments of the northern-end Bosphorus fortifications in early 1915, including one where the ship was attacked by the battle-cruiser Goeben ~ Panteleimon and other warships of the BSF again won the day, seeing her off before she could inflict any serious damage.

Battle of 10th May 1915 ~ After an exchange of fire between the BSF battleship squadron, the *Goeben* and *Breslau* again withdraw. Beginning on 9th May 1915, the Russian squadron attacks Ottoman shipping between Kozluand and Eregli, sinking four steamers and several sailing ships. The *Goeben*, under *Kapitain* Richard Ackermann, immediately puts to sea to intercept the Russian vessels.

Early on the morning of 10th May 1915, a bombardment force consisting of two battleships, with a screen of destroyers and minesweepers, detaches from the Russian squadron in order to attack the Bosphorus forts. The Ottoman torpedo-boat *Numune-i-Hamiyet*, acting as a guard ship at the northern entrance of the Bosphorus Strait sights the BSF bombardment force and radios a warning to *Goeben*. *Kapitain* Ackermann sets the *Goeben* on a course to intercept at 26 knots. As the BSF cruiser *Pamiat Merkuria* spots the *Goeben* and reports it to the fleet, this gives the Russian bombardment group time to break off before being sighted. Cruising 40 km off of the Bosphorus at 5 knots, Admiral Andrei Eberhardt's covering force consists of his flagship *Evstafi*, with battleships *Ioann Zlatoust* and *Rostislav*. Ackermann unaware of this, is soon steaming toward the BSF battleship squadron.

At 07:53 hours, Eberhardt's group meet *Goeben* steaming on a parallel course. Ackermann believes this to be the bombardment detail, though he is confused as to why he is facing three battleships instead of two. He soon realizes his mistake when *Tri Sviatitelia* and *Panteleimon* join the Russian line. The battle-cruiser *Goeben* fires 100 plus 28-cm shells during the engagement, but scoring no hits, causes no damage. One near miss on *Evstafi* sends a cascade of water over the flying-bridge, drenching Admiral Eberhardt and his staff. In return, the Russians land one heavy calibre shell on *Goeben's* forecastle, and another on her forward armoured belt. Outnumbered and outgunned, *Kapitain* Ackermann orders his ship to disengage at 08:12. The BSF pursue the *Goeben* to the north before changing course and returning to Ottoman waters. Though the Ottomans retreat, the damage they suffer is slight. This action makes the Russians wary about dividing their battleship squadron.

From late-1914 the BSF shifts its strategic emphasis to offensive mine-laying of the approaches to the Bosphorus Strait aiming to hinder the *Goeben* and *Breslau* from entering the Black Sea, and to mine Turkish Black Sea ports and coastal shipping routes to paralyse Ottoman shipping routes, and in this way by the end of 1915, Russia secures strategic control of the Black Sea. In 1915 alone, two Ottoman Navy cruisers (including the *Breslau*), one gunboat, and a mine-layer have hit the Russian minefields. *Goeben* also sustains damage from Russian mines, and is put out of action for seven months.

SMS *Goeben*

The Black Sea Fleet was then used ~ not to support any aspect of the Dardanelles campaign, but the land forces of Russian General Nicolai Yudenich facing the Caucasus Front. In August 1915, a Russian submarine and two Russian destroyers attack a Turkish convoy of four transport vessels being escorted by an Ottoman cruiser and two destroyers. The Russian ships sink all four transports without losing a ship. Later in the summer of 1916, after the Ottoman IIIrd Turkish Army under Vehip Pasha, was ordered to re-take Trezibon, the Ottoman forces who began marching along the coast in June were confronted by Russian warships that bombarded these marching troops so accurately and efficiently in destroying their supply columns, that the IIIrd Army were compelled to withdraw.

These examples prove the Russian BSF capable of successfully challenging the best warships of the German-Ottoman Turkish Fleet, and also of being as capable and as effective as the warships of the French and British Navy in targeting and defeating Turkish land targets by naval bombardment. So, the question remains ~ Why, when it has been evidenced that the BSF had the war-winning capability to do so much more, did the Russians do so little during the vital naval phase of the Dardanelles Campaign and contribute so little to the follow-up Gallipoli land campaign?

Furthermore, what was Russia's motive in vetoing the participation of Greek forces twice offered; first by Greek PM Eleutherios Venizelos who offered the entire Greek Navy on 1st March 1915, which would have influenced the Allied forcing the Straits on 18th March, and when Greece later offered three infantry divisions. Russian leaders did not want to see the Greeks (also of the Orthodox faith) making rival claims to their Constantinople Agreement. Russia insisted on sole control of the Straits.

German General Ludendorff stated in his memoirs that he believed the entry of the Turks into the war allowed the outnumbered Central Powers to fight on for two years longer than they would have been able on their own. The war was subsequently extended to the Middle East on the fronts of the Gallipoli Peninsula, Sinai, Palestine, and Mesopotamia. Had the Great War ended in 1916 with an Allied naval victory steaming and shelling thru the Turkish Straits, some of the bloodiest engagements on the Western Front could have been avoided. The U.S.A might not have been drawn from its policy of isolation to intervene in a foreign war. But, the closure of Russia's only ice-free trade route thru the Dardanelles effectively strangled the Russian economy. The sealing off of the Black Sea contributed to revolutionary ferment in Russia, detonated discontent, and condemned Imperial Russia to revolution. The BSF had proved it could prevail, but Russia opted for treachery : waiting and not *doing their bit*. They wasted their chance for the Turkish Straits, failed to shorten the war, and doomed their Empire.

ITALY ENTERS THE WAR ~ 23rd MAY 1915

ENGLAND AND FRANCE (to their new comrade) ~ STICK TO IT!

ROME. The tragic news of the sinking of the *Lusitania* one day before, with its terrible loss of life sends a chill of horror through Italy, as those who have hitherto stood neutral, consider anew to take their place with the war partisans. When the Italian Parliament is gazetted to reassemble on 12th May, the peril of the situation is recognized in Berlin and Vienna, as the subjects of Hapsburg A-H and Imperial Germany begin to plan their swift departure of Italy. Following the secret agreement made by the Allies in the Treaty of London, Italy enters the war on 23rd May 1915, aiming to annex what is known as *Italia Irredenta* or *Unredeemed Italy* ~ the Italian-speaking territories of Tyrol, Trieste, Istria, Zara, and parts of Dalmatia, then still under control of the A-H Empire.

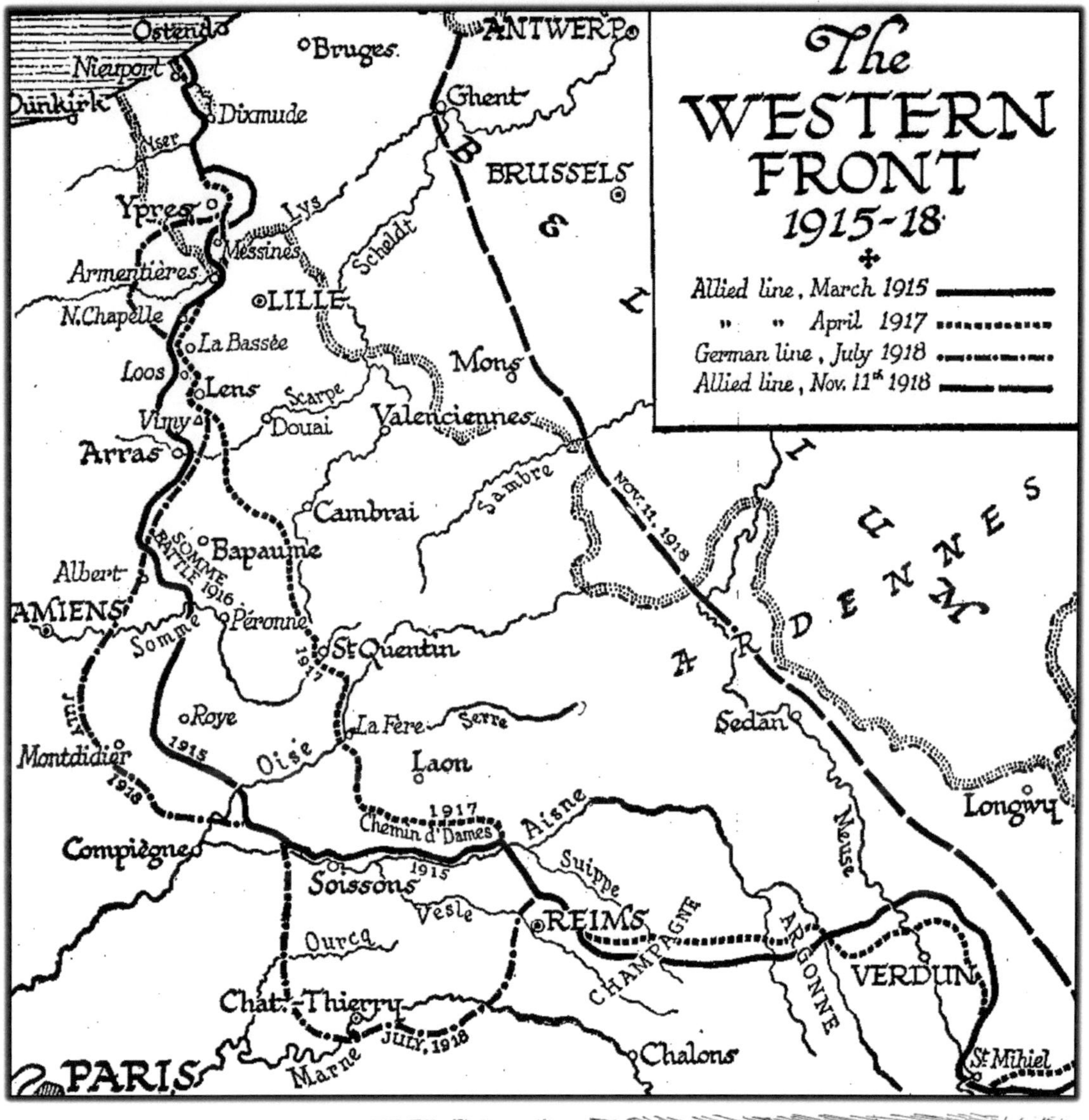
The WESTERN FRONT 1915-18
Allied line, March 1915
" " April 1917
German line, July 1918
Allied line, Nov. 11th 1918
Ostend
Nieuport
Dunkirk
Dixmude
Bruges
ANTWERP
Ghent
BRUSSELS
Yser
Ypres
Lys
Messines
Armentières
Scheldt
N. Chapelle
LILLE
La Bassée
Loos
Lens
Mons
Scarpe
Vimy
Douai
Valenciennes
Arras
Cambrai
Sambre
Bapaume
SOMME BATTLE 1916
Albert
AMIENS
Péronne
Somme
St Quentin
1917
July
Roye
La Fère
Serre
Montdidier
1915
Oise
Laon
1918
Compiègne
1917
Chemin d'Dames
Aisne
1915
Soissons
Suippe
Vesle
REIMS
CHAMPAGNE
ARGONNE
Ourcq
VERDUN
Chat.-Thierry
JULY, 1918
Marne
Chalons
PARIS
St Mihiel
Meuse
Sedan
Longwy
NOV. 11. 1918
ARDENNES
BELGIUM

SAVINSKY ON THE RUPTURE OF BULGARIA

Russian Ambassador to Bulgaria Alexander Savinsky

My appointment to Bulgaria had been decided when I was Ambassador to Sweden, when, in December of 1912 I had come on leave to St. Petersburg, Sazonoff had already told me he was sorry that I was not at Sofia. Coming from the Tsar a few days later, he sent for me and told me in great secret that : His Majesty has decided to transfer you to Bulgaria, however the appointment can only take place at some favourable political moment. The Foreign Minister added : You may consider the matter as definitely settled, but hold it absolutely secret. Besides the Tsar and us two, nobody knows anything about it. I do not deny the delicacy of the task before you. I feel sure that you will find the right tone in your conversations with King Ferdinand and his Ministers. In any case if we had fresh difficulties with Bulgaria, and if I ventured someday to blame you for it, you may always return the shot, for never have I had so many failures with anybody as with the Bulgarians. Our Ministry for Foreign Affairs realized how difficult the political situation was, and this is why, contrary to established custom, I received no written instructions. The verbal directions that given to me were ~ You are to stand aloof from the Bulgarian Government and to wait for its downfall ~ if you can manage to assist in its overthrow without any risk of exposure, that would be still better. After that, negotiations with the Bulgarians can be resumed again. More ample instructions shall be given to you for new questions that may arise ~ if however, you are in doubt and want to discuss viva voce some problem on hand, you may always come up to St. Petersburg for a few days.

Sazonoff's first idea had been to arrive at an understanding with Bulgaria. He realized that in order to win her over to us it was necessary to offer her territorial compensations : Macedonia being the most coveted of all. This is why his original plan was to use with the Serbians, stern language admitting no resistance on their part. How could Serbia refuse to make small territorial concessions useful to the common cause, as well as to Russia; to Russia who was engaged in a war because of them, who supported them and who was dreaming of making of them a Greater Serbia including Bosnia, Herzegovina, Croatia, Slovenia, and having a wide access to the sea?

On 5th August 1915, Sazonoff wired to Belgrade (No. 1684) : It is most urgently needed to put aside petty reckonings and to act deliberately and quickly. The tone of that telegram clearly shows that Sazonoff had no doubt about the Serbians. He instructed Strandtmann, our Charge d'Affaires in Belgrade, to use energetic and authoritative language, the only possible one under the circumstances. Unfortunately, that firmness dwindled to vacillation, partly due to the displeasure of King Ferdinand. The fact did not pass unnoticed by the Serbians, who soon made bold to disregard our suggestions, even to reject them, and use in their negotiations with the Éntente an arrogant language incompatible with our interests. In his telegram of 5th August Sazonoff had quite definitely pronounced himself in favour of the immediate cession to Bulgaria, in return for her assistance, of the eastern part of Macedonia up to the Vardar and, in case of our victory, of the entire Macedonian territory which had been guaranteed to Bulgaria in accordance with the Treaty of 1912. In my opinion it is merely owing to a lack of energy and perseverance in the carrying out of this plan that all that happened later must be ascribed. Replying to Strandtmann's offer, Pasic exclaimed : What, the old frontier of 1912 ~ for nothing in the world! That reply went unchallenged by us, as at the beginning of the war Sazonoff did not admit of the possibility of Serbia, whom we had saved from total destruction, being intransigent enough to create difficulties for us.

Much as the Serbians' irreconcilable attitude has been unreasonable and inexcusable during the entire war, especially so since the war had been precipitated because of them and since they had been assured by the Allies that their interests would be protected, the intransigency of the Serbians is due to the misconceived sympathy displayed towards them by some of our Allies on the grounds of the attack Serbia had been subject to and the gallant defence she opposed ; on the other hand, the London Cabinet since the beginning of the war has acted in the belief that all attempts to win Bulgaria over to the side of the Allies, are in advance doomed to failure.

At that moment, Duke Johann-Albert of Mecklenburg-Schwerin arrived in Sofia. The Duke's personality is well known. Purely German in his convictions, he was often used by the Kaiser in particularly thorny cases. As the Duke of Mecklenburg proceeded with his work and was rather successful, he was assisted by Rosenberg, the Director for Balkan Affairs at the Wilhelmstrasse, who arrived in Sofia to see to the technical side of the arrangement. Being reluctant to give up the game before I had tried every means possible, I went to see Radoslavoff. The long conversation I had with him is summarized in my telegram to Sazonoff of 4th September :

The Duke of Mecklenburg is here and working hard. Radoslavoff admits that the Germans are doing all they can to dazzle the Bulgarians by the most alluring perspectives. Obviously using the arguments of the Germans, he insisted upon the danger that might result for Bulgaria from the fall of Constantinople into Russian hands. He then spoke of the ill-feeling prevailing in Russia against Bulgaria, quoting a phrase concerning territorial concessions to Bulgaria : No compensations to that treacherous State, we shall divide it among its neighbours! To that I replied : Although our public opinion has many reasons to feel embittered against Bulgaria, a frank and decisive action of your country against the Turks still could put everything straight again. Developing my thoughts, I said : It would be sheer madness on the part of Bulgaria to throw herself into the jaws of the wolf. If Bulgaria once more betrays Russia, our public opinion will never forgive it.

The Duke of Mecklenburg's activity was fully supported by the King. The mobilization to be ordered was a current subject of conversation and measures indicating preparatives to war were decreed on the 18th September. Dr. Vladoff of the Macedonian Central Committee, came to see me at the Russian Legation very late one night. Having been enlisted, he wore the reserve uniform and was very upset and nervous. He told me of the measures taken by the Government that established their intentions beyond any possible doubt. He once more entreated me to prevent the catastrophe and warned me that should his visit to me become known, his life would be in danger.

I described these alarming symptoms to Sazonoff in a telegram sent on September 20th 1915 (No. 657) :
The King has thrown off his mask. He no longer tries to conceal his old game nor his recent agreement with the Duke of Mecklenburg. He has personally issued, through the Chief of his Privy Council, mobilization orders to the railways and has answered the last note of the Éntente Powers by decreeing mobilization against Serbia. Radoslavoff declared that the Government had resolved to proclaim a state of armed neutrality, assuring them that the decision bore no aggressive character and amounted to a precautionary measure to meet any emergency. That tale he repeated to me ; All of our neighbours are mobilizing and under arms, while we run the risk of being attacked at any moment. We simply have to take a few protective measures. Do not fear, for this is not war. It is nothing else but a neutralite armee. He smiled in a cunning way when pronouncing these two French words with a Bulgarian accent. Nevertheless, the true meaning of that neutralite armee did not escape me, when I wired Sazonoff on 21st September : Twelve guns of very heavy calibre have been transported to Varna and four to Burgas, and the Bulgarian cavalry have been sent to the Serbian frontier to cover the Bulgarian mobilization.

Documents Diplomatiques se Rapportant a l'entree de la Bulgaria dans la Guerre Europenne ~ being the treaty of alliance between the German Empire, A-H and Bulgaria that fully pledged Bulgaria to an immediate and active alliance with the Central Empires was covertly signed on 6th September 1915, and for almost one month after its signing, King Ferdinand continued to pretend being interested in the Éntente's proposals.

Next morning the boat took me across the Danube, where a Romanian special train was waiting for me. About lunchtime I reached Bucharest. My heart was beating fast when the train was approaching the frontier. To be once more in Russia, to see again my beautiful country to which I had devoted all my forces in trying to make easier for her the struggle with a terrible and cruel enemy! What sublime and entrancing emotions! The first question I asked Sazonoff at our meeting was whether he disapproved of my Bulgarian policy and whether I could have done better or otherwise? Sazonoff : No, my dear friend, I have nothing to reproach you of. I know you have done everything that was humanly possible. But it was hopeless to strive against King Ferdinand's ill will. I know about it myself. Our rupture with Bulgaria remained as unintelligible as ever to the Russian public opinion.

Alexander Savinsky

BULGARIANS ATTACKING A SERBIAN ARMOURED CAR

1914 MODEL PEUGOT 18CV ARMOURED CAR SUPPLIED TO SERBIAN ARMY ~ From a drawing by Fortunino Mantania

Reconnaissance ~ The speed and comparative silence of armoured cars often enables them to carry out routine patrols and distant reconnaissance, and the armoured car holds the field in this role. As the role of armoured cars has developed with the times, so has their design. Armour-piercing munitions have been added to deal with hostile armoured vehicles. Armour has been increased to protect crews, to provide them every prospect of obtaining information, and getting back to report their news. Road Information, whether drivable or boggy saves lives ~ or leads to above situations.

Armoured car crews often supply more detailed reports than aerial observers, and by reaching key tactical point such as a bridge before the enemy can, they often secured routes for consolidation. The role of the armoured car calls for boldness to succeed, and yet towns and villages are among the very worst places to deploy any armoured vehicle. Tanks will forever be the more expensive option for most situations. However, for the same cost a few armoured cars can always be put in the field. From the first pair of Henkart's Minerva tourers, the armoured car has become a distinct weapon in itself, and still protects road warriors by shielding them on fast patrols from snipers and roadside bombs, more so if their design incorporates blast-deflecting design aspects of a sensible V-shaped hull.

THE RUSSIAN FRONT ~ ARMOURED CARS IN ACTION

CLEARING A GERMAN ADVANCED TRENCH ON THE ROAD TO KOVEL

This scene ~ drawn by an artist as it appeared to the naked eye and thru a field-glass ~ has been frequently enacted on the Russian Front, where armoured cars have proved invaluable. The most remarkable feat was performed recently by volunteer Belgian cars crews operating in Galicia. They broke through the Austrian frontline, made a whirlwind tour of the enemy's rear, covering a distance of 370 miles creating panic wherever they appeared, and returned to their own lines without loss. ~ *The Graphic*

FERDINAND OF ROMANIA DECLARES WAR

In 1883, the Kingdom of Romania was ruled by Carol I, and it was he who signed a secret treaty with the Triple Alliance stipulating Romania's obligation to go to war ~ only if A-H was attacked. Romania had joined the Central Powers in order to be safe against possible aggression from Russia. One of the seven articles of the treaty established that Romania would intervene to support A-H in case their Eastern border was attacked. But in the summer of 1914, the treaty became inoperable. Romania remained neutral when the war started, with the contention that A-H itself was the aggressor, and consequently, Romania was under no formal obligation to join it.

In 1915, Lieutenant-Colonel Baron Christopher Thomson, a fluent speaker of French, was sent to Bucharest as British military attaché on the initiative of Lord Kitchener to bring Romania into the war. The Romanian Government signed a treaty with the Allies on 17th August that pledged to declare war on A-H by 28th August. Under a secret treaty signed earlier that month with the Triple Éntente, Romania was promised, in exchange for joining the war, the support of the Éntente for their claim to territory of Transylvania, then part of the A-H Empire but populated by millions of ethnic Romanians. Eventually the Treaty of Trianon transfers Transylvania to the Kingdom of Romania.

Proclamation by King Ferdinand,

Romanians!

The war which for the last two years has been encircling our frontiers more and more closely has shaken the ancient foundations of Europe to their depths. It has brought the day which has been awaited for centuries by the national conscience, by the founders of the Romanian State, by those who united the principalities in the war of independence, by those responsible for the national renaissance. It is the day of the union of all branches of our nation. Today we are able to complete the task of our forefathers and to establish forever that which Michael the Great was only able to establish for a moment, namely, a Romanian union on both slopes of the Carpathians.

For us the mountains and plains of Bukovina, where Stephen the Great has slept for centuries. In our moral energy and our valour lie the means of giving him back his birthright of a great and free Romania from the Tisza to the Black Sea, and to prosper in peace in accordance with our customs and our hopes and dreams.

Romanians!

Animated by the Holy duty imposed upon us, and determined to bear manfully all the sacrifices inseparable from an arduous war, we will march into battle with the irresistible élan of a people firmly confident in its destiny. The glorious fruits of victory shall be our reward. Forward, with the help of God! ~ *FERDINAND*

Prince Hans Heinrich XV of Pless to his wife *Daisy*, Princess Mary Theresa of Pless ;

GROSSES HAUPTQUARTIER
11th November 1915

My Darling Little Wife ~

I had a chance of talking with the Kaiser. He understood everything and told me to tell you so. I was away for a bit the last days. This is also the other reason why I could not see the Kaiser sooner. I stayed two days at Bruxelles, from where I ventured to Antwerp. All was very interesting. Lichnowsky, who was with General von Kluck's army, has now been transferred to General von Mackensen's Staff HQ in Poland. Poor devil. I am sorry for him. But he is too silly. He keeps on telling everybody that he could have prevented the war, and that our Foreign Office is responsible for it. You can imagine that his theory has not had pleasant effects for him. I would like to know your opinion about the future, as far as England is concerned. Be careful what you write. Take care of yourself ~ charming person. Every good wish. *~ Hans of P.*

Dearest Hans,

At the end of January 1916 I was in Munich for a few days and hearing that the Lichnowskys were there I telephoned asking the Prince to come and see me. I knew his wife could not do so as she had just lost her father, Count Maximilian von Arco-Zinneberg. My diary gives the following account of the interview:

January 30th 1916. Munich.

To talk to poor Prince Lichnowsky last night was really sad, and to see him again as I did at tea today. He is frightfully depressed, feeling that the Kaiser and others blame him for the war between England and Germany. He seems to think still that because they were so nice to him and his wife Mechtilde in London that it meant that the political feeling against Germany had absolutely died. I cannot quite think this. The commercial jealousy was always there, and in England they would not disbelieve in the idea of a German invasion and often talked of it. It was like a red rag to a bull.

Lichnowsky said: I have maintained all along, that Austria ought to have accepted the submission of Serbia, as the Serbians had conceded everything, except the absurd stipulation that all her Ministers should be nominated by Austria. If Austria ever made this absurd stipulation ~ the fact has never become publicly known. They insisted upon appointing some of their own. All this could have been arranged and in time the Austrians and Hungarians would have taken up these posts quietly with tact and without fuss.

Lichnowsky also told me something which astonishes me, and which I must find out more about. He said that Germany could have forced Austria to have accepted Serbia's submission and that Berchtold, the Austrian Foreign Minister, would have done anything that the Kaiser and his Government dictated to him. Lichnowsky insists that the whole beginnings of this war the fault of Germany ~ also that if German soldiers had not crossed the frontier, Belgium would have remained neutral. What is a woman to think? When a former Ambassador tells her this: and the Kaiser tells her, as he told me in Pless, that the King of the Belgians refused to remain neutral; that he himself talked on the telephone to King Albert before the war, saying that if Belgium would remain neutral we would not cross the Belgian frontier ~ that the King of the Belgians refused? Bruxelles had publicly declared : Belgium will uphold her neutrality, whatever the consequences. *~ D of P*

BERLIN. U.S. Ambassador James W. Gerard *: At the Chancellor's house in the evenings we had some discussions on the submarine situation and I had several long talks with von Bethmann-Hollweg alone in a corner of the room while the others listened to music or set the mechanical toys in motion. These discussions, without doubt were reported to the Kaiser by the Chancellor. One evening I was asked the question as to what America could do, supposing the almost impossible, that America should resent the recommencement of ruthless submarine warfare by the Germans and declare war. I said that nearly all of the great inventions used in this war had been made by Americans; that the very submarine which formed the basis of our discussion was an American invention, and so were the barbed wire and the aircraft, the ironclad, the telephone and the telegraph, so necessary to trench warfare that had been first developed in our Civil War; and that I believed, if forced to it, American genius could produce some invention which might have a decisive effect in this war. My German auditors said; While you might invent something in America, while you furnish money and supplies to the Allies, you have no men; and the sentiment of your country is such that you will not be able to raise an army large enough to make any impression. I said that; possibly if hostilities once broke out with the Germans, the Germans might force us by the commission of such acts as had aroused England, to pass a law for universal military service. This early proposition of mine was branded by the Germans as absolutely impossible.*

German GHQ Pless Schloss in Silesia

One day we had tea in the garden with the Prince of Pless and motored with him to see some marvelous Himalayan pheasants reared by an old Frenchman, an ex-jailer, who seemed to have a strong instinct to keep something in captivity. The Kaiser's automobile, which he had placed at my disposal, had two loaded rifles standing upright in racks at the right and left sides of the car ~ ready for instant use. On one day we motored, always of course in charge of the officers detailed to take care of us, to the ancient walled city of Rocroy and through the beautiful Ardennes forest east of it, returning via Charleville.

James W. Gerard

General Paul von Hindenburg, Kaiser Wilhelm & General Erich Ludendorff within Pless GHQ

In the German view, there was the possibility of a victory by carrying on the submarine campaign in intensified form : Unrestricted submarine warfare was now the only means left to secure in any reasonable time a victorious end to the war. If submarine warfare on this scale could have a decisive effect, the navy held that in the existing situation it was our military duty to the German nation to embark on this form of warfare.

~ Ludendorff

ADC to Kaiser Wilhelm II ~ Prince Hans Heinrich XV of Pless

Grand-Admiral Alfred von Tirpitz

PLESS CASTLE 31st August 1916. The highest-ranking military and naval commanders of Imperial Germany confer to review all options to bring the war to a timely and triumphant end. The German Navy is convinced that the U-Boat is a war-winning weapon ~ but the opposition of most High Command officers is inflexible, believing it not worth the risk of an American intervention. Previously, Bethmann has succeeded in suppressing Grand-Admiral Alfred von Tirpitz who urges a submarine campaign without caution, but two developments in 1916 make the German decision for unrestricted U-Boat warfare as inevitable as any desperate strategic decision of this sort ever is.

First, the 29th August appointment of Paul von Hindenburg as Chief of Army General Staff with his *alter ego* Ludendorff to the new post of Quartermaster General signifies a strong militarist shift in German strategic decision making. Henceforth, Bethmann who has contained the ultra-militarists with the Kaiser's support will be less able to withstand Paul von Hindenburg when he advises that drastic measures are necessary for their survival.

Second; events of the summer of 1916 ~ the British offensive on the Somme, the increasing economic strangulation of the Allied naval blockade around the German coast, the spread of war weariness, and the growing unreliability of Germany's allies begins to convince Germany's military leaders that they will lose the war unless some way is found to triumph in 1917. With the changing fortunes of the German field armies comes a change in the strategic thinking of Erich von Falkenhayn. *I doubt that Germany's economic resources can sustain the war into 1917, and fear that the morale of the people may collapse if the war continues for more than another year. Obstacles to the use of the U-Boat no longer exist. The situation in the Balkans is secure with the conquest of Serbia and the accession of Bulgaria to the side of the Central Powers. The navy can promise that England will be destroyed within two months after the resumption of unrestricted submarine warfare. Germany can easily sustain the commercial and moral effects of the entry of the United States into the war ~ an event which will probably follow. American hostility will make no difference. Their army improvised from cowboys and roughriders can never face German soldiers. If anything, a declaration of war by the U.S. will help Germany because munitions going to the Allies will be diverted for American use.*

With such influential support, the German Navy renews its push for an all-out U-Boat campaign. Grand-Admiral Tirpitz ; *The United States is already de facto waging economic war against Germany. Formal belligerence will make no difference, because with the U-Boat, the war will soon be won. Without the U-Boat the war might be lost anyway ~ no matter what the United States might do. Even if the United States declares war on Germany, her ability to transmit power will be limited to her available cargo space. Financial support is of no use if the material purchased cannot be shipped. In short, the certainty of bringing England to her knees cannot be weighed against the possibility of complications with neutrals resulting from unrestricted U-Boat warfare.*

This initial Pless Conference fails reach a final conclusion with Bethmann remaining unconvinced that the only way for Germany to win is by the unrestricted use of submarines. *I believe more certainly than ever that the lifting of all restrictions will bring the U.S. into the war, and will herald Germany's defeat. Before England is brought to a state of capitulation ~ and this is the real point at issue ~ she will leave no stone unturned in an effort to overcome her difficulties. She could do much better through better organization, rationing, convoying, and other techniques to improve her situation and will sacrifice the last man and the last penny before admitting that her supremacy on the seas has been destroyed. Men act on the basis of expectations rather than their present reality ~ an American entry would give an enormous moral boost to all our enemies. Their confidence in victory and will to endure would be strengthened. I do agree to an intensification of U-Boat warfare with submarine commanders being permitted to attack armed allied merchant vessels without warning.*

In October 1916, the Kaiser gives his approval for Bethmann to invite the Allies to the peace table, and finally on 12th December, the Chancellor issues his sincere public appeal for a peace conference. Six days after Berlin's appeal, President Wilson issues an appeal of his own calling on all belligerents to state their aims. When the Allies reply to the German appeal with a scathing denunciation and unanimous rejection of Bethmann's peace proposal ~ and to Wilson's appeal with a list of objectives which from Berlin's point of view would constitute a humiliating defeat ~ this is taken as proof in Germany that their enemies will never consent to a peace acceptable to either Berlin or Vienna.

The Kaiser was still under the influence of our peace offer so curtly rejected by the Éntente. At the German GHQ at Pless, I found the prevailing atmosphere largely influenced by the Éntente's answer. Hindenburg and Ludendorff, who were opposed to a démarche for peace, merely remarked to me that a definite victory presented a possibility of ending the war, and the Kaiser said that he had offered his hand in peace but that the Éntente had given him a slap in the face, and there was nothing for it now but war to the uttermost. Count von Czernin

BERLIN ~ U.S. Ambassador James W. Gerard: *The Germans believed that President Wilson had been elected with a mandate to keep out of war at any cost, and that America could be insulted, flouted and humiliated with impunity. In early January of 1917, we began to get rumours of the resumption of ruthless submarine war, and within a few days I was cabling information based not upon absolute facts but upon reports which seemed reliable and which had been collected through the able efforts of our very capable naval attaché, Commander Walter Gherardi. And this information was confirmed by the hints given to me by various influential Germans. Again and again after 6th January, I was assured by Zimmermann and others in the Foreign Office that nothing of the kind was contemplated ~ Now were the German moves in the direction of peace sincere or not?*

I went very far in assuring the Chancellor and other members of the Government of the President's desire to see peace established in the world; but I also impressed on all the members of the Government with whom I came in contact my belief that the election had not in any way altered the policy of the President, and I warned them of the danger to our good relations if ruthless submarine warfare should be resumed.

PLESS 9th January 1917. At a subsequent Imperial Conference at Pless Castle the decision is made. With both sides more convinced than ever ; that the other intends its complete destruction and therefore that victory is the only way to end the war, all strategic considerations drift towards only one possible decision ~ to commence an unrestricted submarine campaign that the German admirals have been pressing for, and which they claim will definitely defeat Great Britain within six months.

This they do, in the full knowledge that it will cause America to enter the quarrel, and in the belief that the conflict will be won long before the United States has time to mobilize. Kaiser Wilhelm now listens to his commander's advice, and prepares to make his fateful decision…

Everyone stood around a large table, on which the Kaiser, pale and excited, leaned his hand. Holtzendorff spoke first from the navy in confidence of victory ; England will lie on the ground in at most six months before a single American has set foot on the continent. The American danger does not disturb me. Hindenburg spoke ; A measure in the reaction of American munitions exports has to be expected. Bethmann with a visible inner excitement ; In view of the recently altered stand of the Supreme Command and the categorical declarations of the admirals as to the success of the measure, I now withdraw my opposition. Kaiser Wilhelm follows their statements with impatience and declares ~ Unrestricted U-Boat war is therefore decided.

From the time when the Chancellor first spoke of peace, I asked him and others what the peace terms of Germany were. I could never get any one to state any definite terms of peace and on several occasions when I asked the Chancellor whether Germany was willing to withdraw from Belgium, he always said, Yes, but with guarantees. Finally in January 1917, when he was again talking of peace, I said, What are these peace terms to which you refer continually? Will you allow me to ask a few questions as to the specific terms of peace? First are the Germans willing to withdraw from Belgium? The Chancellor answered, Yes, but with guarantees. I said, What are these guarantees? He said, We must possibly have the forts of Liége and Namur; we must have other forts and garrisons throughout Belgium. We must have possession of the railroad lines. We must have possession of the ports and other means of communication. The Belgians will not be allowed to maintain an army, but we must be allowed to retain a large army in Belgium. We must have the commercial control of Belgium. I said, I do not see that you have left much for the Belgians except that King Albert will have the right to reside in Brussels with an honor guard. And the Chancellor said, We cannot allow Belgium to be an outpost of England. I said, I do not suppose the English, on the other hand, wish it to become an outpost of Germany, especially as Admiral von Tirpitz has said ; The coast of Flanders should be retained in order to make war on England and America. I continued, How about Northern France? He said, We are willing to leave Northern France, but there must be a rectification of the frontier. I said, How about the Eastern frontier? He said, We must have a very substantial rectification of our frontier. I said, How about Romania? He said, We shall leave Bulgaria to deal with Romania. I said, How about Serbia? He said, A very small Serbia may be allowed to exist, but that is a question for Vienna. A-H must be left to do what she wishes to Italy, and we must have indemnities from all countries and all our ships and colonies back. Of course, rectification of the frontier ~ is a polite term for annexation.

On 22nd January 1917, our President Wilson addressed the Senate referring to his Note of 18th of December, sent to all belligerent governments. In referring to the reply of the Éntente Powers to his Peace Note, he stated; We are that much nearer to the definite discussion of the peace which shall end the present war. He referred to the willingness of both contestants to discuss terms of peace, as follows : The Central Powers united in reply which stated merely that they were ready to meet their antagonists in conference to discuss terms of peace. The Éntente Powers have replied more definitely and have stated, in general terms, with sufficient definiteness to imply details, guarantees and acts of reparation which they deem to be the indispensable conditions of a satisfactory settlement. We are much nearer a definite discussion of the peace which shall end the present war.

The President further referred to a world concert to guarantee peace in the future and said, The present war must first be ended, but we owe it to candour and to a just regard for the opinion of mankind to say that so far as our participation in guarantees of future peace is concerned, it makes a great deal of difference in what way and upon what terms it is ended. He said that the statesmen of the groups of nations at war have stated that it was not part of the purpose they had in mind to crush their antagonists, and he said that it must be implied from these assurances that the peace to come must be a peace without victory.

In the course of his address the President also said: Statesmen everywhere are agreed that there should be a united, independent and autonomous Poland. So far as practicable, every great people now struggling toward a full development of its resources and its powers should be assured a direct outlet to the highways of the sea. Where this cannot be done by cession of territory it can be arranged by the neutralization of direct rights of way; and he closed by proposing that the nations of the world should adopt the Monroe Doctrine, and no nation should seek to explain its policy for any other nation or people.

(A week or two before 31st January) Dr. Solf asked me if I did not think that it would be possible for the United States to permit the resumption of ruthless submarine warfare against Great Britain. He said that three month's time was all that would be required to bring Great Britain to her knees and end the war. And in fact, so cleverly did von Tirpitz and other advocates of submarine war carry on their propaganda that the belief was ingrained in the whole of the German nation that a resumption of this ruthless war would lead within three months to what all Germans so ardently desired ~ peace. It was impossible for any government to resist the popular demand for the use of this illegal means of warfare, because army and navy and people were convinced that ruthless submarine war spelled success and a glorious peace. But this peace of course, meant only a German peace, a peace as outlined to me by the Chancellor; a peace impossible for the Allies and even for the world to accept; a peace which would leave Germany immensely powerful and ready immediately after the war to take up a campaign against the nations of the Western hemisphere; a peace which would compel every nation, so long as German autocracy remained, to devote its best energies and most fruitful periods of life, to preparations for war.

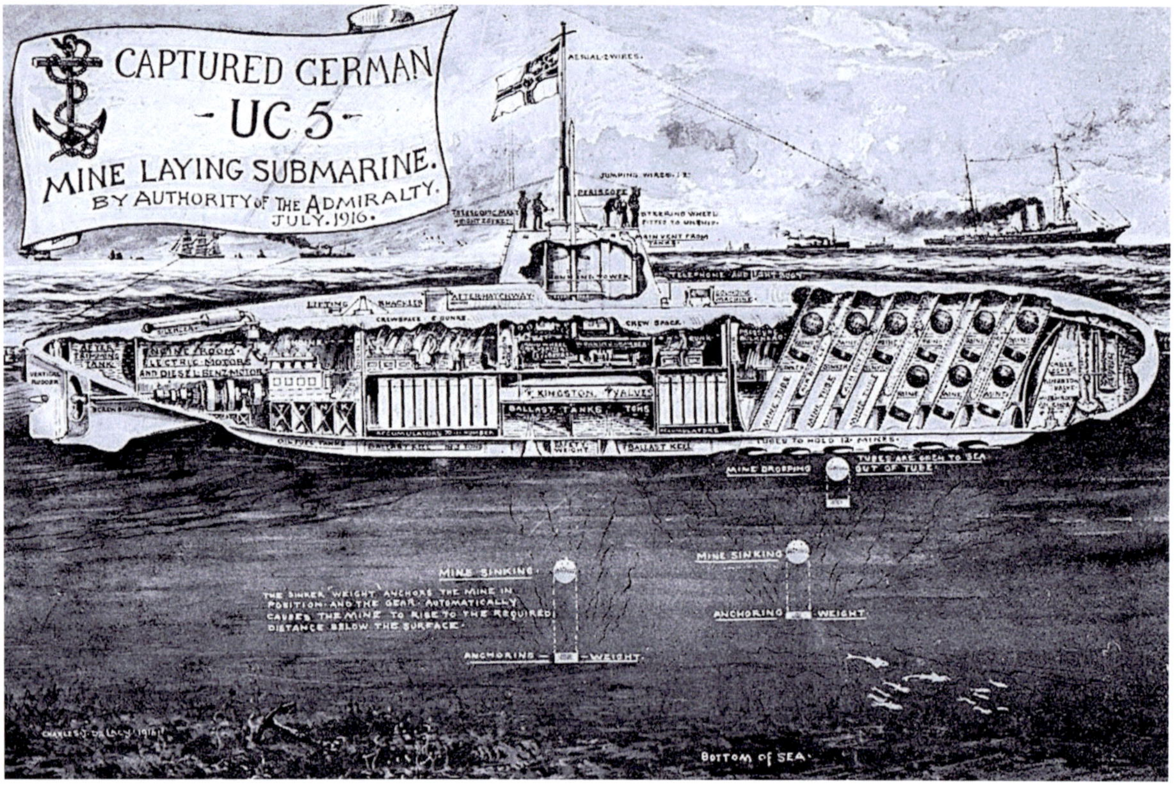

When the U-Boats were unleashed at midnight on 31st January 1917, American Ambassador to Imperial Germany, James W. Gerard was given six-hours' notice. *On 31st January at about four o'clock in the afternoon I received from Zimmermann a short letter of which the following is a copy : The Secretary of State of the Foreign Office, Zimmermann, requests the honor of the visit of his Excellency, the Ambassador of the United States of America, this afternoon at six o'clock in the Foreign Office, Wilhelmstrasse 75/76.*

Pursuant to this letter, I went to the Foreign Office at six o'clock. Zimmermann then read to me in German a note from the Imperial Government, announcing the creation of the war zones about Great Britain and France and the commencement of ruthless submarine warfare at 12 PM that night. I made no comment, put the note in my pocket and went back to the Embassy. It was then about 7 PM and, of course, the note was immediately translated and despatched with all speed to America.

After the despatch of the note I had an interview with the Chancellor in which he criticized both the Peace Note of 18th December as not being definite enough and the President's speech to the Senate of 22nd January; and further said that he believed that the situation had changed, that, in spite of what the President had said in the note, he was now for peace, that he had been elected on a peace platform, and that nothing would happen. Zimmermann at the time he delivered the note told me that this submarine warfare was a necessity for Germany, and that Germany could not hold out a year on the question of food. He said ~ Give us two months of this kind of warfare and we shall end the war and make peace within three months.

Von Bethmann-Hollweg, however, at an interview after the 31st January, said that he had been compelled to take up ruthless submarine war because it was evident that President Wilson could do nothing towards peace. He spoke particularly of the President's speech of 22nd January and said; In that speech the President has made it plain that he considers that the answer of the Éntente Powers to his Peace Note forms a basis for peace, which is a thing impossible for Germany even to consider; and when the President speaks of a united and independent Poland he evidently means to take away from Germany that part of Poland which has been incorporated in the Kingdom of Prussia and give it to this new and independent Kingdom, thereby bringing the Eastern frontier of Germany within two hours by motor from Berlin; and when the President speaks of giving each nation a highway to the sea, he means that the German port of Danzig should be turned over to this new State of Poland, not only taking a Prussian port but cutting the extreme Eastern part of Prussia from the remainder of the country. I said, These objections appeared to me very frivolous. The President, endeavouring to gain peace, has said that all parties were apparently agreed that there should be a peace; that if Germany were fighting a merely defensive war, as she had always claimed, she should be greatly delighted when the President declared that all the weight of America was in favor of a peace without victory, which meant, of course, that Germany should be secured from that crushing and dismemberment which Germany's statesmen had stated so often that they feared. Von Bethmann-Hollweg said that one of the principal reasons why Germany had determined upon a resumption of ruthless submarine warfare was because of this speech of the President to the American Senate.

The unrestricted submarine War Zone proclaimed by Imperial Germany in February 1915 ~ and similarly resumed on 1st February 1917.

Of course, the trouble with this criticism of the President's speech made by the Chancellor is that the orders for the resumption of ruthless submarine warfare had been given before the news of the speech came to Germany.

I had cabled the information collected by Commander Gherardi as to the orders given to submarines before the date of the President's speech, and the night after I received the German note announcing this resumption I was taking a walk after dinner about the snow-covered streets of Berlin. In the course of this walk I met a young German woman of my acquaintance who was on intimate terms with the Crown Princess. She was on her way on foot from the opera house, where she had been with the Crown Princess, to the underground train station and I joined her. I told her of the Ultimatum which, I had received at six o'clock that evening from Zimmermann and I told her that I was sure that it meant the breaking of diplomatic relations and our departure from Germany.

She expressed great surprise that the submarine warfare was set to commence on 31st January and said that weeks before they had been talking over the matter at the Crown Princess's and that she had heard then that the orders had been given to commence it on the 15th. It is certain that the orders to the submarine commanders had been given prior to the 31st and as early as the 15th.

I sincerely believe that the only object of the Germans in making these peace offers was first to get the Allies, if possible, in a conference and there detach some or one of them by the offer of separate terms, or if this scheme failed, then it was believed that the talk about peace would create a sentiment so favourable to the Germans that they might, without fear of action by the United States, resume ruthless submarine warfare against England.

WASHINGTON 31st January 1917. German Ambassador Count Johann von Bernstorff telephones the State Department in the morning to request an interview with Secretary of State Robert Lansing. When von Bernstorff arrives at 4.10 PM he is not walking with his usual confident springy step. Sadly, almost reluctantly he hands Lansing several communiqués from Berlin, saying as he does so, *I have always worked for peace between Germany and the United States and deeply regret the step that my government is about to take.* As von Bernstorff departs, Robert Lansing reads the communiqués;

Berlin deeply appreciates President Wilson's recent efforts for peace and largely subscribe to the principles enunciated in his address to the Senate ~ the Peace without Victory ~ of 22nd January 1917. Germany will cooperate in all efforts to prevent future wars. She has no desire to destroy her enemies or to annex Belgium. However, the Allies have made clear their intention to dismember Austria-Hungary, Germany, and all Turkey. To our wish of reconciliation, they oppose it with their will of destruction. They desire a fight to the bitter end. A new situation has thus been created which forces Germany to new decisions. The Imperial German Government cannot justify before its own conscience, the German people, and history the neglect of any means that might end the war. With its overtures for peace rudely spurned, it is now compelled to continue the fight for existence forced upon it, with the full employment of all the weapons, which are at its disposal.

The German authorities do not doubt that the Washington Government will understand why ~ in view of the brutal methods of the Allies, and their announced intention to destroy the Central Powers ~ Germany has to claim the freedom of action in submarine warfare. After 1st February 1917, German submarines will sink without warning all ships, belligerent or neutral, in a zone around Great Britain, France, and Italy, and in the Eastern Mediterranean. The German Government is confident that this measure will result in that speedy restoration of peace which the United States Government has so much at heart. We grant permission to one American ship to sail weekly to and from Falmouth, England, provided the American Government gives the assurance that these ships carry no contraband and that they are marked with alternating red and white stripes.

After reading the communiqués, Lansing sends them by special courier to the White House with a covering letter to President Wilson, which he has just written. *We are face to face with the gravest crisis presented since the war began. I think that as soon as you have read these papers we should have a conference to determine the course of action to be taken.* Bernstorff next has an aide deliver a special message to Colonel House, Wilson's closest advisor on foreign affairs ~ a telegram from Bethmann Hollweg for Wilson. *The Imperial German Government thanks the President for his recent messages concerning peace, but in view of the recently-announced intention of the Allies to destroy and degrade Germany and her associates ~ the President's message has come too late to prevent the launching of the unrestricted U-Boat campaign on schedule. However, Germany is always ready to do justice to the United States. We beg the President ~ to continue his efforts and we declare ourselves perfectly ready to discontinue the intensive U-Boat warfare as soon as we receive satisfactory assurances that the efforts of the President will lead to a peace which would be acceptable to us.*

WHITE HOUSE. President Wilson played golf on the morning of 31st January and only learned of the German decision for unlimited submarine warfare in the early afternoon from an Associated Press dispatch. He then closeted himself within his office, recalling how he had told Colonel Edward House, *I don't want to go to war over the sinking of any Allied merchantmen.* At 8.45 PM. Secretary of State Lansing arrives and advises, *I strongly urge an immediate severance of relations with Germany ~ failure to do so will cause a massive loss of national prestige.* Wilson, *I am not so sure. I am willing to go to almost any length to avoid belligerency. Nothing can induce me to break relations unless I am convinced that it is the wisest thing to do. However, you should prepare a note severing relations.* Lansing departs at 10.30 PM. Early the following morning, Colonel Edward House arrives carrying Bethmann's personal message for President Wilson. U.S. Secretary of State Lansing: *Prussian militarism is the implacable foe of democracy.* Wilson, *I am not sure that it would be wise to destroy German power and the German nation. I do not want either side to win. Probably greater justice would be done if the conflict ended in a draw.*

Publication of the Imperial German submarine decree in American morning newspapers on 1st February sets off dramatic tremors and shockwaves of excitement across the U.S. . Frank Cobb of the *New York World ~The German announcement is in effect a declaration of war against the United States. Germany has run amuck and the only possible response is an immediate break in diplomatic relations.* Thereafter, Washington appears united in the conviction that relations have to be broken at once.

The Announcement.

German Ambassador von Bernstorff ~ The New York Herald

CONGRESS 3rd February. Wilson severs relations at a joint session of Congress in the afternoon; *The recent German announcement leaves me with no alternative. I have instructed the Secretary of State to inform Bernstorff that relations between the United States and the German Empire are severed. I refuse to believe that Germany will destroy American ships and lives in a submarine campaign. If they do, I will ask for authority to protect peaceful and legitimate American commerce.*

With the ties of diplomatic relations severed between the U.S. and Germany, for the first time the American people now face the real possibility / probability of war ~ depending upon German actions at sea over which they have no control ~ and yet President Wilson is still reluctant to take the first step towards approving the protection of American shipping through armed neutrality. Many in the U.S. still stubbornly hope for peace to somehow still prevail, and one such glimmer of hope that excites President Wilson's hopes that he can resume his peace efforts comes in a peace proposal from the A-H Foreign Minister Count Ottokar von Czernin.

The Imperial and Royal Government is ready to negotiate an honourable peace on a basis of peace without victory. It is now up to the President of the United States to persuade the Allies to negotiate on this basis. Should the President follow this line of conduct, not only the terror of the submarine war, but the war in general could come to a sudden end and Mr. Wilson's name will shine with everlasting letters in the history of mankind.

Wilson relays Czernin's communiqué to London adding, *I know the Teutonic powers desire peace. Austria-Hungary might make peace immediately if she is assured of the continued integrity of her Empire and does not fear radical dismemberment. I am eager to mediate between A-H and the Allies.*
One week later, Lloyd George dashes Wilson's hopes for peace talks declaring;

The Allies are already now conducting secret negotiations with Vienna, and we prefer to have A-H remain in the war, as she is a heavy liability to Germany, and we will neither receive a formal peace offer from Vienna at this time nor give any pledges against dismemberment. The greatest service the United States could render mankind would be to enter the war, because President Wilson would have a full opportunity to heal the wounds of the world and establish a just and lasting peace only if the United States is a fully-fledged belligerent.
(Wilson thereafter plays for time by asking the King of Spain to probe Vienna's position).

BERLIN 3rd February. Gerard: *The President announced to Congress the breaking of diplomatic relations with Germany. News of this, did not reach Berlin until the next day and this Saturday afternoon my wife and I had an engagement to go to the theatre with Zimmermann and Fräulein Marie Anne von Friedlaender-Fuld, a young lady whose father is considered the richest man in Berlin, and who had been married to a John Mitford, a son of Lord Redesdale. Through no fault on the lady's part, there had been an annulment of this marriage, and she is occupying a floor of her own in the grand house of her father and mother on the Pariser-Platz in Berlin. We stopped for her and took her to the theatre. Zimmermann joined us later at the Friedlaender-Fuld Palace. After supper, Zimmermann spoke of the note to America and said; During the past month, this is what I have been doing so often at the GHQ with the Kaiser. I often thought of telling you what was going on as I used to tell you in the old days, but I thought that you would only say that such a course would mean a break of diplomatic relations, and so I thought there was no use in telling you. But as you will see, everything will be all right. America will do nothing, for President Wilson is for peace and nothing else. Everything will go on as before.*

4th February. We had a German who is connected with the Foreign Office and his American wife to lunch, and another German who had been in America, also connected with the Foreign Office. Just as we were going in to lunch someone produced a copy of the B.Z., the noon paper published in Berlin, which contained what seemed to be an authentic account of the breaking of diplomatic relations by America. The Germans looked very sad, and said nothing while I tried to make polite conversation at my end of the table.

5th February. I went over to see Zimmermann, having just received the official despatch from Washington, and told him that I had come to demand my passports. Of course, Zimmermann by that time had received the news and had composed himself. The American correspondents told me that when he saw them on the day before, he had at first refused to say anything and then had been rather violent in his language and had finally shown great emotion. I am sure, from everything I observed, that the break of diplomatic relations came as an intense surprise to him and to the other members of the government, and yet I cannot imagine why intelligent men should think that the United States of America had fallen so low as to bear without murmur this sudden kick.

6th February. The police who have been about our Embassy ever since the commencement of war, were now increased in numbers, and guarded not only the front of the house, but also the rear and the surrounding streets; but there was no demonstration whatever on the part of the people of Berlin. I went out for a walk through most of the principal streets of Berlin, and on my return, I found Count Montgelas, who, with the rank of Minister, was at the head of the department which included American affairs in the Foreign Office. I asked Montgelas why I had not received my passports, and he said that I was being kept back because the Imperial Government did not know what had happened to Count Bernstorff and that there had been rumours that the German ships in America had been confiscated by our government. I said that I was quite sure that Count von Bernstorff was being treated with every courtesy and that the German ships had not been confiscated. I said, I do not see why I have to disprove your idea that Bernstorff is being maltreated and the German ships confiscated. It seems to me it is for you to prove this; and, why don't you have the Swiss Government, which now represents you, cable to its Minister in Washington and get the exact facts? He said ~ Well, you know, the Swiss are not used to cabling. I said, I shall not cable at all. Why do you come to me with a proposed treaty after we have broken diplomatic relations and ask an Ambassador who is held as a prisoner to sign it? Prisoners do not sign treaties and treaties signed by them are worth nothing. After your threat to keep Americans here and after reading this document, even if I had authority to sign it I would stay here until hell freezes over before I put my name to such a paper.

Montgelas seemed rather rattled, and in his confusion left the paper with me ~ something I am sure, he did not intend to do in case of a refusal. Montgelas was an extremely agreeable man and I think at all times had correctly predicted the attitude of America and had been against acts of frightfulness, such as the torpedoing of the Lusitania and the resumption of their ruthless submarine war. I am sure that a gentleman like Montgelas undertook with great reluctance to carry out his orders in the matter of getting me to sign this treaty. Even the most pro-German American correspondents in Berlin, when I told them of Montgelas' threat, showed fine spirit. All begged me not to consider them or their liberty where the interests of America were involved.

As soon as diplomatic relations were broken, and I broke them formally not only in my conversation with Zimmermann of Monday morning but also by sending over a formal written request for my passports on the evening that day, our telegraph privileges were cut off. I was not even allowed to send telegrams to the American consuls throughout Germany giving them their fresh instructions. Mail also was cut off ~ and the telephone. My servants were not even permitted to go to the nearby hotel to telephone. In the meantime, we completed our preparations for departure. We arranged to turn over American interests, the interests of Romania, Serbia and Japan to the Spanish Embassy, and the interests of Great Britain to the Dutch. I have said already that I believe that Spanish Ambassador Polo de Bernabe will faithfully protect the interests of America, and I do believe that Dutch Baron Gevers will fearlessly fight the cause of the British prisoners.

We sold our automobiles… and two beautiful prize-winning saddle horses, one from Kentucky and one from Virginia, which I had brought with me from America. I sold them to the proprietor of the circus in Berlin! The three tons of food which we had brought with us from America we gave to our colleagues in the diplomatic corps ~ to the Spaniards, Greeks, Dutch and the Central and South Americans. I had many friends among the diplomats of the two Americas who were all men of great ability and position in their own country. I think that most of them know only too well the designs against Central and South America cherished by the Pan-Germans.

Friday 9th February. Finally, Mr. Oscar King Davis, correspondent of the New York Times, received a wireless from his editor Mr. Van Anda, telling him that Bernstorff and his staff were being treated with every courtesy and that the German ships had not been confiscated. In the evening our telephone was reconnected, and we were allowed to receive some telegrams and to send open telegrams to the consuls, etc. throughout Germany; and we were notified that we would probably be allowed to leave the next day in the evening.

Saturday 10th February. Always followed by spies, I paid as many farewell visits to my diplomatic colleagues as I was able to see, and I thought that, in spite of the ridiculous treatment accorded us in cutting off the mail and telephone and in holding me for nearly a week, I would leave in a sporting spirit. I therefore had my servant telephone and ask whether Zimmermann and Bethmann would receive me. I had a pleasant farewell talk of about half an hour with each of them and I expressly told the Chancellor that I had come to bid him a personal farewell, not to make a record, and that anything he said would remain totally confidential. At the last moment, the Germans allowed a number of the consuls and clerks who had been working in the Embassy, and the American residents in Berlin, to leave on the train with us; so we were about 120 persons in all on this train, which left the Potsdamer Station at 8.10 in the evening. The time of our departure had not been publicly announced, but there was a large crowd in the station to see us off. All the Spanish Embassy, Dutch, Greeks, and many of our colleagues from Central and South America were there, as was Montgelas from the Foreign Office. As the train pulled out, a number of Americans left in Berlin on the station platform raised a vigorous cheer.

Two officers were sent by the Imperial Government to accompany us on the train; Major von der Hagen, sent by the General Staff the other, a representative of the Foreign Office, Baron Wernher von Ow-Wachendorf. It was quite thoughtful of the Foreign Office to send this last officer, as it was by our efforts that he had secured his exchange when he was a prisoner in England; and he therefore, would be supposed to entertain kindly feelings for our Embassy. I had ordered plenty of champagne and cigars to be put on the train and we were invited to drink champagne with the officers in the dining car; then they joined us in the private salon car which we occupied in the end of the train. The journey was uneventful. Outside some of the stations a number of people stared at the train in a bovine way, but who made no demonstration of any kind.

We went through Würtemburg and entered Switzerland by way of Schaffhausen. The two officers left us at the last stop on the German side. I had taken the precaution before we left Berlin to find out their names, and, as they left us, I gave each of them a gold cigarette case inscribed with his name and the date.

At the first station on the Swiss side, a body of Swiss troops were drawn up, presenting arms, and the Colonel commanding the Swiss Army (there are no generals in Switzerland), attended by several staff officers, came on the train and travelled with us nearly to Zurich. I started to speak French to one of these staff officers, but he interrupted me by saying in perfect English ~ You do not have to speak French to me. My name is Iselin, many of my relations live in New York and I lived there myself some years. At Zurich we left the German train, and were met on the platform by some grateful Japanese, the American Consul and a number of French and Swiss newspaper reporters, thus ended our exodus from Germany.

* * *

Meanwhile, a parallel incident was further ensuring an American involvement in the Great War. In January, the British naval intelligence department designated Room 40 decrypted a secret telegram of an internal diplomatic message from German FM Arthur Zimmermann intended for Berlin's Ambassador to Mexico Heinrich von Eckardt, which had just passed through the German Embassy in Washington en route to Mexico City. Arthur Zimmermann in anticipation of the resumption of unrestricted submarine warfare ~ an act which Berlin supposed would lead to war with the U.S. directed von Eckhardt to approach the Mexican President Venustiano Carranza with an offer to forge a wartime alliance between Germany and Mexico in the event of war between the U.S. and Germany. In return, Germany offered to provide military support for a Mexican attack on the U.S. to annex her so-called *lost territories* of Texas, New Mexico, and Arizona. In addition, von Eckhardt was directed to use Mexican Government channels as a proxy to entice Japan to join the German and A-H cause.

In the interval, the British Director of Room 40 Naval Intelligence, Admiral *Blinker* Hall, was in a delicate situation. Soon after the Iranian Lor tribe Hayat Daoud had chosen to side with Great Britain in March 1915, they captured the German agent Wilhelm Wassmuss and his expedition making their way from the Persian Gulf to Shiraz. Although Wassmuss escaped, his éntourage was secured and their codebooks and propaganda material were handed over to local British authorities who shipped it unopened to London, where *Blinker* Hall unwrapped German Diplomatic Code Book No. 13040. Hall always had to be careful not to reveal the effectiveness of Room 40 decrypts of German codes. Otherwise ~ if mishandled, the discovery of the Zimmermann Telegram had the potential to damage Anglo-American relations. Therefore, before he broke the news to the Americans, *Blinker* Hall held the Zimmermann Note undisclosed for two weeks, hoping that America would declare war on Germany, in response to the German declaration of unrestricted submarine warfare on 1st February. After that failed to occur, *Blinker* Hall disclosed the entire Zimmermann Telegram to the British Foreign Office. The British Ambassador in Mexico was supplied with a copy of the telegram to give the impression that the information had been acquired ~ not through intercepting American communications, but through an agent in Mexico known as *Mr. H.* (The British Ambassador to Mexico Sir Thomas Hohler later identified himself ~ *as the secretive Mr. H responsible for intercepting the inflammatory telegram.*
Thus, the integrity of British intelligence, as well as amiable Anglo-U.S. relations was preserved.

On 19th February, *Blinker* Hall presents Edward Bell, 2nd Secretary of the U.S. Embassy in London and liaison officer to the British, a complete copy of the Zimmermann Telegram *obtained* in Mexico. Bell is furious at the audacity of the Germans to conspire with Mexico to invade the United States. U.S. Ambassador to Great Britain Walter Page, drafts a telegram to Wilson explaining that early in the war the British Government captured a copy of the German code now used for diplomatic messages. On 24th February, Wilson receives Page's telegram. Wilson feels personally betrayed, for he has been negotiating with the Germans to bring an end to the war. After conferring with his closest advisors, Wilson releases the Zimmermann Telegram and its scandalous contents to U.S. papers nationwide. While many Americans remain committed to isolationism, evidenced by Wilson's re-election in 1916 after running on his peace platform and parading the campaign slogan: *He Kept Us Out of the War*, now the Zimmerman Telegram furnishes fresh evidence of severe anti-American aggression:

We intend to begin on the first of February unrestricted submarine warfare. We shall endeavor in spite of this to keep the United States of America neutral. In the event of this not succeeding, we make Mexico a proposal of alliance on the following basis: make war together, make peace together, generous financial support and an understanding on our part that Mexico is to reconquer the lost territory in Texas, New Mexico, and Arizona. The settlement in detail is left to you. You will inform the President of the above most secretly as soon as the outbreak of war with the United States of America is certain and add the suggestion that he should, on his own initiative, invite Japan to immediate adherence and at the same time mediate between Japan and ourselves. Please call the President's attention to the fact that the ruthless employment of our submarines now offers the prospect of compelling England in a few months to make peace. *Signed ZIMMERMANN*

Coupled with the submarine attacks, the American public grow indignant and lose any sympathy for Germany. On 2nd April, President Wilson abandons his policy of neutrality and asks Congress to ; *Declare war on Germany to make the world safe for democracy,* Wilson refers to the Zimmermann Telegram as proof of German perfidy, and this serves as the last straw that pushes the United States into war, when the U.S. joins the Allied Éntente four days later.

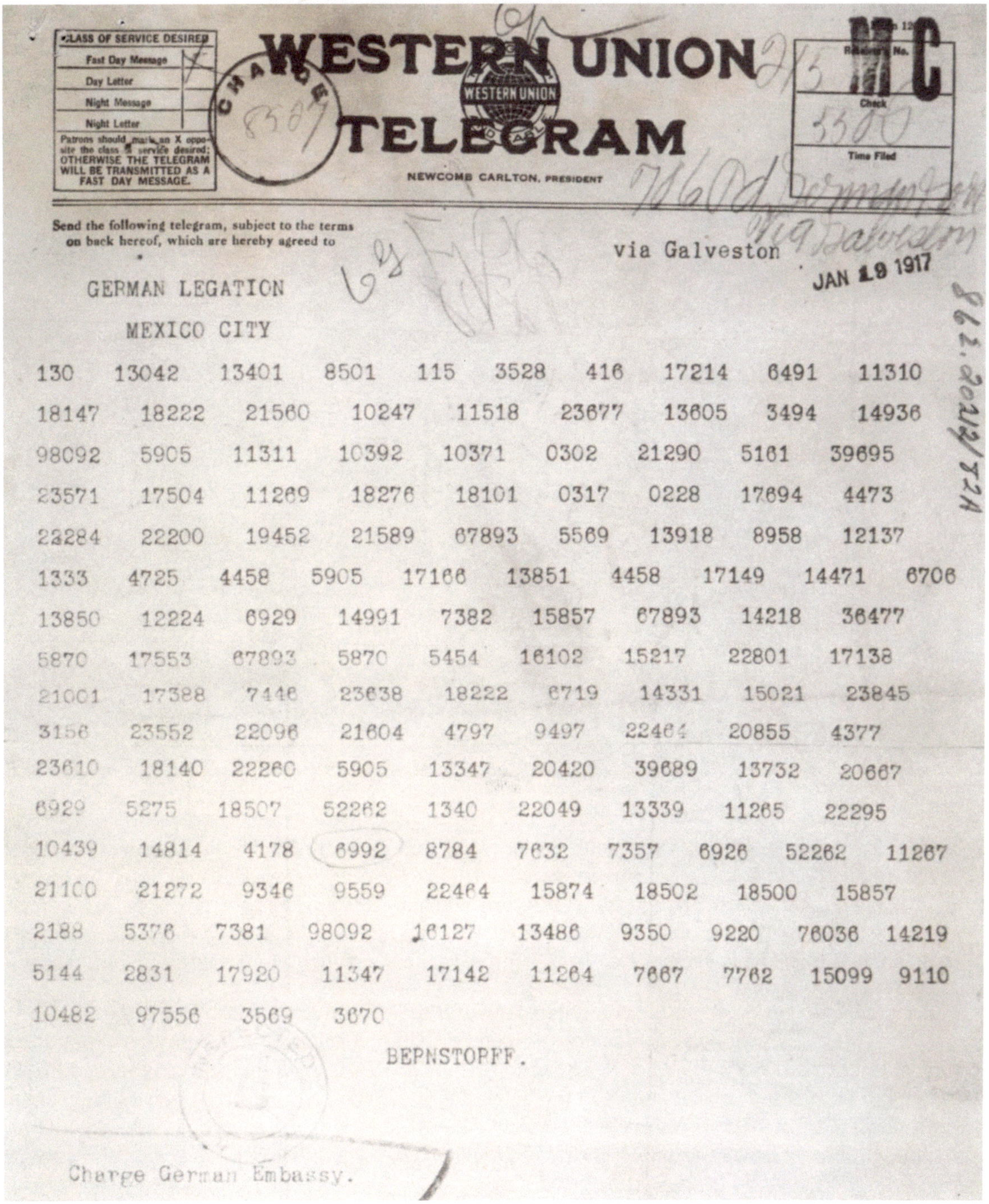

From Berlin, the Zimmerman Telegram was transmitted in Code 13040, used since 1912 to Washington and New York via a wireless Berlin-Stockholm-Buenos Aires-Washington telegraphic route using the American State department wire thru London. (U.S. President Wilson having earlier permitted Berlin to use this line for peace negotiations). Ambassador Count Johann von Bernsdorff then relayed the message to Mexico City using the commercial Western Union telegraph apparatus. It remains unknown whether the Kaiser, Chancellor or Supreme Command knew of or approved the Zimmerman Telegram as all evidence relating to it in the files of the Imperial Cabinet, Foreign Office and Supreme Command were all destroyed soon after the war. In 1919, while Berlin's *Reichstag* Commission of Inquiry probed the causes of American entrance into the war, they asked no questions about this to Foreign Minister Arthur Zimmerman or any of the others. Finally, the German leaders retired to write their memoirs ~ and all remained forever silent on Zimmerman.

Wedding of Archduke Karl & Princess Zita of Bourbon-Parma

Archduke Karl I of A-H

VIENNA. Archduke Karl was still in his twenties, and not until that surprising supper evening at the Belvedere had he expected to become Emperor of the A-H Empire ~ especially as Franz Ferdinand remained in such good health. On 28th June 1914, this all changed with the assassinations in Sarajevo. When Karl heard, his wife Zita observed ; *Though it was a beautiful day, I saw his face go white in the sun.* Karl realized, that he had just become heir presumptive to the Hapsburg Monarchy.

Karl was promoted to General in the A-H Army and took command of 20th Corps for an assault through the Tyrol. The war was personally difficult for Zita, as several of her brothers fought on opposing sides. Prince Felix and Prince René had joined the A-H Army, while Prince Sixtus and Prince Xavier lived in France and chose to enlist in the Belgian Army. When her country of birth, Italy joined the war against A-H in 1915, concerns of the *Italian* Zita arose. The German Ambassador in Vienna, Count Otto von Wedel even reported to Berlin: *The Empress is descended from an Italian princely house... People do not entirely trust the Italian and her brood of relatives.*

At the Emperor Franz Josef's request, Zita and her children left their residence at Hetzendorf and moved into a suite of rooms at Schonbrunn Palace. Here, Zita spent many hours with the old Emperor on both formal and informal occasions, where Franz Josef confided in her, his fears for the future. After Emperor Franz Josef died of bronchitis and pneumonia at the age of 86 on 21st November 1916. Zita later recounted, *I remember the dear plump figure of Prince Lobkowitz going up to my husband, and with tears in his eyes, making the sign of the cross on Karl's forehead. As he did so he said; May God bless Your Majesty ~ It was the first time we had heard the Imperial title used to us.*

BUDAPEST 30th December 1916. Karl and Zita are crowned Emperor and Empress of A-H. Following the coronation in his first public message Karl proclaims ~ *I wish to do everything to banish, in the shortest possible time, the horrors and sacrifices of war and to give back the vanished blessings of peace to my peoples ~ as soon as the vital interests of our faithful Empire and the obstinacy of our enemies will permit.* At the beginning of the reign, with Karl so often away from Vienna, a telephone line is installed from his military GHQ in Baden to the Hofburg to call Zita. Zita has some influence on her husband and discreetly attends audiences with the Prime Minister as she has a keen interest in social policy. Energetic Zita usually accompanies her husband to the provinces and to the front, as well as busying herself with numerous charitable works and regular hospital visits to the war-wounded.

Karl taking his coronation oath December 1916

With the war dragging on towards its fourth year in 1917, Karl's brother-in-law Prince Sixtus of Bourbon-Parma as a serving officer in the Belgian Army, becomes the main mover behind a plan for A-H to make a separate peace with France ~ for Emperor Karl to secretly enter into peace negotiations with France with Sixtus as his intermediary. Sixtus making known his views in politically influential French circles that France's real enemy is Prussia, and that France should regain her boundaries not merely of 1870 ~ but of 1814. He argues that Germany should be dismembered into its constituent kingdoms of Bavaria, Saxony, Hanover, Hesse and the lesser states ; and that Prussia's neighbours should all receive portions of her territory. While depriving the Hohenzollern Kaiser Wilhelm and his line of leadership of Germany, the hope of Prince Sixtus is to preserve the Hapsburg Monarchy. Sixtus suggests that it is in France's best interest to keep the A-H Empire intact as an obstacle to Prussian expansion along the Danube and any future Italian moves in the Mediterranean region. Therefore, Sixtus is in a position to play an important diplomatic role in such a peace proposal.

Emperor Karl initiates contact with Prince Sixtus via neutral Switzerland when he asks his wife, Zita to write a letter to her mother, Duchess Maria Antonia of Bourbon-Parma to invite her brothers, Prince Sixtus and Xavier to Vienna in the hope that they can help in promoting peace negotiations. After Zita visits neutral Switzerland to deliver the letter in person, the Duchess of Bourbon-Parma writes to the Queen of the Belgians via Luxembourg's diplomatic service asking her and her husband Albert to urge the Princes Sixtus and Xavier to travel to neutral Switzerland to see her.

NEUCHATEL 23rd January 1917. The two princes depart the front and arrive in Switzerland a week later for two days of meetings with their sister and mother *incognito.* The Duchess advises her sons that Emperor Karl of A-H wishes to discuss peace terms with them and that all is arranged to conduct them secretly to Vienna. If they cannot go, Karl will send an envoy to them in Switzerland. The brothers choose to wait in Switzerland. Sixtus then reveals to his mother what he believes to be the Éntente's fundamental peace conditions, and conditions for talks that have been agreed to with the French ~ the restoration to France of Alsace-Lorraine, the restoration of the independence of Belgium, independence for Serbia, and the cession of Constantinople to Russia. Karl agrees in principle and drafts a secret letter to Sixtus, *I give the secret and unofficial message that I will use all means and all my personal influence to the President of France.*

PARIS 11th February 1917. The Princes confer with new French FM Jules Cambon to confirm that the French Government at the highest level are indeed interested in Emperor Karl's peace proposal.

NEUCHATEL 13th February. At a second meeting in Switzerland, the Emperor Karl's messenger, his close and trusted friend Count Thomas Erdody, informs Sixtus and Xavier; *The Emperor Karl desires peace and is willing to conclude a secret armistice with Russia and declare his disinterest in Constantinople, return Alsace-Lorraine to France, restore Belgium, and create a south Slav kingdom including Serbia, Montenegro, Bosnia-Herzegovina, and Albania.* Prince Sixtus who is now certain that his brother-in-law has plainly demonstrated his peaceful intentions, farewells Count Erdody who returns to Vienna.

PARIS 5th March. Prince Sixtus during a two-hour interview with French President Raymond Poincaré describes the course of the negotiations thus far. Poincaré: *I feel that the Emperor's secret comments provide a basis for further dealings. I shall communicate the proposals in the greatest secrecy to our two principal allies, Russia and Britain ~ but am concerned about some compensation for Italy, now having entered the war on the Allied side. Italy will undoubtedly assert demands, but Austria will be compensated at German expense. I will ask the King of Belgium to request an extension of your leave from the Belgian Army. Would you consent to travel to Russia for talks with the Tsar ?*
If the Emperor Karl agrees to these conditions, France and her allies will assist A-H in case of German hostility.

GENEVA 19th March. Princes Sixtus and Xavier are met by Count Erdody who now insists; *Instead of merely delivering your messages to the Emperor, I will conduct you personally to Vienna. This time it is absolutely essential that you come to Vienna. The Emperor has told me ~ with all these comings and goings between Vienna and Switzerland too much time is lost. It will end up by drawing attention, and we will be left with no result. If we wish to be successful, we must hurry. One hour of conversation between us will do more to advance peace than 20 letters over six months. The Emperor promises that you will have unhindered entrance to and from Austria and that complete secrecy will be maintained. You will only see the Emperor himself, Zita and Count Czernin.* Erdody reminds Sixtus of all the lives that will be sacrificed if peace is not concluded *Come, I have now prepared everything and can guarantee you on my honour, that there will be no difficulty.* As a final persuasive, Erdody shows Sixtus a letter from his sister Zita the Empress who implores him; *Think of all those unfortunates who live in the hell of the trenches, who die by the hundreds every day and come.* The two Princes, Sixtus and Xavier agree to risk the venture, and in the evening they depart Geneva.
At the border, the guards on orders from the Emperor permit them to pass.

VIENNA 22nd March. Arriving at night, the three men drive to Count Erdody's city residence. The Princes remain there while their host proceeds to deliver a message from Sixtus to the Emperor Karl at Laxenburg Castle, a few kilometers south of the Austrian Capital. At 6 PM the following evening Erdody conducts the two Princes to Laxenburg. The night is dark and heavy snow is falling, as it has for the past two days. Their car stops in a courtyard. Upon receiving the A-H password, the sentinels with their coat collars raised against the icy storm, let them pass. The Princes, Erdody, and a sentinel enter the castle by a side door opening onto a staircase that leads to the private apartment of the Emperor and Empress. Erdody waits on guard outside, while Sixtus and Xavier go into their sister's salon, where they find the royal couple, whom they have not seen since August 1914.

Their reunion is affectionate, but Karl and Zita are grave ~ even sad. With the greetings over, the Emperor opens a serious dialogue with his brothers-in-law. *It is absolutely necessary to make peace, I want it any price. Austria cannot continue to fight for the Kaiser of Prussia. The Foreign Minister must also join our talks.* A few minutes later Czernin enters. Tall and lean, his manner is completely reserved. The conversation remains formal and cold, despite Karl's many efforts to introduce a warmer tone. Czernin is evasive and as he makes no proposals, he clearly has no enthusiasm for peace discussions. Sixtus: *I need to insist on precise terms.* Karl: *I will give them to you tomorrow evening.* After being in the room for less than an hour Czernin departs. Karl accompanies Czernin and returns shortly to say that Czernin will meet the Princes at Erdody's house the next day, when he expects the Foreign Minister's suspicions will give way to a more cordial attitude. The Princes depart Laxenburg Castle.
The snowstorm is over, and Vienna is silent under a blanket of snow.

24th March. Czernin is more co-operative, yet hesitates for A-H take a first step towards peace, strongly insisting on maintaining their current level of secrecy, fearing the consequences should Berlin discover Vienna's bid for peace. Czernin has cause for such a fear, as he is scheduled to meet German policy makers in Berlin in two days. As Czernin departs, Erdody implores him not to waste time,
You know that we cannot hold out indefinitely! (Czernin ignores the Sixtus Affair in his memoirs).

That same evening ~ and again in the greatest secrecy, Sixtus and Xavier return to Laxenburg. Emperor Karl hands to Sixtus a letter he has written saying; *Here are the fuller particulars as promised.* They talk awhile... and then accompanied by Erdody, the Princes leave for Geneva, and then Paris.

PARIS 30th March 1917. Arriving in Paris, the Princes are handed a letter advising them that President Poincaré wishes to see them the following morning. At their third meeting, Sixtus submits to the President of France the letter Karl handed to him at Laxenburg a week earlier. In this letter, written in French in his own hand, the Emperor addresses himself to the points put to him by Sixtus.

I promise to use all my influence to support the just claims of France in regard to Alsace-Lorraine.
As for Belgium, she must be re-established in her sovereignty, retaining the whole of her African possessions, without prejudice to the compensations she may receive for the losses she has sustained.
Serbia, too, will be re-established in her sovereignty, and as a token of our good will we are willing to grant her a just and natural access to the Adriatic Sea, as well as liberal economic concessions.
Austria-Hungary requires in return, that Serbia shall for the future abandon all connection with and shall suppress every society of which the political object is the disintegration of the Monarchy, and particularly the Narodna Obrana or Black Hand ~ responsible for the assassination of the Archduke Franz Ferdinand in 1914.
Finally on the question of the Straits. The revolutionary events that are taking place in Russia compel me to withhold my ideas in regard to her until a legal and compact government shall be established there.

Poincaré and Jules Cambon ~ also attending the meeting raise questions about the Emperor's attitude towards Russian claims to Constantinople. They also ask Sixtus for more precise details than Karl has supplied regarding France's recovery of the frontiers of 1814, and the compensations Belgium can expect. After receiving reassurances from the Prince, Poincaré makes it clear that he is aiming not for an armistice, but for a separate peace with A-H to severely weaken Germany. President Poincaré, on the whole finds Karl's proposals favourable enough for transmission to Great Britain.

FOLKESTONE 11th April 1917. British PM David Lloyd George is shown the Emperor's letter. He believes that Italy should be made aware of the possibility of peace with A-H. The French disagree, stating that the negotiations with Karl should remain completely secret, lest the plan unravels fast. In time, Paris gives their approval for British PM Lloyd George to arrange a meeting with the Italian Premier and Foreign Minister Baron Sonnino to discuss peace ~ without mentioning Karl's offer.

PARIS 12th April 1917. Sixtus has his fourth interview with Poincaré, who reveals the French involvement with Lloyd George. Sixtus is justified in feeling extremely anxious, as he is keenly aware of the intense hostility between A-H and Italy. He insists that Karl's peace proposal must not be divulged to the Italians who will endanger the peace negotiations and possibly also leak it to Berlin. On 19th April, Lloyd George meets Sixtus at St Jean de Maurienne to stress the need to obtain Italian agreement to a separate peace with A-H. Sixtus knowing this will ruin everything properly refuses. Lloyd George cautiously raises the possibility of A-H peace negotiations with Italy. Baron Sonnino is vehemently opposed to this course. He wants territorial gains for Italy ~ the Trentino, Trieste, Dalmatia and the Adriatic islands ~ which he knows can be obtained by a complete military victory, (otherwise he cannot justify his own role in bringing Italy into the war on the side of the Éntente). Lloyd George gives in to the obstinate Italians that effectively ends Sixtus' attempt to negotiate peace.

Sixtus continues his efforts, even meeting Lloyd George in London about Italy's demands on A-H but Lloyd George cannot persuade his generals that Britain should make a separate peace with A-H. Jules Cambon then concludes; *There is no good in carrying on negotiations which can only end in deadlock.* At a time when Karl rightly worries that he may lose his throne, Zita manages an impressive personal achievement at this time by stopping German plans to send aircraft to bomb the home of the King and Queen of Belgium. Prince Sixtus continues trying to save his brother-in-law Emperor Karl of A-H, and to buttress the old monarchial order of Europe against the overwhelming new force of mass politics.

(In April 1918, after A-H Foreign Minister Count Ottokar Czernin delivers a dramatic speech attacking incoming French PM Georges Clemenceau ~ *France has been the main obstacle to peace.* Clemenceau is so incensed that he has Emperor Karl's letter of 24th March 1917 published that leads to Count Czernin's resignation, and subsequently appears to place the life of Prince Sixtus in peril).

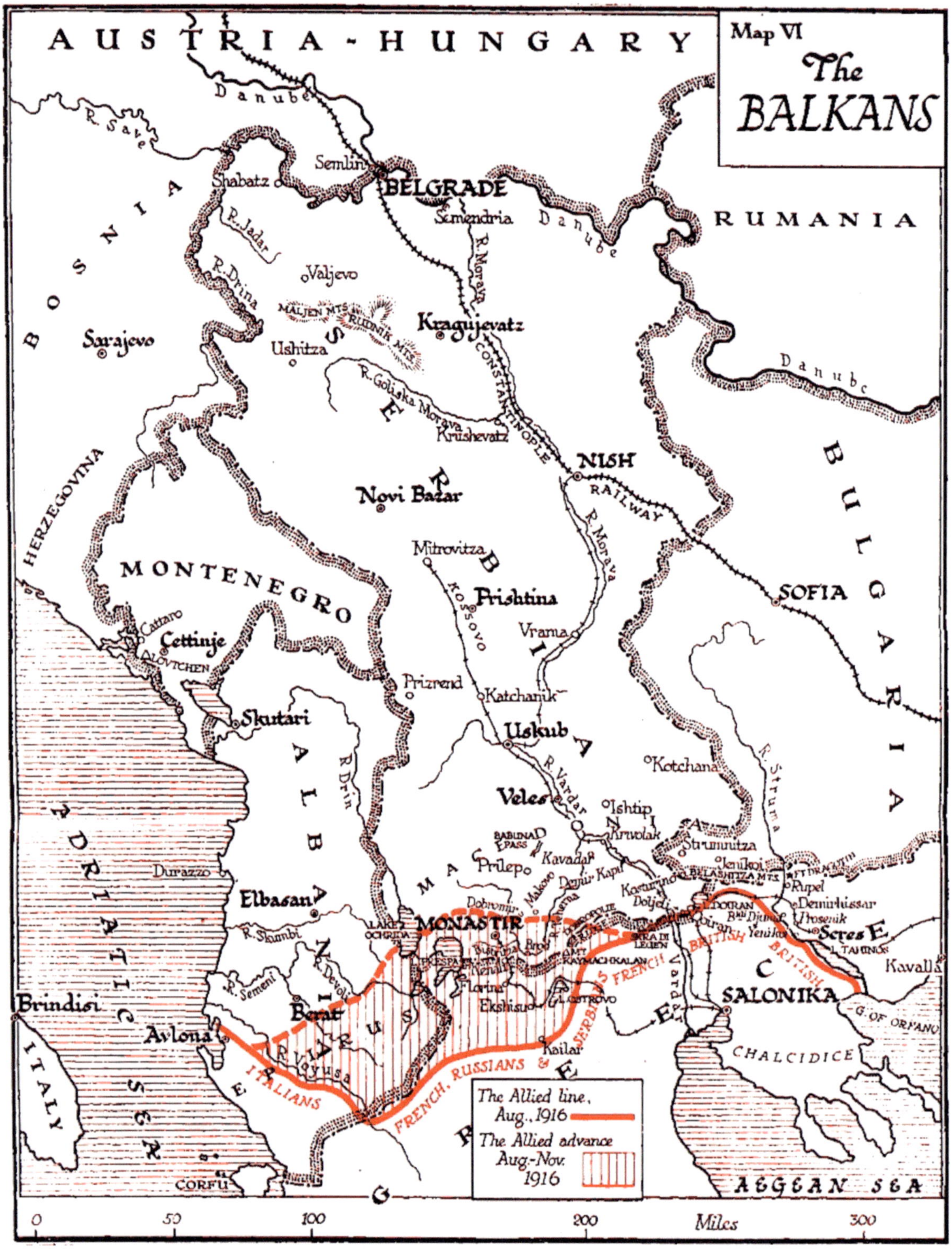

SALONIKA. Following the invasion of Serbia by A-H and German forces that drive the Serbian Army from Serbia into Albania, and from there to the island of Corfu ~ from Salonika, the *Black Hand* argue for an attack on Bulgaria. Crown Prince Alexander and Pasic reject the idea of such an attack, and dispute leads to open conflict between the Serbian Government and the leaders of the *Black Hand*. However, as so many prominent *Black Hand* officers have fought and perished in harsh campaigns, including the recent fighting on the Salonika Front, the *Apis* faction is diminished in influence.

Pasic and his Minister of the Interior ~ former Serbian Minister of Education ~ Ljubo Jovanovic, seek to resolve the dispute by dismissing *Apis* and any officers associated with him from within the Serbian Army. Crown Prince Alexander and Minister of War General Bozidar Terzic decide that *Apis* and his collaborators should be put on trial, for their end has become politically desirable to Pasic and those who hope of making a separate peace between Serbia and their opponents.

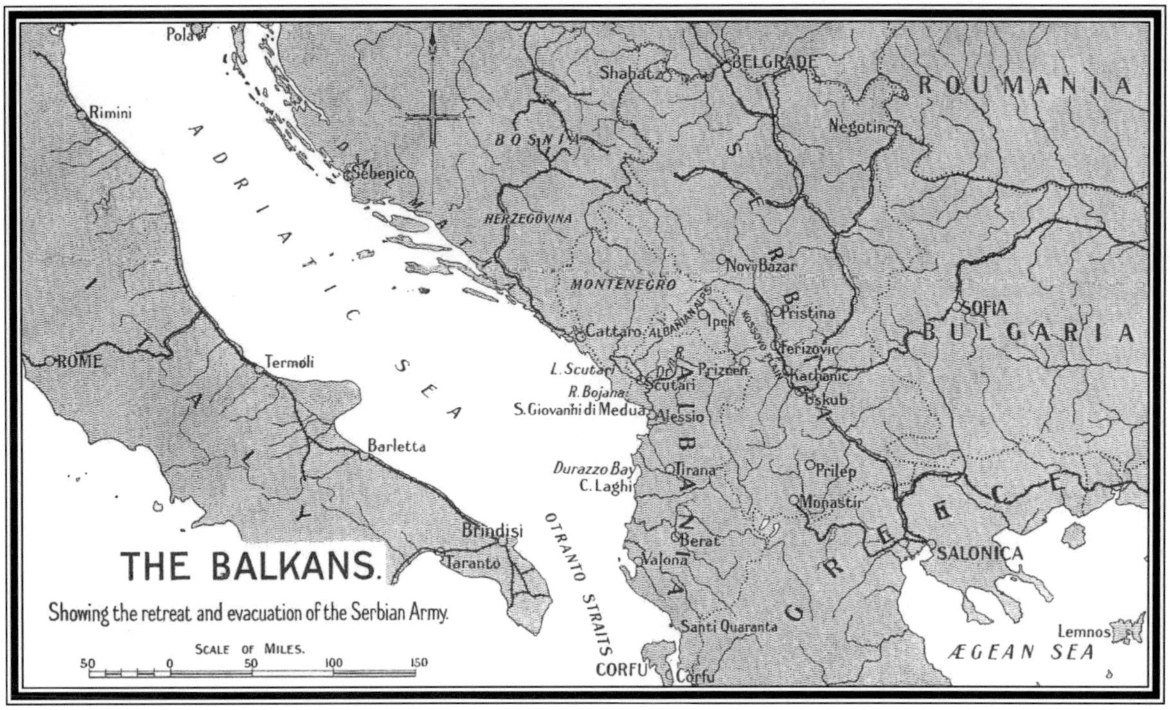

In endless postwar attempts to assign responsibility for the war to this or that European Empire, the assassination itself is often lost sight of. It is worthwhile now to recall the setting and circumstances of the murders and re-investigate the unsolved questions that still puzzle historians. While few condone the assassination as a political act, in considering this early phase of the 1914 crisis, we find the fundamental question in Austro-Serbian relations ~ whether Austria had the better right to defend her existence or Serbia to liberate her brethren? The spot where Franz Ferdinand was shot at the intersection of Franz Josef Street and the Appel Quay, has become a shrine for Bosnian nationalists to the assassin Gavrilo Princip, who hoped to make it possible to unite all South Slavs from Bosnia-Herzegovina, Serbia, Montenegro, Macedonia, Croatia, Slovenia and Dalmatia. At times, the old bridge crossing the river at the end of Franz Josef Street has been renamed the Princip Bridge.

Gavrilo Princip first had the idea of murdering the Archduke before Easter 1914, when he learned that Franz Ferdinand would visit Sarajevo. At once Princip decided that this was the opportunity he and other young Serbs had been waiting for, and it is believed that Princip secured this information via Colonel Dragutin *Apis* Dimitrijevic of the Serbian General Staff, who had received his information from his top agent in Vienna (Rade Malobabic) before any word of it ever appeared in the press.

Princip along with two of his Bosnian friends, Chabrinovic and Grabezh, the son of an Orthodox priest, got in touch with Ciganovic, another Bosnian. He was an employee of the Serbian State Railway and a member of the *Black Hand.* He in turn sounded out Major Voija Tankosic, a close collaborator of Dimitrijevic. Tankosic had connections with the Serbian State Arsenal at Kragujevac, and secured supplies of revolvers, ammunition and bombs, and through the *Black Hand* he had access to (Russian supplied) money. Although at the initial Sarajevo trial in October 1914, Chabrinovic asserted that Tankosic and Ciganovic were Masons and that it was the Masons who were determined upon the assassination, this was asserted to deflect attention from Dimitrijevic and the *Black Hand,* whose Dimitrijevic and Tankosic (with Russian attachés Artamonov and Verkovskii) initiated the plot.

As for the question of the responsibility of the Serbian Government ; once it was established by the 1914 Minister for Education Ljuba Jovanovic that PM Pasic knew of the plot itself, he cannot be absolved of complicity. Never of his own volition did Pasic open judicial enquires into the part any subjects of Serbia might have had with its preparation on Serbian soil. Fearing the consequences to himself and his government ~ overthrow of the later or even his own assassination by the *Black Hand,* with whom he was not on good terms. Instead, Pasic waited for Vienna to present proof and make demands, which were soon forthcoming in the A-H 48-hr. ultimatum/note of investigative demands to Serbia on 23rd July 1914. Pasic thereafter remains silent when enquires are made of him.

In December of 1916, the Prince Regent Alexander and officers loyal to him in the Serbian régime in exile move to eradicate the clique headed by *Apis*. They arrest *Apis* and other surviving *Black Hand* leaders, accusing them of treason and the attempted murder of the Serbian Prince Regent Alexander. Therefore, on 15th March 1917, a Serbian Court Martial on the French-controlled Salonika Front indicted *Apis* and the others. There is some reason to suspect however that this alleged plot against Prince Alexander was in part a mere pretext for silencing *Apis* forever from the fear on the part of Paşic that he might reveal to the world the truth about the Belgrade Cabinet's part in the murder plot that gave rise to the Great War, and thus reveal the Serbian Government's own guilty knowledge of the plot. After all, Nikola Pasic was well known for his role in the gradual building up of the moral and material forces of Serbia as a means for the eventual liberation and union of all Serbs in a powerful state, such as the Italian unification in the preceding generation. With this main aim in view, Pasic founded the Radical Party in 1881, and asserted as pursuant in its organ *Samouprava ; The people's welfare and freedom at home, and the country's independence and unification with the parts of Serbdom abroad. Foreign Policy; There must be organized a way of helping the divided and unliberated parts of Serbdom, as well as of keeping alive the sense of our national unity in the Serb provinces which, being far away are exposed to the influence of foreign elements.* In other words, discontent must be kept alive in the Serb provinces of the Turkish and Hapsburg Empires until the future war of liberation should join them to a Greater Serbia.

Much new information concerning the *Black Hand* came to light in the official record of the Salonika Trial of 1917. This thick volume; officially published in Salonika in 1918, was later withdrawn from circulation and suppressed as far as possible because it contained so much material damaging to the reputation of the Serbian Government of 1914. It now seems almost impossible to acquire a copy. *According to one account, the edition was destroyed by fire in Salonika, according to another it was suppressed. Nevertheless, I found a copy in Zagreb in 1928. Unfortunately, the book has not been translated into any Western language* ~ Bernadotte Everly Schmitt. However, it contains much information about the activities of the *Black Hand* before 1914, and about those members who participated in the plot to assassinate the Archduke. This trial throws much valuable light on the preparations within Serbia that culminated in the assassinations, yet few outside of Austria and Germany ever gave it any attention.

Four of the defendants confessed to their roles in the assassination. In his written confession submitted to the Officer's Court at Salonika, Colonel *Apis* enlightens them. *I enlisted Rade Malobabic, Serbian Military Intelligence's top undercover operative, to organize for me a new intelligence network in Austria-Hungary ~ in agreement with the Russian Military Attaché Artamonov, who personally met with Rade in my presence. After Rade began work, believing that Austria was preparing for war with us, I thought that with the removal of the heir to the throne, Franz Ferdinand, the military party he headed would lose strength, thus removing or at least delaying the danger of war somewhat. For that purpose, I hired Malobabic to organize Franz Ferdinand's murder upon his arrival to Sarajevo. I decided on this definitely only after Artamonov had assured me that Russia would not leave us undefended if Austria-Hungary attacked us. Malobabic carried out my instructions, organized, and carried through the assassination. Its chief participants were all in my service and received small honoraria which I sent them through Malobabic. I received money for this from Artamonov, since the Serbian General Staff did not have credit to conduct this intensified work.*

The alarming aspect of *Apis's* statement : *in agreement with the Russian Military Attaché Artamonov* reveals the official complicity and financial contribution of Russia in the 1914 Sarajevo assassinations. When Artamonov was elsewhere asked ~ *Was yours the hand behind the hands that launched the war? While I admit to my close co-operation with Apis, I could not be consulted beforehand about the assassination as I was far away from Belgrade on leave in Switzerland and then in Italy, well before the crime was perpetrated, having left Assistant Military Attaché, Captain Alexander Verkhovskii in charge at the Embassy in Belgrade.* Years later, the Principal of the St Petersburg Archaeological Institute, Trydar Burzynski conferred; *The Sarajevo assassination was prepared with the support of the acting Russian Military Attaché in Belgrade, Captain Verkhovskii, a young man I had known for years and who told me sincerely the truth about the origins, preparation and execution of the plot.*

With plenty to hide, it comes as no surprise to learn that the Russian collection of diplomatic papers from 1914 has much missing. During 1917, an unknown person from the Russian Foreign Ministry removed from the archives, all of Nikola Hartwig's dispatches from Belgrade for May-July 1914. Another with a tale to tell did not live long ~ after previous Russian Ambassador Isvolsky began to write his biography in Biarritz, he was found dead… slumped over his desk, pen in hand, before he finished his first volume ~ all of Isvolsky's papers and telegrams from 1914 then vanished.

Apis on the left ~ with Ljubomir Vulovic on the right

Major Voijaslav Tankosic

KAFANA ALBANIJE ~ the Okhrana Coffee house in Belgrade where the assassination plan was at times discussed

Apis and others of the Black Hand at the Salonika Trial

Apis

During the Salonika Trial, the internal Serbian situation in the spring of 1914 becomes clearer. The *Black Hand* of *Apis* undertook a campaign against PM Pasic of the Serbian Radical Party, and from early May of 1914, the priority question that came to dominate internal Serbian politics was; whether the army or civilian government should control the *New Serbia* area won in the Balkan Wars? If the military at this time, still strongly under the influence of officers who belong to the *Black Hand*, were allowed to administer half of all Serbian territory, the army could move to challenge the civilian government for control of the entire state. Pasic, determined to maintain civilian rule in Macedonia, took measures against the *Black Hand*. Opposition groups then grasped the unstable opportunity to launch a simultaneous attack on Pasic's government. Although forced to resign briefly on 2nd June, Nikola Pasic was able to return to power on 11th June. The assembly was next dissolved on 24th June, (four days before the assassinations) with new elections scheduled. King Petar now in bad health, stepped down for his son Crown Prince Alexander to become Prince Regent of the Kingdom of Serbia. In this crucial period of internal turmoil, the three Bosnian students, with Serbian arms and acting with the knowledge and approval of *Apis* then crossed into Bosnia. During the subsequent week, the three travelled to recruit more supporters. While Serbia was still in a state of political crisis, Pasic learned about the actions of the students. Aware of their intentions, he discussed the matter with his ministers, but declined to notify any A-H officials.

Colonel Ljubomir Vulovic, the Commander of the Serbian Frontier Service testified as follows: *In 1914, on my official trip from Loznica to Belgrade, I received a letter at the General Staff signed by Marshal Putnik, Serbia's top military officer, noting that agents of Malobabic would come and a teacher so I could send them into Bosnia. Because of that, I went to Loznica and very soon afterwards sent Rade into Bosnia. Soon thereafter occurred the Sarajevo assassination of Archduke Franz Ferdinand.* On the eve of Malobabic's execution he confessed to a priest. *They ordered me to go to Sarajevo when the assassination was to take place and when everything was over, they ordered me to come back and fulfill other missions, and then there was the outbreak of the war.* (This confirms Putnik's involvement in the *Black Hand* plot). Rade Malobabic arrived in Sarajevo on the eve of the assassination to give final consent for the operation to Danilo Ilic.

Colonel *Apis'* confession (kept secret until 1953) sealed his fate. In wishing to remove *Apis* as a political rival, the Salonika régime was also anxious to show the then likely victors, A-H and Germany that it has dealt summarily with the Serbian schemers and assassins of Sarajevo. *Apis* understands : *As the Chief of the Intelligence Department of the General Staff, I emphasized once to you that perhaps I erred in writing openly that I carried out the Sarajevo assassination. I would say now, on the basis of these hearings that this is the main reason why I will be killed. I believed that the name, Black Hand should be interpreted that every member must be prepared to give his life to the unification of Serbdom.*

After confessing, *Apis* and eight of his associates are sentenced to death and the Serbian Army is authorized to execute Rade Malobabic, Ljubomir Vulovic, and *Apis*. In justifying their executions, Pasic wrote: *Dimitrijevic Apis besides everything else admitted he had ordered Franz Ferdinand to be killed. And now who could reprieve them? As far as Apis is concerned, he admitted to the court that it was he who had organized the Sarajevo assassination which gave Austria the excuse to declare war against us.*

Salonika Officer's Prison 12th June 1917. *Apis* is in suspense about the Prince Regent's decision... Often that evening, *Apis* calls Prison CO Captain Milan Stojkovic to his cell to ask about the verdict, but Stojkovic can tell him nothing. *Apis: No decision will come this night.* Captain Stojkovic: *He had little for supper and went to bed early, so when I came to him with the Priest at 11 PM, he was sleeping very soundly. I awakened him, he roused himself, and then he sat on the bed.* Stojkovic then introduced Serbian Orthodox Priest Father Zdravko Paunovic, after which Stojkovic left. Father Paunovic recorded the event ;

The room is large, the door open, well lighted by a lamp on a table in the middle of the room. Dragutin did not notice our entrance. He was lying on the bed with his face toward the ceiling. As Captain Stojkovic preceded me, he went up to him and said: Colonel, sir! Apis jumped up hastily, offered me his hands, brought me a chair, and urged me to be seated. Apis: Why have you come so late? So they have convicted us, but what our heads and lives will give them, I don't know. Do you know if the Crown has confirmed our sentences and whether they will shoot all of us and when, or have some at least been reprieved? Surely the verdict was approved before they let you come to us tonight. That means we'll be shot tomorrow... Have you been to see my other friends?
Father Paunovic: I have seen Vulovic and Malobabic. Apis: That means we three are finished!

The fewer the better… that is how I calculated. Thus, it is death! That means I should have died in the war, but God did not grant that, wishing me to perish thus instead, probably because the interests of Greater Serbia demand it. All right, if it is to be so, then I will die calmly and easily. Believe me, sir! Our intentions, desires and activities were totally patriotic. We spilled blood and carried out the revolution of May 29th 1903. And when we saw that we had been mistaken, then we thought we had to do everything in secrecy. Since the cause was almost lost because some comrades betrayed us, secrecy and caution became even more essential… But I never believed Pasic and the Radical Party would oppose us. The Radicals even approved the May 1903 coup initially. This toppled the Obrenovic rulers, brought the Karadjordjevic dynasty to the throne, and the Radicals to power.

Father Paunovic: *Excuse me Dragutin, I must now visit the other convicted officers… First, do you wish to confess and to take communion? Apis: I have not received communion for a long time; nor will I now. I have harmed no one.* The Priest delivers his sermon, and persuades *Apis* to confess and to take communion. Around 11.30 PM *Apis* summons Captain Stojkovic who then sends his deputy, Lieutenant Josif Protic, who had earlier fought under *Apis*: *Ah, there you are, dear Protic!* Protic: *The Captain is busy, but he will come with us to the place of execution.* Protic gently informs *Apis. We will soon be leaving for the execution. Apis*: *Well, thus my dear fellow, as you see I have lost the game. It is all over. Look, you can see I am now as quiet as a child, see I'm not at all worried. Please let us smoke. I'm smoking a lot since it is for the last time.* Lighting Protic's cigarette, *Apis* asks: *Now there is something you can do for me. Greet Ljuba Jovanovic… Tell him that he has won the game, but let him watch out since if he has erred, his children will soon atone for it. I blame him alone for being abandoned by my people and friends. I erred Protic, but I never expected them to sign my death warrant… But you'll see. I shall die quietly, peacefully.*

As the condemned ~ *Apis* leader of the broken *Black Hand,* Rade Malobabic once Serbian Army Intelligence's best undercover operative, and Colonel Ljubomir Vulovic, former commander of the Serbian Frontier Service are driven to their execution, *Apis* confirms his rôle in the plot to their driver. *Now it is clear to me and clear to you too, that I am to be killed today by Serbian rifles solely because I organized the Sarajevo outrage. The thing is over. The Military Court has condemned me ~ and my two friends to death.*

When eventually they reach the place of execution, they wait for the final proceedings to begin. With the first signs of dawn appearing in the east, Lieutenant Protic looks at his watch. It is 3.20 AM. From the distance comes a faint rumble through the night. *Apis*: *There, the Salonika Front does not rest. The dead man of Sarajevo has long arms ~ he reaches from the grave, and he catches us all… one after the other. I only envy the men who will see victory with their own eyes. And if I had only experienced one thing, I would have to say that life has been beautiful. Never, not even over there in eternity, will I forget that day in December of 1914 when we entered recaptured Belgrade. The people in the streets cried and gasped for joy as King Petar drove through the streets, and his car rolled over all the banners abandoned by the Austrians and into his palace.*

Apis smiles because he's that way, because he is a giant, a hero, a lion. Apis smiles because he knows they'll shoot him. You don't know such people. When they smile, they are the most dangerous. Colonel C.A. Popovic

Finally at 4.30 AM *Apis* enquires ~ *Is it time to say farewell?* Embracing Father Paunovic, *Apis* takes the priest's hand to kiss it. When the priest resists, *Apis* pleads; *Please permit me to kiss you, as I would father and mother.* Paunovic lets him kiss his hand, and the other condemned do likewise. Holding the whimpering Malobabic by the hand, *Apis* turns: *This is a good Serb and an honourable man. He is not guilty.* Looking at his friend ~ *Only be brave Rade!* ~ Malobabic falls sobbing on *Apis's* neck, and presses his face on his chest. Three accompanying officers lead the condemned to the three graves.
Apis standing before the right-hand grave, smokes his last cigarette ~ *This grave is too small for me.*
The soldiers carry the howling Rade Malobabic, who grimly gripes ~ *I want to see the sun!*
Apis: *Let them do it Rade ~ the law requires it.*
The three are bound to their poles and blindfolded in their graves.
Captain Stojkovic: *How Apis looked then, bound to his pole, the kerchief not covering his eyes was poignant. That facial expression, sad and questioning, with a terrible glassy stare.*
After his eyes were bound, Apis shouted ~ Long live Greater Serbia!
From the neigbouring grave Vulovic ~ Greet our children when you go!
Captain Stojkovic raised his sabre, to command: F-I-R-E ! The first salvo killed Rade Malobabic outright.
Apis came loose from his pole, his head supported on the left of his grave. Blood poured from his frame, but Apis still lived. Vulovic moved his arm… After another volley, the doctor confirmed their deaths.
The graves were filled in, the Priest Paunovic intoned a requiem, and the soldiers marched off.

Nicholas II of Russia with the Imperial Romanov Family (left to right)
Olga, Maria, Nicholas II, Alexandra, Anastasia, Alexei, and Tatiana.

PETROGRAD (formally St Petersburg) February 1917. After days of violent street demonstrations the Provisional Government orders Russian troops to open fire. The next day these troops mutiny. The Tsar abdicates after he hears that Moscow has also fallen to the revolution. Following the February Revolution, Nicholas, now no longer a monarch and addressed with contempt by sentries as *Nicholas Romanov*, is reunited with his family under house arrest and confined to Alexander Palace.

April. Emperor Karl of A-H learns of a German plan to knock Russia out of the war by sending Lenin and others, then in exile in Switzerland, back to Russia to ferment a more severe revolution. Karl blocks the plan, *I refuse to act toward the Russian people in a way that would be unfair and irresponsible,* refusing to allow the train carrying Lenin and his éntourage to cross the secure Austrian frontier. Rebuffed, the German plotters send the train through neutral Sweden instead.

Alexander Kerensky's Provisional Government evacuates the Romanov family to Tobolsk, to protect them from the rising tide of revolution. There they live in the former governor's mansion in considerable comfort. After Lenin's Bolsheviks come to power in October of 1917, the conditions of their imprisonment grows stricter, talk of putting Nicholas on trial more frequent, and the sentries scrawl lewd drawings on the fence to offend his daughters. On 1st March 1918, the family is placed on soldier's rations, having to give up butter and coffee, and having to part with eleven devoted servants. When Lenin's Bolsheviks accept the German-dictated peace of Brest-Litovsk, Russia is out of the war.

May 1918. Revolutionaries decide to move the Romanovs to the dull Russian city of Ekaterinburg, where the family is imprisoned with a few retainers in Ipatiev House ~ *The House of Special Purpose.* At midnight on 17th July, Bolsheviks acting on orders from Moscow, and led by Yakov Yurovsky, order the Romanov's physician to awaken the sleeping family and ask them to put on their clothes. The family is told that they are to be photographed to prove to the people that they are still alive. The family are arranged appropriately and some men under the command of Yurovsky enter the room…

Having gone down to the room, I ordered them to stand along the wall. Obviously, at that moment they did not imagine what awaited them. Alexandra Feodrovna said; There are not even chairs here. Nicholas was carrying Alexei in his arms. Then I ordered a couple of chairs. On one of them, to the right of the entrance, almost in the corner, Alexandra Feodrovna sat down. The daughters stood next to her, to the left of the entrance. Beside them Alexei was seated in the armchair. Behind him the others stood. Nicholas now stood opposite Alexei. At the same time, I ordered the men to go down and to be ready in their places when the command was given. Nicholas had put Alexei on the chair and stood in such a way, that he shielded him. Alexei sat in the left corner, and I said to Nicholas; Your royal and close relatives inside the country and abroad are trying to save you, but the Soviet of Workers Deputies has resolved to have you shot. He asked ~ What? What? ~ and turned to Alexei.

At that moment, I shot him and killed him outright. He did not get time to face us to get an answer. Disorganized firing began. The room was small, but everybody came in and carried out the shooting. But many shot through the doorway. Bullets began to ricochet because the wall was brick. Moreover, the firing intensified when the victim's shouts arose. I managed to stop the firing, but with great difficulty.

A bullet, fired by somebody in the back, hummed near my head and grazed over the finger of somebody. When the firing stopped, it turned out that all of the daughters, Alexandra Feodrovna and Alexei too, were alive. I think they had fallen from fear or maybe intentionally, and so they were alive. Then we proceeded to finish the shooting. Alexei remained sitting petrified. I killed him. They shot the daughters but did not kill them. Then Yermakov resorted to a bayonet, but that did not work either. Finally, they killed them by shooting them in the head. Only in the forest did I discover the reason why it had been so hard to kill the daughters and Alexandra.

After the shooting, it was necessary to carry away the corpses, but it was a comparatively long way. Somebody came up with an idea of stretchers. We did not think about it earlier. Having confirmed they were dead, we put sheets on them and carried them out. It was discovered that traces of blood would be everywhere. I said to get some smooth woolen military cloth immediately and put some of it onto the stretchers and then line the truck with it. When they had removed the first corpse somebody said; Someone has taken some valuables.

Then I understood that evidently there had been valuables in the things that they had brought with them. I stopped the removal immediately, assembled the men, and demanded the valuables be returned. After some denial, two men returned the valuables they had taken. After I threatened the looters with shooting, I removed those two and ordered Comrade Nikulin to escort the bodies, having warned him about valuables. I first collected everything ~ the things they had taken and other things as well ~ and I sent all of it to the commandant's office.

At three o'clock we left. I took several minutes, as I did not know where the corpses were meant to be buried. Comrade Filipe Goloshchyokin was walking back and forth near the house, because he was anxious about how everything would turn out. Yermakov drove us somewhere at the Verkh-Isetsky Works. I heard only shouts; We thought they would come here alive, but it turns out they are dead. Where we stopped several people were stretching out girl's blouses. We discovered again that there were valuables and they were taking them.
I ordered that men be posted to keep anyone from coming near the truck.

Only at daybreak did we come to the famous gully. Several steps from the mine where the burial had been planned, peasants were sitting around the fire, having spent the night at the hayfield. Those damned valuables! Yes, it was too much! I had to disperse the people. We had to cordon the place off at some distance, and we did it. Besides that, I sent an order to the village to keep everybody out, explaining that I ordered the men to turn back anybody to the village and to shoot any stubborn, disobedient persons if that did not work. Having done all of this, I ordered the men to load the corpses and to take off the clothes for burning, to destroy absolutely everything they had, to remove any incriminating evidence if the corpses were somehow discovered. I ordered bonfires.

When we began to undress the bodies, we discovered the daughters had on bodices of diamonds and other precious stones. Those were not only places for valuables, but protective armor at the same time. That is why neither bullets nor bayonets got results. Greed was so great that on Alexandra Feodrovna, there was an enormous piece of round gold wire, turned out as a sheer bracelet. All the valuables were ripped out immediately. Several diamonds were handed over to me the next day by Comrades who had found them there. How did they overlook the other valuables? They had enough time for it. Most likely they simply did not figure it out. The valuables had been collected, the things had been burned, and the naked corpses had thrown into the mine. The pit was filled up and covered. At 6 in the morning, I assembled everybody and stated the importance of the work completed. I warned everybody to forget the things they saw ~ and never speak about them with anybody. In the evening of 19th July 1918, I went to Moscow with my report. Yakov Yurovsky

* * *

THE AUSTRIAN FERMENT

Kaiser Wilhelm : Shove like mad, Karl! Remember Nicky!
We mustn't let our skeleton get out of the cupboard, as Russia did!

Wilton Williams ~ *London Opinion* ~ 2nd February 1918

28th September 1918. The Allied offensive triggers a process of collapse inside Imperial Germany. Ludendorff spends most of the day in an incoherent rage. In the evening he tells Paul von Hindenburg that Germany needs to seek an armistice, as it is no longer possible to win the war on the battlefield. The German spring and summer offensives once designed to win the war before the Americans could arrive in numbers, now face an increasingly large numbers of U.S. troops taking part in the fighting, as well as British Empire and French troops with armoured support demonstrating their war-winning ability to now force their way through the strongest of German defensive lines on the Western Front.

29th September. The crisis for Germany spreads. The Kaiser visits his GHQ at Spa (S.E. of Liége) and is told that victory is no longer possible. On the same day, Bulgaria begins armistice negotiations. There is also an increasing amount of unrest on the home front. Sacrifices are now seen as intolerable.

3rd October. The Kaiser appoints Prince Max von Baden as the new Chancellor of Germany, along with a more representative government, it having been decided that the only way to gain a good peace is to transform Germany into a democracy. Kaiser Wilhelm decides to grant an amnesty to free numerous political prisoners, including Karl Liebknecht who promptly calls for revolution.

Ludendorff advises that the Western Front can collapse at any time, and that a ceasefire has to be negotiated without delay, suggesting that the request for the ceasefire comes from a new government. Prince Max has Hindenburg admit in writing that there is no chance of forcing a peace on the Allies. After the request for a ceasefire goes out, the Kaiser makes this proclamation to the German Army;

For months past the enemy, with enormous exertions and almost without pause in the fighting, has stormed against your lines. In the struggle, often without repose, you have had to persevere and resist a numerically far superior enemy. Therein lies the greatness of the task which has been set for you and which you are fulfilling. Troops of Germany are doing their part and are heroically defending the Fatherland on foreign soil.
The eyes of those at home rest with pride and admiration on the deeds of the Army and Navy.
Hard is the task. I express to you the thanks of myself and the Fatherland.

The collapse of the Macedonian Front has occurred in the midst of the hardest struggle. In accord with our allies, I have resolved once more to offer peace to the enemy, but I will only extend my hand for an honourable peace. We owe that to the heroes who have laid down their lives for the Fatherland, and we make that our duty to our children. Whether arms will be lowered still is a question. Until then we must not slacken. We must, as hitherto, exert all our strength tirelessly to hold our ground against the onslaught of our enemies. The hour is grave, but trusting in your strength and in God's gracious help, we feel ourselves to be strong enough to defend our beloved Fatherland.

WILHELM

VIENNA 14th October. Karl I issues a proclamation to change the structure of the A-H Empire. Poland is granted full independence to join their ethnic kin in Russia and Germany as a Polish state. The remainder of the A-H Empire is transformed into a federal union of four parts: German, Czech, South Slav, and Ukrainian; each to be governed by a federal council. Trieste will have a special status. But mere autonomy for the nationalities is no longer enough. A Czechoslovak provisional government joins the Allies in mid-October, and an independent South Slav state is declared on 29th October 1918. This effectively ends any efforts to keep the A-H Empire together.

BERLIN 24th October. Ludendorff issues and withdraws a proclamation denouncing Prince Max. One copy is leaked, and on 27th October, Ludendorff is ordered to resign. The Kaiser returns to Spa GHQ from where he plans to use the army to restore order in Germany. With defeat clearly imminent the sailors of the German High Seas Fleet are ordered to prepare their warships to put to sea to seek a final battle with the Royal Navy. Not surprisingly, with the war approaching its conclusion, many sailors take unauthorized leave and indicate their refusal to put to sea from Kiel and Wilhelmshaven.

WASHINGTON 28th October. U.S. President Wilson initially refuses to agree to the ceasefire. His notes seem to indicate that Kaiser Wilhelm II remaining as head of state is a particular obstacle. Ebert does not favour exchanging the monarchy for a republic, but like many others, he is worried about the danger of a socialist revolution, which seems more likely with each day passing. After the German constitution is changed to transfer power to the Reichstag, the majority parties of the Reichstag are satisfied with the state of affairs, and proceed with the issues of negotiating an armistice.

WILHELMSHAVEN 30th October. A dramatic confrontation between officers and sailors of the German High Seas Fleet sparks a revolutionary uprising that spreads over parts of Germany during the next week. Against the backdrop of a country falling into anarchy, the SPD led by Friedrich Ebert demands a more powerful voice in the cabinet, an extension of parliamentarism to the state of Prussia and the renunciation of the throne by both Kaiser Wilhelm and his oldest son, Crown Prince Wilhelm. Ebert who had favoured retaining the monarchy under another ruler now tells Prince Max von Baden;
If the Kaiser does not abdicate, social revolution is inevitable. I do not want it ~ I hate it like sin.
Spartacist revolutionaries and activist trade unions now prepare to revolt in Berlin.

SPA ~ Count von Moltke: *At 4 PM on October 30th we arrived. Minister of the Interior Dr. Drews presented himself to the Kaiser for the purpose of requesting abdication. The Kaiser heard him thru and replied; What ~ you as a Prussian officer, who have sworn your oath of loyalty to your king, you come before me with such a proposition! I will describe to you the chaos you invite. I abdicate! Immediately all German dynasties fall! The army is left without a leader! The front dissolves and the soldiers stream back across the Rhine. The traitors throughout the country get together burning, plundering, murdering and the enemy assists in nefarious work! I have no idea of abdicating! The King of Prussia dares not to betray Germany, least of all in this hour of her greatest need. I also have sworn my oath ~ and I shall keep it!*

Elsewhere, the Ottoman Turkish Empire surrenders. Now Germany's only remaining loyal ally, the Austro-Hungarian Empire, is in the process of dissolving. One by one, the nationalities of the diminishing A-H Empire proclaim their independence. Little remains of Karl's Hapsburg Monarchy and Karl's political future becomes uncertain. Austrian Prime Minister Heinrich Lammasch, advises;
In an impossible situation, your best course is to ~ temporarily, give up your right to exercise sovereign power.

3rd October. King Ferdinand of Bulgaria abdicates.
2nd November. Collapse of Austrisn Front in Upper Italy.
3rd November. Austrian cease-fire comes into broad effect.
4th November. The Kaiser's brother Prince Henry of Prussia, Naval CO at Kiel is forced to flee. Hindenburg then contacts the Allied Supreme Commander Foch to commence armistice negotiations.

SPA 5th November. Count Moltke : *In the early morning when we arrived, we learned of the revolt in Kiel and of the demand for the abdication of the Kaiser. Railroad connection with Germany began to be irregular and broken. Soon it ceased altogether. No train crossed the Rhine. Telegraphic and telephonic connection with Berlin became difficult. It appeared to be under some sort of censorial control and confidential messages could no longer be trusted. Finally, we recoursed to our air service. Two airships flew to Berlin, but none returned.*

BERLIN 9th November. The revolution reaches Berlin. Companies are targeted with strikes. Worker's and soldier's councils are created and important buildings occupied. As the striking masses march on the centre of Berlin, the SPD, now anxious of losing its influence on the revolution, announces that it is resigning from the government of Prince Maximilian von Baden who has failed to convince Kaiser Wilhelm II at the German Army GHQ at Spa, in Belgium of the need to abdicate. Wilhelm has resigned himself to the loss of the Imperial crown, but hopes to remain King of Prussia. When Prince Max fails to convince him of the unreality of this belief, he untruthfully announces;
The Kaiser Wilhelm has abdicated and the Crown Prince has agreed to relinquish his right of succession.

Shortly after, the leadership group of the SPD arrives at the chancellery, Ebert asks Prince Max to hand over the government to him. After a short meeting of the cabinet, the false Chancellor resigns and in an unconstitutional move, hands his office to Ebert, who thus becomes Chancellor of Germany.

Ebert's first action as Chancellor is to issue some proclamations asking the people to remain calm, stay off the streets and to restore peace and order. This fails to work. Ebert then has lunch with Scheidemann at the Reichstag, and when asked to do so, refuses to speak to the masses now outside. Scheidemann seizes his chance to forestall whatever the Spartacist leader Liebknecht is telling his followers at the Berliner Stadschloss ~ Scheidemann proclaims Germany a republic. Ebert is furious;
You have no right to proclaim a Republic! Such a decision must to be left to an elected national assembly.
Even if that decision might be the restoration of the monarchy.

THE LEAGUE OF ABDICATIONS

Ex-King of Greece : Hullo, Ferdie! Seen anything of Wilhelm?
Ex-King of Bulgaria : He's somewhere behind. He'll join us a bit later.

After his abdication, Ferdinand returns to live in Germany. Having managed to salvage much of his fortune, he is able to live in some style ~ he sees his exile as one of the hazards of kingship. *Kings in exile are more philosophic under reverses than ordinary individuals; but our philosophy is primarily the result of tradition and breeding, and do not forget that pride is an important item in the making of a monarch. We are disciplined from the day of our birth and taught the avoidance of all outward signs of emotion. The skeleton sits forever with us at the feast. It may mean murder, it may mean abdication, but it serves always to remind us of the unexpected. Therefore, we are prepared and nothing comes in the nature of a catastrophe. The main thing in life is to support any condition of bodily or spiritual exile with dignity.*
If one sups with sorrow, one need not invite the world to see you eat.

On November 9th ...the German Army was not beaten, but its strength had dwindled and the enemy had fresh masses in readiness for a new attack. The conclusion of the armistice was directly impending. At this moment of the highest military tension revolution broke out in Germany; the insurgents seized the Rhine bridges, important arsenals, and traffic centres in the rear of the army, thereby endangering the supply of ammunition and provisions, while the supplies in the hands of the troops were enough only to last for a few days. The troops on the lines of communication and the reserves disbanded themselves, and unfavourable reports arrived concerning the reliability of the field army proper. The complete collapse of Germany was inevitable, and civil war would have been added to fighting with the enemy without ~ who would doubtless have pressed on.

The Kaiser could have betaken himself to the fighting troops, in order to seek death at their head in a last attack; but the armistice so keenly desired by the people, would thereby have been postponed, and the lives of many soldiers uselessly sacrificed. The peaceful return home of the Kaiser was no longer to be thought of and could only have been enforced at the head of loyal troops. Finally, the Kaiser might leave the country. He chose this course in agreement with his advisers, solely in the hope that he could thereby best serve the Fatherland, save Germany further losses, distress, and misery, and restore to her peace and order. Paul von Hindenburg
German Army Chief of Staff

Prince Max von Baden's announcement of Kaiser Wilhelm II's intended abdication :

The Kaiser and King has decided to renounce the throne.

The Imperial Chancellor will remain in office until the issues connected with the abdication of the Kaiser; the renouncing by the Crown Prince of the throne of Imperial Germany and of Prussia, and the setting up of a regency have been settled. For the regency, he intends to appoint Deputy Ebert as Imperial Chancellor, and he proposes that a bill shall be brought in for the immediate promulgation of general suffrage and for a constitutional German National Assembly, which will finally settle the future form of government of the German Nation and those peoples which might be desirous of coming within the Empire.

10th November. To maintain control of events, Ebert cleverly manages to ensure that a majority of the newly elected worker's and soldier's councils come from the ranks of their own SPD supporters. Ebert receives a call from Chief-of-Staff Wilhelm Groener, who offers to cooperate with him. Groener promises Ebert the loyalty of the military in exchange for his demands for; a fight against Bolshevism, an end to soldier's and worker's councils, a national assembly and a return to a state of law and order. The Imperial Government and the German Army then proceeded to place the blame for their defeat on the new democratic government ~ verified when Ludendorff's successor, Wilhelm Groener wrote; *It was just fine with me when the German Army and Army High Command remained as guiltless as possible in these wretched truce negotiations, from which nothing good could be expected.* Groener also described the dramatic scene when the Kaiser became irreversibly aware that the Hohenzollern monarchy was over; *He said nothing, just looked from one to the other, with an expression first of amazement, then piteous appeal, and then ~ just a curious wondering vagueness. He said nothing, and we took him as if he were just a little child and sent him to Holland.* Count Detlef von Moltke, who also recorded this event wrote: *At 10 AM, the Kaiser was advised by his counselors to go to neutral territory to avoid further shedding of blood and civil war. He was now called upon to make his decision under the feeling that he had been abandoned, by his own people, and betrayed by his highest counselor. What should he do? Which way should he turn after both the Front and the Fatherland had been closed to him? After agonizing self-deliberation, he chose Holland as his place of refuge.*

With the German Government actively seeking an armistice and the Spartacist revolutionaries growing in strength daily, the massed calls for the Kaiser to abdicate grow in intensity. But Wilhelm at first instead expresses a preference to lead his armies back into Germany from the Western Front. Following GHQ advice that his troops cannot be relied upon not to harm him, the Kaiser disposes of this option, and explains his final difficult decision in a letter to his son, Crown Prince Wilhelm.

My Dear Boy,

After the Court Chamberlain had informed me that he could no longer guarantee my safety at the GHQ, and that the troops also were no longer trustworthy, I resolved after a severe mental struggle to leave the army, which has collapsed, and go to Holland. I advise you to stick to your post until the conclusion of the armistice.
In Berlin, two Governments under the leadership of Ebert and Liebknecht, are fighting against each other.
I hope to see you again in happier times. Your faithful and deeply affected father ~ *WILHELM*

Before the war, during one of Queen Wilhelmina's many visits to Potsdam, as the Kaiser was reviewing his famous Prussian Guards, when he asked the young Queen ~ *Are they not magnificent?* Wilhelm supposed that he could impress the Queen of a relatively small Kingdom by also telling her; *My guards are seven feet tall and yours are only shoulder-high to them.* Wilhelmina smiled and replied ; *Quite true your Majesty, your guards are seven feet tall. But when we open our dikes, the water is ten feet deep!*

10th November. Kaiser Wilhelm II departs Imperial Germany to go into exile in the Netherlands. Since Wilhelm has not yet signed any abdication papers, Germany legally remains a monarchy until he eventually chooses to sign his formal abdication ~ nevertheless Wilhelm hands over supreme command of the German Army to Paul von Hindenburg and leaves for exile in the Netherlands, and in doing so, effectively leaves Germany without a head of state. In the dramatic turmoil of this day, Friedrich Ebert's acceptance of the harsh terms of the Éntente for a truce, is almost unnoticed.

The night passed quietly but I doubt if a single eye was closed in sleep during its entire course. After 4 AM we assembled in the dining car. The Kaiser came in apparently self-possessed and calm, gave us all a friendly shake of the hand, as usual. During the breakfast, we learned the shameful terms of the armistice. At 5 AM the train started for the Dutch frontier. It had a guard of four soldiers in each car, since it had to pass through places occupied by mutinied troops. Ten minutes after, we halted at the station of La Reide. In the darkness, the Kaiser left the train, and stepped into the automobile provided to take him across the Dutch frontier. The rest of us continued in the train thru Liége and ruined Vise. At 7 AM the train stopped as it reached the Dutch frontier. As a parting greeting, the last German sentinel called out some coarse words. Our car was uncoupled and we waited for the Dutch engine to take us across the frontier. It came at 10 AM, and drew us into neutral Holland.

When we reached the first Dutch station we saw the Kaiser, who had preceded us in an automobile, walking up and down the platform. In great depression, we presented ourselves before him. The Dutch Government had been made acquainted with the decision of the Kaiser by its Consul in Bruxelles. The Kaiser had also telegraphed to Queen Wilhelmina asking her permission to enter her kingdom as a private gentleman. The Kaiser was received at the frontier by the Dutch Major van Dyl who looked out for our protection against the public, as the place was filled with hostile Belgian deserters. In the morning, the German Consul at Maastricht had a number of Dutch officers present themselves before the Kaiser. We learned that the Queen had placed Castle Amerongen, the home of Count Godard Bentnick at the service of the Kaiser. Our departure from the frontier for the Castle was the next day. It was a depressing, shameful journey. At every station thousands of people were gathered, greeting us with shouting, whistling, and cursing. They threatened us, made signs of choking and hanging us. In such manner was our poor Kaiser received on Dutch soil. Count Detlef von Moltke

On the 11th hour of the 11th day of the 11th month ~ German Centre Party deputy Mathias Erzberger signs the Armistice Agreement in Compiégne to end the Great War. Eric Ludendorff disguises himself with a false beard and blue glasses to travel to the home of his brother, astronomer Hans Ludendorff who lives in Potsdam. A week later, he boards a steamer from Copenhagen for temporary exile in Sweden where he will write his memoirs, and then return to Weimar Germany in the spring of 1919.

On 11th November 1918, the same day as the armistice ends the Great War between the Allies and Germany, Emperor Karl of Austria issues a proclamation ~ *I recognize the Austro-Hungarian people's right to determine the form of the state and relinquish every participation in the administration of the State. I also release all officials from their oath of loyalty to me.* The Imperial family then exit Schonbrunn for Castle Eckartsau, east of Vienna. On 13th November, Karl issues a similar proclamation for Hungary. Although it is cited as an abdication, that actual word is never mentioned in either proclamation. Indeed, Karl deliberately avoids using the word *abdication*, hoping that his people will recall him. Privately Karl has no doubt that he sees himself as rightful Emperor. *I did not abdicate and never will. I see my manifesto of 11th November as the equivalent to a cheque, which a street thug has forced me to issue at gunpoint, and I do not feel bound by it in any way whatsoever.*

Instead, an uneasy truce-like situation ensues and persists until late March 1919, when Karl leaves for Switzerland. As his train departs Austria on 24th March, Karl confirms his claim of sovereignty. *Whatever the national assembly of Austria has resolved with respect to these matters since 11th November is null and void for me and my house.* Once in Switzerland, Karl acquires Chateau de Pangrins near Geneva. The politicians of the new Austrian Parliament, anxious from Emperor Karl's departure without an explicit abdication, pass the Hapsburg Law on 3rd April 1919 to permanently prohibit Karl and Zita from returning to Austria. Other members of the House of Hapsburg are also banished from Austria, unless they renounce all intentions of reclaiming the throne and accept the status of ordinary citizens. Another law passed the same day, abolishes all noble titles within Austria.

On 25th January 1919, the Paris peace-conference appoints a commission to investigate the question of responsibility for the war. Among the 15 members of this commission are two Americans: Robert Lansing, who is elected chairman and James Scott. After two months, the commission renders a report[1] in which the exclusive guilt of Germany, A-H, and their allies is recorded. On 4th April 1919, the American delegates emphasize the responsibility of Germany for the war and declare:
Its originators should be held up to the execration of humanity.

They also publish several diplomatic documents from July 1914 ~ one of which, the Wiesner Report, has not been known up to this time. The American delegates publish the following passage of the *first* Wiesner Report, originally sent from Sarajevo in July 1914 : *There are no proofs, not even a surmise that the Serbian Government knew of the organization of the plot or of its preparation or the supply of weapons. There are clues instead, which make this appear as out of the question*. This is intended to appear extremely damaging to Austria, since their Note of 23rd July 1914, had accurately accused Serbia of complicity. The American delegates announce this as the essential part of the report.

But the complete text published by the Austrian Government later that same year, gives a very different impression. For Wiesner had further stated that the Greater Serbian propaganda against the integrity of A-H had been carried on from Belgrade with the cognizance of the Serbian Government : that the plot had been hatched in Belgrade, and that the assassins had received help from Serbian military officials ~ especially from Major Voija Tankosic ; who had supplied them with bombs, pistols and ammunition, and got them secretly over the border. While it is never asserted that the American delegates willfully suppressed these facts, a Slav official of the 1919 Austrian Government had handed the Wiesner Report to the Americans, and he purposely only supplied them with the *essential passage* from which they were misled to subsequently quoted.

While Wiesner was making his investigations at Sarajevo, the English weekly *John Bull* published an article on 11th July 1914 entitled *The Murdered Archduke: Complicity of the Serbian Government* which created a great sensation. The editor, Horatio Bottomley wrote: *About eight months ago Serbia instituted a Secret Service Bureau at their London Legation. When the Legation moved in April 1914, from Belgrave Mansions Hotel to Queen's Gate, it was decided to tear up and burn a large number of documents which it was considered unwise to preserve. But one document of the most incriminating character, and relating to the projected murder of F.F. ~ Franz Ferdinand, fell into the hands of John Bull. It was in cipher on the official letter-paper of the Legation, with its embossed address, and bore the date 5th April 1914.*

Bottomley reproduced a facsimile of it in the private code of the Secret Service. Translated into English the document read: *For the total elimination of F.F. the sum of £2000 sterling paid as follows: £1000 on your arrival in Belgrade by the hands of Mr. G. and the rest, £1000, on finishing the work paid as above. The sum of £200 for expenses and to pay agents, etc., before you leave here.* Horatio Bottomley asserted that ; *The Serbian Secret Service were actively at work, at the Legation, plotting the foul deed.* On that same day, the Serbian Legation declared its intention to begin legal proceedings against *John Bull* ~ an intention it never dared to carry out. While Horatio Bottomley was sentenced for embezzlement to seven years penal servitude (a detail detrimental to his reliability). Edith Durham, one of the foremost authorities on Balkan affairs wrote that she believed Horatio Bottomley to have stated the truth against Serbia.

The A-H Government was well aware of the political agitation carried on by Serbians through a number of secret societies, the most prominent which was the *Narodna Odbrana* (National Defence). It also had knowledge of the *Black Hand,* but thought it merely a branch of the *Narodna Odbrana* and this explains why the Austrians, in their charges against the Serbians, never mentioned the *Black Hand.* Since the assassinations at Sarajevo, a great deal of light has been brought to bear upon this pernicious secret society whose members, as even some Serbians themselves declare, provoked the Great War.

Serbian officer *Commandant* Dobrivoi Lazarevitch stated ; *The Black Hand originated in the gang of 57 officers and officials who, in June, 1903, murdered their King and Queen. The new King Petar, having been put on the throne by these men had no power over them. Even Crown-Prince Alexander had to join the society. Many members of the Government belonged to it, among them all Ministers of War after 1903, and Pasic.*

1 Georges Clemenceau as President of the peace-conference, refused to let the German delegates have Commission reports, as they were ~ *of an internal nature.* But, probably through an indiscretion, the report soon after appeared in the Berlin press. Its authenticity was never denied.

During the Balkan Wars the *Black Hand* showed its strength, and was largely responsible for the atrocities seen by the British eyewitness, General Christopher Birdwood Thomson who gave a vivid picture of affairs in Serbia in 1913: *Pasic only retained office by submission to forces independent of the Government : the foreign policy of Serbia was dictated by the Russian Hartwig. Serbia's internal affairs were at the mercy of factions and secret societies ; of these the most influential was a society known as the Black Hand, which included among its members some of the ablest men in the country, whose patriotism was beyond dispute, but who all had the vices of their virtues in their eyes to give free rein to the unbridled appetites of ignorant, shortsighted men intoxicated by success. An orgy of wanton destruction was permitted, if not encouraged, by the Serbian staff that stained indelibly the heroism and the endurance of Serbian soldiers.*

When the Carnegie Commission investigated the atrocities of the Balkan belligerents in 1913; *Traces of the Black Hand were found everywhere, for instance; The population at Uskub called their station the Black House, from the name of the League itself ~ the Black Hand. The worst crimes were committed by this organization, known to be under powerful protection. It was of distinct advantage to the Serbian Government to have under its hand an irresponsible power like this, which soon became all-powerful and could be disowned. Our records are full of depositions which throw light on the activities of these legalized brigands.*

Thomson wrote from Belgrade in April 1914 : *The Serbian General Staff are talking and planning war. King Petar has retired and his second son Alexander has taken over the regency (Crown-Prince George having been forced to renounce his right to the throne in 1909, because he killed his valet). The country is in a state of excitement. The Austrian heir to the throne Franz Ferdinand, is feared; for his plan to form an autonomous South-Slav State within a federation of the Dual Monarchy, and this means of course, the destruction of all Greater Serbian aspirations.*

With A-H and Italy pledged to an autonomous Albania, and having refused to let Serbia have a port of its own on the Adriatic Sea, the *Black Hand* gained many new adherents from the Serbian Army at that time. Made bitter by helplessness and disappointment, these men believed that the *Black Hand* society alone stood up for Serbia's rights, and so they joined the ranks of the rogue organization. Major Voija Tankosic, mentioned in the Wiesner Report as one of the more conspicuous members of the *Black Hand* took under his personal protection Bosnian students who came to Belgrade to study. Many of these students were radicalized with nihilistic/nationalistic ideas and easy to recruit. *Commandant* Dobrivoi Lazarevitch : *When the papers published the news that the Archduke Franz Ferdinand of Austria was coming to Sarajevo, the members of the Black Hand considered this a superb piece of good luck; for they had been preparing to kill Franz Ferdinand for a long time.*

When the news of the murder reached the Serbian capital, the Black Hand were jubilant and the whole city triumphed. In Cetinje, the Archduke's murder was accepted unhesitatingly as Serb work. The first thing I did in London was to send back to King Petar the Order of St. Sava he had bestowed upon me, with a letter telling him I had heard the attack upon Austria freely discussed the previous year, and that I considered him and his people guilty of the greatest crime in history. Edith Durham

There is further evidence to prove that the Serbian régime was well aware of what was going on; as the Serbian Diplomat, Bogitchevitch wrote; *There is good reason to believe that Prime Minister Pasic would have liked to prevent the deed. It was too early from his viewpoint, because Serbia had not recuperated sufficiently from the late wars; and Russia too, needed more time to prepare. But Pasic found himself in a painful dilemma ~ if he exposed the plot, he ran the danger of exposing his own connection with these criminal societies. Therefore, he contented himself with giving the Serbian Ambassador in Vienna orders to privately warn the A-H Government, but this was never carried out. After the murder, Pasic was still in the same dilemma, which explains why the Serbian authorities did not move a finger to call the Serbian instigators to account, although two days after the assassination they received from Police HQ of the prefecture of Belgrade the following astounding document which was addressed to the Minister of the Interior ~ I have the honour to inform you, that yesterday evening three individuals came to Svetolic Savic, owner of the paper Balkan, and asked him on behalf of Major Tankosic not at any price to publish anything in his paper regarding any connections and relations of the assassin Cabrinovic with the Belgrade personalities, their associates. Moreover, he should not write anything in his paper likely to compromise any Serbian or else something serious might happen to him. In letting you know this, I report to you, that I have taken measures to procure the names of these individuals.*

On the day this Belgrade police document was written, von Storck of the Austrian Legation in Belgrade telegraphed to Vienna : *I have addressed to M. Gruitch, secretary to the Minister of Foreign Affairs, the question appropriate to the moment, to inquire what measures the police have already taken, or intend to take, to follow up the traces of the crime which are notoriously spread through Serbia. He replies that up till now the police have not occupied themselves with the affair.* Edith Durham's reaction; *The consummate impudence of this remark needs no comment. Up to the time of the Austrian ultimatum of 23rd July 1914, a period of more than three weeks, the Serbian Government took no steps whatever to clear up the conspiracy.* Dr. Bogitchevitch condemned this frivolous attitude and explained, *Serbia had received the assurance of Russian help. Without Russia's backing, Serbia would never dared to provoke Austria-Hungary, and that the accusations contained in the Austrian Ultimatum were based on facts ~ but Russia made Serbia's cause her own.* Former Italian Premier Francesco Nitti similarly wrote ; *The Russian policy in Serbia could be called truly criminal. Russia was Europe's biggest danger ~ its greatest menace.*

Journalist Leopold Mandl, who enjoyed exceptional connections with Serbian politicians asserts; *Pasic did not think A-H would make the murder of the Archduke the cause of a punitive expedition into Serbia. He and other Ministers were ready to accept the Austrian Ultimatum. In this case, the King was to abdicate and Pasic to resign. At the same time the Black Band was to be utterly done away with. This society on the other hand, planned a rebellion against the Government and the murder of Pasic in case the ultimatum were accepted. But telegrams from Russia on 25th July completely changed the situation: Pasic, that old atheist crossed himself and with tears in his eyes embraced the Russian Charge d'Affaires, exclaiming ~ Long live the Tsar!*

In September 1914, the French Ambassador in St Petersburg Maurice Paléologue noted an interesting conversation with Count Witte. Witte: *The war is a stupid adventure, which should be brought to an end as soon as possible, since it might benefit only France and England.* To this Paléologue retorted : *Allow me to recall to you, that if the world is today given over to blood and flames, it is for a cause in which Russia was pre-eminently interested, a cause eminently Slav, which concerned neither France nor Britain.* To this Count Witte answered : *Serbia should have been allowed to receive the chastisement it deserved.*

The Serbian consul in the Russian Black Sea port of Odessa Marko Zemovic wrote in a booklet published in 1915 (with the complete consent of Russian censorship) : *The helpless little Slav state Serbia dared to shake the heavily-armed European peace to its very foundations. The South Slavs forced little Serbia into the war with Austria-Hungary by provoking the deed in Sarajevo. The calculations of the South Slavs, at the time considered naive realized themselves only too well. The shots at Sarajevo set the whole world ablaze.*

After the 1918 armistice, Serbian officials hurried to Vienna and insisted on being given all the documents connected with the trial of the assassins. Edith Durham's comment upon this action was ;

They suppressed at the earliest possible moment evidence of Serbian guilt. Had they found any flaw in that evidence, any sign of a miscarriage of justice, we may be certain they would have trumpeted it to the world. In the summer of 1919 the Serbians exhumed the bodies of the murderers. An altar was erected at the spot where the Archduke and his wife fell, and a funeral mass was celebrated ~ for the souls of the assassins. At this ceremony the Serbian Government was officially represented, and the city streets were lined with soldiers. The murderers are now regarded as national heroes throughout Serbia.

Edith Durham
Foreign Affairs
October 1919

In spite of all this evidence, the commission came to a truly remarkable conclusion. *A crime, committed by an Austro-Hungarian subject, on Austro-Hungarian territory, can in nowise compromise Serbia.*

The commission sat for two months. A great deal of the material here quoted had already been published. Besides, General C.B. Thomson, who had been attached to the Serbian Staff during the Balkan Wars and knew the country well, was at that very time in Versailles as a member of the Supreme War Council of the Allies. Edith Durham might also have very easily been consulted : (that she was a recognized authority on Balkan affairs is evidenced by the fact that Sir Edward Grey had asked her in the autumn of 1915 to undertake a political mission to Northern Albania. But any testimony of Gen. Thomson and Edith Durham would have been in contradiction to the conclusions of the Paris Commission, as Edith Durham indeed later wrote her views on the origin of the war :

On Sunday, June 28th 1914, I was having tea with Mr. and Mrs. Lamb, when we saw Admiral Troubridge climbing the hill towards us. He came into the house very hot, and said almost at once: I have come to tell you our wireless has picked up a bit of a message. The Archduke Franz Ferdinand has been murdered at Sarajevo. Just that! My first idea was ~ They have done what they said they would last year. They have begun in Bosnia.

In the autumn of 1915, I received a telegram from Sir Edward Grey suggesting that I and a few others who knew the land should go to Albania and recruit the tribesmen to our side. The frontier could thus have been held, and the Serbian debacle prevented in all probability. But to do this it was necessary to guarantee to the Albanians the independence of their land. It seems Russia refused consent, and the plan was dropped.

I was in Egypt when a Reuter telegram announced that the Austrians had taken the Lovtchen, occupied Cetinje, and appointed the Bulgarian Vuletitch as Mayor. I guessed at once this was my old friend Vulco of the Grand Hotel. His son-in-law, Rizoff, who had to leave Rome, where he was working on pro-German propaganda, was now the Bulgarian Ambassador in Berlin. There was something truly Balkanic in the surrender of Cetinje, arranged by the Grand Hotel and his son-in law, which appealed to my sense of humour. I soon learnt my guess was true. The Fates willed that I should meet a Montenegrin official. Last time we met during the Balkan war I had vituperated him about the cutting off of noses. Now in a strange land, we were old friends.

Tell me, said I, what happened? The Austrians cannot really have taken the Lovtchen. One does not march troops up two thousand feet of rocks under guns, when one can walk in by the back door. Cheerfully he replied: You have been up the Lovtchen yourself. Frankly, we welcomed the Austrians, even with enthusiasm. A small detachment on the road had not been warned, and fired. Otherwise nothing occurred. Yes, Vulco is Mayor! All your old friends still remain, Yanko Vukotitch, and all! You know our Montenegrins. They are funny dogs. For me this war is the struggle of the Slav and the Teuton for the only unexploited lands in Europe. We always knew it would come. But in the past, we had never reckoned that England would range herself with Russia and permit her to take Constantinople. That would mean the end of Romania, Bulgaria, Serbia, of us, and of you too, if you are not careful. Therefore we ranged ourselves with Austria. Those who have travelled in Austria know that the Slavs there are richer, better educated, and better off than we poor devils of Serbia and Montenegro. In return for the taxes they pay they get roads, schools ~ what you will. Our taxes all run out of the breech pockets of those Two Families of Petrovic and Karageorgevic. I can tell you those Two Families will never return. Who is responsible for killing the Archduke? Certainly ~ those mad students of Belgrade. You remember how they tried to kill King Nikola? Well! The Serbs wanted war. Now they have got it, let us hope they are content. Politics are all cochonnerie. As for me, I have had enough and I wash my hands of them.

The Montenegrin opinion of the Sarajevo murders is corroborated by several facts. The Serb students refuged in London had postcards printed of the murderer Princip, on which he was described as a national hero! One said to me: Yes, it is a pity so many people were killed. But you see the plan succeeded and Greater Serbia has been made. He seemed to think it the object of the war. Another told a friend of mine that bomb throwing had been taught at Shabatz, and a Serb officer boasted to Lady Nina Boyle, when she was doing Serbian relief work, that he was one of the men who taught the murderers to shoot. He took their photographs from his pocket, and called on her to admire how well he had taught them.

The bombs used, like those prepared for King Nikola, came from Krajujevac. The assassins told in great detail at their trial that they had been supplied with weapons, and taught to use them, by a Serbian railway employee Ciganovic, and by Major Tankosic a known member of the Narodna Odbrana and of the Black Hand. And he was in constant touch with the Belgrade students at the Zelenom Vjencu eating-house. A Serb student, who himself had frequented this place, told me that Princip was chosen because he was so far advanced in tuberculosis he could not live long in any case. He saw him just before he left for Sarajevo, looking very ill indeed. He described that when the news of the murders arrived, three hundred Bosnian students rushed through Belgrade shouting and singing and led by a Montenegrin playing the gusle. Did not the police try to stop them? No, why should they? And no arrests were made? Oh, no.

The attitude of the Serb Government was precisely that it knew nothing and therefore it made no inquiry. Nor, during the whole three weeks that elapsed before the A-H ultimatum, did the Serb Government do anything to clear up the matter and mitigate Vienna's just anger. One can deduce that war was expected and intended. Time, the revealer of all secrets, will one day unveil the whole of this one.
Meanwhile, I am glad that the Order of St. Sava is not in my house

ARMISTICE DAY CELEBRATIONS LONDON

King George V and Queen Mary passing thru London's Armistice Day celebrations in Trafalgar Square

Colonel Edward House

Sir Edward Grey

The Great War ended in 1919, with the signing of the Treaty of Versailles in the Hall of Mirrors : the same symbolically significant place where Otto von Bismarck celebrated the French defeat of 1871 by proclaiming the unification of German states that signaled the emergence of Imperial Germany. When the Treaty of Versailles was finalized and handed to German representatives on 7th May 1919, Berlin declared the Treaty unfair and attempted several changes. The allies responded by threatening the continuation of hostilities unless the German representatives signed within 24 hours, in default of which Allied troops would march. Berlin being fully aware that their now fractured forces could not form an effective defense, signed the Treaty on 28th June 1919.

With immediate condemnation of the Versailles Treaty from Berlin, the German people found their acknowledgement of defeat, as much as the treaty terms themselves, extremely hard to accept. The terms of the treaty that caused the most resentment within Germany were the loss of territory, Article 231 ~ *The War Guilt Clause* which was directed solely at Germany, and demands of reparations. The ratified reality of the fragmentation of Imperial Germany saw the loss of 13.5% of their former territory under the terms of the Treaty, within which seven million German citizens were put under the jurisdiction of foreign nations. The main aspects of this were: France to acquire Alsace-Lorraine and the coalmines in Saar for 15 years. The modern state of Poland to be re-created from West Prussia, Posen, and Upper Silesia ~ with the formation of the Polish Danzig Corridor to access the North Sea. Belgium received Eupen, Malmèdy, Moresńet, and St. Vith. Denmark acquired Northern Schleswig. Czechoslovakia received the border area near Troppau (present day Opava). For 15 years, the Allies took control of the industrial Rhineland area. Germany also lost all her colonies, and merchant vessels.

I am leaving Paris, after eight fateful months, with conflicting emotions. Looking at the conference in retrospect, there is much to approve and yet much to regret. It is easy to say what should have been done, but more difficult to have found a way of doing it. To those who are saying that the treaty is bad and should never have been made and that it will involve Europe in infinite difficulties in its enforcement, I feel like admitting it. But I would also say in reply that empires cannot be shattered, and new states raised upon their ruins without disturbance. To create new boundaries is to create new troubles. The one follows the other. While I should have preferred a different peace, I doubt very much whether it could have been made, for the ingredients required for such a peace as I would have, were lacking at Paris. Edward House

The consequences of Versailles Treaty Article 231 would be far reaching. The *War Guilt Clause* triggered lasting resentment within German society that resulted in prolonged humiliation and rage. The German people did not understand why blame was so falsely and unjustly placed upon them and their nation, and this directly served to create the core conditions for the rise of an enraged German Nazi faction whose fanaticism led directly to World War Two and the further deaths of 20 million. Therefore, it is essential to take an overview of the *Question of War Guilt* ~ with all applicable evidence.

GERMANY after the PEACE TREATY of 1919

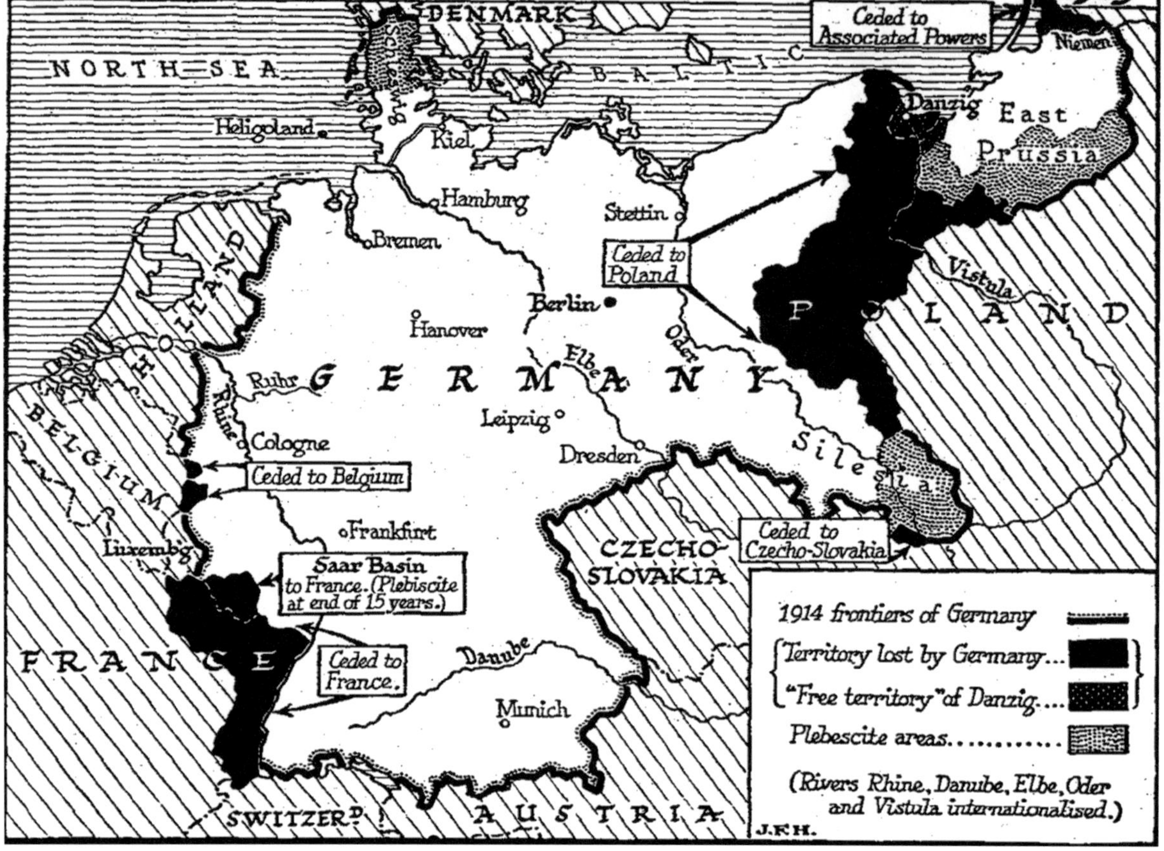

The BREAK-UP of AUSTRIA-HUNGARY

ITALY

as she would be under the terms of the Treaty of London.

Redeemed Lands

The black sections indicate the territory which is to be restored to Italy.

General Viktor Artamonov
Russian Military Attaché to Belgrade

Serge Sazonoff
Russian Foreign Minister 1910-16

R-L : General Artamonov (Russian uniform) Baron Hartwig & Serbian Crown Prince ~ with 16th Slava Regiment in Nish.

THE QUESTION OF WAR GUILT

Once Great Britain so deeply involved herself in the alliance of the Triple Éntente that military support for France and Russia became a point of honour, the policy of Vienna and Berlin passed from a period of passivity to practical activity. One reason why war came, therefore appears to be that at some point in 1913, Berlin succumbed to the momentum of the militarists, whom up to that time the diplomats had held in restraint. From their initial intent of peace, the defensive alliances of Europe then transformed into group designers for war and began to array their intrigues against each other.

While there is still little agreement on the precise causes of the Great War, this enduring division of opinion perfectly reflects the situation in the summer of 1914 when Europe's leaders evaluated the assassination of Franz Ferdinand and its consequences from significantly different perspectives. As suspicious misunderstanding of the interests of the rivals contributed so much to wider operations, it remains essential to review the underlying issues to appreciate the role of Europe's diplomats and strategists, whose strong opinions were so influenced by their own Imperial passions and prejudices.

Russia's urgency for the domination of the approaches to the Mediterranean, and its precipitancy for the hegemony of the Slav world permeated all Russian foreign policy with a strength that varied but never vanished. The determination of prominent Russian militarists and diplomats to acquire possession of the Straits at the premeditated cost of a European war has been documentarily proven.

Some Russians even appraised the presence of Franz Ferdinand before the assassination. When Tsar Nicholas II and his family visited Romania to hold urgent talks with King Carol I on bilateral relations and the situation in the Balkans ~ the Russian official Evgeni de Schelking noted :

On 1st June 1914, Tsar Nicholas had an interview with King Carol I of Romania at Constanza. Sazonoff was convinced that if Archduke Franz Ferdinand were out of the way, the peace of Europe would not be endangered. Four weeks later, *Franz Ferdinand was out of the way.*

Economic rivalry, national ambitions and antagonisms, and newspaper incitements all played a role in bringing on the Great War, but it is doubtful if all these together could have led to war if it had not been for the assassination of Franz Ferdinand. This was the flame, which caused the rapid and complex succession of events, which culminated in a world war ~ and for this Serbian nationalism was ultimately responsible. S.B. Fay

The case of Austria-Hungary against Serbia and Russia in 1914 is far stronger than most suppose. Pasic and the Belgrade Cabinet did nothing to inform A-H of the known *Black Hand* assassination plot.

The 1914 Wiesner Report shows that the Austrians were certain that the assassins had come from Belgrade into Bosnia with the *Black Hand* connivance of Serbian Frontier Service border guards; that the A-H *Evidenzbureau* had detected and decoded telegrams that revealed further development of Russo-Serbian intrigues in the Balkans to confirm their awareness that Sazonoff in St Petersburg, and Hartwig in Belgrade, were pursuing an authorized policy to fracture the A-H Empire.

Our conviction at that time ~ a conviction in no way weakened by the subsequent course of events ~ was that the continuance of the Austro-Hungarian Monarchy was unthinkable without a definitive elimination of the Serbian menace. The danger to the life of our nation was bound to the fact that Russian diplomacy immediately assumed its moral direction and thus assured it a powerful backing. This sprang less from altruistic motives for the development of the Balkan States ~ than from what was their true aim, to win over the Balkans and the eventual control of Constantinople and the Straits. On 23rd April 1913, Sazonoff wrote to Hartwig in Belgrade : Serbia's promised land lies in the present-day domain of Austria-Hungary. Energetic work should be devoted to preparing for future conflict. Make this clear to the Serbs. Count Leopold von Berchtold

Wilhelm II made great efforts to turn the Archduke to a friendly attitude to us. His death was a tragedy and its results disastrous; for they have brought about for Austria after four years of war, the fate that Russia, the real author of the tragedy intended ~ true, Russia ruined herself in the process. General Erich Ludendorff

Aya Sofia, Constantinople ~ Gaspare Fossati

A breach in the peace of Europe is what Russia had in view because they wanted Constantinople, and therefore war is what was wrought. To the militarists of Russia, Sazonoff agreed at the conference of February 1914 that Russia must seize the Straits and could only do so through a European War, therefore with Russia's motive in supporting the Balkan Slavs and in seeking to appropriate the Straits for their own imperialist expansion ~ and not for their own self-preservation ~ fresh promises of Russian support spurred on the Serbian extremists who fought to fracture A-H, and whilst these Serbian extremists sought national freedom for themselves, they also intended to acquire Albania and several other neighbouring lands to create a Greater Serbia. Thus they share substantial war-guilt with the militarist clique in Russia headed by Grand Duke Nicholas, who sought to take advantage of a wider war to seize the Straits, and to capture Constantinople as their new capital *Tsargrad*. Therefore, all efforts for peace were bound to fall against the strong will to war of the Russian militarists.

No one quite realizes yet that Austria's quarrel with Serbia might have remained a local disturbance on the Balkan Peninsula, if Russian statesmen had not seized this opportunity for beginning a Great War. No country in the world had more reason to avoid a European war than Austria-Hungary. But we could not remain neutral onlookers when a small state like Serbia began working with bombs, assassins, and treacherous propaganda with the intention of separating our southern states from the Hapsburg Monarchy. Sazonoff in his memoirs wrote : Russia kept the peace in 1908 in order to save Europe from a war, and because she had not recovered financially or military from her defeats of the Russo-Japanese War. In 1914, the situation had changed. Russia had restored her finances and army, and there was no necessity for saving the peace of Europe. Count Alexander Hoyos

With the Russian élite, now lacking a motivating self-confident belief in both their own society and government (at this time particularly pronounced within British and German society) they came to realize that public opinion in Europe would not sanction her posturing as the protectress of Slavs in her tussle with Vienna over ongoing Balkan issues. From this perspective, Vienna dragged Berlin into the July Crisis and drifted toward a wider war which Vienna did not want ~ but was willing to risk ~ in her determination to put an end to the political threat from the conspirators for a Greater Serbia, who progressively menaced the very existence of the Hapsburg Monarchy. Even so, the Austrians had many plausible and some weighty arguments. Germany considered herself the friend and protector of the Ottoman Empire, and having committed herself to vast enterprises in Turkish territories ~ these efforts would be ruined if Belgrade's dominance prevailed in the Balkans and Russian influence prevailed in Constantinople. Unless something was done, the Triple Alliance would cease to exist.

In the decision taken in Vienna and backed by the militarists of Berlin for action in Serbia, the Austrians were much misinformed of the true strength of the Russian military ~ and this gave cause for Austrian Count Adalbert von Sternberg to blame the traitor Colonel Alfred Redl for the Great War. *This scoundrel denounced every single Austrian spy. Redl betrayed our secrets to the Russians and hindered our learning the Russian secrets through spies. Thus the existence of 75 Russian divisions ~ more than the entire Austrian Army remained unknown to Germany and us in 1914. Hence our desire for war, and our early defeat. Had we a clear view of things, our generals would not have pushed the court dignitaries to war.* Similarly, the head of the German Secret Service *Arbeit III,* Colonel Walter Nicolai concurred that the chiefs-of-staff of the Central Powers would have advised in July 1914 more strongly against war had they known what Redl knew (but had thoroughly suppressed) regarding the actual strength of the Russian Army.

Germany endeavoured to localize the conflict between A-H and Serbia. All other European Governments (particularly Great Britain) save one took the same attitude. Russia alone asserted that she had to be heard in a settlement of the matter. Thus the danger of a European crisis raised its threatening head. Bethmann Hollweg

Serbia wantonly played with fire. The undermining work which had been carried on for years from Belgrade in the south-eastern portions of the Habsburg Monarchy, with the unmistakable object of revolutionizing and detaching them, relying on the goodwill if not express approval and encouragement from the great Slav Power, did not shrink from the most abominable methods, had become so great a danger that, after repeated warnings, which had no effect, Austria-Hungary found it necessary to counteract these subversive activities with all her available energy. The Sarajevo murder was startling evidence of the urgent necessity for crushing the rebellion at its centre. Hartwig, the soul of the Russo-Serbian machinations against A-H did not survive his fatal work, but it survived him, to the misfortune of his country and of the world. And after the first outbreak of the fire she herself had laid, Russia defended political immorality, and unsheathed the sword. Baron Wilhelm von Schoen

Tsar Nicholas with his children and Cossack officers

I am fully convinced that great times are coming for your reign and Russia. We must give a strong country to our son and dare not be weak for his sake. Don't let things slip through your fingers and leave it to him to build all over again. Tsarina Alexandra

Lighthouse of Fenerbahce on north coast of Marmara

It is not entirely accurate that the Kaiser gave Vienna *carte blanche* to do as they pleased in Serbia. When the July Crisis was at its most acute, Berlin succeeded in bringing Vienna to declare expressly that it laid no claim to any Serbian soil, that it would not impair the Sovereignty of Serbia, and only proposed a temporary military occupation of Serbian territory. Germany restored direct conversations between Vienna and St Petersburg, and then advocated in Vienna the acceptance of the mediation desired by Grey. After such preventative pressures from Berlin failed, Bethmann wired to Berchtold: *While we are quite ready to fulfill our treaty obligations, we must refuse to let ourselves be drawn into a world conflagration by Austria-Hungary, owing to the later ignoring our advice.* Although it is evident that Berlin had initially given their approval of support to Vienna, the Kaiser believed that a wider war could be avoided, for he supposed it would suffice for A-H to carry out a prompt artillery bombardment of Belgrade and then to quickly occupy it ~ but the Austrians took *six weeks to act!*

Germany profoundly disapproved of Vienna's delay in taking firm decisive steps against Serbia. Berlin felt that the quicker the A-H Army struck, the less the danger of a European war as a result of Franco-Russian intervention. Furthermore, every day of delay would provide more time for Europe to recover from the first shocks of the assassination and this would diminish the universal sympathy Austria could exploit in her action against Serbia. Hence, Berlin implored Vienna to move swiftly against Belgrade to reduce both the possibility and probability of a Franco-Russian intervention that could trigger wider European conflict. But Russia had so continually stirred Serbian policy and encouraged Serbia into action against Austria through their support of *Black Hand* extremists that Russo-Balkan nationalism would detonate Europe into war. Because of the persuasive pressure Berlin faced from Vienna in responding to Russo-Serbian provocateurs, it can be evidenced that Germany attacked from the precarious perspective of a preventative war. Furthermore, if Berlin had not supported Vienna, the ever-increasing Russo-Serbian influence would continue to fracture A-H and ruin German commerce in the East by blocking access to their heavily invested *Berlin-Baghdad Railway.*

The conclusion of many Germans appears to have been that while a war should not be forced; no favourable opportunity which provided a prospect of defeating Russia (whilst there was yet time) should be neglected. The chance of a swift win from a Balkan issue that would bring Russia in was not in the eyes of many Prussian militarists in 1914 a reason for retiring, but a reason for going forward. The Kaiser who had long resisted such arguments, had great faith in his power of influencing the monarchs of other lands and guiding them on the path of peace, however from the beginning of 1914, rapid change set in. The reason for this change, beyond doubt was alarm about Russian ambitions. The dual Austro-German belief was that in a few years' time, if nothing happened in the interval, Imperial Russia with her new strategic railways and swifter mobilization could overwhelm them. Ultimately, the grinding logic of the Alliance system presented all with the choice of standing by their partners ~ or sharing their downfall ~ and whatever sacrifices of traditional interests or policies might be required for this purpose, it had to be made lest a far worse thing befell ~ the downfall of Empires.

Whereas the 1905 dated *Schlieffen Plan* had previously assumed that Russia would take at least six weeks longer than Germany and France to fully mobilize their forces, due to Russia's then smaller railroad infrastructure and military capacity following the depletory Russo-Japanese War in 1905, during the years leading up to 1914, the gap between Russia and the West had rapidly closed. Massive French bank loans helped advance Russian railroads closer and closer to the German border, to construct new factories, to expand their Baltic Fleet, and to re-equip an augmented Russian Army. Therefore, German strategic thinkers during 1914 were wary of the military progress of Russia in both the numbers of its standing army, and the alarming progress of their railway construction across Poland toward the German frontier. With these French-financed railways soon able to move massive numbers of Russian troops to within 160km of Berlin, German planners who knew the Russian Army was rapidly rearming and expanding, estimated that by 1917, Imperial Russia would be sufficiently strong that Germany and A-H combined could no longer hope to defeat it. *Now or never* was the call of the German warlords in their air of fatalism, and their powerful influence reflected that assumption.

Whereas formerly the German Government had aggressive intentions; they are now genuinely alarmed at the military preparations in Russia, the prospective increase in her military forces and particularly at the intended construction, at the insistence of the French Government and with French money, of strategic railways to converge on the German frontier. Germany is not afraid, because she believes her army to be invulnerable, but she is afraid that in a few years hence she might be afraid ~ Germany is afraid of the future. Sir Edward Grey

With German plans reflecting their assumption that by 1917 Russian rail growth and rearmament would reach the point where Russia could mobilize and attack Germany before *Plan 1914* could achieve victory in the west, waiting would make a pre-emptive German strike against Russia suicidal. From this scenario presented to them by the warlords of Berlin; Bethmann and Kaiser Wilhelm finally became men who accepted the burden of what they were about to unleash with utmost reluctance, feeling that they were trapped in a lesser-of-two-evils choice, and yet believed that it now needed to be done, even if it led to a convulsion of civilization. Thereupon, the warlords of Berlin made their leap into the dark, with their decision that risked undermining the dominance of their Prussian ruling class via defeat, to avoid the worse fate of their perception of a future Germany facing a modernized Russia of overwhelming size and expansionist intent. Then to ensure the collapse of Tsarist Russia, German and Turkish forces simply closed the Straits to disable Russia ~ economically and militarily.

While the Chief of the German General Staff, General von Moltke, was ready to risk war based on his assumption that an Austro-Serbian war could be localized (thus giving Britain no interest for war), German Army strategists further fatally supposed that Russia would stay out as she was not ready, but if Russia did enter the war, the German Army could defeat her after defeating France in the west. However, as *Plan 1914* required a German passage through Belgium into France, this made it quite improbable for Great Britain to stand aside; and with the British Empire intervening, a broadening European war rapidly became a wider global war. Thus, the capable diplomats who had previously and successfully managed to localize two recent Balkan Wars could not localize the quarrel of 1914. During the July Crisis, the Éntente were willing to submit the Austro-Serbian quarrel to international mediation (as was Serbia herself) but bolstered by the hawks of Berlin, Vienna resolutely refused. Since it was unlikely that A-H would have attacked Serbia without the backing of Berlin, the militarists of Germany and A-H therefore bear substantial responsibility for the Great War, as they do for sending their triple ultimata to Russia, Belgium, and to France in particular, with the furtive and fabricated pretext that French aircraft had bombed the German cities of Karlsruhe and Nuremberg.

The Tsar had proposed to the Kaiser that the Austro-Serb dispute should be settled by the Hague Tribunal. The refusal of a Conference by Germany, though it did not decide British participation, did in fact decide the question of peace or war for Europe, and sign the death warrant of those killed in this war. Sir Edward Grey

Whenever Helmuth von Moltke as head of the German Army and his A-H counterpart in Vienna Conrad had pressed for a preventive war, they had been kept in check by their respective monarchs. Helmuth von Moltke often mocked and derided the Kaiser as *Wilhelm the Timid ~ the brave coward,* during the earlier war-scares from which Kaiser Wilhelm declined to consider armed action from 1908-12, when he judiciously avoided force to pursue more peaceful approaches. Similarly in 1914, when informed of his friend Franz Ferdinand's assassination, Wilhelm's unswerving response was to cancel his nautical vacation and return to Berlin, *to take things in hand and preserve the peace in Europe.* Yet previous derision stirred the Kaiser's resolve not to be seen as weak ~ *This time, I shall not give in.* One of the principal causes of the Great War being the German High Command's mistaken belief that neither Russia nor Britain (and still less France) were willing to intervene, is the fundamental reason why Kaiser Wilhelm II (at least initially) had advised Vienna to assume an uncompromising attitude.

Austria, in fact, had finally yielded, and that she herself had at this point good hopes of a peaceful issue is shown by the communication made to you on 1st August by Count Mensdorff, to the effect that Austria had neither banged the door on compromise nor cut off conversations. Unfortunately, conversations at St Petersburg and Vienna and were cut short by the transfer of the dispute to the more dangerous ground of a direct conflict between Germany and Russia. Sir Maurice de Bunsen to Sir Edward Grey

On 29th July 1914, once Russia ordered a partial mobilization of their armed forces, Nicholas II assured Wilhelm II that their partial mobilization did not actually mean war : but Russia's Pan-Slavic Foreign Minister Sazonoff and General Nikolai Yanuskevich were committed to a full mobilization, (having finally cajoled Nicholas to agree to it on 30th July). Wilhelm appealed to Nicholas to halt Russia's mobilization or it would mean war. These appeals to Nicholas were futile on account of the adamant insistence upon mobilization by Yanuskevich; who after the war testified to acting as he did from Pan-Slavic principles ~ *knowing the Tsar being of mixed Danish and German origin would not share. I resolved to smash my telephone and generally adopt measures which will prevent anyone from finding me for the purpose of giving any contrary orders which would again stop our mobilization.*

With Russia mobilizing, Germany asserted they had no other option, but to do the same due to their offensive-defensive *Plan 1914* ~ the invasion of Belgium and France. *War by timetable* between the European Empires commenced the moment when Nicholas II signed off on Russia's full mobilization. Had Tsar Nicholas continued to stand his ground, peace in Europe might well have been saved. From St Petersburg's central telegraph office, commencing at 6 PM, General Sergei Dobrovoloski began directing the Tsar's General Mobilization orders to be sent throughout Imperial Russia during the night and the red call-up placards were posted up in the towns and villages of the Tsar's Empire. When Friday 31st July dawned, there was no way back for Tsar Nicholas, nor for Europe.

The Tsar misjudged his next act, when he expressed the true Russian state-of-preparedness in a personal telegram to the Kaiser that exposed the truth about Russia's mobilization ~ with underlining. *The military measures which have now come into force were decided five days ago for defence reasons because of Austria's preparations. I hope from all my heart that these measures won't in any way interfere with your part as mediator which I value much. We need your strong pressure on Austria to come to an understanding with us.*

This was an unwise indiscretion to disclose to Wilhelm II, who instinctively sensed danger ;
The Tsar has clearly been playing me along, to get almost a week ahead of us in mobilization. I cannot agree to any more mediation. It is only a manoeuvre, in order to hold us back and increase the start they already have. My work is at an end!

By degrees, the people of Imperial Germany who were taught from childhood to regard their Empire as something beyond and above them, to be reverenced and implicitly obeyed and honestly believed Germany was the Super-State ~ that they were in the Kaiser's words ~ *the very salt of the earth.* The moral effect of all this self-glorification on the continuum of German opinion and policy was not surprising, since nothing clouds the judgment so much as vanity. Whilst this statement can be culturally transposed for application to our wider cast of characters in the Imperium of Empires, it remained an underlying flaw of the German psyche and foreign policy decisions of their ruling élite.

A false appraisement of oneself is certain to lead to a false opinion of one's neighbours. Always somewhat unsympathetic and therefore unintelligent of others, the Germans have been led by their pride on the one hand, by their desires on the other, into the grossest miscalculations as to the mind of foreign nations, and even as to the political conditions of other countries. Never has it proved more true that the wish is father to the thought.

The Germans had been led to believe that France was utterly decadent; that Russia would collapse at a blow; that Britain would never fight, and that, if she did, her colonies would desert her, India would revolt, and the British Empire would crumble to pieces. There is no end, in fact, to the delusions and false hopes which have been nourished in the minds of Germans, as a result of this habitual glorification of themselves and depreciation of their neighbours. Their vanity has destroyed their judgment.

Their judgment, like their diplomacy, was never so much at fault as in the summer of 1914, when the assassination of the Austrian Archduke Franz Ferdinand and his wife at Sarajevo ; the capital of the Balkan province of Bosnia once part of the ancient Kingdom of Serbia was seized upon by the warlords as Germany's golden opportunity to hack out her destiny by force of arms. At any rate, she would be sure of Austria-Hungary on her side, whatever Italy, the third partner of the Triple Alliance, might decide to do in view of the fact that this compact was solely of a defensive nature, and dissolved if war resulted from their own aggressive action.

Every effort, therefore, was made by Germany both then and afterwards to prove that the sole responsibility for the war rested with the Allies in general and Great Britain in particular; that Germany, in short, had unsheathed her sword solely in self-defence. It was not difficult to convince the German people that they were fighting for their lives against a ring of jealous and unscrupulous enemies. They knew that France had never ceased to nurse her hopes of revenge for Alsace-Lorraine ; that Russia had set her hand to the freeing of the Slavs, and would block the path of Germany to Constantinople and the East; and that Britain had causes of friction on account of naval supremacy and commercial rivalry. Hence it was not difficult for the Berlin authorities to persuade all Germany that Britain had deliberately egged on France and Russia to goad the Germanic Powers to a conflict, while she lured Germany to her doom under the cloak of peace, meaning all the time to deal the treacherous blow which she struck on 4th August, 1914.

Professor G.W. Prothero
King's College, Cambridge

When Imperial Germany declared war, most Germans believed that Vienna could not let the assassinations go unpunished. If A-H had been allowed to fracture, Germany would become isolated and vulnerable. Berlin had not expected Russia to suddenly intervene in the Austro-Serbian quarrel, and Russia's decision to fully mobilize had come as a complete surprise to the German camp ~ to them Russia's mobilization was the true cause of the war. Foreign Secretary Gottlieb von Jagow asked; *Were we supposed to wait until the Russian Army had marched to our frontier, ready to overrun our country? Until our existence was simultaneously threatened from east and west by an annihilating superiority of power?* Bethmann Hollweg's summation was ; *We thought the Russians would shrink back before making the terrible last step : mobilization ~ and the Russian mobilization was critical, because of the Schlieffen-Plan 1914. The world cataclysm was conjured up, not only through the collective guilt of nations, but also through the collective guilt of men and parties within them. War could have been avoided only by a rapprochement with England ~ but after we had decided for a policy with A-H, we could not then desert her in that crisis.*

The immediate causes of the war came from three actions. First : the Serbian prompted *Black Hand* assassination plot that succeeded in the Sarajevo assassination. Second : the refusal of Belgrade to help Vienna with their police investigation of the assassinations. Third : Russia's decision to support Serbia. As the Austrian case is a sound one, Germany cannot be blamed for supporting it ~ except in their brutal invasion of Belgium pursuant to manœuvre against France, in which case Britain intervenes.

The fact of most importance however, was that our advance into Belgium was responsible for England's decision to intervene. But the violence done to Belgium was not only a strategical and political error, it was also a breach of international law, as the competent authorities had to admit from the first. It was both wrong and dishonourable, it exposed us to the contempt of the world, and furnished our enemies with weapons with which they fought us no less effectively than by force of arms. The hardships of the campaign and of the occupation did all and more than was required to lash our enemies' hatred into fury. It may be quite correct, from a military point of view, to carry the horrors of war successfully into the enemy's country, but crushing a weak country, protected by sacred treaties, is a crime against which all the world's conscience revolts, and for which it demands reparation. The war was the disastrous result of antagonisms between the Powers, engendered by elementary national impulses, and of there being no mastermind to settle them peaceably. Baron Wilhelm von Schoen

And so it persists, that the world puts most war-guilt upon the Kaiser and Imperial Germany, whereas the proximate causes of the *Black Hand, Mlada Bosna,* and the precipitate mobilization of Russia are all deemed lesser contributory causes of the Great War than Berlin's support for Vienna. This is apparent, because the Great War did not erupt because diplomacy failed ~ it detonated on the world because of the agreement between the leaders in Vienna and Berlin to bring about a limited war against Serbia ~ in the self-deluded hope and fatal misjudgment that Great Britain would stay out. While some say the leaders of Britain might have averted hostilities by clarifying their position earlier, this was prevented by the British Cabinet disunity at the time, and if we disregard British domestic political obstacles to an alternative course ~ London's efforts were all for peace, rather than action.

In my opinion, an outspoken threat from England ~ a sudden roar of the British Lion ~ would not have made Wilhelm II draw back. The Kaiser would have risked losing all prestige in the eyes of his subjects, if at the bidding of the Anglo-Saxon he had refused to go further ~ Germany barks, but does not bite ~ was the current saying abroad, and this naturally tended to exasperate the Germans. An ominous warning from the lips of Sir Edward Grey would only have served to precipitate the onslaught of the Kaiser's armies. Baron N.E. Beyens

Great Britain's non-committal policy up to the last appeared the only wise course to adopt in the perilous circumstances prevailing, Sir Edward Grey's balanced pacifism had helped to solve what had seemed an equally grave crisis in the Balkans in 1913, and there is little doubt that it would have prevailed in 1914 but for the myopic militarism on the Continent. British foreign policy in the fateful hours before the German declaration of war, was described by Sir Winston Churchill as :

Not to encourage France and Russia into ambitious courses by any assurance that we should take their side: not to let Germany suppose we were afraid or unable to fight her: to try somewhere, somehow, to get both sides to parley round a table. To this must be added the Admiralty policy ~ Ready ~ And had there not been a deliberate overriding intention to make war on the part of Germany, to which intention Kaiser William II had consciously and resolutely given his support, Sir Edward Grey would unquestionably have succeeded.
But against the ram of a stubborn purpose nothing could avail.

To the leaders of Great Britain who had hoped to guarantee Belgian neutrality, but had agreed to war in 1914 when this was not forthcoming ~ David Lloyd George assessed their decision as follows : *If Germany could not go through Belgium, an attack on France would have enabled Great Britain to remain neutral, at least for a while. France would be defending a heavily fortified 250-mile frontier instead of the 500 miles of French and Belgian frontier. In the former case, France would have had no difficulty holding off the Germans, and Great Britain would have had time to build up its land forces. Then with its Fleet mobilized and a million men under arms, it would have been able to force a peaceful settlement. But the invasion of Belgium put an end to these possibilities. Then our Treaty obligations were involved.* Sir Edward Grey further explained ; *Fear has created more history than it is usually given credit for. The real reason for going to war was that if we did not stand by France and stand up for Belgium against this aggression, we should be isolated and discredited, and hated; and there would be before us nothing but a miserable and ignoble future.*

By late 1918, Count István Tisza knew that his life was in peril in Hungary after three failed assassination plots. Advised to leave the country, he said ~ *I wish to die upright, the way I have lived,* and often carried a pistol. The fourth, and successful assassination attempt came on 31st October 1918. He was at his residence, Roheim Palace on the edge of Varosliget Park. In the evening, three armed soldiers entered Roheim and confronted Tisza in the hallway, waiting with his revolver in his hand. The soldiers told Tisza to drop his weapon. He refused. His wife and niece, the Countess Almassy were told to leave. They refused. One soldier accused Count Tisza ; *It is your fault millions have perished! You brought about this war!* Tisza : *No!* The soldiers fired. Tisza's last words ~ *I am dying. It has to be…*

After the deaths of ten million by 1918, Germany signed an Armistice to conclude the Great War ; the greatest tragedy being that they did not lose the war in its first months, instead of protracting it for four years. At 11 AM on 11-11-1918 ~ the Armistice of Compiègne ended the fighting of the Great War.

Prince Max von Lichnowsky who spent his remaining years trying to clear his name wrote : *Bethmann and Stumm did not to listen to me, because they disliked me. They could not accept the advice of a man they regarded as a rival. Only he who knows the truly fabulous incompetence of our leading statesmen will understand that such a policy was indeed possible ~ without their desiring war!*

In 1926, two men strolling along promenade of Merano; formerly part of the A-H province Tyrol, but Italian since the end of the Great War. Prince Max Lichnowsky was unknowingly approaching Count Leopold von Berchtold who when they met said ; *I know that you might not want to speak to me, but I have something important to tell you. I would never have taken so hard a line on Serbia if I had not been so constantly pressured to do so by Berlin. Berlin's actions, gave me no reason to suppose that Germany opposed a general war.* Having spoken, Count Leopold von Berchtold walked on. He continued to maintain that he could not have acted in any way differently, as he told A-H Ambassador in Rome Count Mérey ; *With Belgrade implacable, Russia organizing the Balkan League against A-H, and with France financing them both ~ would it not be suicide to wait for one's opponents to choose their moment? I believed I had no choice but to attack Serbia if we were to save the Hapsburg Monarchy ~ and that the War would never have happened if Russia had not suddenly mobilized, and Germany had not declared war in response.*
Having spoken, Count Berchtold walked on.

Accordingly, ultimate war-guilt falls upon those militarists and diplomats who either deliberately set out to ignite a limited war that got out of hand or for various reasons, then failed to prevent the situation from becoming a wider war. In the end, it appeared their statecraft would or could aspire to nothing higher than staking their future on the chances of war ~ a fate that was already hanging over the world like a storm, thus Europe went down to disaster in the delusion that their politicians could convey some timely conviction to certain countries to revise their ambitious attitudes towards war.

The true culprits in this long list of assassinations and atrocities are those who mislead public opinion to take advantage of the people's ignorance to raise disquieting rumors and sound the alarm bell, inciting their country and consequently other countries into enmity. The real culprits are those who by interest or inclination, declare constantly that war is inevitable, end by making it so, and assert that they are powerless to prevent it. The real culprits are those who sacrificed the general interest to their own personal interest that they so little understood, and who hold up to their country a sterile policy of conflict and reprisals. In reality, there is no way out for countries except by union and conciliation. Baron Paul-Hénri d'Estournelles de Constant

~ THE END ~

Der Letzte Mann (*The Last Man*) : Hans Bohrdt

EPILOGUE

The decisions of the July Crisis of 1914 did more to reshape our modern world than any other event, with a few dozen men, from a few Continental Chancelleries bringing about the Great War. The outcome of this catastrophic conflict included ; the dramatic collapse of Imperial Russia with the rise of the revolutionary Soviet Union, the ascent to world power status of the United States, the dismemberment of the Ottoman Turkish Empire with the abolition of its inherent Caliphate which witnessed the formation of the modern Middle East whose recurrent dramas we still hear about daily.

As one of the causes of the Great War was Russian foreign policy, so an unequal relationship in the parallelism form of the Soviet Union forced itself upon other realms resulting in the suppression and subordination of their economies and territory ~ of which A-H and Germany had been so wary. Whether the principal impetus for Soviet imperialism was cultural or political, their provocateurs, agents, and military detachments sought decisive political influence in any unstable part of the world, and so Soviet Russian foreign policy became interested in the turbulent Caribbean island of Cuba.

Barbara Tuchman, granddaughter of the U.S. Ambassador to Constantinople Henry Morgenthau, mentioned from an early age that the *Belgian Twins* book of Belgium's violation had influenced her. Years later she published *The Guns of August* in 1962, to win the Pulitzer Prize for General Non-Fiction. While U.S. President John F. Kennedy was reading Tuchman's *Guns of August* he was absolutely stunned to see Chancellor Bethmann's response when asked how the war began, *Ah, If only one knew!* When President Kennedy himself was facing the possibility of global Armageddon during the Berlin Crisis of 1961, he asked General Curtis Le May ~ *What will happen if the Soviet Union does not back off?* The General replied ~ *We will order a nuclear strike!*

Kennedy refused to recognize the U.S. military policy of a threatened nuclear response, all too similar to the hair-triggered volley of war ultimata issued in the July Crisis of 1914 in that drift to war. Perceiving disaster, J.F.K. immediately removed the nuclear option from the military as a response, and Bethmann's ominous words inspired him to say; *If this planet of ours is ever ravaged by nuclear war, if the survivors of that devastation can then endure the fire, poison, chaos and catastrophe ~ I do not want one of these survivors to ask another ; How did it all happen? and receive the incredible reply ~ Ah, If only one knew!* In the midst of the crisis in that fateful October of 1962 he told his brother Robert, *I am not going to follow a course which will allow anyone to write a book about this time, and call it ~ The Missiles of October.*

Kennedy having studied history at Harvard, and having served in the U.S. Navy during WWII, knew of the successful 1914-18 Royal Navy blockade of Germany, so his equally successful decision was to impose a naval *quarantine* around Cuba to avoid escalating the 1962 Cuban Crisis of his era. President Kennedy also remembered from the failed Bay of Pigs invasion of Cuba the year before, that the military are not always correct. When the Cuban Missile Crisis tested J.F.K., he resolved not to allow global catastrophe on his watch from a situation in which nations found themselves enmeshed in unwise self-destructive policies of time-limited ultimata and reflected auto-responses similar to the July Crisis when European Empires opted for hair-trigger diplomacy when facing severe problems. When such policies are as tumultarily and generally adopted as they were in July to August of 1914, it is only a matter of time until someone somewhere, causes the trigger to be pulled and all fall to disaster from the swift unfolding of miscalculations that rapid escalations often have in abundance.

With the destiny of nations hidden from us, no one can reveal what our future shall bring forth. We can at least look back through the past, which as the creator of the present contains the causes of events that are now taking place and of those that are to come. In the past lie the seeds of the errors, which weigh us down and have all but destroyed some nations, however as the present shows results, so the past reveals to us the source of the present. One reason why the Romans were so great was because they were aware of the finest qualities of their rivals and adopted their methods as required. We should keep this lesson in mind ~ for those of long ago, still offer valuable lessons for us today.

THE KAISER AND HIS NEW CHANCELLOR

THE DRAMATIC MEETING IN THE PALACE GARDEN ON THE SPREE

Theobald von Bethman-Hollweg has succeeded Prince von Bulow as Imperial Chancellor, President of the Ministry and Minister for Foreign Affairs. The announcement formally appeared in the Imperial Gazette of July 14, but a far more picturesque statement was made in the little garden at the rear of the old Palace at Berlin, where Kaiser Wilhelm II was seen by his people promenading with the six-foot-three newcomer, who towered over his Imperial Majesty. The conservation lasted three-quarters of an hour, while all Berlin stared through the railings. The New Chancellor was a college reading chum of the Kaiser at Bonn University, and until the other day was Imperial Home Secretary.

THE GRAPHIC ~ 24th July 1909

Aksakal, Mustapha *The Ottoman Road to War in 1914* Cambridge University Press 2008
Albert I: *The War Diaries of Albert I, King of the Belgians* Kimber, London 1954
Albertini, Luigi *The Origins of the War of 1914* Oxford University Press 1952
Arthur, Sir George *Life of Lord Kitchener* 3 Vols. Macmillan, London 1920
Aspinall-Oglander, Brigadier General C.F. *History of the Great War: Gallipoli*
Asquith, H.H. *The Genesis of the War* Cassell and Company, London 1923
Asquith, H.H. *Memories and Reflection 1852-1927* Little & Co. Boston 1928
Barnard, Charles Inman *Paris War Days* Brown and Company, Boston 1914
Barnes, H.E. *In Quest of Truth and Justice: De-bunking the War Guilt Myth* Arno NY 1928
Basily, Nicolas de *Nicolas de Basily, Diplomat of Imperial Russia* Stanford, California 1973
Beaverbrook, Lord *Politicians and the War 1914-16* Thornton Butterworth, London 1928
Berchtold, Count Leopold *Russia, Austria and the World War. Contemporary Review* 1928
Bernhardi, Freidrich von *Germany and the Next War* Longmans Green, NY 1914
Bethmann Hollweg, Theobald von *Reflections on the War* Butterworth, London 1920
Berghahn, V.R. *Germany and the Approach of War in 1914* St. Martin's, New York 1993
Beyens, Baron Eugéne Napoléon *Germany Before the War* Thomas Nelson, London 1916
Bobroff, R.P. *Roads to Glory ~ Imperial Russia and the Turkish Straits* Taurus, London 2006
Bradford, Ernle *The Great Betrayal ~ Constantinople* White Lion, London 1967
Buchan, John *Greenmantle* Hodder & Stoughton, London 1916
Buchanan, Sir George *My Mission to Russia & Other Diplomatic Memories* London 1913
Buchanan, Meriel *The Dissolution of an Empire* John Murray, London 1932
Bülow, Prince Bernhard von *Memoirs of Prince von Bülow* Brown, London 1932
Buxton, Noel and Charles *The War and the Balkans* Allen & Unwin, London 1915
Cambon, Paul *Correspondence 1870-1924.* 3 vols. Grasset, Paris 1940-46
Carter, Miranda *Three Emperors, Three Empires and the First World War* London 2009
Charykov, N.V. *Glimpses of High Politics* Allen and Unwin, London 1931
Churchill, Winston *The World Crisis 1911-1914* Scribner & Sons, New York 1928
Clark, Christopher *The Sleepwalkers: How Europe Went to War in 1914* London 2014
Dedijer, Vladimir *The Road to Sarajevo* Simon & Schuster, London & NY 1966
Deutsch, H. *Sidelights on the Redl case: Russian Intelligence on the eve of the Great War (Intelligence and National Security 4, No. 4)* 1989
Douglas, Rachel *June 28 : A red-letter day for Serbs ~ EIR* #13, March 24 1989
Durham, Mary Edith *Twenty Years of Balkan Tangle* London, Allen & Unwin 1920
Dumitru, Colonel Laurentiu-Cristian *Preliminaries of Romania's Entering the WWI*
Dugdale E.T.S. *German Diplomatic Documents* Harper and Brothers, London 1931
Earle, Edward M. *Turkey, The Great Powers, and the Baghdad Railway* New York 1923
Evans, R.J.W. and Hartmut von Strandmann *The Coming of the First World War* 1990
Evera, Stephen van *The Cult of the Offensive and the Origins of the First World War* 1984
Fay, Sidney Bradshaw *The Kaiser's Secret Negotiations with the Tsar, 1904-1905*
Fay, Sidney Bradshaw *The Origins of the World War* MacMillan, New York 1928
Fischer, Ben. B *Okhrana ~ Paris Operations of the Russian Imperial Police* C.I.A. 1997
Fischer, Fritz *Germany's Aims in the First World War* Norton, New York 1967
Fischer, Fritz *War of Illusions: German Policies from 1911-1914* Norton N.Y. 1975
Fromkin, D. *Europe's Last Summer: Who Started the Great War in 1914* Knopf NY 2004
Gainham, Sarah *The Hapsburg Twilight: Tales from Vienna* Athenaeum, NY 1979
Gauss, Christian *The German Emperor* Charles Scribner's Sons, New York 1915
Geiss, Immanuel, *July 1914. The Outbreak of the First World War* Batsford, NY 1967
Gerard, James W. *My Four Years in Germany* Doran, New York & London 1917
Gerard, André *A New German Empire ~ The Story of the Baghdad Railway* London 1911
Gibson, Hugh S. *A Journal from our Legation in Brussels* Doubleday, New York 1917
Goschen, Edward *The Diary of Edward Goschen* Royal Historical Society, London 1980
Grey of Fallodon *Twenty-five Years 1892-1916* Hodder & Stoughton, London 1925
Grey, Sir Edward *Speeches on Foreign Affairs 1904-1914* Allen & Unwin, London 1931
Gwinner, Arthur van *The Berlin-Baghdad Railway ~ Nineteenth Century* June 1909
Haldane, Viscount Richard *Before the War* Funk & Wagnalls, NY & London 1920
Hamann, Otto *The World Policy of Germany, 1890-1912* New York & London 1927
Hamilton, Richard, and Herwig Holger *Decisions for War 1914-17* Cambridge 2004
Helmreich, E.C. *The Diplomacy of the Balkan Wars* Harvard University Press 1938
Henig, Ruth *The Origins of the First World War* (3rd Edition) Routledge, London 1989

Herrmann, D.G. *The Arming of Europe and the Making of the First World War* Princeton 1997
Hewitson, Mark. *Germany and the Causes of the First World War* Bloomsbury Publishing 2014)
Hillgruber, Andreas *Germany and the Two World Wars* Harvard University Press 1981
Hobson, R. *Imperialism at Sea: Naval Ideology of Sea Power, and the Tirpitz Plan* 2002
Holquist, Peter *Making War, Forging Revolution, Russia's Continuum of Crisis* 2002
Hopkins *The Baghdad Railway ~ Journal of the United Service Institution, India* 1909
Howard, Michael *The Outbreak of World War I* Boston, Houghton Mifflin, 1997
Hull, I.V. *The Entourage of Kaiser Wilhelm II, 1888-1918* Cambridge University 1982
Isvolsky, Alexander *Recollections of a Foreign Minister* Doubleday, New York 1921
Jannen, William Jr. *The Lions of July ~ Prelude to War 1914* Presidio, Novato, CA 1997
Jastrow, Morris *The War and the Berlin-Baghdad Railway* Lippincott, Philadelphia 1917
Jovanovic, Ljuba *The Blood of Slavdom* Belgrade 1924 (translated *National Review*) 1925
Kalmykov, A.D. *Memoirs of a Russian Diplomat* Yale University Press, New Haven 1971
Kazemzadeh, F. *Russia and Britain in Persia 1864-1914 : A Study in Imperialism* 1968
Keiger, J.F.V. *France and the Origins of the First World War* St. Martin's Press, NY 1983
Kennedy, P.M. *The Rise of the Anglo-German Antagonism, 1860–1914* Allen & Unwin 1980
Kennedy, P.M. *The War Plans of the Great Powers, 1880–1914* Allen & Unwin, NY 1979
Keyes, Sir Roger *The Narrow Seas to the Dardanelles 1910-15* Butterworth, London 1934
Kisch, Egon Erwin *The Raging Reporter ~ The Case of Colonel Alfred Redl* Purdue 1997
Kleinmichel, Countess Marie *Memories of a Shipwrecked World* Brentano, London 1923
Knutsen, Torbjørn *The Rise and Fall of World Orders* Manchester University Press 1999
Kronenbitter, Rita *The Paris Okhrana 1885-1905* CIA Historical Review Program 1993
Kurenberg, Joachim von *The Kaiser; A Life of Wilhelm II, Last Emperor of Germany* 1955
Langer, W.L. *The Franco-Russian Alliance* Harvard University Press 1929
Lawrence, T.E. *Seven Pillars of Wisdom* Jonathon Cape, London 1935
Lichnowsky, Prince Karl Max *Heading for the Abyss* Payson & Clarke, NY 1928
Lloyd George, David *War Memoirs* Ivor Nicholson & Watson, London 1934
Lieven, D.C.B. *Russia and the Origins of the First World War* St. Martin's Press 1983
Lynch H.F.B. *The Baghdad Railway ~ Fortnightly Review* March & May 1911
Ludendorff, General Erich *My War Memories 1914-18 Vol. 1* Hutchinson, London 1920
Ludendorff, General Erich *The General Staff and its Problems* Hutchinson, London 1921
MacMillan, Margaret *The War That Ended Peace: The Road to 1914* Random House, 2013
McMeekin, S. *The Berlin-Baghdad Express ~ Germany's Bid for World Power* Penguin 2010
Miller, G. *Straits : British Policy towards and the Origins of the Dardanelles Campaign* 1997
Milshtein, M. *The Colonel Redl Affair ~ War History Journal #1* (Russian Language) 1966
Morgenthau, Henry *Ambassador Morgenthau's Story* New York, Doubleday 1918
Morgenthau, Henry *Secrets of the Bosphorus* London, Hutchinson 1918
Morris, A.J.A. *The Scaremongers ~ The Advocacy of War and Rearmament* Routledge 1984
Mossolov, A.A. *At the Court of the Last Tsar* (Trans. E. Dickes) Methuen, London 1935
Murray, Gilbert *The Foreign Policy of Sir Edward Grey 1906-1915* Clarendon Oxford 1915
Napier, H.D. *Experiences of a Military Attaché in the Balkans* Drane, London 1924
Neiberg, M.S. *Dance of the Furies: Europe and the Outbreak of World War I* Harvard 2011
Nekludoff, A.V. *Diplomatic Reminiscences before the World War* John Murray, London 1920
Nikolai, Colonel Walter *The German Secret Service* Paul Stanley & Co. Ltd. London 1924
Paléologue, Maurice *An Ambassador's Memoirs, Last Ambassador to the Russian Court* 1923
Paléologue, Maurice *Journal 1913* Librairie Plon, Paris 1947
Palmer, Alan, *Twilight of the Hapsburgs ~ The Life and Times of Emperor Franz Josef* 1997
Pears, Edwin *The Baghdad Railway ~ Contemporary Review* November 1908
Petrovich, M.B. *A History of Modern Serbia* Harcourt Brace, New York 1976
Pless, Daisy Princess of, *From my Private Diary* London 1931
Pozzi, Henri *Black Hand Over Europe* The Francis Mott Company, London 1935
Poincaré, Raymond *The Memoirs of Raymond Poincaré* Doubleday, London 1926
Poincaré, Raymond *The Origins of the War* Cassel & Co. Ltd. NY, London 1922
Porch, Douglas *The French Secret Services* Farrar, Strauss and Giroux, NY 1995
Price, Crawford *Serbia's Part in the War* Fisher Allen & Unwin, London 1922
Prothero, G.W. *German Policy Before the War* E.P. Dutton, New York 1916
Purves, J.G. & D.A. West *War and Society in the 19th Century Russian Empire* Toronto 1972
Radziwill, Princess Catherine *Behind the Veil at the Russian Court* Cassell & Co., London 1914
Radziwill, Princess Catherine *France from Behind the Veil: 50 Years of Social & Political Life* 1914

Radziwill, Princess Catherine *The Austrian Court from Within* Frederick Stokes & Co. NY 1916
Radziwill, Princess Catherine *Sovereigns and Statesmen of Europe* Cassell & Co., London 1916
Radziwill, Princess Catherine *Secrets of Dethroned Royalty* John Lane & Company, NY 1920
Radziwill, Princess Catherine *The Intimate Life of the Last Tzarina* Harper, NY London 1928
Radziwill, Princess Catherine *Nicholas II: The Last of the Tsars* Harper, NY London 1931
Remak, Joachim *The Origins of World War I, 1871–1914* Harcourt, Fort Worth 1995
Rohl, J.C.G. *Wilhelm II Into the Abyss of War and Exile 1900-1941* Cambridge 2014
Rosen, Baron Freidrich *Forty Years of Diplomacy* Volume 1 London 1922
Rosen, Baron Friedrich *Oriental Memoirs of a German Diplomatist* London 1930
Rumbold, Sir Horace *The War Crisis in Berlin July-August 1914* London 1940
Sazonoff, Serge *Fateful Years: 1909-1916* Frederick A. Stokes, New York 1928
Savinsky, A.A. *Recollections of a Russian Diplomat* Hutchinson, London 1917
Schelking, Baron Eugene de *Suicide of a Monarchy* MacMillan, Toronto 1918
Schilling, M.F. *How the War Began in 1914* Allen & Unwin, London 1925
Schindler, J.R. *Redl ~ Spy of the Century?* Essay published in the ;
International Journal of Intelligence and Counterintelligence 18 – No. 3 2005
Schmitt, B.E. *The Annexation of Bosnia 1908-09* Cambridge 1937
Schoen, Wilhelm von *The Memoirs of an Ambassador* London 1922
Sixte de Bourbon *L'Offre de paix séparé de l'Autriche* Librairie Plon, Paris 1920
Sontag, R.J. *European Diplomatic History 1871-1932* New York & London 1932
Sontag, R.J. *Germany and England: Background of Conflict, 1848-1894* London 1938
Steiner, Zara *Britain and the Origins of the First World War* MacMillan Press 1977
Seymour, Charles (Editor) *The Intimate letters of Colonel House* Benn 1928
Stevenson, David *The First World War and International Politics* Oxford U.P. 1998
Stevenson, David *Cataclysm: The First World War as a Political Tragedy* Oxford 2004
Stevenson, David *War by Timetable : The Railway Race Before 1914* NY 1999
Stokes, G *Serbian Documents from 1914 ~ Journal of Modern History* 1976
Strachan, Hew *The First World War Vol. 1 : To Arms* Oxford University 2001
Tabouis, Genevieve *The Life of Jules Cambon* Jonathon Cape, London 1938
Taylor, A.P.J. *War by Timetable : How the First World War Began* MacDonald 1969
Taylor, Edmund *Fall of the Dynasties ~ The Collapse of the Old Order 1905-22,* 1963
Thaden, E.C. *Russia and the Balkan Alliance of 1912* Pennsylvania 1965
The Kaiser's Daughter, Memoirs of Princess Viktoria Luise London 1977
Thomson, C.B. *Old Europe's Suicide, or the Building of a Pyramid of Errors* 1922
Tirpitz, Grand-Admiral Alfred von, *My Memoirs* Hurst & Blackett, London 1919
Tompkins, R.O. *Anglo-Russian Diplomatic Relations* North Texas University 1975
Treadway, J.D. *The Falcon and the Eagle ~ Montenegro and Austria Hungary* 1983
Tuchman, Barbara *The Guns of August.* Macmillan, New York 1962
Tuchman, Barbara *The Zimmermann Telegram* Viking, New York 1958
Turner, E.R. *Causes of the Great War ~ American Political Review* 1915
Turner, L.C.F. *Origins of the First World War* Norton New York 1970
Usher, R.G. *The Story of The Great War* MacMillan & Co. New York 1919
Vasilyev, A.T. *The Okhrana : The Russian Secret Police* G. Harrap London 1930
Vivian, H. *A Royal Serbian Tragedy* (& Chandomille Mijatovich) London 1906
Viviani, René *As We See It. France and the Truth About the War* London 1923
Vopicka, Charles *Secrets of the Balkans : Seven Years of a Diplomatists life in*
The Storm Centre of Europe Chicago Rand McNally & Company 1921
Westwood, J.N. *A History of Russian Railways* Allen & Unwin, London 1964
Whitlock, Brand *The Letters and Journal of Brand Whitlock* Appleton, NY 1936
Wiesner Dr. Friedrich *Austria's Life and Death Struggle Against Irredentism* 1928
Wilhelm II, German Emperor *The Kaiser's Memoirs* Harper, NY & London 1922
Wilhelm, Crown Prince: *My War Memories* Thornton Butterworth, London 1922
Williamson S. *Austria-Hungary and the Origins of the Great War* Macmillan 1991
Wilson, Sir Arthur *The Final Report of the Dardanelles Commission* H.M.S.O. London 1916
Witte, Sergei *The Memoirs of Count Witte ~* Doubleday & Co. New York 1912
Woods, H. Charles F.R.G.S. *The Cradle of the War* Brown & Co. Boston 1918
Woods, H. Charles F.R.G.S. *The Danger Zone of Europe* Brown & Co. Boston 1918
Young, H.F. *Prince Lichnowsky and the Great War* University of Georgia Press, Athens Ga., 1977
Young, H.F. *The Misunderstanding of 1st August ~ Journal of Modern History #48,* December 1976

THE CRISIS AT CONSTANTINOPLE.

~INDEX~

INDEX

CAROLINE ISLANDS 8 267

INDEX

INDEX

INDEX

AERIAL ENGAGEMENTS

AND THE USE OF AIR POWER FROM THE GREAT WAR TO THE PERSIAN GULF WARS

This revealing book of winning tactics is filled with fascinating illustrations taken from older sources. For example, when dealing with Gotha bombing raids on London during World War One, he has pertinent illustrations of the B.E.2c night-fighter and Handley Page 0/400 heavy bomber. There are also three antique aviation-related advertisements for Sphinx ignition-spark plugs, a Wright Seaplane, and Burberry Air-Warm (flying coat). These reflect the style of the time and put the reader into that era. MacLean has done a commendable job. Coverage of the Arab-Israel and the Indian-Pakistan conflicts are very thorough.

Capt. J.F. O'Connell, USN (Ret.)
National Air and Space Museum
Washington, D.C.

Beware of the Hun in the Sun!

This book takes a slightly different approach to military aviation history. The author studies evolving aerial engagements from aeronauts in the French Revolution, US Civil War, and Franco-Prussian War to the present. An interesting concept ; the author then discusses the critical aerial reconnaissance missions that occurred in the first days of The Great War. As the book progresses through WWI's aerial innovations, it pays homage to leading pilots and lesser-covered activities, such as the Zeppelin Campaign, naval aviation, and the Middle Eastern successes of 1918. Similar treatment follows for later conflicts, including the Spanish Civil War, WWII, the Korean, Vietnam, and Middle Eastern Wars, in addition to the Falklands War, as well Desert Shield, Desert Storm, and on-going campaigns around the Persian Gulf. This work is well illustrated with interesting contemporary images of training diagrams, advertisements, recruiting posters, etc. Readers interested in the evolution of aerial warfare will find this book of interest.

~ Over the Front

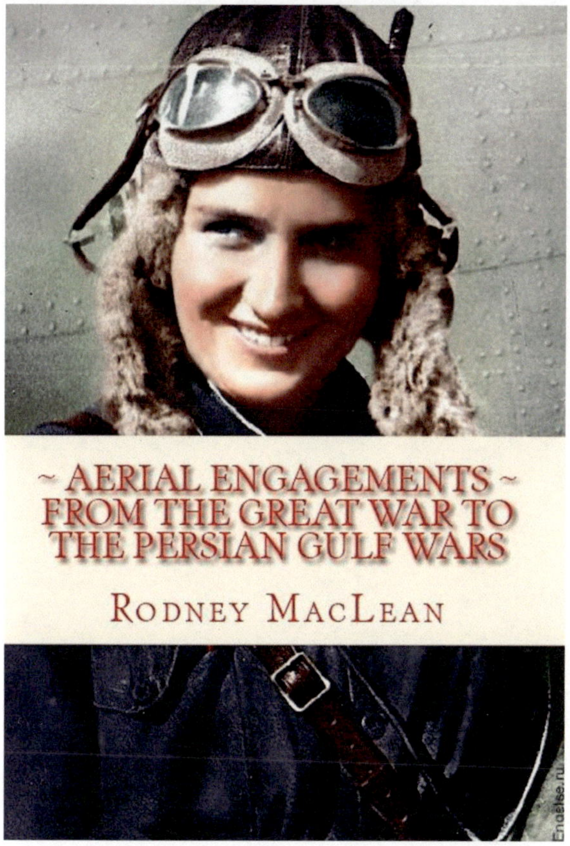

This volume is, in nutshell, how to win a war in the air as told by those who either took the fight to the enemy or defended against him across the centuries of manned flight. From the wicker basket of a French war balloon floating above the battles of Mauberge and Flerus in 1794, through to the first aerial bombing by the Austrians targeting Venice in 1849 ; from the World Wars, Vietnam, Korea, the Six Day War, and Falklands to Desert Storm. It is also a study of the refinement of tactics which, although have always been fluid, have had the same template since Boelcke and the Knights of the Air jousted over the trenches. These men (and women) had to be calculating, able to wring the very last drop of performance out of their mounts and apply a fleeting advantage ruthlessly. Those that survived continued to write the tactics and strategies for their successors far into the future. The subject is touched on often in the multitude of books on aircraft and air war campaigns, but not to any great depth and certainly not all encompassing as this work is. It takes on the challenge of enticing the modern reader to understand the mindsets of those who were living and flying at a time different from our own ~ we have to fly a circuit in their cockpit!

There are plenty of famous names here ~ Garros, Immelmann, Boelcke and his Dicta, Lufbery and his Circle, Mannock and Ball, von Richthofen, Guynemer, Nungesser, and Deullin ~ whose tactical principles were absorbed into the training of the US Army Air Service in 1917 ~ to name a few of the Great War's prominent tacticians. Discover how they influenced the outcome of a moment that meant survival or oblivion in a time when every passing day confirmed the growing importance of aviation in war. In a different war, with a different level of technology, combat formations had to be modified too. This was apparent to Luftwaffe ace Werner Mölders who took Boelcke's array of a pair of fighters and added a second element, this formation known as the finger four which is the standard fighter flight formation today. This is a classic example of a calculating visionary applying the lessons learned from the past and incorporating them into the now.

Technology would also force a rethink : the closing speed of the new Me-262 jet fighters on lumbering bombers meant a standard attack was impractical with their decreased reaction times. Tactics had to stay abreast of weapon innovation, pilot training proving more critical than ever ~ the technically brilliant more often than not undone by the tactically sharp. The book also features the document created by Greg Boyington in early 1944 detailing his thoughts on Combat and Strategy. This is a very interesting read, with much authentic first-hand material and a wealth of contemporary accounts untainted by revisionist historians. Significant aerial action is well reconstructed to preserve a sense of period and reveal to the reader the most significant and harrowing engagements of all time.

~ Classic Wings

GERMAN AVIATORS OF THE GREAT WAR

Kapitainleutant Heinrich Mathy ~ CO Zeppelin L 31

Oberleutnant Max Immelmann ~ 15 victories

Oberleutnant Werner Voss ~ 48 victories

Leutnant Baron Walter von Bülow ~ 28 victories

BOMB DROPPING APPARATUS OF GERMAN AVIATORS

HOW THE ROLAND BOMB-DROPPING APPARATUS IS WORKED

THE ROLAND BOMB-DROPPING APPARATUS CONSISTS OF THREE PARTS ~ THE BOMB TUBES, THE TUBE COVERS, AND THE PEDAL BOARD. THE BOMB TUBES ARE ARRANGED IN A PLATE LET INTO THE FLOOR OF THE FUSELAGE, A LITTLE BEHIND THE PILOT'S SEAT. THE PEDAL BOARD IS OPERATED BY THE FOOT OF THE OBSERVER. THE EXPLOSIVE CHARGE IS PLACED IN A STREAMLINE CASING ~ IN THE NECK OF THIS CASING IS A CIRCUMFERENTIAL GROOVE, AND WHEN THE BOMB HAS BEEN PULLED UP INTO THE TUBE IT IS ARRANGED THAT A FORK ENGAGES THIS GROOVE IN THE NECK, ALLOWING THE BOMB TO DROP. IN ORDER TO OBVIATE MISTAKES, THE PEDALS AND THE TUBE COVERS ARE PAINTED IN CORRESPONDING COLOURS.

AMERICAN AVIATORS OF THE PACIFIC THEATRE WWII

LCDR John S. Thach ~ CO of VF-3 Squadron

Commander David McCampbell ~ 34 kills

LT Alex Vraciu, showing his Philippine Sea tally for 19th June 1944

Major Gregory *Pappy* Boyington ~ 28 kills
CO USMC VMF-214 *Black Sheep Squadron*

Major *Dick* Bong, 40 kills, poses with *Marge*, his P38J-15 *Lightning* with a lovely photo of his fiancée, Marjorie-Ann affixed to its nose

LCDR *Jimmy* Doolittle

Dauntlesses over Midway Atoll ~ 1942

If you end up in front… your only salvation may lay in how quickly you get your nose down and get into this Diving Spiral. Maybe you can get some airspeed before you get hit too badly. Become familiar with this manœuvre for it is useful regardless of the type of aircraft being flown. Your best protector with an aircraft at six o'clock is G force. You need speed to get G's ~ thus, the diving spiral. Keep the G's on and keep the nose down so you can keep enough airspeed to make it rough for the shooter. Watch the aircraft behind you. If you can see the belly of his aircraft, he's beginning to pull lead and may be about to fire. Tighten up your turn immediately.

Captain Frederick C. *Boots* Bleese USAF
Basic Principles of Offense ~ No Guts, No Glory

FLAME-OUT APPROACH

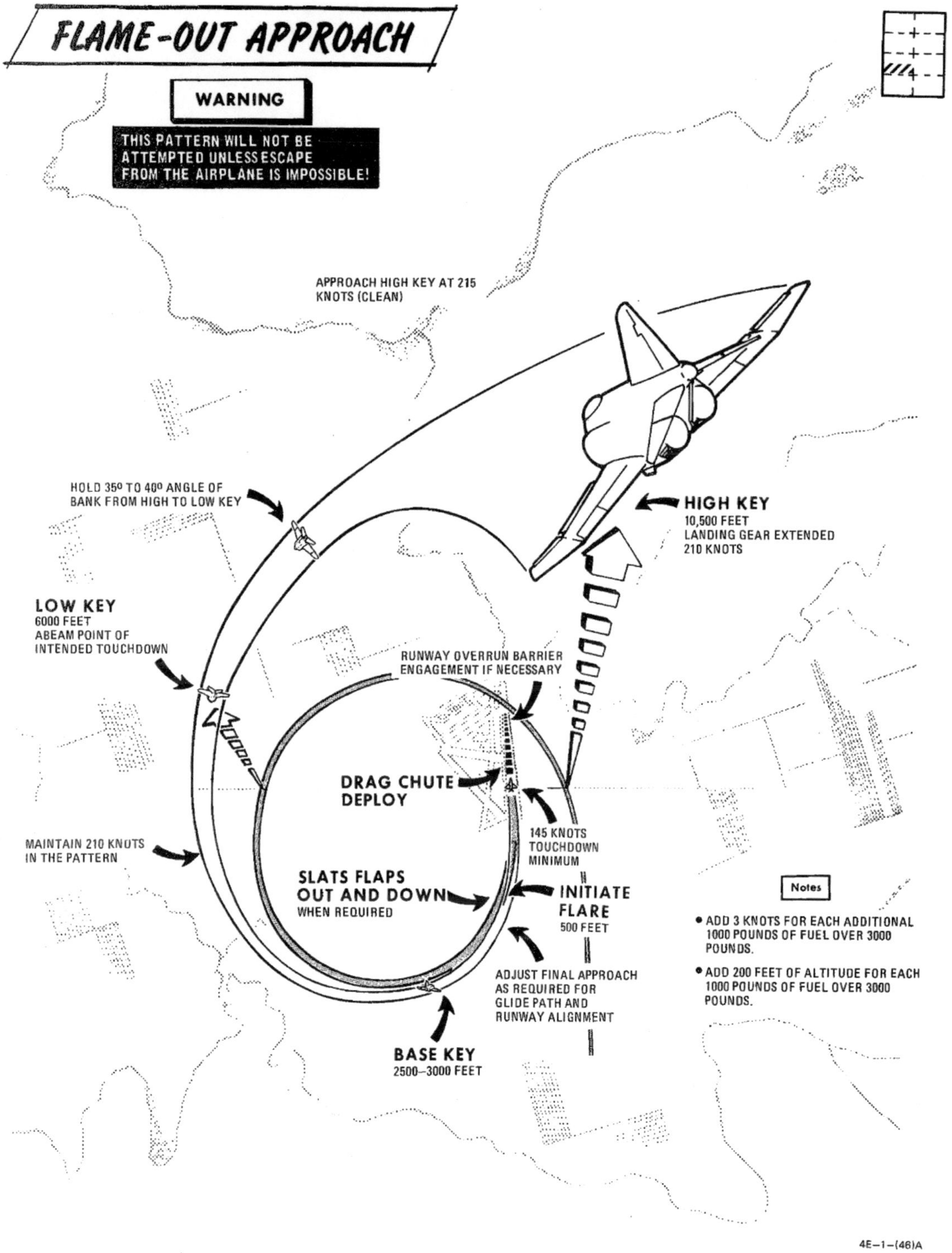

Figure 3-10

ARCHIVES

d'Histoire et Militaire

ADELAIDE

~ to contact the author ~

20 Reedie Street
Henley Beach 5022
Australia

Made in the USA
Monee, IL
17 March 2026

46257779R00240